Molecular Physics

Personal Study Notes
Atomic, Molecular, And Nuclear Physics

Every Thing I learned In Modern Physics
From High School To Post-Doctorate

By

Mohamed F. El-Hewie

TABLE OF CONTENTS

CONSTANTS

Avogadro number .... $N_A = 6.02296 \cdot 10^{26}$ 1/kmole

Faraday constant .... $F = 9.64914 \cdot 10^{7}$ C/kmole

Elementary charge .... $e = 1.6020310^{-19}$ C

Planck constant .... $h. = 6.62517 \cdot 10^{-34}$ J $\cdot$ s

.... $\hbar = 1.05443.10^{-34}$ J $\cdot$ s

....$= 6.5817 \times 10^{-16}$ eV $\cdot$ s

Boltzmann constant .... $k = 1.3805. \, 10^{-23}$ J/deg $= 8.617X \times 10^{-5}$ eV/deg

Velocity of light in vacuo.........$c = 2.997925 \cdot 10^{8}$ m/sec

Electric constant ....$g_o = 8.854 \cdot 10^{-12}$ F/m

Weak interaction constant........$g = 10^{-62}$ J.m^3

Bohr magnetos ....$\mu_B = 9.2732 \cdot 10^{-24}$ J/T

Electron magnetic moment....... $M_e = \mu_B$

Nuclear magneton .... $\mu_N = 5.051 \cdot 10^{-27}$ J/T

Proton magnetic moment......... $\mu_p = .2.7928 \cdot \mu_N$

Neutron magnetic moment....... $\mu_n = 1.913148 \cdot \mu_N$

Classical electron radius .... $r_e = 2.818. \, 10^{-15}$ m

Radius of action (range) of nuclear forces ....$R \approx 1.5 \cdot 10^{-15}$ rn

Hydrogen atom radius (first Bohr radius) ....$R(H^1) = 0.529172. \, 10^{-10}$ m

Mass of mass unit ....$m_\mu = 1.6603 \cdot 10^{-27}$ kg

Energy equivalent of mass unit .... $m_\mu c^2 = .931.4$ MeV

Conversion of electron volts into joules ....1 eV $= 1.60203 . \, 10^{-19}$ J

Atomic mass of carbon C^{12} ....$A_r = 12.000000$ (standard)

Neutron atomic mass ....$A_r = 1.008665$

Proton atomic mass .... $A_r = 1.007276$

Electron atomic mass..............$A_r = 0.00054859$

Atomic mass of hydrogen atom ....$A_r = 1.007825$

Mass of hydrogen atom . . ..$M(H^1) = 1.67. \, 10^{-27}$ kg

Unit of measurement of nuclear cross-sections ....1 barn $= 10^{-28}$ m^2

Temperature of Maxwellian distribution at average energy of 1 eV

....$T = 7737$ °K

CHAPTER 9

THE WAVE NATURE OF MATTER

Introduction. With the recognition of the quantum theory at the beginning of the twentieth century, physicists were obliged to admit a dual nature, wave and particle, for *radiant energy, as* we have seen in the preceding chapter. A similar situation arose with *matter,* when in 1924 the French physicist, Louis de Broglie, put forward the bold suggestion that matter which is ordinarily considered as made up of discrete particles molecules, atoms, protons, electrons and like--might exhibit wavelike properties under appropriate conditions. This means that *matter, like radiation, has a dual nature.*

Such a novel idea brought no little consolation and encouragement to those who were attempting to reconcile existing concepts with the double aspect of radiation. For like Einstein's mass-energy relation of the theory of relativity, De Broglie's concept of "matter waves" was able to bridge over the almost insuperable gulf which Classical Physics had created between the two fundamental forms in which Nature manifests herself, viz., matter and radiation, and arrive at a certain unity between them, since both matter and radiation could be conceived to possess a dual nature of the same type. But the synthesis was not to be as easy as it appeared at first sight.

At any rate, scientists, encouraged by the important suggestion of De Broglie, set themselves to elaborate it by means of precise mathematical theories as well as check it by delicate experimental researches. Thus arose a very interesting and important but difficult chapter in Modern Physics which comprises the phenomenon of *diffraction of material particles,* such as electrons, protons, etc., forming a solid experimental proof of the existence of waves associated with matter and the theories worked out along different lines such as *the quantum mechanics* of Heisenberg and the *wave mechanics* of Schrödinger, wherein the "matter wave" concept not only received a firm mathematical basis and thereby became extremely useful in the interpretation of many a difficult problem in Atomic Physics but also led to a strange conclusion known as the *Principle of Uncertainty* that shook Classical Physics to its very foundations.

The aim of the present chapter is to give a brief outline of the main features of this important branch of Atomic Physics. After stating De Broglie's concept of matter waves we shall describe some of the experiments carried out on the diffraction of electrons and other material particles. We shall then study the theoretical treatment of the problem given by Heisenberg and Schrödinger, without entering into all the complicated mathematical details involved. Finally we shall consider the revolutionary Principle of Uncertainty that has resulted from the analysis of the wave nature of matter.

DE BROGLIE'S CONCEPT OF MATTER WAVES

Louis de Broglie, attempting to develop a theory of radiation in terms of light quanta or photons, was led to the new conception of matter waves by the following considerations

1. Symmetry requirement. According to this principle, the two fundamental forms, matter and energy, in which Nature manifests herself, must be mutually symmetrical.

Since radiant energy has been shown to possess a dual nature, wave and particle, matter also must possess the same dual nature, particle and wave.

2. The close parallelism between mechanics and optics, as regards their fundamental laws governing motion, momentum, mass and energy, appeared to De Broglie as a direct index of the essential similarity between matter and radiation, required in the evolution of the now concept of matter waves. For, the *Maupertuis' principle of least action* in mechanics and *Fermat's principle of least time* in optics implied similar conditions which were very suggestive. According to the former, a moving particle chooses always that path for which the action is minimum, *i.e.,* the integral of the momentum over the path is a minimum and is analytically expressed as

$$\delta \int_{P_1}^{P_2} (mu)\, ds = 0$$

According to the latter, a light ray chooses always that path for which the time of transit is a minimum as represented by the relation

$$\delta \int_{P_1}^{P_2} \mu \ ds = 0$$

This close analogy of the two principles belonging to two different branches of Physics argued to the probability of the behavior of matter as a wavelike entity under suitable circumstances. Just as radiation, ordinarily treated as a wave, has to be supplemented with a particle characteristic for a satisfactory explanation of observed optical phenomena, *so also* material particles, considered in mechanics as having a corpuscular structure alone, might have to be supplemented with a wave aspect for a full understanding of their behavior.

3. Bohr's theory of atomic structure. In 1913 Bohr basing his analysis on the quantum theory was able to give a very satisfactory explanation of the structure of electronic orbits in the atom and of the origin of spectral lines. According to this interpretation, an electron was shown to remain in an orbit of definite size for a considerable time without radiating energy. Further

the orbits available to the electron were rigorously selected by quantum rules from many orbits permitted by classical mechanics and it turned out that their radii were proportional to the square of integral numbers.

Thus the stable non-radiating orbits of the electron in the atom are governed by integer rules. De Broglie wishing to find a satisfactory explanation for these restrictions of the quantum theory argued that since

the only phenomena involving integers in Physics were those of interferences and modes of vibration of stretched strings, both of which imply wave motion,

the electrons in the privileged orbits could not be regarded simply as material particles but a certain intrinsic periodicity should also be assigned to them.

These reflections led Louis de Broglie to make bold to suggest in his doctorate thesis the new idea of matter waves, for which he was later (in 1929) honored by the scientific world with the award of the Nobel Prize. In his thesis he wrote that

there is an intimate connection between waves and corpuscles not only in the case of radiation, but also in the case of matter.

A moving particle of matter has always got a wave associated with it and the particle is controlled by the wave in a manner similar to that in which a photon is controlled by waves. To study the path of a beam of monochromatic radiation we use the wave theory, while to calculate the amount of energy transactions of the same beam we have recourse to the photon or quantum of energy $h\nu$. In a similar way, the electrons are particles, that is to say, their charge, mass and energy are observable in particle form; but

if we want to find the path of a beam of electrons and whether and how it is reflected by objects, we must treat it as though it were a beam of waves.

It is to be noted that the energy is carried by the electrons and not by the waves associated with them. In other words, whatever might vibrate in the matter wave it is not a strain in a real medium possessing energy.

De Broglie waves. These matter waves conceived by De Broglie are called "De Broglie waves" for obvious reasons. Since the essential feature of any wave is its wavelength, De Broglie next directed his attention on this quantity and derived an expression for the wavelength of the matter waves using the general equation of a standing wave system and the principles of the theory of relativity.

De Broglie wavelength. Picturing a material particle, such as an electron, as a standing wave system in the region of space occupied by the particle, let the quantity that undergoes periodic changes giving rise to matter waves be ψ. Its value at any instant t_o at a point x_o, y_o, z_o, in the immediate vicinity of the particle is given by $\psi = \psi_o \sin 2\pi\nu_o t_o$, where ψ_o is the amplitude at the point chosen and ν_o the frequency as observed by an observer at rest with respect to the particle.

Suppose now the particle is given a velocity v along the X -axis. In order to represent the variation of ψ under this new condition, we have to apply the *transformation equation* of relativity:

$$t_0 = \frac{t - vx/c^2}{\sqrt{1 - v^2/c^2}}$$

$$\therefore \quad \psi = \psi_0 \sin \frac{2\pi v_0 (t - vx/c^2)}{\sqrt{1 - v^2/c^2}}$$

Comparing this with the standard equation of wave motion

$$y = A \sin \{ (2\pi/T)(t - x/u) \}$$

where A is the amplitude, T the periodic time and u the velocity of the wave along the X-direction, we see that

$u = c^2 / v$ and
$1/T = v = v_0 / (1 - v^2/c^2)^{1/2}$

From Einstein's mass-energy relation

$$h v_0 = m_0 c^2 \quad \text{or} \quad v_0 = \frac{m_0 c^2}{h}$$

$$\therefore \quad v = \frac{m_0 c^2}{h(1 - v^2/c^2)^{1/2}}$$

$$\text{Since} \quad m = \frac{m_0}{\sqrt{1 - v^2/c^2}}, \quad v = \frac{m c^2}{h}$$

Now the wavelength of the matter wave

$$\lambda = \frac{\text{velocity}}{\text{frequency}} = \frac{u}{v} = \frac{c^2/v}{mc^2/h} = \frac{h}{mv}$$

Thus we get an expression for the De Broglie wavelength, which translated in physical language means that a material particle of mass *m* moving with a velocity *v* has a wave associated with it, whose wavelength is given by the ratio of Planck's constant *h* to the momentum *mv* of the particle.

The expression for λ could have been more easily derived on the analogy of radiations as follows:

On the basis of relativity principle the momentum *p* of a photon of energy hv is given by hv/c. as we have already seen. Hence $p = hv/c = h/\lambda$ or $\lambda = h/p$

Similarly, the wavelength of matter waves is given by h/p. If the mass of the material particle is *m* and its velocity v, then $p = mv$, therefore, $\lambda = h/mv$

But the first method of derivation indicates certain important features of the matter waves which the second does not. Thus, it can be noticed that with a material particle in motion two different velocities are involved, viz., one referring to the mechanical motion of the particle, represented by v and the other to the propagation of the associated wave represented by *u.* The two are connected by the relation $u = c^2/v$.

Particle velocity and group velocity. It can be shown that the particle velocity v is the same as the group velocity, *i.e.,* the velocity of a group or system of waves as follows:

The *group velocity* ω of a system of waves is given by

$\omega = u - \lambda \, du/d\lambda$

where λ is the wavelength of an individual wave in the system and u the velocity of that wave, frequently called the *phase velocity*. In dispersive media where $du/d\lambda$ is positive, $\omega < u$. In free space $du/d\lambda$ is zero and $\omega = u$.

The above relation may be written as

$$w = \lambda^2 \left(\frac{u}{\lambda^2} - \frac{1}{\lambda} \frac{du}{d\lambda} \right)$$

$$= - \lambda^2 \frac{d}{d\lambda} \left(\frac{u}{\lambda} \right) = - \lambda^2 \frac{d\nu}{d\lambda}$$

where ν is the frequency of the wave.

$$\therefore \quad \frac{1}{w} = - \frac{1}{\lambda^2} \frac{d\lambda}{d\nu} = \frac{d}{d\nu} \left(\frac{1}{\lambda} \right) \qquad \dots (1)$$

If E and V represent the total and potential energies of the particle respectively,

$$\tfrac{1}{2}mv^2 = E - V \quad \text{or} \quad v = \{ 2 (E - V)/m \}^{1/2} \qquad \dots (2)$$

Using the relation for the De Broglie wavelength: $\lambda = h/mv$,

$$1/\lambda = mv/h = 1/h \cdot \{ 2m (E - V) \}^{1/2} \qquad \dots (3)$$

Substituting this value of $1/\lambda$ in (1),

$$1/w = d/d\nu \{ 1/h \sqrt{2m (E - V)} \}$$

Replacing E by $h\nu$, since $E = h\nu$,

$$1/w = 1/h \cdot d/d\nu \{ 2m (h\nu - V) \}^{1/2}$$

$$= 1/h \cdot \tfrac{1}{2} \{ 2m (h\nu - V) \}^{1/2} \cdot 2mh$$

$$= m/\{ 2m (E - V) \}^{1/2}$$

$$= \{ m/2 (E - V) \}^{1/2} = 1/v$$

Therefore, the group velocity (ω) = the particle velocity (v)

From this result, a very significant conclusion may be drawn, viz., that

a material particle in motion is equivalent to a group of waves or a "wave packet,"

as it is called. The wave packet, formed by the superposition of a number of waves and traveling with the velocity of the particle, behaves very much like a corpuscle. But a serious difficulty at once arises against this concept of wave packet, since such a wave packet is, in general, very soon dissipated. For example, if we produce a wave crest at any point on a smooth surface of water, it is not long before it spreads out and disappears.

Again, since the particle velocity v is less than that of light c, it follows that

the velocity of propagation u of the individual waves is greater than c as seen from the relation $u = c^2/v$. The *De Broglie waves are not of the same kind as those associated with radiation*, since the former have a velocity greater than that of light, while the latter have always the same constant velocity of light.

This shows that the *De Broglie waves are not of the same kind as those associated with radiation*, since the former have a velocity greater than that of light, while the latter have always the same constant velocity of light. The individual waves forming the wave packet may be considered as possessing an average velocity, the "phase velocity," which seems to have no physical significance since it has a value greater than c. These facts render the interpretation of matter waves very difficult. We shall come back to this point a little later.

It is to be noted that the expression for the De Broglie wavelength ($\lambda = h/p$) combines corpuscular and undulatory concepts in a very intimate way, since λ the wavelength has a clear-cut meaning only in wave motion and p the momentum is most naturally associated with the moving particle.

The wavelength to be expected for the waves associated with different material particles, such as electrons, protons, atoms, molecules, etc. can be readily calculated from the relation $\lambda = h/mv$. For instance, for the electrons of the cathode *rays* we have the relation $(1/2)mv^2 = eV/300$, where m is the mass of the electron, v its velocity which is assumed much below that of light, e the electronic charge in *e.s.u.* and V the P.D. between the electrodes of the discharge tube expressed in volts.

From it we get

$$m^2v^2 = \frac{meV}{300} \times 2$$

$$\text{or} \quad mv = \sqrt{\frac{meV}{150}}$$

$$\therefore \quad \lambda = \frac{h}{mv} = h\sqrt{\frac{150}{meV}}$$

Substituting the values of $h = 6.55 \times 10^{-27}$, $m = 9 \times 10^{-28}$ gm. and $e = 4.77 \times 10^{-10}$ e.s.u.,

$$\lambda = \sqrt{\frac{150}{V}}\ 10^{-8}\ \text{cm.}$$

When the P.D. is 150 volts, the wavelength of the waves associated with the electron subjected to that field is equal to 1 A°. If the P.D. is 15,000 volts, λ is 0.1 A° . These values lie within the range of X-ray wavelengths, so that it should be possible to check them by experiments similar to those used with X-rays.

With very high speed electrons, whose velocity is comparable to that of light, relativistic correction will have to be applied. When this is carried out

$$\lambda = \sqrt{\frac{150}{V}}\left(1 + \frac{\alpha}{2}\right)^{-1/2}, \text{ where } \alpha = \frac{eV}{300\,m_0c^2}$$

Substituting the known values e, m_0, and c^2

$$\lambda = \sqrt{\frac{150}{V}}\ \frac{10^{-8}}{\sqrt{1 + 9.836 \times 10^{-7}\,V}}\ \text{cm.}$$

The correction factor is evidently very small, except for very large values of V. It amounts to only 5% with 50,000 volt-electrons and 8.5% with 200,000 volt-electrons.

The wavelengths of atom-waves, molecule-waves, even of waves for larger masses, can be calculated in a similar way. For a given velocity v, the larger the mass m the shorter is the wavelength λ. From this we see that, in practice, it is not easy to measure the wavelengths of heavy particles.

These are results derived from De Broglie's concept of matter waves. If then a beam of electrons does have undulatory characteristics, it should also exhibit the phenomenon of reflection, refraction, polarization and diffraction as in the case of a beam of light. Further, since the calculated wavelengths of electron waves are of the same order as X-ray wavelengths, it should be possible to produce clear diffraction effects with a beam of electrons using crystal gratings, as with X-rays. All these conclusions have been experimentally tested and found true. Thus the ingenious idea of matter waves, proposed by De Broglie, has been put on a firm experimental basis. We shall now consider some of these experimental researches.

> The wavelengths of electron waves are of the same order as X-ray wavelengths and electron diffraction is achievable by crystal gratings.

EXPERIMENTAL STUDY OF MATTER WAVES

The material particles, which readily lend themselves to experimental investigation of matter waves, are the electrons; for, not only are they easily produced in fairly intense beams with a definite velocity, like the cathode rays, but also on account of the smallness of their mass the wavelengths involved have magnitudes which could be measured by methods used for X-rays.

Clinton Joseph Davisson (October 22, 1881 – February 1, 1958)

Lester Halbert Germer (October 10, 1896 – October 3, 1971)

THE EXPERIMENTS OF DAVISSON AND GERMER

The electron waves predicted by De Broglie were first experimentally and rather accidentally detected in 1927 by two American physicists, Davisson and Germer, who succeeded also in measuring the De Broglie wavelengths for slow electrons by diffraction methods. As frequently happens, an experimental accident was to result in a new discovery. Davisson and Germer were studying the reflection of electrons from a nickel target and accidentally subjected the target to such heat treatment that it was transformed into a group of large crystals. As a result, the reflection of electrons became anomalous, *i.e.,* instead of decreasing continuously from the angular position of maximum reflection, the reflected intensity showed striking maxima and minima. This unexpected result called to mind X-ray diffraction from crystals and made them suspect that a beam of electrons might be diffracted from crystals like X-rays, which would mean that electrons behave like waves under certain circumstances. This was a very important point that deserved verification. So they prepared a target consisting of a single crystal of nickel and carried out the following researches which established that in fact a beam of electrons could be reflected, diffracted and even refracted.

Apparatus. Their experimental arrangement is shown diagrammatically in Fig. 179.

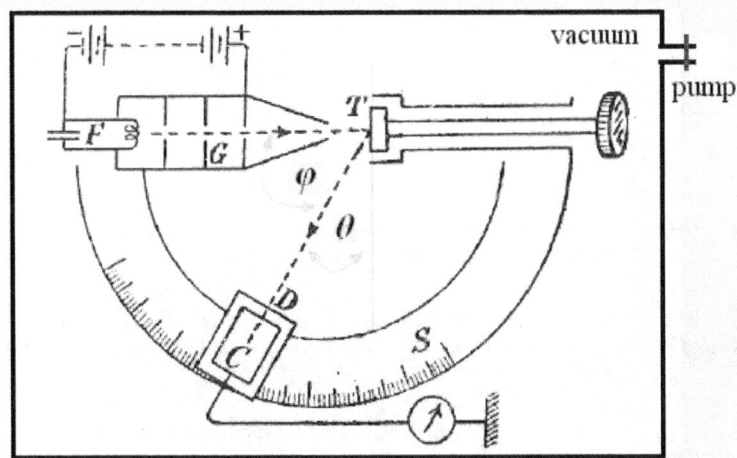

Fig. 179. Apparatus of Davisson and Germer used for diffracting electrons.

The electron beam is produced from an *electron gun* G. This contains a tungsten filament F heated to dull red, when electrons are emitted by thermionic action. The electrons are then accelerated in an electric field of known P.D. and collimated by suitable slits so that a fine parallel pencil emerges. This pencil is directed to fall on a large single crystal of nickel, known as *the target* T which is capable of rotation about an axis parallel to the axis of the incident beam, the purpose of which will be made clear presently. The electrons are *reflected* from the crystal in different directions, the angular distribution being measured with a Faraday cylinder called the *collector* C, which is connected to a sensitive galvanometer and can be moved along a graduated circular scale, S, so that it is able to receive the reflected electrons at all angles between 20° and 90°. The collector has two walls insulated from each other. A retarding potential is applied between the inner wall and the outer D so that only the fastest electrons, i. e., those possessing nearly the incident velocity but not the secondary slow elections excited by collisions with atoms may enter the collector and be detected by the galvanometer. The accelerating potential used ranged from about 30 to 600 volts and the retarding potential was nine-tenths of the accelerating voltage. The whole apparatus was completely enclosed, highly evacuated and degassed.

The nickel crystal which is known to be of the face-centered cubic type is cut so as to have a smooth reflecting surface parallel to the lattice plane (1, 1, 1), *i.e.*, perpendicular to one of the diagonals of the cube. By rotating the crystal about the axis stated above, any azimuth of the crystal can be presented to the plane defined by the incident beam and the beam entering the collector.

Experimental procedure. The experiment was conducted in the following two different ways:

(a) *With the beam of electrons at 'normal' incidence on the surface of the crystal*, when a diffraction effect from the surface layer acting a plane grating was produced. For each azimuth of the crystal, a beam of low voltage electrons was made to fall normally on the surface of a crystal, the collector war moved to various positions on the scale S and the galvanometer current at each position noted. The current, which was a measure of the intensity of the diffracted beam of electrons, was plotted against the angle between the incident beam and the beam entering the collector, known as the *"colatitude"*. The observations were repeated for different voltage electrons and several curves were drawn as shown in Fig. 180.

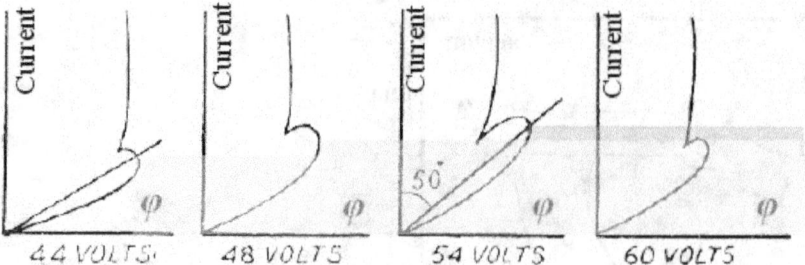

Fig. 180. Current vs. Colattitude at various voltages at azimuth A.

It is seen that a "bump" begins to appear in the curve for 44 volt-electrons. With increasing voltage the bump moves upward and attains its greatest development in the curve for 54 volts at a colatitude of 50°. At higher voltages the bump gradually diminishes, there being hardly any trace of it at about 68 volts.

The bump in its most prominent state of development offers a convincing evidence for the existence of the electron waves. For, according to De Broglie's theory, the wavelength for a beam of 54 volt-electron is given by

$$\lambda = \sqrt{\frac{150}{54}} \, A^\circ = \sqrt{2.8} \, A^\circ = 1 \cdot 66 \, A^\circ.$$

Now according to experiment, we have a diffracted beam at a colatitude of 50°. Applying the well known relation of a plane reflection grating, $n\lambda = d \sin \theta$, n referring here to the first order, d being equal to 2.15 A°, as given by crystallographic analysis,

$\lambda = 2.15 \sin 50° = 1.65 \, A°$.

This excellent quantitative agreement is very important, showing as it does, that a beam of electrons does really possess wavelike characteristics. The above result was obtained with the A azimuth and the same was verified with two other azimuths, B and C.

(b) With the beam of electrons falling 'obliquely' upon the crystal, when a diffraction effect from a space-lattice, *i.e.,* from successive parallel layers of atoms in the crystal, analogous to Bragg's X-ray diffraction, was produced, as was indicated by the existence of *regular selective reflections* depending upon the velocity of the incident electrons. If the electrons were simple corpuscles one cannot explain these selective reflections at definite electron velocities. On the other hand, if it is assumed that electrons have waves associated with them, their wavelengths can be calculated using Bragg's formula $n\lambda = 2d \sin \theta$, appropriate to the case. The values so found agree with those calculated by the De Broglie relation; $\lambda = - (150/\mathbf{V})^{1/2}$.

If the electron gun and the collector are fixed, the glancing angles of incidence and reflection can be kept constant. Then varying the electron velocity, the galvanometer current is measured for each value of velocity. Plotting the current values against the

corresponding electron velocities or accelerating voltages, a curve with several sharp maxima are obtained as shown in Fig. 181.

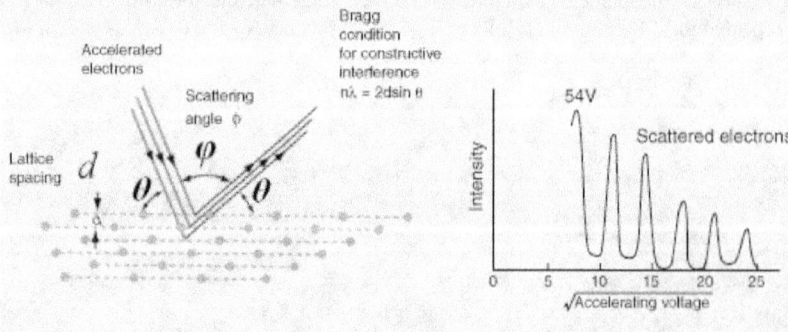

Fig. 181.

The different maxima correspond to the various orders. From the curve the accelerating voltages which produce maximum reflection in the different orders can be found and hence λ calculated using, De Broglie's relation. On the other hand, with the known values of the glancing angle θ and the grating space d, using Bragg's relation, λ is found. The two values are found to agree.

Refraction of electrons. The results obtained in the second series of experiments of oblique incidence are such as to call for the phenomenon of *refraction also* of a beam of electrons. For, Davisson and Germer found
that the agreement between the values of λ obtained from experiment using the relation $n\lambda = 2d \sin \theta$ and from De Broglie's theoretical equation, $\lambda = (150/V)^{1/2}$ was not perfect, the former being systematically less than the latter.

An explanation of this discrepancy was first given by Eckart and Bethe who suggested that the electron beam is refracted as it enters the crystal. In the case of X-rays we have seen that such an effect takes place, which modifies the Bragg's relation into

$$n\lambda = 2d \, (\underline{\mu}^2 - \cos^2 \theta)^{1/2}$$

μ being the refractive index for X-rays entering the crystal. This relation presumably holds also for the electron-waves.

But what is μ and how does refraction arise in the case of the electrons?

This point has been treated in a very interesting manner by Eckart and Bethe. Consideration of the thermionic emission of electrons shows that the electrons must do work in escaping from a metal surface. Hence we may say that a metallic crystal is at a positive potential ΔV above that of the surrounding. Consequently, if a pencil of V volt-electrons is incident at an angle i on the surface of a crystal, then on entering it, it is further accelerated through ΔV volts and is refracted. Let r be the angle of the refraction and v_1, v_2 the velocities of the electrons outside and inside the crystal. Then

$$\tfrac{1}{2} \, mv_1{}^2 = eV$$
$$\tfrac{1}{2} \, mv_2{}^2 = e \, (V + \Delta V)$$

Since the electric field parallel to the surface does not change, the components of the velocities parallel to the surface are equal.

$$\therefore \qquad v_1 \sin i = v_2 \sin r$$

$$\text{or} \qquad \frac{\sin i}{\sin r} = \mu = \frac{v_2}{v_1} = \sqrt{\frac{V + \Delta V}{V}}$$

It is to be noted that μ for electrons is greater than unity.

Davisson and Germer verified the above expression for μ with their experimental data. Using the formula $n\lambda = 2d \, (\mu^2 - \cos^2 \theta)^{1/2}$ they calculated μ for various values of λ. The results were found to agree with those calculated from the relation

$$\mu = \sqrt{\frac{V + \Delta V}{V}}$$

putting ΔV for nickel as 18 volts. They also found that μ approaches unity as the speed of the electrons and hence **V** is increased, as is to be expected from the theory. Other workers, such as G. P. Thomson and Rupp, have also obtained evidence of refraction of electron beams and have verified the above relation.

George Paget Thomson (3 May 1892- September 10, 1975)

THE EXPERIMENTS OF G. P. THOMSON

 In 1928, G. P. Thomson in Scotland extended the research on electron waves to high speed electrons ranging from 10,000 to 50,000 volts, diffracted by very thin metallic films. He used a method analogous to the Debye- Scherrer powder method of X-ray analysis of crystals.

Apparatus. His experimental arrangement is shown in Fig. 182.

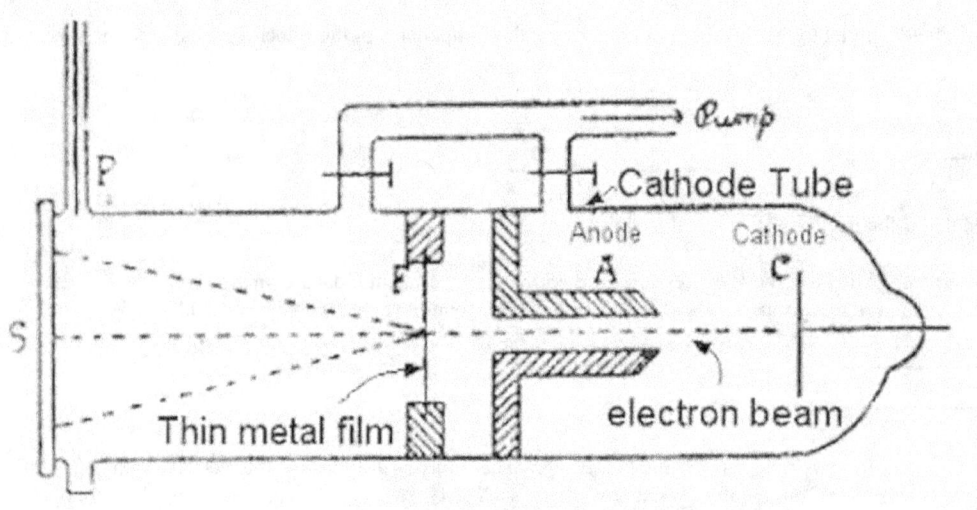

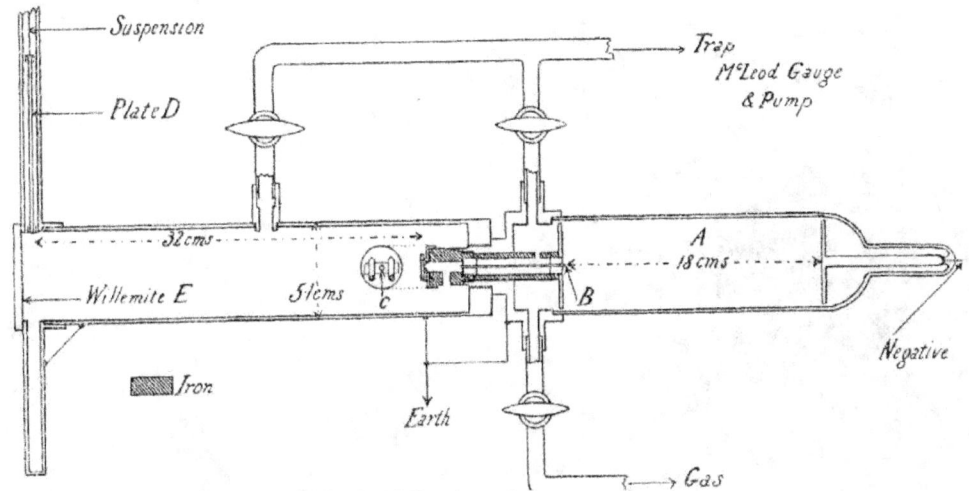

Fig. 182. G.P. Thomson's apparatus for the diffraction of electrons.

A beam of cathode rays is produced in a discharge tube AC by means of an induction coil. The rays are passed through a diaphragm tube A to obtain a fine pencil of electrons which is then allowed to fall upon a very thin metallic film F of gold, aluminum, etc. The film should be extremely small in thickness, of the order of 10^{-6} cm., to obtain good results. Special techniques are used to obtain such extremely thin films. Some are produced by thinning down commercial foils by means of suitable solvents, while others by "sputtering" the metals on some base which can be then dissolved away. P is a photographic plate which can be slided down into position to receive the pencil of electrons after it has traversed the film. S is a fluorescent screen which can be used instead of the photographic plate for visual examination of the result obtained by the passage of electrons through the foil. The camera part FP of the apparatus is exhausted to a high vacuum while air is allowed to leak into the discharge tube section through a needle valve. Since the only connection between the camera and the discharge tube is through the diaphragm A, it is possible to have the camera at a low pressure and yet maintain the discharge tube sufficiently soft to give a beam of the required voltage. The current from the induction coil is rectified by a kenotron and several smoothing condensers are connected in parallel with the discharge tube.

Experimental procedure. Allowing a pencil of electrons of known velocity to fall on the photographic plate after traversing the thin foil and developing the plate, a symmetrical pattern consisting of concentric rings about a central spot is obtained very much like that produced by X-rays in the powdered crystal method, as seen from the following photo.

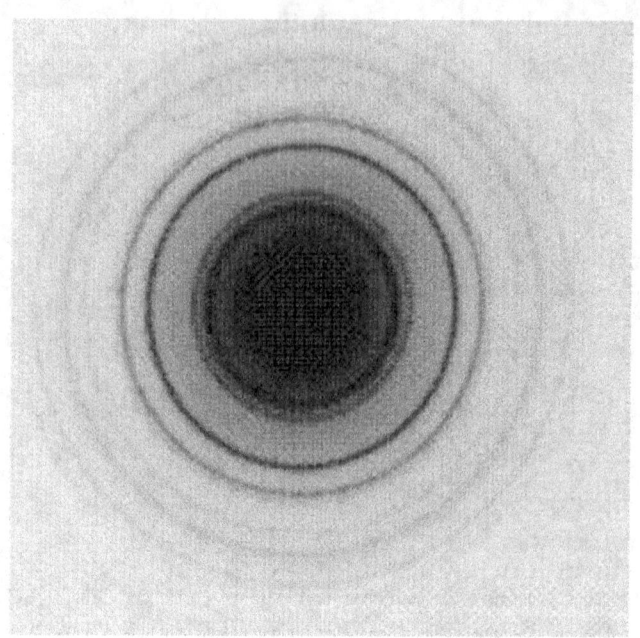

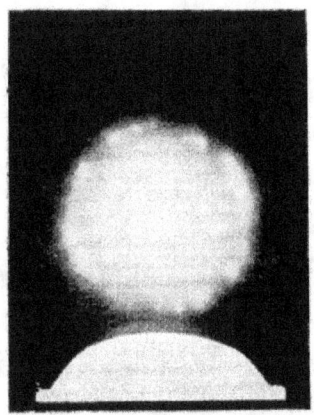

FIG. 1.—Aluminium.

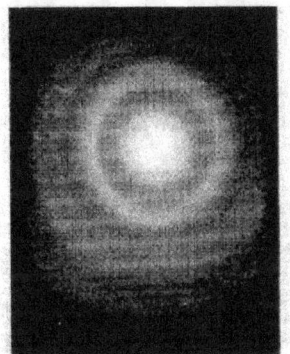

FIG. 4.—Gold.

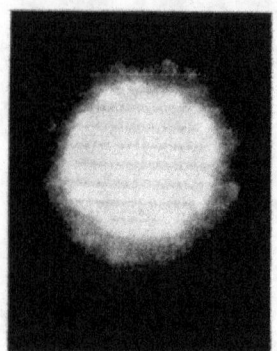

Fig. 2.—Aluminium.

Fig. 5.—Celluloid.

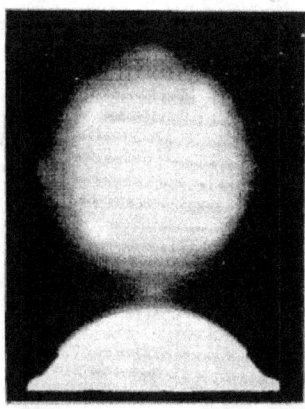

FIG. 3.—Aluminium.

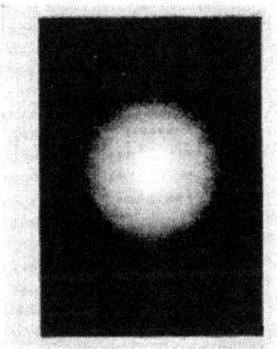

FIG. 6.—Film X.

To make sure that the pattern is produced by the diffracted electrons and not by secondary X-rays generated by the electrons in their passage through the foil, the cathode rays in the discharge tube are deflected by means of magnetic field when it is found that the whole pattern as observed on the fluorescent screen S shifts correspondingly, which cannot happen if X-rays are responsible for the pattern.

Further, on removing the film F, the pattern disappears, showing that the presence of the film is essential. Clearly then this experiment demonstrates in a striking manner that the electron pencil behaves as waves, since diffraction patterns can be produced only by waves.

The quantitative verification of the De Broglie equation can be made as follows:

As with the Debye-Scherrer powdered crystal method for X-rays, in the polycrystalline film there will be some crystals set at the correct angle to give a Bragg reflection. If there are enough crystals distributed at random the result of such reflections will be a series of rings arising from the intersection of the cones of diffraction with the photographic plate. Let AB be the incident beam passing through the film at B and let BE be a beam which has suffered a Bragg reflection in some small crystal in the film at B and falls at the point E on the photographic plate at a distance R from the central point C (Fig. 183).

21

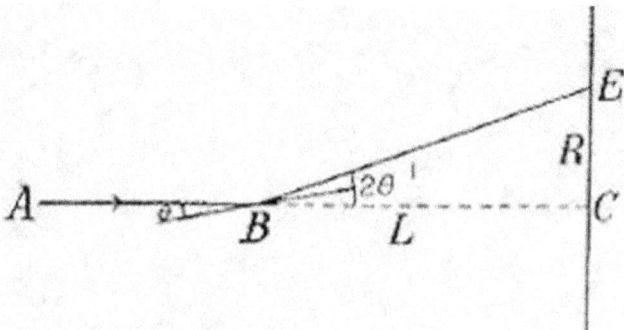

Fig. 183

Let the distance BC from the film to the plate be L. The angle CBE = 2θ, where θ is given by the Bragg relation $n\lambda = 2d \sin \theta$.

Now R $= L \tan 2\theta = L.2\theta$, since 2θ is small.
But from $n\lambda = 2d \sin \theta = 2\,d.\,\theta,\ \theta = n\lambda/2d$.
Therefore R $= L\, n\lambda/d$

Substituting De Broglie's value of λ where relativistic correction is included on account of the high speed electrons used,

$$R = \frac{nL}{d}\sqrt{\frac{150}{V\,(1 + \alpha/2)}}$$

$$\therefore\ RV^{1/2}\,(1 + \alpha/2)^{1/2} = \frac{nL\,\sqrt{150}}{d} = \text{constant.}$$

If D be the diameter of the diffraction ring considered, R=D/2 so that $DV^{1/2}(1 + \alpha/2)^{1/2}$ is still a constant. As a first approximation, neglecting relativity correction, $DV^{1/2}$ = constant. This relation has been verified by Thomson using different voltage electrons and measuring the diameter of a given ring in the pattern.

Alternately, by calculating the grating spaces d from the other readily measured quantities in the relation

$$d = \frac{nL}{R}\sqrt{\frac{150}{V\,(1 + \alpha/2)}}$$

in the case of various metals and comparing them with those found by means of X-rays, De Broglie's law was verified again within the limits of experimental error.

For example, in one of the experiments, where an aluminum foil was used to diffract the electrons, the photographic plate was at a distance of 25 cms. from the foil. When the applied voltage V was 30,000 volts the diameter D of the first ring in the diffraction pattern obtained was 2.45 cms. When V was made equal to 56,000 volts, the diameter D of the same order ring was found to be equal to 1.8 cms. These data show that $DV^{1/2}$ is a constant.

Aluminum belongs to the cubic system of crystals and the crystal face involved in the above experiment has been identified to have the Miller indices (2, 2, 0) for the first order ring. With this additional finding, the side of the unit cube for aluminum can be estimated as follows:

We have already seen that for crystals of the cubic system, the distance d between successive lattice planes is related to the side a of the unit cube by

$$d = \frac{a}{\sqrt{h^2 + k^2 + l^2}},$$

when h, k, l are the Miller indices of the concerned face.

Considering only the first order ring,

$$d = \frac{\lambda L}{R} = \frac{2\lambda L}{D} \qquad \frac{2\lambda L}{D}$$

$$\therefore \qquad a = \frac{2\lambda L}{D} \sqrt{h^2 + k^2 + l^2}$$

Taking the experimental data, viz., L = 25 cms., V = 56,000 volts, D = 1.8 cm. and for high speed electrons

$$\lambda = \sqrt{\frac{150}{V}} \cdot \frac{10^{-8}}{\sqrt{1 + 9 \cdot 836 \times 10^{-7} \, V}} \text{ cm.},$$

$$a = 2 \times \sqrt{\frac{150}{56,000}} \times \frac{10^{-8}}{\sqrt{1 + 9 \cdot 836 \times 10^{-7} \times 56,000}} \times \frac{25}{1 \cdot 8} \times \sqrt{2^2 + 2^2 + 0^2}$$

$$= \frac{2 \times \sqrt{150} + 10^{-8} \times 25 \times \sqrt{8}}{\sqrt{56,000} \times \sqrt{1 \cdot 055} \times 1 \cdot 8} = 3 \cdot 959 \times 10^{-8} \text{ cm.}$$

This value of *a* is in good agreement diffraction experiments.

Diffraction at glancing incidence
G. P. Thomson also obtained diffraction patterns with fast electrons by reflection both from single crystal and from polycrystalline surfaces at small angles. In general the patterns obtained with a single crystal showed a number of separate spots as in the adjacent photo, while with polycrystalline surfaces the patterns were rings similar to those obtained by transmission through thin foils.

Further experimental researches on the diffraction of electrons

Kikuchi in Japan, in 1928, obtained some remarkable photographs of the diffraction of high speed electrons by mica. The electrons in his experiments could be accelerated through a P.D. of 10 to 85 KV and made homogeneous by magnetic separation. These were then made to pass through a very thin mica film about 10^{-6} cm. and the diffracted pattern photographed. The pattern obtained was quite similar to that given by dimensional grating, a number of spots arranged in three parallel rows making $60°$ with each other as the following photo, obtained by Springer in Germany.

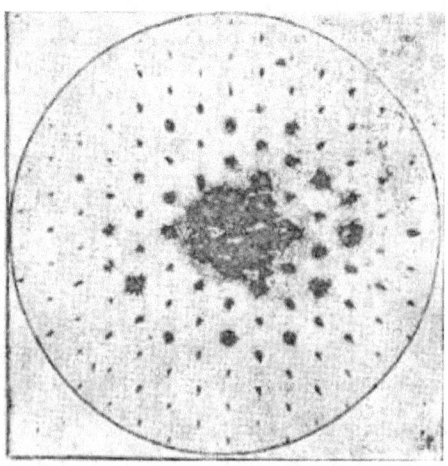

Diffraction of electron by thin mica (45,000 volts)

From the study of the pattern, between the diffracting elements could be calculated assuming the wavelength as given by De Broglie's formula. This gave values lying between 5.1 A° and 5.23 A° with an average value 5.18 A°. Crystallographic studies on the other hand, give the value as 5.17 A°. The close agreement between the values argues once again to the existence of electron waves and also to the correctness of De Broglie's formula.

Rupp in Germany succeeded in 1929 in diffracting slow electrons (150-300) volts) by means of a ruled grating with 1300 lines per cm. The wavelength obtained from his experimental data agreed within two per cent with that predicted by De Broglie's theory.

J.V. Hughes in 1934 was able to verify De Broglie's law for electrons with velocities comparable with that of light.

Practical applications of electron diffraction

The phenomenon of electron diffraction has now passed far beyond the stage of academic interest and is being used to extend the methods of X-ray diffraction to the study of the inner structure of materials, particularly in the structure of thin films of surfaces and of complex molecules.

In the study of surface structure, electrons have two inherent advantages over X-rays, viz. they penetrate less deeply and their interaction with the atoms of the body under test is more intimate than in the case of X-rays. A further practical advantage of electron diffraction is the high intensity of the diffraction pattern so that only very short exposure times are necessary for photographic purposes, whereas much longer times are required when X-rays are used.

G. P. Thomson has used electron diffraction to investigate surface layers of unknown composition. The nature of the patterns produced enables the crystalline structure of the surface to be determined.

Buhl and Rupp have shown that reproducible patterns are obtained when electrons are reflected from oil surfaces.

Dauvillier has used the transmission method of diffraction in the structural analysis of organic materials prepared in thin films. Jenkins, Murison, Langmuir and others have used electron diffraction in the study of lubrication by graphite, grease and oils. Modern electron microscopes are usually fitted with an electron diffraction unit as a very useful complement.

Polarization of electron waves

The close analogy between the electron waves and X-rays has naturally led investigators to search for polarization in these waves, chief among whom were Davisson, Germer and Rupp. The method used is similar to that applied in the case of X-rays, viz., to

produce a double reflection of the electron beam on two similar surfaces and examine whether the intensity of the second reflected ray depends upon the azimuth of the second reflector with respect to the first.

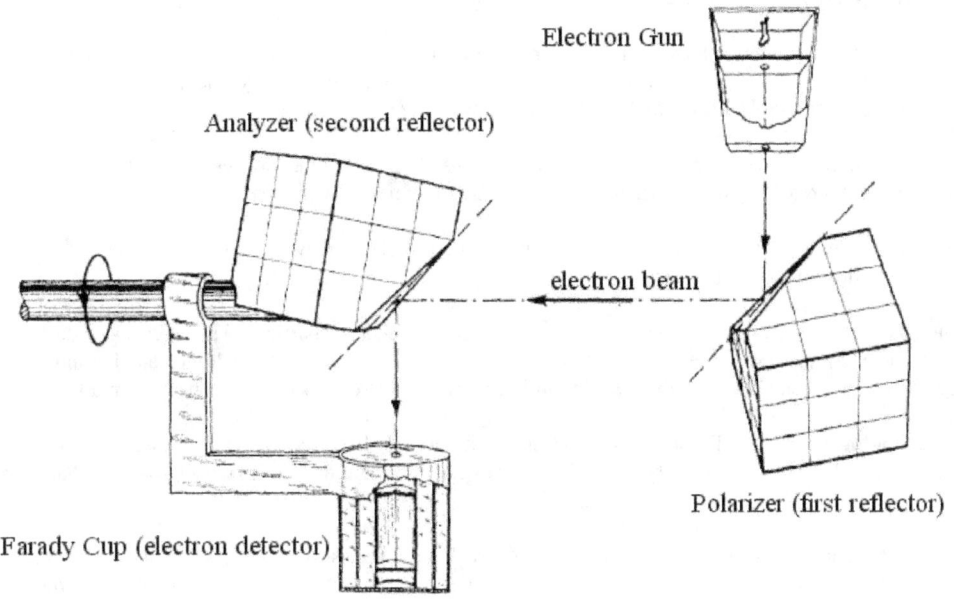

It was found that there was no variation in intensity greater than 0.5 per cent. On the whole, the polarization of the electron waves appears doubtful,

From the theoretical point of view the results obtained are more promising. For, Dirac in 1928, by an ingenious application of the principles of relativity to wave mechanics, established that the electron possessed a magnetic moment and spin. From this finding, it could be argued to the possibility of polarization of electron waves, since just as the polarization of light waves can be attributed to the spin of the photon, so also the spin of the electron should cause polarization of electron waves.

In 1992, SLAC (Stanford Linear Accelerator Center) commissioned the Polarized Electron Source that generates the polarized electrons by illuminating a photocathode with circular polarized light. The photoinjector also delivers the unpolarized electron beam. The new GaAs source for the SLC produces highly reliable polarized-electron beam to the level of 80-86%. polarization.

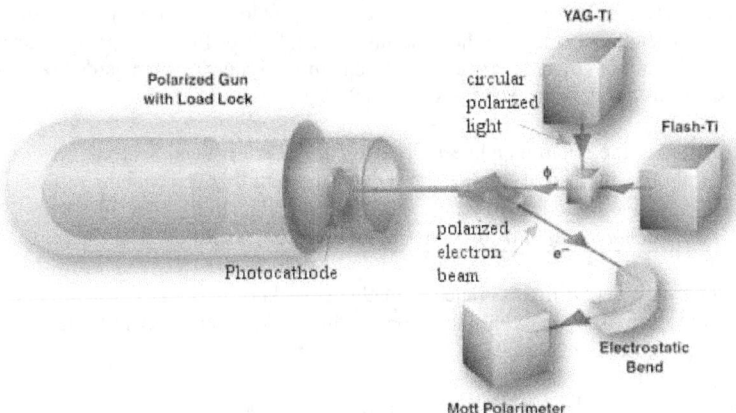

Wave nature of neutral particles, such as free atoms and molecules

Since De Broglie's theory applies to any material particle, experiments designed to show the wave characteristics of free atoms and molecules have been tried by several workers. Special difficulties have to be overcome in producing diffraction effects with atoms and molecules, which do not arise in working with electrons.

First of all, as the atoms and molecules are much heavier than electrons, the De Broglie wavelength will be much smaller for them and it will therefore be more difficult to detect diffraction effects with them.

Secondly, the free atoms or molecules possess a Maxwellian distribution of velocity, whereas for diffraction experiments, a beam of uniform velocity is desirable.

Thirdly, neutral atoms or molecules are very much harder to detect than are charged particles. However, Dempster in 1927 and Estermann and Stern in 1930 succeeded in obtaining clear diffraction effects with hydrogen and helium by suitably reflecting them from crystal gratings, thus lending full support to De Broglie's general concept of matter waves.

It may be noted here that for bodies in bulk such as those we deal with in daily life, a cricket ball for instance, the wavelengths are far too small to be observable, being of the order of 10^{-24} A^o. But there is no reason to doubt that even these have a De Broglie wavelength associated with them.

It is clear from the experiments described above that some sort of wave must be associated with all material particles whatever their nature. This fundamental fact leads to **a** new form of mechanics different from the classical one when the fine structure details of matter are to be considered. This new mechanics must be able to combine valid aspects of the wave theory of light, the quantum theory of radiation and the dynamics of particles in motion; it should also be able to throw more light on the nature of the matter waves which, according to the simple analogical conception of De Broglie, can just be conjectured as a kind of wave packet, not of the same type as that of radiation, but without further physical significance. We shall, therefore, now pass on to the study of the new type of mechanics which the matter waves called for.

THEORETICAL STUDY OF MATTER WAVES

One of the reasons which made De Broglie have recourse to the idea of matter waves was the concept of the stable and non-radiating electronic orbits restricted by quantum rules to integral numbers in the Bohr theory of atomic structure as we have already mentioned. In the same theory, on the assumption that energy is radiated as spectral lines when the electron jumps from one quantum orbit to another and that the frequency of the emitted radiation is strictly governed by a law known as Bohr's frequency condition, great advance was made in the analysis of optical and X-ray spectra. But after a time it was realized that Bohr's picture of the atom was not complete. Spectral lines were observed which could not be made to correspond to any known electronic orbits and special orbits had to be provided, often with very little reason other than that the spectrum seemed to demand them. With the experimental analysis of fine structure of spectral lines and anomalous Zeeman effect etc., the Bohr atom model was being more and more frequently disregarded because of its increasing inability to account satisfactorily for the observed facts.

In 1925, therefore, even before the wave nature of the electron had been discovered, Heisenberg and Schrödinger, taking up the clue of matter waves of De Broglie as a possible solution that might successfully overcome the difficulties met with in experimental spectroscopy, developed independently and from two different directions a new mechanics. The one of Heisenberg is known as *quantum, mechanics* and that of Schrödinger wave *mechanics.*

The quantum, mechanics of Heisenberg involves a very difficult branch of mathematics known as matrix calculus and it is well nigh impossible to describe it in simple terms. Hence, within the scope of this book, we can only indicate the *main* line of argument of Heisenberg and that too qualitatively.

The wave mechanics of Schrödinger is more readily understood and lends itself to a more or less pictorial representation. We shall therefore deal with it somewhat more in detail. But here also, applications to particular cases involve fairly complex mathematical solutions and hence only the results of such solutions will be made use of, wherever necessary, in the study of Atomic Physics. It may be noted that Schrödinger subsequently showed that the quantum mechanics of Heisenberg and his own wave mechanics were mathematically equivalent and led to the same results. Both methods have been fruitful and are included in the general term, the *New Quantum Theory.*

QUANTUM MECHANICS

Heisenberg started with a fundamental principle, already emphasized strongly by Einstein, that it is meaningless for Science to talk about things unless we can observe and measure them. In other words, scientific enquiry should contain only "observable" quantities and strictly abstain from "unobservable" quantities. In Heisenberg's view,

Bohr's theory partially fails because it deals with quantities which entirely elude observation, such as electronic orbits, motion of the electron round the nucleus, electronic jumps from one orbit to another, etc.

Hence if a logically consistent system of atomic mechanics is to be built up, only physically observable entities should be introduced, such as, spectral lines, wavelengths, frequencies, intensities of the light emitted by the atom, etc.

Considering therefore only the observable frequencies of the spectral lines which could be represented by paired quantum numbers, Heisenberg arranged them in a square array as

$$
\begin{array}{cccc}
\nu_{11} = 0 & \nu_{12} & \nu_{13} & \cdots \\
\nu_{21} & \nu_{22} = 0 & \nu_{23} & \cdots \\
\nu_{31} & \nu_{32} & \nu_{33} = 0 & \cdots
\end{array}
$$

In mathematics such an arrangement is called a •*matrix* while the individual members characterized by two indices the *elements of the matrix.* In a similar manner, for the intensities of the lines, a second matrix can be formed with the amplitudes of the "virtual resonators" associated with the various frequencies, since intensity is given by the square of the amplitude. Thus:

$$
\begin{array}{cccc}
A_{11} & A_{12} & A_{13} & \cdots \\
A_{21} & A_{22} & A_{23} & \cdots \\
A_{31} & A_{32} & A_{33} & \cdots
\end{array}
$$

Heisenberg next basing himself on experimental facts, such as Ritz's combination principle, established that if two vibrational factors, say

$$
A_{nk} = e^{2\pi i \nu_{nk} t} \qquad \text{and} \qquad A_{km} = e^{2\pi i \nu_{km} t}
$$

are multiplied, a third vibrational factor is obtained belonging to the same array. From this he deduced a general multiplication rule, viz., the product of two square arrays is also a square array of the same type. This empirical rule is identical with one which has long been known in mathematics as the rule for forming the product of two matrices. Hence the mathematical theory of matrices can be legitimately applied to Heisenberg's square arrays.

Now we come to the central point of Heisenberg's theory. A representative matrix of the above type is associated with every physical quantity. We can thus form a coordinate matrix, momentum matrix, etc., and then calculate with these matrices in practically the same manner as in classical mechanics. But there is one essential distinction between matrix and classical mechanics, viz., that
- when matrices are introduced as momenta *(p)* and coordinates *(q),*
- *their product is not commutative;* analytically this means that $pq \neq qp$, *unlike* in classical mechanics. The difference $(pq — qp)$ has a definite meaning, being itself a matrix, which is usually different from zero.

Heisenberg and a great number of other workers have shown that the whole problem of atomic spectra is solved, if $pq - qp = h/2\pi i$, where *h* is Planck's constant and $i = \sqrt{-1}$. This replaces the quantum condition in Bohr's theory and is known as *the commutation law* of Heisenberg.

MATRICES ALGEBRA

(a) Addition and subtraction

These operations require that the matrices concerned must be of the same order. If A and B are two matrices of the same order with elements a_{ik} and b_{ik} respectively then their sum $A + B$ is defined as a matrix C whose elements $c_{ik} = a_{ik} + b_{ik}$. Clearly C has the same order as A and B. For example if

$$A = \begin{bmatrix} 1 & -2 & 3 \\ 1 & 0 & 2 \end{bmatrix} \quad \text{and} \quad B = \begin{bmatrix} 2 & 3 & -1 \\ 0 & 2 & 3 \end{bmatrix}$$

then

$$C = A + B = \begin{bmatrix} 3 & 1 & 2 \\ 1 & 2 & 5 \end{bmatrix}$$

Two matrices of the same order are said to be conformable for addition and subtraction. Subtraction of matrices is defined in a similar way in that the difference of two matrices A and B of the sane order is a matrix D whose elements $d_{ik} = a_{ik} - b_{ik}$. For example using the matrices A and B in the previous example we have

$$D = A - B = \begin{bmatrix} -1 & -5 & 4 \\ 1 & -2 & -1 \end{bmatrix}$$

From the definition it follows that $A + B = B + A$
Similarly we have $A + (B + C) = (A + B) + C = A + B + C$

addition and subtraction of matrices is both commutative and associative.

(b) Equality of matrices

Two matrices A and B with elements a_{ik} and b_{ik} respectively are only equal if they are of the same order, and if all their corresponding elements are equal (i.e., $a_{ik} = b_{ik}$)

(c) Multiplication by a number (or scalar)

The result of multiplying a matrix $A = [a_{ik}]$ by a number k (real or complex) is defined as a matrix B whose elements b_{ik} are k- times the elements of A For example, if

$$A = \begin{bmatrix} 1 & 2 & 0 \\ 3 & -1 & 2 \end{bmatrix} \quad \text{then} \quad kA = \begin{bmatrix} k & 2k & 0 \\ 3k & -k & 2k \end{bmatrix}$$

From this definition it follows that the distributive law of elementary algebra holds also for matrices in that

$$k (A + B) = kA + kB$$

Furthermore

if we define $kA = Ak$, then multiplication by a number is commutative. It is important to state that k must be a number and not for instance another matrix.

(d) Matrix multiplication

The definition of matrix multiplication A and B can only be multiplied together to form their product AB if the number of **columns** of A is equal to the number of **rows** of B. A is then said to be conformable to B for multiplication.

Suppose A is a matrix of order (**mxp**) with elements a_{ik} and B is a matrix of order (**pxn**) with elements b_{ik}. Then their product AB is a matrix G of order (**mxn**) with elements c_{ik} defined by

$$c_{ik} = \sum_{s=1}^{p} a_{is} b_{sk}$$

For example, if **A** and **B** are (3x 2) and (2x 2) matrices, respectively, given by

$$A = \begin{bmatrix} a_{11} & a_{12} \\ a_{21} & a_{22} \\ a_{31} & a_{32} \end{bmatrix} , \quad B = \begin{bmatrix} b_{11} & b_{12} \\ b_{21} & b_{22} \end{bmatrix}$$

then the product $C = AB$ is a (3×2) matrix defined as

$$C = \begin{bmatrix} a_{11} b_{11} + a_{12} b_{21} & a_{11} b_{12} + a_{12} b_{22} \\ a_{21} b_{11} + a_{22} b_{21} & a_{21} b_{12} + a_{22} b_{22} \\ a_{31} b_{11} + a_{32} b_{21} & a_{31} b_{12} + a_{32} b_{22} \end{bmatrix}$$

Clearly $AB \neq BA$, since the orders of the matrices representing the two products are different. This non commutative property of matrix multiplication appears even when **A** and **B** are such that their two products **AB** and **BA** are matrices of the same order.

Matrix multiplication is non commutative, i.e., $AB \neq BA$, either when the orders of the matrices representing the two products are different or when **A** and **B** are such that their two products **AB** and **BA** are matrices of the same order.

However, apart from this non commutative law of multiplication,

Provided that the products are defined, matrices satisfy the distributive and associative laws of multiplication

(i) $(A + B) C = AC + BC$
(ii) $(AB) C = A (B C)$

Although matrix multiplication is distributive with respect to addition,

the failure of the commutative law means that the usual formulae for real numbers concerning special products and factorizing are not in general valid for matrices,

for example

$(A+B)^2 = (A+B)(A+B) = A^2 + AB + BA + B^2$
$(A+B)(A-B) = A^2 - AB + BA - B^2$

(i) Rank of a matrix

A matrix is of rank r if there exists at least one non zero determinant of order r while all determinants of order higher than r are zero.

The existence of a non - zero determinant of order r implies the existence of at least one non - zero determinant of every order less than r. The rank of a matrix is of fundamental importance in the study of the linear dependence of equations such as those encountered in quantum mechanics.

(ii) Row vector

A set of **n** quantities arranged in a row is a matrix of order (**1x n**). Such a matrix is usually called a row matrix or row vector and is denoted by

$$[A] = \begin{bmatrix} a_1 & a_2 & a_3 & \ldots & a_n \end{bmatrix}$$

(iii) Column vector

A set of **m** quantities arranged in a column is a matrix of order (**mx1**). Such a matrix is called a column matrix or column vector.

(iv) Null matrix

Any matrix of order (**mxn**) with all its elements <u>equal to zero</u> is called a null matrix of order (**mxn**)

(v) Square matrix

A square matrix has the same number of rows as columns.

(v) Unit matrix

The unit matrix is a diagonal matrix with all its diagonal elements <u>equal to unity</u>. It is usual to denote such matrices by the letter **I**; for example the (3x 3) unit matrix is

$$I = \begin{bmatrix} 1 & 0 & 0 \\ 0 & 1 & 0 \\ 0 & 0 & 1 \end{bmatrix}$$

In general, if **A** is a square matrix of order (**mxm**) and **I** is the unit matrix of the same order then

I A = A I = A

Multiplying any matrix by a unit matrix leaves the matrix unchanged provided the product is defined.

(vi) Determinant of a matrix

The determinant of a matrix **A** is only defined when **A** is square and is then just the determinant of the matrix elements. It is usually denoted by |**A**| or det **A**.
For example if

$$A = \begin{bmatrix} 1 & 0 & 1 \\ 2 & 3 & 1 \\ 3 & 1 & 2 \end{bmatrix}$$

then

$$|A| = \begin{vmatrix} 1 & 0 & 1 \\ 2 & 3 & 1 \\ 3 & 1 & 2 \end{vmatrix} = -2$$

When a square matrix is such that its determinant is zero it is called a singular matrix (otherwise it is non - singular or regular).

For example, the matrix

30

$$\begin{bmatrix} 1 & 2 & 1 \\ 3 & -1 & 3 \\ 5 & 7 & 5 \end{bmatrix}$$

is singular since its determinant vanishes in virtue of two columns being identical.

(vii) Diagonal Matrix

A square matrix with zero elements everywhere except in the leading diagonal is called a diagonal matrix.

In other words, if a_{ik} are the elements of a diagonal matrix we must have $a_{ik} = 0$ for $i \neq k$.

A typical (4 x 4) diagonal matrix is

$$A = \begin{bmatrix} 1 & 0 & 0 & 0 \\ 0 & 3 & 0 & 0 \\ 0 & 0 & 5 & 0 \\ 0 & 0 & 0 & 2 \end{bmatrix}$$

Clearly all diagonal matrices of the same order commute under multiplication.

(viii) The Transposed Matrix

In the case of determinants we have seen that interchanging rows and columns leaves the value of that determinant unaltered. However,

if the rows and columns of a matrix are interchanged a new matrix called the transposed matrix is obtained.

For example if A is a (3 x 2) matrix given by

$$A = \begin{bmatrix} a_{11} & a_{12} \\ a_{21} & a_{22} \\ a_{31} & a_{32} \end{bmatrix}$$

then its transpose (denoted by A') is the (2 x 3) matrix

$$A' = \begin{bmatrix} a_{11} & a_{21} & a_{31} \\ a_{12} & a_{22} & a_{32} \end{bmatrix}$$

$$\{A\}'\{A\} = [A][A]' = a_1^2 + a_2^2 + a_3^2$$

In general we see that

if **A** is of order (**mxn**) then **A'** is of order (**nxm**), and hence **A** and **A'** are conformable to both products **A A'** and **A' A** (i.e., both products exist but are of different orders unless **A** is square).

We now show that if **A** and **B** are two matrices conformable to the product **AB** = **C**, then

C'= (AB)' = B' A'

Suppose **A** is of order (**mxp**) with elements a_{ik} and **B** is of order (**pxn**) with elements b_{ik}. Then **C** is a matrix of order (**mxn**) with elements c_{ik} given by:

$$c_{ik} = \sum_{s=1}^{p} a_{is} b_{sk}$$

Consequently

$$c'_{ik} = \sum_{s=1}^{p} a_{ks} b_{si} = \sum_{s=1}^{p} b'_{is} a'_{sk}$$

and therefore

C' = B' A'

Similarly we may show that

(ABC)' = C' B' A'

In other words,

in taking the transpose of matrix products the order of the matrices forming the product must he reversed.

(ix) Complex and Hermitian Matrices

If **A** is any matrix of order (**mxn**) with complex elements a_{ik}, then the complex conjugate A* of A is found by taking the complex conjugates of all the elements.

For example, if

$$\mathbf{A} = \begin{bmatrix} 3+2i & 4 & -i \\ 5 & 2-i & 1+i \end{bmatrix}$$

Then

$$\mathbf{A}^* = \begin{bmatrix} 3-2i & 4 & i \\ 5 & 2+i & 1-i \end{bmatrix}$$

Clearly $(A^*)^* = A$

A Hermitian matrix is a square matrix which is unchanged by taking the transpose of its complex conjugate,

that is, A is Hermitian if

$$(\mathbf{A}^*)' = \mathbf{A}$$

For example if

$$\mathbf{A} = \begin{bmatrix} 1 & 1+i \\ 1-i & 3 \end{bmatrix}$$

Then

$$\mathbf{A}^* = \begin{bmatrix} 1 & 1-i \\ 1+i & 3 \end{bmatrix}$$

And consequently

$$(A^*)' = \begin{bmatrix} 1 & 1+i \\ 1-i & 3 \end{bmatrix} = A$$

It is obvious that the diagonal elements of a Hermitian matrix are always real.

If a square matrix A is such that $(A^*)' = -A$, it is called a skew-Hermitian.

(x) The Adjoint and Reciprocal matrices
If **A** is the square matrix

$$\begin{bmatrix} a_{11} & a_{12} & \cdots & a_{1n} \\ a_{21} & a_{22} & \cdots & a_{2n} \\ \cdot & \cdot & \cdots & \\ \cdot & \cdot & \cdots & \\ a_{n1} & a_{n2} & \cdots & a_{nn} \end{bmatrix}$$

then its adjoint (usually denoted by adj **A**) is defined as the transpose of the matrix of its cofactors.

The adjoint (usually denoted by adj **A**) of the square matrix **A** is the transpose of the matrix of its cofactors.

In other words, if A_{rs} is the cofactor of the element a_{rs} (i.e., the value of the determinant formed by deleting the row and column in which a_{rs} occurs and attaching a plus or minus sign in accordance with the known rule), then the matrix of the cofactors is the square matrix **B** (of the same order as **A**) where

$$B = \begin{bmatrix} A_{11} & A_{12} & \cdots & A_{1n} \\ A_{21} & A_{22} & \cdots & A_{2n} \\ \cdot & \cdot & \cdots & \\ \cdot & \cdot & \cdots & \\ A_{n1} & A_{n2} & \cdots & A_{nn} \end{bmatrix}$$

Consequently

$$\text{adj } A = B' = \begin{bmatrix} A_{11} & A_{21} & \cdots & A_{n1} \\ A_{12} & A_{22} & \cdots & A_{n2} \\ \cdot & \cdot & \cdots & \\ \cdot & \cdot & \cdots & \\ A_{1n} & A_{2n} & \cdots & A_{nn} \end{bmatrix}$$

We now consider the product

$$A \, (\text{adj } A) = \begin{bmatrix} a_{11} & a_{12} & \cdots & a_{1n} \\ a_{21} & a_{22} & \cdots & a_{2n} \\ \cdot & \cdot & \cdots & \cdot \\ \cdot & \cdot & \cdots & \cdot \\ a_{n1} & a_{n2} & \cdots & a_{nn} \end{bmatrix} \begin{bmatrix} A_{11} & A_{21} & \cdots & A_{n1} \\ A_{12} & A_{22} & \cdots & A_{n2} \\ \cdot & \cdot & \cdots & \cdot \\ \cdot & \cdot & \cdots & \cdot \\ A_{1n} & A_{2n} & \cdots & A_{nn} \end{bmatrix}$$

$$= \begin{bmatrix} |A| & 0 & 0 & \cdots & 0 \\ 0 & |A| & 0 & \cdots & 0 \\ 0 & 0 & |A| & \cdots & 0 \\ \cdot & \cdot & \cdots & \cdot \\ 0 & 0 & 0 \, 0 & \cdots & |A| \end{bmatrix} = |A| \, I$$

where $|A|$ is the determinant of A and I is the unit matrix of the same order as A
It follows that the matrix A^{-1} defined by

It follows that the matrix A^{-1} defined by

$$A^{-1} = \frac{\text{adj } A}{|A|}$$

has the property that $AA^{-1} = I$

and in this sense it is referred to as the reciprocal (or inverse) matrix of A. It is easily verified that

multiplication of matrices and their inverses is commutative in that $A \, A^{-1} = A^{-1} \, A = I$
Clearly only nonsingular square matrices have reciprocals, since A^{-1} is undefined when $|A| = 0$

We shall now show how a non - singular square matrix can be transferred from one side of a matrix equation to the other side. Suppose A is a non - singular square matrix and suppose that B and C are two matrices such that

$AB = C$,

we shall now show that

$B = A^{-1} \, C$

For $A^{-1} \, C = A^{-1} \, A \, B = I \, B = B$

Thus A occupies the first position on L. H. S. and A^{-1} occupies also the first position on R. H. S.
Again suppose that

$BA = D$,

We shall show that

$B = D \, A^{-1}$

For

$D A^{-1} = BA A^{-1} = B I = B$

Thus **A** occupies the second position on L. H. S, and A^{-1} occupies also the second position on R. H. S. Hence we have the following rule:

> *If a non - singular matrix is transferred from one side of a matrix equation to the other, then its inverse must occupy in the new side a position similar to that which the matrix occupied in the original side.*

(xi) Pre Division and Post Division

Let **A** be a non-singular square matrix of order **n** x **n** and **B** is a matrix. We are now in a position to give meaning to **B** ÷ **A** by finding the matrix **P** or **Q** such that **A P = B** or
Q A = B.

If **A P = B** then $P = A^{-1} B$. This implies that **B** has **n** rows. This is called a <u>pre-division.</u>

If **QA = B** then $Q = B A^{-1}$. This implies that **B** has **n** columns. This is called a <u>post-division</u>

If **B** is a square matrix of order (**n** x **n**) then both **P** and **Q** exist but they are in general different.

(xii) Orthogonal matrices

A square matrix **A** is said to he orthogonal if **A' A = I**
Therefore,

$AA' = A I A^{-1} = A A^{-1} = I$

(xiii) The characteristic equation of a matrix

> If **A** is a square matrix of order (**n** x **n**) and **I** is the unit matrix of the same order, then the matrix **B = A** − λ**I** is called the characteristic matrix of **A**, λ being a parameter.

For example

$$A = \begin{bmatrix} 1 & 0 & 1 \\ 0 & 2 & 0 \\ 1 & 1 & 3 \end{bmatrix}$$

Then

$$B = \begin{bmatrix} 1 & 0 & 1 \\ 0 & 2 & 0 \\ 1 & 1 & 3 \end{bmatrix} - \lambda \begin{bmatrix} 1 & 0 & 0 \\ 0 & 1 & 0 \\ 0 & 0 & 1 \end{bmatrix}$$

$$= \begin{bmatrix} 1-\lambda & 0 & 1 \\ 0 & 2-\lambda & 0 \\ 1 & 1 & 3-\lambda \end{bmatrix}$$

> $|\mathbf{B}| = |A - \lambda I| = 0$
> is called the characteristic equation of **A** and is in general an equation of the **n**th degree in λ. The **n** roots of this equation are called the characteristic or latent roots (or eigen values) of **A**.

(xiv) Consistent and Inconsistent systems of Equations

A set of equations that have at least one common solution is said to be a consistent set of equations. A set for which there exists no common solution is called an inconsistent set.

The question of consistency is frequently of practical importance in the quantum theory. For example, there are often more conditions than there are variables. This leads to a system in which there are more equations than there are unknowns. It is important to have a method for testing whether all the conditions can be satisfied simultaneously.

Consider a system of **m** linear equations in **n** unknowns,

$$a_{11} x_1 + a_{12} x_2 + \ldots + a_{1n} x_n = h_1$$
$$a_{21} x_1 + a_{22} x_2 + \ldots + a_{2n} x_n = h_2$$
$$\cdots \cdots \cdots \cdots , \cdots \cdots \cdots \cdots$$
$$a_{m1} x_1 + a_{m2} x_2 + \ldots + a_{mn} x_n = h_m$$

where at least one $h_1 \neq 0$

From the matrix of coefficients, *namely*

$$\begin{bmatrix} a_{11} & a_{12} & \cdots & a_{1n} \\ a_{21} & a_{22} & \cdots & a_{2n} \\ \cdot & \cdot & & \cdot \\ a_{m1} & a_{m2} & \cdots & a_{mn} \end{bmatrix}$$

Let its rank be **r**. Form the augmented matrix by adjoining to the matrix of the coefficients the column in the right band side of the given equations, i.e.,

$$\begin{bmatrix} a_{11} & a_{12} & a_{1n} & \cdots & h_1 \\ a_{21} & a_{22} & a_{2n} & \cdots & h_2 \\ \cdot & \cdot & \cdot & & \cdot \\ a_{m1} & a_{m2} & a_{mn} & \cdots & h_m \end{bmatrix}$$

Let its rank be **r'**.

Then, we have the following fundamental rule

I) Unique solution: if $r = r' = n$
II) More than one solution: if $r = r' < n$

In this case we can give **n − r** of the unknowns arbitrary values and then express the remaining **r** unknowns in terms of these. The **r** unknowns, which are expressed in terms of the others must be associated with some non vanishing determinant of order **r**

III) No solution if **r' > r**

Since r can not exceed r' since all determinants in the matrix of coefficients are included in the augmented matrix.

1. THE CORPUSCULAR CONCEPT OF MATTER

The derivation of the mathematical scheme of the quantum theory is based on the following four pillars:

- the wave theory fields;
- the particle or corpuscular picture of light;
- the available empirical facts; and
- the correspondence principle.

The correspondence principle which is due to Bohr, postulates a detailed analogy between the quantum theory and the classical theory appropriate to the mental picture employed. This analogy does not merely serve as a guide to the discovery of formal laws; its special value is that it furnishes the interpretation of the laws that are found in terms of the mental picture used.

We commence with a derivation of the mathematical structure of quantum mechanics from the corpuscular analogy.

Classical equation of motion of a particle

The fundamental equations of classical mechanics for a system of f-degrees of freedom may be written in the so-called "canonical" form,

$$\dot{p}_k = -\frac{\partial H}{\partial q_k}, \qquad \dot{q}_k = \frac{\partial H}{\partial p_k}, \quad (k = 1,2,...,f) \ldots\ldots\ldots\ldots(1)$$

Where
$q_1, q_2, q_3, \ldots, q_f$ are the generalized coordinates,
$p_1, p_2, p_3, \ldots, p_f$ their conjugate momenta, and
H the Hamiltonian function corresponding to the total energy of the system. Its spectrum is the set of possible outcomes in relation to the time-evolution of a system.

$\dot{p}_k$ and $\dot{q}_k$ are the time derivatives of the momenta and coordinates which corresponds to the energies and velocities of the particle for each k-degree of freedom.

Equation (1) describes the state of motion of a particle according to Newton's law where the rates of changes of momenta; p's and of displacements, q's, are determined by the changes of the total energy through H. This is the classical Newtonian invariable of energy and momentum which requires that once motion is initiated, it never vanishes, but rather converts into different kinds of energies. That is energy is invariable

When H does not depend explicitly on the time as in steady state conditions, the energy equation

$$H(p,q) = W \ldots\ldots\ldots\ldots\ldots\ldots\ldots\ldots\ldots\ldots\ldots(2)$$

where W, the total energy, is a constant, follows at once.

Multiply periodic coordinates and momenta

For simplicity it may be assumed that the system is multiply periodic, in which case any coordinate q_k and momenta p_k as a function of the time may be written as a Fourier series, that is, as a sum of harmonic terms in the form

$$q_k \sum_{\tau_1=-\infty}^{+\infty} \sum_{\tau_2=-\infty}^{+\infty} \cdots \sum_{\tau_f=-\infty}^{+\infty} q_{\tau_1,\tau_2,\ldots,\tau_f}^{(k)} \, e^{2\pi i(\tau_1 v_1 + \tau_2 v_2 + \ldots \tau_f v_f)t} \quad \ldots\ldots\ldots\ldots\ldots(3a)$$

$$p_k \sum_{\tau_1=-\infty}^{+\infty} \sum_{\tau_2=-\infty}^{+\infty} \cdots \sum_{\tau_f=-\infty}^{+\infty} p_{\tau_1,\tau_2,\ldots,\tau_f}^{(k)} \, e^{2\pi i(\tau_1 v_1 + \tau_2 v_2 + \ldots \tau_f v_f)t} \quad \ldots\ldots\ldots\ldots\ldots(3b)$$

Where,

$q^{(k)}_{\tau_1, \tau_2, \ldots\ldots\ldots f}$ are amplitudes independent of the time and
$v_1, v_2, \ldots v_f$, are the fundamental frequencies of the motion.

Equation (3) is a pure mathematical device that links Newton's law which is depicted in equations (1) and (2) with the wave picture of particles. Because, equation (3) shows that

> the definite boundaries of a material particle could be described as a Fourier summation of waves whose resultant delineates the discreteness of a material body.

The resultant of summation of waves entails the destructive and constructive interference of waves which are governed by the state of energy.

Similar expressions involving the same frequencies may be written for any function of the p_k and q_k.

Canonical transformation of force and velocity to action-angle variables

By a canonical transformation—that is, one which leaves invariant the form of equations (1)—it is possible to introduce a. new set of canonical conjugates J_k, w_k, known as action-angle variables." These are essentially defined by the following properties:

The Hamiltonian H depends on the J_k only and the w_k are related to the fundamental frequencies of the motion by equations of the form

$$\omega_k = v_k t + \beta_k$$

where the β_k are constants. In these variables the equations of motion (1), given above as

$$\dot{p}_k = -\frac{\partial H}{\partial q_k}, \qquad \dot{q}_k = \frac{\partial H}{\partial p_k}, \qquad (k = 1,2,\ldots,f)$$

Can be written in the form

$$\dot{p}_k = -\frac{\partial H}{\partial \varpi_k}\frac{\partial \varpi_k}{\partial t}\frac{\partial t}{\partial q_k}, \qquad \dot{q}_k = \frac{\partial H}{\partial J_k}\frac{\partial J_k}{\partial t}\frac{\partial t}{\partial p_k}, \qquad (k = 1,2,\ldots,f)$$

Which gives

$$\dot{J}_k = -\frac{\partial H}{\partial \varpi_k} == \frac{\dot{p}_k \dot{q}_k}{v_k}, \qquad \dot{\varpi}_k = \frac{\partial H}{\partial J_k} = v_k, \qquad (k = 1,2,\ldots,f) \ \ldots (4)$$

Equation (4) is a canonical transformation that leaves invariant the form of equations (1) where the canonical conjugates J_k and w_k, "action-angle variables" replace the p_k and q_k. Evidently, J_k has the units of (energy x time) as it represents the energy invariable and ω_k is dimensionless as it represents the angular coordinate.

Restraining the Fourier harmonics to the frequencies of spectral lines

According to classical electrodynamics the frequencies of the spectral lines emitted by an atom will be the frequencies of the harmonic terms in equation (3) and the amplitudes will determine the corresponding intensities.

As we see from equations (3) and (4),

Heisenberg has already concluded that the particles oscillate through their coordinates q_k's and that such oscillations determine the magnitude and frequencies of the spectral lines of the emitted radiations from those very particles.

According to the correspondence principle, there must exist a close relationship between the mechanics of classical particles as outlined above and the mechanics of the quantum theory. For the latter we must therefore seek a set of equations analogous in form to the equations of classical theory, but which also take account of certain well-established empirical facts of atomic physics. Primary among these are the following:

1. *The Rydberg-Ritz combination principle*
The observed spectral frequencies of an atom possess a characteristic term structure. That is, all the spectral lines of an element may be represented as the differences of a relatively small number of terms. If these terms are arranged in a one-dimensional array $T_1, T_2, \ldots$, the atomic frequencies form a two-dimensional array

$$\nu(nm) = T_n - T_m, \qquad\qquad (5)$$

from which follows at once the combination principle

$$\nu(nk) + \nu(km) = \nu(nm). \qquad\qquad (6)$$

Equation (6) shows how Bohr's correspondence principle combined *Rydberg-Ritz* finding of the rules governing spectral frequencies had led Heisenberg to matrix formulation since both the rules of matrix algebra and those of *Rydberg-Ritz* were congruent. The emitted radiation from atomic stimulation depends on the differences between orbits or levels of the excited electrons in such manner that greater energy is released or absorbed depending of the order of the electronic shells and the direction of jumps, toward the nucleus or away from the nucleus, as follows.

$$E(\text{photon}) = h\nu = \frac{hc}{\lambda} = E_f - E_i = (-13.6\ \text{eV}))(\frac{1}{n_f^2} - \frac{1}{n_i^2})$$

When *f* and *i* denote the final initial order of the orbits whose integer numbers are denoted by *n*. As we see from the above relation, the values of *n*'s determine with the radiation is emitted (positive sign of *E*) or absorbed (negative sign of *E*).

$$\nu(nk) + \nu(km) = \nu(nm)$$

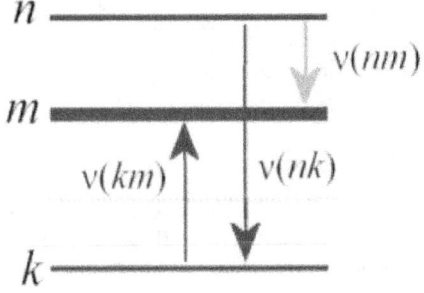

2. *The existence of discrete energy values*
The fundamental experiments of Franck and Hertz on electronic impacts show that the energy of an atom can take on only certain definite discrete values, $W_1, W_2, \ldots\ldots$

3. The Bohr frequency relation

The characteristic frequencies of an atom are related to its characteristic energies by the equation

$$\nu(nm) = \frac{1}{h}\left(W_n - W_m\right) . \tag{7}$$

We shall now sketch the deduction of the fundamental equations of the new quantum mechanics, following the program outlined above. It should be distinctly understood, however, that this cannot be a deduction in the mathematical sense of the word, since the equations to be obtained form themselves the postulates of the theory. Although made highly plausible by the following considerations, their ultimate justification lies in the agreement of their predictions with experiment.

Deviation from classical mechanics

A profound modification, not only of classical dynamics, but of classical kinematics, is evidently necessary if the simple experimental facts mentioned above are to be incorporated in the foundations of a new theory.

In the classical theory all possible motions of the coordinates may be built up by addition from Fourier terms

of the kind contained in equation (3), and these may be termed the "kinematic elements," since the quantities with which the theory deals, and in particular the energy, can be expressed in terms of them. Their amplitudes and frequencies are functions of continuously variable constants of integration as well as of the integers $\tau_1, \tau_2, ..., \tau_f$, which determine the order of the harmonics.

This is indirect contradiction to the existence of only discrete values of the atomic energies and frequencies and, in fact, to the very existence of sharply defined spectral lines.

Similar elements must be assumed in quantum mechanics if a correspondence is to be preserved between the two theories.

To assure the existence of discrete energy values at the outset, the elements will be taken to be functions of integers. Corresponding to the Rydberg-Ritz combination principle, a dependence on two sets of integers is required,

while the f-fold character of the classical harmonics suggests that each set contain f integers. We therefore postulate elements of the form

$$q\left(n_1 .. n_f ; m_1 .. m_f\right)e^{2\pi i\nu(n_1 .. n_f; m_1 .. m_f)t} , \tag{8}$$

in which the complexes $(n_1, n_2, ..., n_f)$ and $(m_1, m_2, ..., m_f)$ replace the single integers n and m in an easily understandable way. Furthermore,

the amplitudes and frequencies are assumed to be directly those which are given by aspectral analysis of the emitted radiation, so that the new theory may be described as a calculus of observable quantities.

The frequencies $\nu(n_1, n_2, ..., n_f ; m_1, m_2, ..., m_f)$ are therefore assumed to have the term structure (5); they accordingly obey the combination principle (6).

Postulating a practical Fourier series of coordinates fitting experimental findings

There can clearly be no question of the addition of such elements to form a Fourier series as in the classical theory; there must, however, be an analogue to the representation of a coordinate by such a series. A sufficiently general and flexible method is afforded by taking simply the ensemble of all elements of the form (8) as the entity which, in the quantum theory of the particle picture, replaces mathematically the classical representation of a coordinate given in equation (3). The ensemble may be written as a matrix,

$$\left\| q\left(n_1 .. n_f ; m_1 .. m_f\right)e^{2\pi i\nu(n_1 .. n_f; m_1 .. m_f)t} \right\| ,$$

that is, as an infinite quadratic array, ordered according to the integers n_i, m_i, which take on all real values. The new kinematics is accordingly based on a matrix representation of the coordinates, with

$$q_k = \left\| q_k(nm)e^{2\pi i\nu(nm)t} \right\| \qquad (9)$$

corresponding to q_k. As here, the complexes $(n_1, n_2, ..., n_f)$ and $(m_1, m_2, ..., m_f)$ will, in general, be replaced by single letters n and m. For the momenta p_k a similar matrix representation is assumed, with the same frequencies, as is the case in classical Fourier series.

Defining the mathematical operations of the postulated quantum theory

Such a representation is, however, meaningless both mathematically and physically until properties and rules of operation for the matrices have been defined. The correspondence principle must be our guide here.

Requirement of real values of physical variables

In the first place, the classical expression (3) must have a real value; since the terms are complex this can be the case only if for each term there occurs the conjugate imaginary.

This will also be true of the elements of the matrix (9) if we assume

$$q_k(mn) = q^*_k(nm)$$

since by (6),

$$\nu(mn) = -\nu(nm).$$

The asterisk denotes the conjugate imaginary. Matrices with this type of symmetry are called Hermitian and in the quantum theory all coordinate matrices are assumed to be of this kind.

In the quantum theory, all coordinate matrices are Hermitian, that is for each term there occurs the conjugate imaginary such that:

$$q_k(mn) = \text{conjugate imaginary } [q_k(nm)]$$

The time derivative $\dot{q}_k$ of any coordinate is represented classically by the Fourier series whose terms are the time derivatives of those of the series representing q_k. Hencefor the quantum-theory matrices

$$\dot{q} = \left\| 2\pi i\nu(nm)q(nm)e^{2\pi i\nu(nm)t} \right\| \quad \dots\dots\dots\dots\dots\dots\dots (10)$$

which is again a Hermitian matrix of the form (9).

It must be possible in the quantum theory to answer such elementary kinematical questions as the following.

Given the matrices representing, say, a momentum p and a coordinate q, what matrices represent pq, pq, and in general any function of p and q?

Addition operation

In the case of <u>addition</u>, the answer is obvious from the classical analogue. Since
the sum of two Fourier series of the form (3) is again a series of the same kind and with the same frequencies, but with amplitudes which are the sums of the component amplitudes,
we must expect for the elements of the quantum-theory matrices

$$(p+q)(nm) = \| [p(nm) + q(nm)] e^{2\pi i \nu(nm)t} \| .$$

Multiplication operation

The rule for <u>multiplication</u> is defined from similar considerations with, however, a characteristic difference from classical multiplication, due to the fact that the quantum frequencies obey the Rydberg-Ritz combination principle. The product of two Fourier series in the classical theory may be written as the double sum

$$pq = \sum_{\sigma} \sum_{\sigma'} p_\sigma q_{\sigma'} e^{2\pi i [(\sigma+\sigma')\nu]t} ,$$

where σ replaces the complex σ_1, σ_2. . σ_f and $[(\sigma + \sigma')\nu]$ stands for $(\sigma_1 + \sigma'_1)\nu_1 + \ldots + (\sigma_f + \sigma'_f)\nu_f$.

To write this again in the form of equation (3) terms of the same frequency must be collected, i.e., those for which $\sigma + \sigma' = r$, giving

$$pq = \sum_{\tau} (pq)_\tau e^{2\pi i [\tau\nu]t} ,$$

Where

$$(pq)_\tau = \sum_{\sigma} p_\sigma q_{\tau - \sigma} . \qquad\qquad (11)$$

In the quantum theory the matrix representing pq must be an ensemble made up of terms $p(nm)e^{2\pi i \nu(nm)t}$ and $q(nm)e^{2\pi i \nu(nm)t}$.

A matrix of the type (9) is again obtained if all elements with the same frequency are added together, i.e., those for which

$\nu(nk) + \nu(km) = \nu(nm)$

by the combination principle (6). The new amplitudes are therefore taken to be

$$pq(nm) = \sum_{k} p(nk)q(km) , \qquad\qquad (12)$$

and the elements are then $pq(nm)e^{2\pi i \nu(nm)t}$

This is the well-known mathematical rule for the multiplication of matrices or tensors, and justifies the use of these terms here. As is obvious from equation (12),

$pq(nm) \neq qp(nm),$

so that

multiplication in the quantum theory is non-commutative--a result of great importance for the further development.

By means of the rules for addition and multiplication a meaning is given to any function $x(p , q)$ of the coordinate and momentum matrices, at least in so far as the function may be expressed as a power series. The elements of the function x will always be of the form

$x(nm)e^{2\pi i v(nm)t}$ and the array of frequencies $v(nm)$ will always be the same for a given atomic system. Hence a matrix is sufficiently well represented by its amplitudes $x(nm)$ alone, the exponential terms being understood.

The customary definitions and conventions of the theory of matrices are adopted in the quantum theory.

(i) Equality of two matrices means equality of corresponding elements.

(ii) The unit matrix is defined as the matrix whose diagonal elements are all unity and whose non-diagonal elements are zero. It is conveniently written

$$I = \left\| \hat{\delta}_{nm} \right\| ,$$

where

$$\delta_{nm} = \begin{cases} 1 \text{ when } n = m , \\ 0 \text{ when } n \neq m . \end{cases}$$

(iii) The reciprocal x^{-1} of a matrix x is the matrix satisfying . the equations

$$x^{-1}x = xx^{-1} = 1 .$$

(iv) The transpose $\tilde{x}$ of x is the matrix $\|x(mn)\|$ obtained by interchanging the rows and columns of x.

We are now in possession of

the elements of a quantum algebra, in which it is readily seen that all the rules of ordinary algebra remain valid with the exception of the commutative law.

Thus if x, y, and z represent any functions of the dynamical variables they obey, in the quantum theory, the rules of matrix algebra:

$$x + y = y + x ,$$
$$x(y + z) = xy + xz ,$$
$$x(yz) = (xy)z ,$$
$$(x + y) + z = x + (y + z) ,$$

but, in general,

$$xy \neq yx .$$

So far the Planck constant h, which must play a fundamental role, has not been introduced into the theory. Its appearance proves to be closely related to the non-commutative relation of the variables which forms so striking a contrast to the classical theory.

Poisson bracket

In fact, it has been found by Dirac that in the quantum theory the expression

$(2\pi i/h)(xy - yx)$

is the analogue of the Poisson bracket

$$[xy] = \sum_{k=1}^{f} \left(\frac{\partial x}{\partial q_k} \frac{\partial y}{\partial p_k} - \frac{\partial y}{\partial q_k} \frac{\partial x}{\partial p_k} \right)$$

in classical mechanics. The invariance of this expression with respect to canonical transformations of the p_k and q_k is well known. In order to make plausible this significant connection it will be shown that

in the limiting region where the integers n and m are large compared to their differences, there is asymptotic agreement between the matrix elements of $(2\pi i/h)(xy—yx)$ and the harmonic elements of the classical bracket expression $[xy]$.

It is first necessary, however, to state more exactly the connection between the matrix elements and the Fourier amplitudes.

Introducing Planck constant "h" and the discreteness of energy values

It will be recalled that in the theory of stationary states, which formed a preliminary stage in the development of the present quantum mechanics,

the existence of only discrete energy values is attained through the fixation of "stationary" classical motions.

If these are defined from among the continuum of possible motions by the equations

$$J_k = n_k h \qquad (k = 1, 2, \ldots, f) , \qquad\qquad (13)$$

where the J_k are the action variables and the n_k integers, the Bohr frequency condition (7) then appears as the analogue of the classical relation

$$\nu_k = \frac{\partial H}{\partial J_k} .$$

For since H is a function of the n_k only by equations (4), $\partial H/\partial J_k$ may be written

$$\frac{\partial H}{\partial J_k} = \lim_{a_k = 0} \frac{H(n_1 \ldots n_f) - H(n_1 \ldots n_k - a_k, \ldots, n_f)}{a_k h} ,$$

And in the limiting region where the n_k are very large compared to the a_k,

$$\nu(n_1 \ldots n_f; m_1 \ldots m_f) = \frac{1}{h} \left[H(n_1 \ldots n_f) - H(n_1 - a_1, \ldots, n_f - a_f) \right]$$

$$\sim a_1 \frac{\partial H}{\partial n_1} + \ldots + a_f \frac{\partial H}{\partial n_f}$$

$$= a_1 \nu_1 + \ldots + a_f \nu_f .$$

There is therefore asymptotic agreement in this region, which may be briefly referred to as that of large quantum integers, between the spectral frequency

$\nu(n_1, n_2, \ldots, n_f; m_1, m_2, \ldots, m_f)$

and the harmonic

$$(n_1\text{-}m_1)v_1 + \ldots + (n_f\text{-}m_f)v_f$$

in the $(n_1, n_2, \ldots, n_f)$ of $(m_1, m_2, \ldots, m_f)$ stationary state.

Since the harmonic elements of the matrices of quantum mechanics represent the spectral lines this suggests a general coordination between the matrix element

$q(n_1, n_2, \ldots, n_f; n_1\text{-}a_1, \ldots, n_f-a_f) \exp[2\pi i v(n_1, n_2, \ldots, n_f; n_1\text{-}a_1, \ldots, n_f-a_f) t]$

and the harmonic $(a_1 \ldots a_f)$ in the $(n_1 \ldots n_f)$ stationary state.

More briefly,

$$q(n, n-a)e^{2\pi i \nu(n, n-a)t} \quad \text{corresponds to} \quad q_a(n)e^{2\pi i [a\nu]t} \quad (14)$$

in the region of large quantum numbers. This coordination is further justified by the approximate agreement found empirically in this region between the intensities calculated classically from the Fourier amplitudes $q_a(n)$ in the stationary states and the intensity of the spectral line $v(n, n$ -a).

> The indices **n** and **m** of the matrix elements thus correspond to the quantum numbers of two stationary states, while the diagonal elements (**n** = **m**) correspond to the stationary states themselves.

With the aid of the coordination (14) the above-mentioned correspondence with the Poisson brackets is readily shown. The (nm) element of $(2\pi i/h)(xy\text{-}yx)$ maybe written as a sum over α and β of terms of the form

$(2\pi i/h) \{x(n, n\text{-} a) y(n\text{-} a, n\text{-}a\text{-}\beta) - y(n, n\text{-} \beta)x(n\text{-}\beta, n\text{-} a\text{-}\beta)\}$,

where
$a - \beta = n - m$.

On adding and subtracting $y(n\text{-} a, n\text{-}a\text{-}\beta)x(n\text{-}\beta, n\text{-} a\text{-}\beta)$ this becomes

$$\left(\frac{2\pi i}{h}\right)\{[x(n, n-a) - x(n-\beta, n-a-\beta)]y(n-a, n-a-\beta)$$

$$-[y(n, n-\beta) - y(n-a, n-a-\beta)]x(n-\beta, n-a-\beta)\} \ .$$

Now in the region of "large quantum numbers" where $\alpha, \beta << n$,

$$x(n, n-a) - x(n-\beta, n-a-\beta) \sim h\beta \frac{\partial x_a(n)}{\partial J} ,$$

And

$$y(n-a, n-a-\beta) \sim \frac{1}{2\pi i \beta}\frac{\partial y_\beta(n-a)}{\partial w} \sim \frac{1}{2\pi i \beta}\frac{\partial y_\beta(n)}{\partial w}$$

since the harmonics of **y** are of the form $y_\beta(n)e^{2\pi i \beta \omega}$ by equations (4).

Hence the foregoing matrix element is approximately

45

$$\sum_{a+\beta=n-m} \sum_{k=1}^{f} \left[\frac{\partial x_a(n)}{\partial J_k} \frac{\partial y_\beta(n)}{\partial w_k} - \frac{\partial y_\beta(n)}{\partial J_k} \frac{\partial x_a(n)}{\partial w_k} \right] ,$$

The summation necessarily extends into the region where the quantum numbers are not large compared to their difference; hence for numerical agreement the matrix elements far removed from the diagonal must be assumed negligible, since they correspond to high harmonics in the classical theory. The formal agreement, which is of most importance here, is, of course, unaffected which by the rule (12) for the multiplication of Fourier amplitudes is the (n-- m) harmonic of [xy], expressed in terms of the action-angle variables.

In the classical theory the Poisson brackets of canonically conjugate variables p_k and q_k satisfy the relations

$$[p_k \; q_l] = \left\{ \begin{array}{l} 1 \text{ when } k=l \\ 0 \text{ when } k\neq l \end{array} \right\} , \quad [p_k, \; p_l]=0 , \quad [q_k, \; q_l]=0 .$$

The analogous relations will therefore be assumed for conjugate variables in the quantum theory, that is

$$\left. \begin{array}{l} p_k q_l - q_l p_k = \left\{ \begin{array}{l} \dfrac{h}{2\pi i} \; 1 \text{ when } k=l \\ \\ 0 \quad \text{ when } k\neq l \end{array} \right\} \\ \\ p_k p_l - p_l p_k = 0 , \\ \\ q_k q_l - q_l q_k = 0 . \end{array} \right\} \qquad (15)$$

These "exchange relations," by means of which h is introduced into the equations, are of fundamental importance for quantum mechanics. They correspond to the quantum conditions of the theory of stationary classical motions, but whereas these conditions could be applied only to a multiply periodic system, the present exchange relations must be regarded as generally valid for any motion. In fact, as will appear later, they are necessary in order to give meaning to the problem of integration of the equations of motion, which will now be established.

The canonical equations (1) of the classical theory, if expressed in terms of the Poisson brackets, become

$$\dot{p}_k = [H p_k] , \qquad \dot{q}_k = [H q_k] .$$

The simplest assumption is to take over these equations formally into the quantum theory, replacing the Poisson brackets by their quantum analogues. We therefore assume the equations of motion in the quantum theory to be

$$\left. \begin{array}{l} \dot{p}_k = \dfrac{2\pi i}{h} \left(H p_k - p_k H \right) , \\ \\ \dot{q}_k = \dfrac{2\pi i}{h} \left(H q_k - q_k H \right) . \end{array} \right\} \qquad (16)$$

46

Clearly the equations (15) and (16) are not independent of each other. Strictly speaking, it is only permissible to assume equation (15) to be true at a single instant of time. The exchange relations at any other time must then be determined by the solution of equations (16); however, a calculation shows that equations (15) are really independent of the time.

Summary of corresponding equations between classical and quantum theory

I. Classical equations of motion

(1) Equations of motion for a system of f-degrees of freedom:

$$\dot{p}_k = -\frac{\partial H}{\partial q_k}, \qquad \dot{q}_k = \frac{\partial H}{\partial p_k}, \quad (k = 1, 2, \ldots, f) \ldots\ldots\ldots\ldots(1)$$

(2) The Hamiltonian H in steady state conditions:

$$H(p,q) = W \ldots\ldots\ldots\ldots\ldots\ldots\ldots\ldots\ldots\ldots\ldots\ldots(2)$$

(3) Harmonic for coordinate q_k and momenta p_k for multiply periodic system:

$$q_k \sum_{\tau_1=-\infty}^{+\infty} \sum_{\tau_2=-\infty}^{+\infty} \cdots \sum_{\tau_f=-\infty}^{+\infty} q^{(k)}_{\tau_1,\tau_2,\ldots,\tau_f} e^{2\pi i(\tau_1\nu_1+\tau_2\nu_2+\ldots\tau_f\nu_f)t} \ldots\ldots\ldots\ldots\ldots(3a)$$

$$p_k \sum_{\tau_1=-\infty}^{+\infty} \sum_{\tau_2=-\infty}^{+\infty} \cdots \sum_{\tau_f=-\infty}^{+\infty} p^{(k)}_{\tau_1,\tau_2,\ldots,\tau_f} e^{2\pi i(\tau_1\nu_1+\tau_2\nu_2+\ldots\tau_f\nu_f)t} \ldots\ldots\ldots\ldots\ldots(3b)$$

(4) Canonical transformation from (q,p) to (J, ω)
$\omega_k = \nu_k t + \beta_k$

$$\dot{J}_k = -\frac{\partial H}{\partial \varpi_k} == \frac{\dot{p}_k \dot{q}_k}{\nu_k}, \qquad \dot{\varpi}_k = \frac{\partial H}{\partial J_k} = \nu_k, \quad (k = 1, 2, \ldots, f) \ldots\ldots(4)$$

(5) Transitions between energy states are commutative in all combinations
xy = yx

II. Quantum equations of motion

(1) The Rydberg-Ritz combination principle

$$\nu(nm) = T_n - T_m , \qquad (5)$$

That gives

$$\nu(nk) + \nu(km) = \nu(nm) . \qquad (6)$$

(2) The Bohr frequency relation

$$\nu(nm) = \frac{1}{h}(W_n - W_m) . \qquad (7)$$

(3) Planck's quantization of energy requirement

$$J_k = n_k h \qquad (k = 1, 2, \ldots, f), \qquad (13)$$

(4) Consequences of Rydberg-Ritz combination principle

$$q(n_1, n_2, \ldots n_f; m_1, m_2, \ldots m_f) e^{2\pi i \nu (n_1, n_2, \ldots n_f; m_1, m_2, \ldots m_f) t} \quad \ldots \ldots \ldots \ldots \ldots (8)$$

$$q_k = \| q(nm) e^{2\pi i \nu (nm) t} \| \ldots \ldots \ldots \ldots \ldots \ldots (9a)$$

$$p_k = \| p(nm) e^{2\pi i \nu (nm) t} \| \ldots \ldots \ldots \ldots \ldots \ldots (9b)$$

(5) Requirement of real values of physical variables
$$q_k (mn) = q^*_k (nm)$$

$$\overset{\circ}{q} = \| 2\pi i \nu (nm) q(nm) e^{2\pi i \nu (nm) t} \| \ldots \ldots \ldots \ldots \ldots (10)$$

(6) Transition of states of energy are non-commutative

$xy \neq yx$
and
$(2\pi i / h)(xy - yx)$

In the limiting region where the n_k are very large compared to the a_k, this approximates to Poisson's bracket

$$[xy] = \sum_{k=1}^{f} \left(\frac{\partial x}{\partial q_k} \frac{\partial y}{\partial p_k} - \frac{\partial y}{\partial q_k} \frac{\partial x}{\partial p_k} \right)$$

(7) Equations of motions (obtained by replacing the Poisson brackets by their quantum analogues):

$$\left. \begin{aligned} \dot{p}_k &= \frac{2\pi i}{h} \left(H p_k - p_k H \right), \\ \dot{q}_k &= \frac{2\pi i}{h} \left(H q_k - q_k H \right). \end{aligned} \right\} \qquad (16)$$

(8) The indices n and m of the matrix elements correspond to the quantum numbers of two stationary states, while the diagonal elements ($n = m$) correspond to the stationary states themselves.

The physical constraints on the mathematical formulation of the quantum theory

The formal basis of the new mechanics is now completed;

for any physical application, however, the form of the Hamiltonian corresponding to the special dynamical problem must be known.

It is in general sufficient, in the spirit of the correspondence principle, to assume the same form as in the classical theory. The ambiguity as to the order of factors in a product which may occur here seldom arises; when it does, special considerations suffice to determine the correct form.

The law of the conservation of energy and the Bohr frequency condition are not contained explicitly in the postulates of the theory; it is therefore necessary to show that they may be derived from them.

We commence by forming a diagonal matrix $\mathbf{W}$ with elements

$$W(nm) = \begin{cases} T_n h & \text{when } n = m \\ 0 & \text{when } n \neq m \end{cases} \qquad (17)$$

where the T_n are the term values of equation (5). The time derivative of any quantity x may be expressed in terms of this matrix by the equation

$$\dot{x} = \frac{2\pi i}{h}(Wx - xW), \qquad (18)$$

since the (nm) element of $(2\pi i/h)(\omega x - x\omega)$ is

$$\frac{2\pi i}{h}\sum_k [W(nk)x(km) - x(nk)W(km)] = 2\pi i(T_n - T_m)x(nm)$$

$$= 2\pi i \nu(nm)x(nm) = \dot{x}(nm)$$

by equation (10). From equation (18) and the equations of motion (16) it follows that
$Wp - pW = Hp - pH$
And
$Wq - qW = Hq - qH$, or

$$(W - H)p = p(W - H), \quad (W - H)q = q(W - H). \quad (18')$$

That is, the matrix $W - H$ "commutes" with both p and q, and it is readily shown that it therefore commutes with any function of p and q that can be represented as a power series. In particular it commutes with H, so that

$$(W - H)H - H(W - H) = WH - HW = 0, \qquad (19)$$

which, by equation (18), means

$$\dot{H} = 0, \qquad (20)$$

expressing the conservation of energy.

Equation (20) gives for the elements of H the infinite set of equations $\nu(nm)H(nm) = 0$. If $\nu(nm) = 0$ only when $n = m$, all the non-diagonal elements of H are zero and H is necessarily a diagonal matrix. In this case, the system is said to be "non-degenerate." It may happen, however, that $\nu(nm) = 0$ for $n \neq m$; the corresponding elements of H are then undetermined and H is not necessarily diagonal. The system is then said to be "degenerate."

It follows further from equation (18') that

$$(W_n - H_n)p(nm) = p(nm)(W_m - H_m),$$

$$(W_n - H_n)q(nm) = q(nm)(W_m - H_m);$$

i.e.,

$$W_n - H_n = W_m - H_m$$

for any value of n and m. Therefore

$$H = W + C,$$

where C is the unity matrix, multiplied by an arbitrary constant. It is most convenient to put

$$H = W \ . \qquad\qquad (21)$$

The mathematical apparatus belonging to the particle picture has been outlined above. Its physical interpretation is discussed in detail elsewhere, but the two most important rules follow naturally at this point from the correspondence principle.

1. The time average of a quantity represented as a Fourier series is given by the terms independent of t. Hence, for a non-degenerate system, the diagonal elements of the matrix representing any variable give the time averages corresponding to the various stationary states.

2. The radiation process, when the particle picture is used, may be regarded as the emission of photons with the spectral frequencies $v(nm)$ accompanied by a simultaneous transition of the atom from the initial state with energy W_n to the final state with energy W_m, $(W_n > W_m)$.

The intensity (rate of emission of energy) may then be represented statistically as

$A(nm)hv(nm)$

where $A(nm)$ is the probability of spontaneous transition from state n to state m with emission of a photon. On the other hand, the classical theory gives for the average intensity corresponding to the τth harmonic

$(2/3) \, (e^2/c^3)(2\pi)^4 [\tau v]^4 \, |r_\tau|^2. \ 2$

Where er is the vector dipole moment of the electrons (r is the vector with components

$$x = \sum_k q_k^{(x)}, \qquad y = \sum_k q_k^{(y)}, \qquad z = \sum_k q_k^{(z)},$$

Where

$$q_k^{(x)}, q_k^{(y)}, q_k^{(z)}$$

being the rectangular coordinates of the electrons.

On equating the expressions of the two theories and replacing Fourier terms by matrix elements we obtain for the transition probability

$$A(nm) = \frac{1}{hv(nm)} \frac{2}{3} \frac{e^2}{c^3} [2\pi v(nm)]^4 |r(nm)|^2 \cdot 2 \ . \qquad (22)$$

The justification of this second rule is not obvious since the Maxwell theory also requires reconsideration. However, equation (22) determines only the time average of the emitted radiation, and it has been shown that the Maxwell theory is competent to furnish this information exactly.

2. THE TRANSFORMATION THEORY

The mathematical scheme of quantum mechanics has been derived in §1 in a way which displays its analogy to classical mechanics; it is not, however, as yet in an easily usable form. In this section it will be shown that the solution of a dynamical problem in the quantum theory is equivalent to the principal axis transformation of a Hermitian form or tensor.

This provides the basis for a practicable method of solution and shows the consistency of the conditions imposed.

Suppose a set of Hermitian matrices p_k, q_k can be found which are independent of the time, satisfy the exchange relations, and make $H(p,q)$ a diagonal matrix. The dynamical problem is then solved, for if the matrices are provided with the time factors

$exp\ [(2\pi i/h)(H_n - H_m)t,$

where H_n and H_m are the diagonal elements of H, it is readily seen that the equations of motion (16) are satisfied. If $p_k^{(o)}$, $q_k^{(o)}$ is any set of matrices satisfying the exchange relations, the transformations

$$p_k = S^{-1} p_k^{(o)} S , \qquad q_k = S^{-1} q_k^{(o)} S , \qquad (23)$$

where S is any matrix, give a new set likewise satisfying the exchange relations. This is seen algebraically on substituting equations (23) in the exchange relations for the new variables; in a similar way it is easily proved that if f is any function of the $p_k^{(o)}$ and $q_k^{(o)}$ that can be written as a power series, then

$$f(p_k, q_k) = f(S^{-1} p_k^{(o)} S , S^{-1} q_k^{(o)} S) = S^{-1} f(p_k^{(o)}, q_k^{(o)}) S . \quad (24)$$

Since special Hermitian matrices satisfying the exchange relations can be found, the problem reduces to that of finding a transformation function S such that

$$S^{-1} H(p_k^{(0)}, q_k^{(0)}) S = W , \qquad (25)$$

where W is a diagonal matrix.

The transformations (23) are analogous to the canonical transformations of classical mechanics; but they have also a geometrical interpretation of great importance if the matrices of the quantum theory are interpreted as tensors in a unitary space of infinitely many dimensions (Hilbert space). This not only furnishes an analytical method of representing the transformations (23) and equation (25) but also provides a convenient language for the physical interpretation of the theory. For present purposes a purely abstract formulation will suffice.

Let $u_1^{(o)}$, $u_2^{(o)}$,...., be an infinite set of unit orthogonal vectors. The space used is that of all vectors

$$t = \sum_n t_n^{(0)} u_n^{(0)} ,$$

where the components $t_n^{(o)}$ are complex numbers. A tensor q then expresses a linear relation between two vectors according to the equations

$$t = qs, \text{ or } t_n^{(0)} = \sum_m q^{(o)}(nm) s_m^{(0)} .$$

Consider now a transformation from the foregoing coordinate system $U_o(u_1^{(o)}, u_2^{(o)}, ...)$ to a new coordinate system $U(u_1, u_2, ...)$, the new vectors being given in terms of the old ones by the linear equations

$$u_n = \sum_m S(mn) u_m^{(0)} . \qquad (26)$$

The components t_n of any vector t and $q(nm)$ of any matrix q in the new system are then given by the equations

$$t_n^{(0)} = \sum_m S(nm)t_m \ , \tag{27}$$

$$q(nm) = \sum_{k,\,l} S^{-1}(nk)q^{(0)}(kl)S(lm) \ , \tag{28}$$

where $\mathbf{S}^{-1}$ is the matrix of the transformation

$$t_n = \sum_m S^{-1}(nm)t_m^{(0)}$$

inverse to equation (27). [S is assumed to be non-singular.] Of special importance are the so-called "unitary" transformations, i.e., those which leave invariant the quadratic form

$$\sum_n t_n t_n^*$$

which is the analogue of distance in unitary space. It is readily verified that for such unitary transformations

$$\sum_k S(nk)S^*(mk) = \sum_k S(kn)S^*(km) = \delta_{nm} \ ,$$

which means that

$$S^{-1} = \tilde{S}^*, \quad \text{or} \quad S\tilde{S}^* = \tilde{S}^*S = 1 \ . \tag{29}$$

They are the analogue in unitary space of rotations of rectangular coordinate systems in real, three-dimensional space.

It is now seen that equations (23) are of precisely the form of equations (28), by virtue of the rule (12) for quantum multiplication; p_k, q_k may therefore be regarded as the same matrices or tensors as $p_k^{(0)}$, $q_k^{(0)}$ expressed in a new coordinate system U, the new coordinates being related to the coordinates in the original system U_o by equations (27). Equation (25) then expresses the condition on the transformation matrix S that in the new system the tensor H is in the diagonal form—i.e., the coordinate vectors of the new system are the principal axes of H. It is sufficient to consider only unitary transformations [S satisfying eq.(28)] since under these conditions it is well known that the principal axis transformation problem, at least for finite matrices, always has a solution.

A word is necessary as to the notation. In general it is not expedient to distinguish matrices in different coordinate systems by new symbols; they are more conveniently characterized by using a distinguishing letter for the indices of the components in each coordinate system. Different numerical values of the indices will be indicated by primes; thus $p(l'l'')$, say, represents the components of p in the "l" system and $p\,(a'a'')$ the components in another "a" system of coordinates. The first of equations (23), for example, is to be written

$$p(a'a'') = \sum_{l'} \sum_{l''} S^{-1}(a'l')\,p(l'l'')\,S(l''a'') \ .$$

The indices of the transformation matrix S then refer naturally to different coordinate systems.

The solution of a quantum-mechanical problem given by the equations of motion (16) and the exchange relations (15) thus reduces to the problem of the principal axis transformation of the Hermitian matrix H. It remains to state briefly the method of solution, which is a well-known one. The equation (25) may be written

$$HS - SW = 0 , \tag{30}$$

which gives for the elements of S the equations

$$\sum_{l''} H(l'l'')S(l''a') - \sum_{a''} S(l'a'')W(a''a') = 0$$

$$\left(\begin{array}{l} l' = 1, \ 2, \ \ldots \ . \\ a' = 1, \ 2, \ \ldots \ . \end{array} \right) ,$$

or, since W is diagonal, an infinite set of homogeneous linear equations

$$\sum_{l''} H(l'l'')S(l''a') - S(l'a')W_{a'} = 0 \quad (l' = 1, \ 2, \ \ldots \ .) , \tag{31}$$

for the determination of the elements of any column of the matrix $S(l'a')$. The $W_{a'}$'s, which appear as parameters, are also determined, and, in fact, independently of the $S(l'a')$, since the equations (31) will have a solution when and only when the determinant of the left-hand member is zero, that is, when the $W_{a'}$'s are solutions of the algebraic equation

$$\begin{vmatrix} H(11) - W & H(12) & H(13) & . & . \\ H(21) & H(22) - W & H(23) & . & . \\ H(31) & H(32) & H(33) - W & . & . \\ . & . & . & . & . \\ . & . & . & . & . \end{vmatrix} = 0 . \tag{32}$$

The roots $W_{a'}$, of this equation are thus characteristic values of equation (30) or equations (31) and are always real. They are the diagonal elements of W and therefore give the energy levels of the system; when the roots of equation (32) are multiple, the system is degenerate, for there is then coincidence of frequencies by equation (7).

To each $W_{a'}$, corresponds a characteristic solution $C_{a'}S(1a')$, $C_{a'}S(2a')$, ..., of equations (31) and hence a column of the matrix S, the arbitrary constant $C_{a'}$ occurring because of the homogeneity of the equations (31). In case the system is not degenerate it is readily seen that any two characteristic solutions are orthogonal to each other, i.e.,

$$\sum_{l'} S(l'a')S^*(l'a'') = 0 \quad \text{when } a' \neq a'' .$$

The relation (29) is thus satisfied for the non-diagonal elements. It may also be satisfied for the diagonal elements by proper choice of the $C_{a'}$, although this "normalization" obviously determines only the absolute magnitude of the $C_{a'}$. There is therefore always an undetermined factor of absolute magnitude one common to the elements of each column of S. In case of degeneracy there is a further indeterminateness, but equation (29)may always be satisfied.

From the transformation function S the coordinates and momenta which form the solution are given by equations (23). In the preceding it has been tacitly assumed that theorems for finite matrices and sets of equations are true for the infinite ones of quantum mechanics. This may be directly justified only under certain conditions, but the more rigorous treatment shows that the results of the formal treatment above are essentially correct. There is one important distinction, however,

in the case of infinite matrices: The characteristic value "spectrum" may contain a continuous sequence of values as well as the discontinuous one hitherto exclusively considered. In the case of the energy this accounts for the existence of continuous optical spectra.

The occurrence of continuous characteristic values also means that in certain coordinate systems the elements of the matrices will have continuously variable indices, or indices discontinuous in a certain range and continuous in another. Our matrix relations must accordingly be extended to include this case. The methods of Dirac will be used for this purpose; though somewhat formal in character they have the advantage of great clarity and may be rigorously justified in all cases which occur practically.

In the first place sums must be replaced by integrals in a range where the indices are continuously variable, the elements becoming functions of two sets of variables. Thus when the range is wholly a continuous one the product rule, for example, becomes

$$pq(nm) = \int dk \; p(nk) q(km) \; ,$$

while in the case of mixed ranges there will occur a sum and an integral. To represent the unit matrix in the continuous case Dirac has introduced a function $\delta(\xi)$, corresponding to δ_{nm} defined by the following properties:

$\xi \, \delta(\xi) = 0$,

so that $\delta(\xi) = 0$ for $\xi \neq 0$

$\delta(-\xi) = \delta(\xi)$ ....(33),
and

$$\int_{\xi_1}^{\xi_2} \delta(\xi) d\xi = 1 \; , \qquad\qquad (34)$$

when the value zero lies between ξ_1 and ξ_2. It is thus a function with a singularity at $\xi_1 = 0$ and is only possible as the limit of a sequence of functions. From the foregoing properties it follows readily that

$$\int_{-\infty}^{+\infty} f(\xi) \delta(a-\xi) d\xi = f(a) \; , \qquad\qquad (35)$$

$$\int_{-\infty}^{+\infty} f(\xi) \delta'(a-\xi) d\xi = f'(a) \; , \qquad\qquad (36)$$

where $f(\xi)$ is any regular function and $\delta'(\xi) = (d/d\xi)\delta(\xi)$. Equation (35) results from an integration by parts. Furthermore, since

$$\int_{-\infty}^{+\infty} \delta(a-\xi)\delta(\xi-b) d\xi = 0$$

when $a \neq b$ and

$$\int db \int \delta(a-\xi)\delta(\xi-b) d\xi = \int \delta(a-\xi) d\xi \int \delta(\xi-b) db = 1 \; ,$$

$$\int_{-\infty}^{+\infty} \delta(a-\xi)\delta(\xi-b) d\xi = \delta(a-b) \; , \qquad\qquad (37)$$

since the integral has all the properties of the δ-function of $a - b$.

54

The elements of the unit matrix in the continuous case may be expressed in terms of the δ-function, for $\delta(a'-a'')$ has, by equation (37), the property that

$$\int \delta(a'-a''')x(a'''a'')da''' = x(a'a'') \ . \qquad (38)$$

Hence

$$\mathrm{I}(a'a'') = \delta(a'-a'') \ .$$

A diagonal matrix with continuous indices is one of the form $q(a'a'')\ \delta(a'-a'')$. The extension to multiple indices causes no difficulty; the unit matrix, for example, becomes

$$\mathrm{I}(a'a'') = \delta(a_1'-a_1'')\delta(a_2'-a_2'') \ . \ . \ . \ . \ \delta(a_f'-a_f'')$$

and may again be written simply $\delta(a'-a'')$.

For the quantum theory those coordinate systems in which quantities other than the energy take the diagonal form are also of importance. In such a system it often proves convenient to replace the indices of all matrices by corresponding diagonal elements of matrices which are diagonal in that system. Rows and columns are thus designated by characteristic values of the matrices which define the coordinate system. This is equivalent to replacing quantum numbers by the energies of the corresponding stationary states in a system of one degree of freedom; by the energy and, for example, the angular momentum in a system of two degrees of freedom, etc.

In general, if the matrices $x_1, x_2, \ldots, x_f$ have the diagonal form, the matrix elements of q will be written

$$q(x'x'') = q(x_1'x_2' \ . \ . \ . \ . \ x_f' \ ; \ x_1''x_2'' \ . \ . \ . \ . \ x_f'') \ ,$$

the primed letters denoting characteristic values of the corresponding matrices; in particular, the diagonal matrices x, when the indices are continuous, have the form

$$x(x'x'') = x'\delta(x_1'-x_1'')\delta(x_2'-x_2'') \ . \ . \ . \ . \ \delta(x_f'-x_f'') \ . \quad (39)$$

The question naturally arises as to what matrices can simultaneously have the diagonal form in a given coordinate system. The answer is well known from the theory of Hermitian forms, and is highly significant for the quantum theory:

Any set of matrices all of which commute with any other of the set can be simultaneously brought to the diagonal form by a unitary transformation.

Thus it will always be possible to find a coordinate system in which the position coordinates $q_1, \ldots q_f$ are diagonal, but if the exchange relations are satisfied the momenta $p_1 \ldots p_f$ cannot also have the diagonal form.

Without proceeding further with the development of matrix mechanics as a formal mathematical calculus, we shall simply mention some results of fundamental importance. When applied to the case of a linear harmonic oscillator, it is found that such an oscillator can never be devoid of energy, its energy values being given by the general expression $(n+1/2)h\nu$, where n is any integer including zero.

A linear harmonic oscillator can never be devoid of energy, its energy values being given by the general expression $(n+1/2)h\nu$, where n is any integer including zero.

This brings out the difference between the old quantum theory proposed by Planck and used by Bohr and the new quantum theory developed by Heisenberg.

According to the former, the energy of a system is assumed to be always an integral multiple of $h\nu$ with zero energy in the lowest state.

According to the latter, the energy of a system is a half integral multiple of $h\nu$ with $1/2\ h\nu$ in the lowest state and hence there is no state with zero energy.

Surprising though this conclusion of the new quantum theory may be, it is strikingly confirmed by many experimental facts, such as specific heats of solids at low temperatures, molecular spectra, etc.

But, when Heisenberg's equation is applied to ordinary radiation in empty space, the latter being treated as made up of a number of separate free vibrations of a given frequency v, it is found that the energy of the radiation, which is equal to the sum of the energies of the component vibrations, is an integral multiple of $h\nu$, quite in accordance with Planck's idea.

Thus Heisenberg's relation shows that all radiations can be regarded as consisting of indivisible photons each of energy $h\nu$; it brings therefore atomicity in the wave picture of radiation and thereby achieves the equivalence of the wave and particle pictures.

Practical application of quantum mechanics

1. The Simple Harmonic Oscillator
Simple harmonic oscillations occur in diatomic as well as more complex molecules, in the motion of the atoms or ions in a crystal, and in electromagnetic radiation, to say a few.

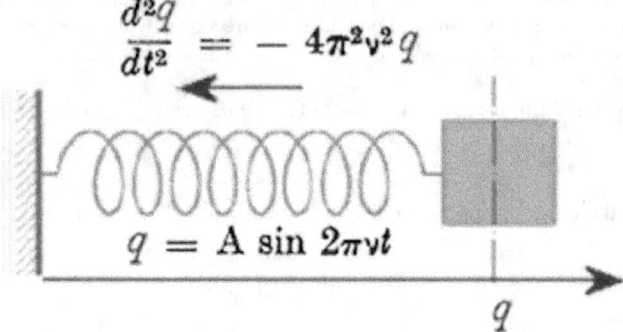

The equation of displacement-time of a harmonic oscillator is
$$q = A \sin 2\pi\nu t$$

$$\frac{d^2q}{dt^2} = -4\pi^2\nu^2 q$$

Restoring force is given by

$$-m\frac{d^2q}{dt^2} = 4\pi^2\nu^2 mq$$

Potential energy is given by

$$V = \int_0^q 4\pi^2\nu^2 m \, q dq = 2\pi^2\nu^2 m \, q^2$$

The Hamiltonian energy of a linear oscillator is of the form
$$H = p^2/(2m) + cq^2.$$

Here,
$p = m \, dq/dt$, the momentum

q is the coordinate,
$c = 2\pi^2\nu^2\, m\, q^2$

We need to determine the energy levels of the eigenvalues of the oscillator by solving its simultaneous equations with the matrix method described above.

These are uniquely determined by the above formula for the energy, and the non-commutation rule $qp—pq = ih$

Let the square of the angular frequency of the oscillations, ω, be

$\omega^2 = 2c/m$

Since
$p = m\, 2\pi\nu\, A \cos(2\pi\nu\, t)$ and
$q = A \sin(2\pi\nu\, t)$,

we can write

$$p = (m\hbar\omega)^{\frac{1}{2}} P,$$
$$q = (\hbar\omega/2c)^{\frac{1}{2}} Q,$$

so that

$$H = \tfrac{1}{2}(P^2+Q^2)\hbar\omega$$

The eigenvalues of the operator $(1/2)(P^2 + Q^2)$ represent the energy levels of the oscillator and hence are the sought solution.

Solution
We need to find matrix or a set of equations such that the non-commutative law holds true. That ensures that we are only picking the proper values of oscillations that conform to the Hamiltonian of the oscillator. The non-commutative rule gives

$$qp-pq = (m\hbar^2\omega^2/2c)^{\frac{1}{2}}(QP-PQ)$$
$$= \hbar(QP-PQ).$$

Thus

$$QP-PQ = i.$$

Consider the infinite matrix $[A_{kl}]$ defined by

$$A_{kl} = \begin{cases} 0, & l \neq k+1 \\ \alpha k^{\frac{1}{2}}, & l = k+1 \end{cases}$$

(where α is a complex number of modulus 1) for $k, l = 1, 2, 3, \dots$.

The Hermitean conjugate of $[A_{kl}]$ is the matrix $[A_{kl}]^*$ with

$$A_{kl}^* = \begin{cases} 0, & k \neq l+1 \\ \alpha^* l^{\frac{1}{2}}, & k = l+1 \end{cases}$$

where α^* is the complex conjugate of α, and $\alpha\alpha^* = 1$. Writing A_{kl} and A_{kl}^* explicitly

$$A = \alpha \begin{pmatrix} 0 & 1 & 0 & 0 & \vdots \\ 0 & 0 & 2^{\frac{1}{2}} & 0 & \vdots \\ 0 & 0 & 0 & 3^{\frac{1}{2}} & \vdots \\ & \cdots \cdots \cdots & & \vdots \end{pmatrix}$$

$$A^* = \alpha^* \begin{pmatrix} 0 & 0 & 0 & \vdots \\ 1 & 0 & 0 & \vdots \\ 0 & 2^{\frac{1}{2}} & 0 & \vdots \\ 0 & 0 & 3^{\frac{1}{2}} & \vdots \\ & \cdots \cdots & & \vdots \end{pmatrix}$$

Considering the above described matrix algebra we find that

$$AA^* = \begin{pmatrix} 1 & 0 & 0 & \vdots \\ 0 & 2 & 0 & \vdots \\ 0 & 0 & 3 & \vdots \\ & \cdots \cdots & & \vdots \end{pmatrix}$$

And

$$A^*A = \begin{pmatrix} 0 & 0 & 0 & 0 & \vdots \\ 0 & 1 & 0 & 0 & \vdots \\ 0 & 0 & 2 & 0 & \vdots \\ & \cdots \cdots \cdots & & \vdots \end{pmatrix}$$

Thus,

$$AA^* - A^*A = 1.$$

Then, if

$$A = (Q + iP)/2^{\frac{1}{2}}, \qquad A^* = (Q - iP)/2^{\frac{1}{2}},$$

one has

58

$$AA^* = \tfrac{1}{2}(Q^2 + P^2) - \tfrac{1}{2}i(QP - PQ),$$
$$A^*A = \tfrac{1}{2}(Q^2 + P^2) + \tfrac{1}{2}i(QP - PQ)$$

by subtraction,

$$QP - PQ = i,$$

as required above. Thus, if

$$Q = (A + A^*)/2^{\frac{1}{2}},$$
$$P = i(A^* - A)/2^{\frac{1}{2}},$$

the commutation rule is automatically satisfied. It is worth noticing that Q and P, defined in this way, are both Hermitean, and therefore qualify as observables. Also

$$\tfrac{1}{2}(AA^* + A^*A) = \tfrac{1}{2}(P^2 + Q^2)$$

$$= \begin{bmatrix} \frac{1}{2} & 0 & 0 & 0 & \vdots \\ 0 & \frac{3}{2} & 0 & 0 & \vdots \\ 0 & 0 & \frac{5}{2} & 0 & \vdots \\ 0 & 0 & 0 & \frac{7}{2} & \vdots \\ \cdots & \cdots & \cdots & \cdots & \vdots \end{bmatrix}$$

From this formula it can be seen that the eigenvalues of $(1/2)(P^2 + Q^2)$ are
1/2, 3/2, 5/2, .. etc,
The energy levels of the harmonic oscillator, i.e., the eigenvalues of H, are therefore

$$E^{(n)} = \tfrac{1}{2}(2n - 1)\hbar\omega,$$

where n is a positive integer, and ω is the angular frequency of the classical theory.

The measurement of the energy is bound to yield one of these eigenvalues. Thus energy can be gained and lost by the oscillator only in integral multiples of the interval $h\omega$ between successive levels. Also, the state of lowest energy of the oscillator has energy $(1/2)h\omega$, not zero as in the classical theory.

The complex number α appearing in the definition of A is arbitrary, so far as the determination of the eigenvalues is concerned, but is needed to satisfy the commutation rule

$$AH - HA = i\hbar A,$$

postulated by Heisenberg. As

$$H = (A^*A + \tfrac{1}{2})\hbar\omega,$$
$$AH - HA = (AA^* - A^*A)A\hbar\omega$$
$$= A\hbar\omega.$$

2 The Energy Levels of the Hydrogen Atom

Consider a hydrogen atom consisting symmetric positive ion in interaction with an electron. Denoting the charges of the positive ion by **Ze** and the electron by —**e**.

The Coulomb energy of the two particles at distance **r** is —**Ze²/r**, in electrostatic units. If the masses and momenta are m_1, $\mathbf{p}_1$ for the electron and m_2, $\mathbf{p}_2$ for the positive ion, the total energy is

$$H = \mathbf{p}_1^2/(2m_1) + \mathbf{p}_2^2/(2m_2) - Ze^2/r.$$

The kinetic energy can be separated into translational, rotational and vibrational energies, so that total energy is expressed in the form

$$H = \mathbf{P}^2/(2M) + (p_r^2 + r^{-2}\mathbf{L}^2)/(2m) - Ze^2/r.$$

The translational energy $\mathbf{P}^2/(2M)$ commutes with H; so does $\mathbf{L}^2$ (which commutes with *any* scalar); therefore H, $\mathbf{P}^2/(2M)$, and $\mathbf{L}^2$ have common eigenvectors.

If ψ is a common eigenvector, and E, T and $l(l+1)\hbar^2$ are the corresponding eigenvalues, we have

$$H\psi = [T + (p_r^2 + l(l+1)\hbar^2/r^2)/(2m) - Ze^2/r)]\psi = E\psi,$$

i.e.,

$$2m(H-T)\psi = [p_r^2 + l(l+1)\hbar^2/r^2 - 2c/r]\psi,$$

where $c = m\,Ze^2$.

So, to determine the energy levels, we have to find the eigen-values of the linear operator

$$A = p_r^2 + l(l+1)\hbar^2/r^2 - 2c/r.$$

According to the theory, the eigenvalues are the set of numbers $a^{(j)}$ defined recursively by

$$A = \theta_1^*\theta_1 + a^{(1)},$$
$$A_{j+1} = \theta_j\theta_j^* + a^{(j)},$$
$$A_j = \theta_j^*\theta_j + a^{(j)},$$

where, in case of ambiguity, the greatest value of $a^{(j)}$ determines the choice of θ_j at each stage.

In this application, the form of A suggests we should take

$$\theta_j = p_r + i(a_j + b_j/r),$$

where a_j and b_j are real numbers, to be determined.

Then we shall have

$$\theta_j^*\theta_j = [p_r - i(a_j + b_j/r)][p_r + i(a_j + b_j/r)]$$
$$= p_r^2 + (a_j + b_j/r)^2 + ib_j[p_r,\ r^{-1}]$$
$$= p_r^2 + a_j^2 + 2a_jb_j/r + (b_j^2 - b_j\hbar)/r^2,$$

60

$$\theta_j \theta_j{}^* = p_r{}^2 + (a_j + b_j/r)^2 - ib_j[p_r, \ r^{-1}]$$
$$= p_r{}^2 + a_j{}^2 + 2a_j b_j/r + (b_j{}^2 + b_j\hbar)/r^2.$$

Since (comparing the two expressions for A)

$$\theta_1{}^*\theta_1 + a^{(1)} = p_r{}^2 - 2c/r + l(l+1)\hbar^2/r^2,$$

we see that we must choose a_1 and b_1 so that

$$a_1 b_1 = -c, \quad b_1(b_1 - \hbar) = l(l+1)\hbar^2,$$

and $a^{(1)}$ will then be given by

$$a^{(1)} + a_1{}^2 = 0.$$

There are the possibilities: either

(i) $b_1 = -l\hbar$. when $a_1 = c/(l\hbar)$ and $a^{(1)} = -c^2/(l\hbar)^2$, or
(ii) $b_1 = (l+1)\hbar$. when $a_1 = -c/[(l+1)\hbar]$ and $a^{(1)} = -c^2/[(l+1)\hbar]^2$

As the second possibility gives the greater value of $a^{(1)}$, this is chosen, i.e., we take $b_1 = (l+1)\hbar$,

Also, since (comparing the two expressions for A_{j+1})

$$\theta_{j+1}^*\theta_{j+1} + a^{(j+1)} = \theta_j\theta_j{}^* + a^{(j)},$$

we must choose a_{j+1} and b_{j+1} so that

$$a_{j+1}b_{j+1} = a_j b_j, \quad b_{j+1}(b_{j+1} - \hbar) = b_j(b_j + \hbar),$$

And $a^{(j+1)}$ is then given by

$$a^{(j+1)} + a_{j+1} = a^{(j)} + a_j{}^2.$$

There are again two possibilities; either
(i) $b_{j+1} = -b_j$ when $a_{j+1} = -a_j$ and $a^{(j+1)} = a^{(j)}$. or
(ii)) $b_{j+1} = b_j + \hbar = \dots = b_1 + j\hbar = (l+1+j)\hbar$.

$$a_{j+1}b_{j+1} = a_j b_j = \dots = a_1 b_1 = -c,$$

$$a^{(j+1)} + a_{j+1}^2 = a^{(j)} + a_j{}^2 = \dots = a^{(1)} + a_1{}^2 = 0,$$

i.e.,

$$b_j = (l+j)\hbar,$$

$$a_j = -c/[(l+j)\hbar] \quad \text{and}$$

$$a^{(j)} = -c^2/[(l+j)\hbar]^2.$$

The first alternative is obviously unacceptable, so the eigenvalues E of the energy are given by

$$2m(E-T) = -c^2/[(l+j)\hbar]^2,$$

where l is a non-negative integer and j is a positive integer.

For a hydrogen atom at rest, $T = 0$ and

$$E = -\frac{mZ^2e^4}{2(l+j)^2\hbar^2}.$$

3 The Deuteron

The deuteron is a bound state of a proton and neutron. As the neutron is uncharged, there are no electrostatic forces involved, and one is concerned with only the short range nuclear forces. Owing to complexities arising from the spin of the two particles, pair production within the nucleus, etc., the nuclear forces are quite complicated in detail, but to a good approximation the interaction energy of the proton and neutron is represented by Hulthen's potential

$$V(r) = -\frac{g\mu^2}{e^{\mu r}-1},$$

where g is a known constant and μ is the mass of the π-meson, multiplied by $c/\hbar$, where c is the velocity of light.

From the small electrical quadrupole moment of the deuteron it is inferred that the eigenvalue zero of $\mathbf{L}^2$ predominates, though there is a small probability of the eigenvalue $6\hbar^2$ which will be neglected here.

The energy of a deuteron at rest may therefore be assumed to be

$$H = p_r^2/(2m)+V(r),$$

where m, the reduced mass, is about *one half* the proton mass. So we have to investigate the eigenvalues of

$$A = p_r^2-c\mu^2/(e^{\mu r}-1),$$

where $c = 2mg$.

This time we take

$$\theta_j = p_r+ia_j+ib_j/(e^{\mu r}-1)$$

and find

$$\theta_j{}^*\theta_j = p_r{}^2 + [a_j + b_j/(e^{\mu r}-1)]^2 + ib_j[p_r,\ (e^{\mu r}-1)^{-1}]$$
$$= p_r{}^2 + a_j{}^2 + 2a_j b_j/(e^{\mu r}-1) + b_j{}^2/(e^{\mu r}-1)^2$$
$$- b_j \hbar\mu\, e^{\mu r}/(e^{\mu r}-1)^2$$
$$= p_r{}^2 + a_j{}^2 + (2a_j - \hbar\mu)b_j/(e^{\mu r}-1)$$
$$+ b_j(b_j - \hbar\mu)/(e^{\mu r}-1)^2.$$

Similarly,

$$\theta_j\theta_j{}^* = p_r{}^2 + a_j{}^2 + (2a_j + \hbar\mu)b_j/(e^{\mu r}-1)$$
$$+ b_j(b_j + \hbar\mu)/(e^{\mu r}-1)^2.$$

since

$$A = \theta_1{}^*\theta_1 + a^{(1)},$$

We find (rejecting $b_1 = 0$)

$b_1 = \hbar\mu$,
$(2a_1 - \hbar\mu)b_1 = -2c$, and
$a^{(1)} + a_1{}^2 = 0$.

Thus
$a_1 = (1/2)\hbar\mu - c/(\hbar\mu)$ and
$a^{(1)} = -[c/(\hbar\mu) - (1/2)\hbar\mu]^2$

This is the negative binding energy of the deuteron.

WAVE MECHANICS

Schrödinger in 1926 directly started with De Broglie's idea of matter waves and developed it into a rigorous mathematical theory which has received the name of wave mechanics. The essential feature of this theory is the incorporation of the expression for the De Broglie wavelength into the general classical wave equation. By this means a wave equation for a moving particle is derived, which is known as *Schrödinger's fundamental -wave equation.*

Schrödinger's fundamental wave equation
According to the De Broglie theory, a particle of mass *m* and moving with a velocity *v* has associated with it a wave system of some kind, of wavelength $\lambda = h/mv$. Though we have no knowledge of what it is that vibrates, we can indicate it by ψ, periodic changes in which are responsible for the wave system.

Supposing a system of stationary waves to be associated with the particle and referring the particle to the Cartesian coordinate system, at any point x, y, z in the immediate vicinity of the particle, ψ undergoes periodic changes, its value at any instant *t* being given by

$$\psi = \psi_0 \sin 2\pi \nu t \qquad\qquad ...(1)$$

where ψ_o is the amplitude at the point considered, independent of *t* but a function of s, y, z, and ν the frequency.

The differential equation of this wave motion can be written in the classical way as

$$\frac{\partial^2 \psi}{\partial t^2} = v^2 \left(\frac{\partial^2 \psi}{\partial x^2} + \frac{\partial^2 \psi}{\partial y^2} + \frac{\partial^2 \psi}{\partial z^2} \right) \qquad \ldots(2)$$

$$= v^2 \nabla^2 \psi$$

Where

$$\nabla^2 \psi = \frac{\partial^2 \psi}{\partial x^2} + \frac{\partial^2 \psi}{\partial y^2} + \frac{\partial^2 \psi}{\partial z^2} ,$$

∇^2 being the Laplacian operator.

From equation (1)

$$\frac{\partial^2 \psi}{\partial t^2} = - 4\pi^2 \nu^2 \psi = - \frac{4\pi^2 v^2 \psi}{\lambda^2}$$

Since, frequency ν = velocity v/ wavelength λ.

Substituting this in equation (2)

$$- \frac{4\pi^2 v^2 \psi}{\lambda^2} = v^2 \nabla^2 \psi$$

Or

$$\nabla^2 \psi + \frac{4\pi^2}{\lambda^2} \psi = 0 \qquad \ldots(3)$$

So far the treatment is general. The wave mechanics concept is now introduced on replacing λ by *h/mc* from De Broglie's theory.

With this change, equation (3) becomes

$$\nabla^2 \psi + \frac{4\pi^2 m^2 v^2}{h^2} \psi = 0$$

Now if **E** is the total energy of the particle and **V** its potential energy, its kinetic energy

$$\tfrac{1}{2} mv^2 = E - V,$$

from which we get

$$m^2 v^2 = 2m (E - V)$$

If this is substituted in the wave equation derived above, we have

$$\nabla^2 \psi + \frac{8\pi^2 m}{h^2} (E - V) \psi = 0 \qquad \ldots(4)$$

This is known as Schrödinger's fundamental wave equation *with respect to space*. It is to be noted that the time factor does not explicitly enter into this amplitude equation, as is to be expected in a stationary wave system. The quantity ψ is usually referred to as a wave *function*. The potential energy V is, in general, a function of the coordinates.

Schrödinger's equation, with respect to time can be derived as follows:

In the most general form of wave motion, the energy density at any point is found to be proportional to the sum of the squares of two quantities. For instance, in the electromagnetic wave, energy density is found to be proportional to (E^2 +H^2), where E and H are the electric and magnetic vectors respectively. To describe, in a similar manner, the state of the De Broglie wave, let us take two functions *f(x,y,z)* and *g(x,y,z)*. The functions are, however, to be assumed *scalar*, since in a stream of material particles, say electrons, there is no property similar to polarization of electromagnetic waves (that was the thought then). Hence, as long as the spin of the electron is not considered, the electron waves can be treated as scalar waves, with properties analogous to sound waves in a gas, which consist of periodic variations of pressure, a scalar quantity.

Assuming that the electron density in a De Broglie wave is proportional to (f^2+g^2), let the units in which the wave displacement is measured be so chosen that the electron density is equal to (f^2+g^2). Further, since the electron density can be legitimately assumed to be constant, (f^2+g^2) is a constant. This means that *f* and *g* should have a phase difference of $\pi/2$ in a plane wave, so that if

$f = a \cos 2\pi(vt - x/\lambda)$,
then
$g = a \sin 2\pi(vt - x/\lambda)$.

Now, instead of two separate real functions, a single complex function is more conveniently used, and in the present case, the only way of doing this is to put
$\psi = f + i\,g$

Equation (1) may, therefore, be written as

$$\psi = \psi_0 \cos 2\pi(vt - z/\lambda) + i\psi_0 \sin 2\pi(vt - z/\lambda)$$
$$= \psi_0\, e^{2\pi i(vt - z/\lambda)}$$
$$= \psi_0\, e^{2\pi i vt} \cdot e^{-2\pi i \bar{v} z}, \ (\text{where } \bar{v} = 1/\lambda) \qquad \ldots (5)$$

In equation (4) replacing **E** by *hv* and multiplying throughout by $h^2/8\pi^2 m$, we get

$$\nabla^2\psi(h^2/8\pi^2 m) + hv\psi - V\psi = 0 \qquad \ldots (6)$$

To eliminate v, differentiating (5) with respect to time,

$$\partial\psi/\partial t = 2\pi i v\psi_0\, e^{2\pi i vt} \cdot e^{-2\pi i \bar{v} z}$$
$$= 2\pi\, i\, v\, \psi$$
$$\therefore\ v = \frac{\partial\psi}{\partial t} \cdot \frac{1}{2\pi\, i\, \psi}$$

Substituting this value of v in (6),

65

$$\nabla^2\psi \frac{h^2}{8\pi^2 m} + \frac{\partial\psi}{\partial t}\cdot\frac{h}{2\pi i} - V\psi = 0$$

$$i.e., \qquad -\frac{h}{2\pi i}\cdot\frac{\partial\psi}{\partial t} = \frac{h^2}{8\pi^2 m}\nabla^2\psi - V\psi$$

$$or \qquad \frac{ih}{2\pi}\cdot\frac{\partial\psi}{\partial t} = \frac{h^2}{8\pi^2 m}\nabla^2\psi - V\psi \qquad\qquad \dots (7)$$

This is the Schrödinger's equation containing the time factor. It is unique among the differential equations of mathematical physics, as it includes the imaginary quantity $i=\sqrt{-1}$. But for this i, the equation would resemble that for the flow of heat in a solid conductor. The motion of matter waves combines, in fact, some of the features of the three familiar types of wave motion, viz., the vibrations of a string, sound waves in a gas and electromagnetic waves.

Applications

In the application of wave mechanics to different problems, the appropriate values for the potential energy V are substituted in the fundamental equation (4), which is then solved by a suitable mathematical device. We shall here consider only a few important cases of a general nature.

1. The harmonic oscillator

The equation of motion of a linear harmonic oscillator is

$$x = A\sin 2\pi\nu t$$

$$\frac{d^2 x}{dt^2} = -4\pi^2\nu^2 x$$

Restoring force is given by

$$-m\frac{d^2 x}{dt^2} = 4\pi^2\nu^2 mx$$

Potential energy is given by

$$V = \int_0^x 4\pi^2\nu^2 mx\,dx = 2\pi^2\nu^2 mx^2$$

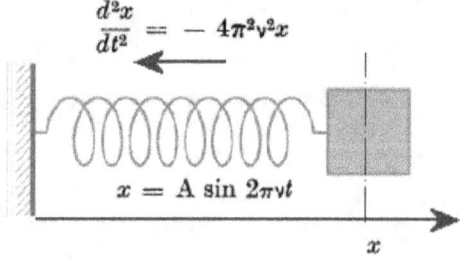

$$\frac{d^2 x}{dt^2} = -4\pi^2\nu^2 x$$

$$x = A\sin 2\pi\nu t$$

Substituting this value of V in Schroedinger's wave equation

$$\frac{\partial^2\psi}{\partial x^2} + \frac{8\pi^2 m}{h^2}(E - 2\pi^2\nu^2 mx^2)\psi = 0.$$

Using a new variable k, where k is given by $k = xb^{1/2}$ and $b = 4\pi^2\, mv/h$ (a constant),

$$\frac{\partial^2\psi}{dx^2} = \frac{\partial}{\partial x} \cdot \frac{\partial\psi}{\partial x} = \frac{\partial}{\partial x} \cdot \frac{\partial\psi}{\partial k} \cdot \frac{\partial k}{\partial x} = \frac{\partial}{\partial k} \cdot \frac{\partial\psi}{\partial x} \cdot \frac{\partial k}{\partial x}$$

$$\therefore \qquad \frac{\partial^2\psi}{\partial x^2} = \frac{\partial}{\partial k} \cdot \frac{\partial\psi}{\partial k} \cdot \left(\frac{\partial k}{\partial x}\right)^2 = \frac{\partial^2\psi}{\partial k^2} \cdot \left(\frac{\partial k}{\partial x}\right)^2$$

Since $k = xb^{1/2}$.

$$\frac{\partial k}{\partial x} = \sqrt{b} \ \text{ or } \ \left(\frac{\partial k}{\partial x}\right)^2 = b$$

$$\frac{\partial^2\psi}{\partial x^2} = b \cdot \frac{\partial^2\psi}{\partial k^2}$$

$$\therefore \qquad b\frac{\partial^2\psi}{\partial k^2} + \frac{8\pi^2 m}{h^2}\left(E - 2\pi^2 v^2 m \cdot \frac{k^2}{b}\right)\psi = 0$$

Putting

$$\frac{8\pi^2 mE}{h^2} = a,$$

we have

$$b\frac{\partial^2\psi}{\partial k^2} + (a - bk^2)\,\psi = 0$$

or

$$\frac{\partial^2\psi}{\partial k^2} + \left(\frac{a}{b} - k^2\right)\psi = 0$$

On the assumption that ψ should be finite, continuous and single valued at $k = 0$, the proper solutions of this equation can be shown to be as follows:

For $a/b = 1$, $\psi = (1/2)\exp(-k^2/2)$,
For $a/b = 3$, $\psi = (2k)\exp(-k^2/2)$,
For $a/b = 5$, $\psi = (4k^2 - 2)\exp(-k^2/2)$,
For $a/b = 7$, $\psi = (8k^2 - 12k)\exp(-k^2/2)$,

Thus the proper values of a/b are given by $(2n+1)$, where n is an integer and the values of ψ corresponding to these are known as the *characteristic eiyen-functions.*

Substituting for a/b,

$$\frac{a}{b} = \frac{8\pi^2 mE}{h^2} \times \frac{h}{4\pi^2 mv} = \frac{2E}{hv} = 2n + 1$$

$$\therefore \qquad E = \left(n + \tfrac{1}{2}\right)hv$$

These values of the total energy represent the quantum states of the harmonic oscillator. Thus, wave mechanics leads to the same result as quantum mechanics with reference to the energy states of a linear harmonic oscillator and *zero-point energy* which is not zero but $(1/2)h\nu$.

The characteristic functions for which $n = 1, 2, 3$ in the case of a harmonic oscillator are indicated in Fig. 184.

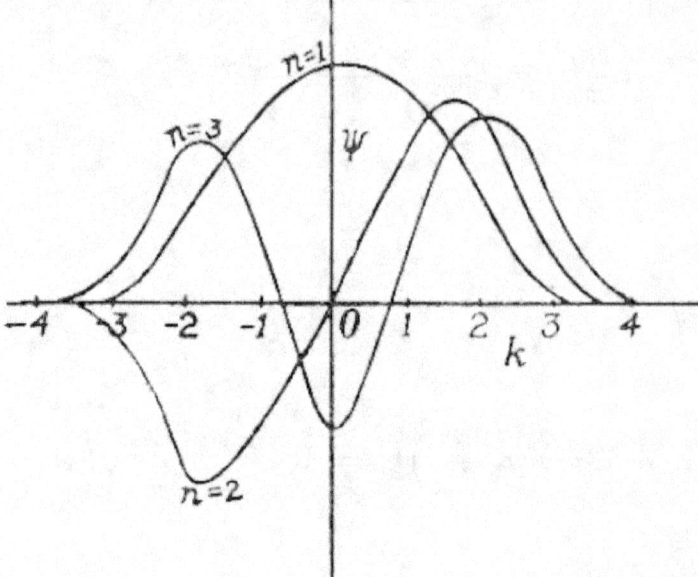

Fig. 184

The X-axis represents the variable k which is a function of x, while the Y-axis the normalized wave function ψ. It can be seen that the wave amplitudes become zero even for small values of x. This means that the associated wave packet is found only in the immediate neighborhood of the particle, in accordance with the initial assumptions; the wave amplitudes have a finite and rather indefinite extent in space, which indicates that the *wave packet behaves very much like a particle with* fuzzy *edges*.

2. The rotator
We shall now consider another important case to which wave mechanics has been applied and which gives results different from the older quantum theory but in complete accord with experimental facts, *viz.,* a *simple rotator*. By a rotator is understood a dumb-bell shaped body consisting of two spheres connected together by a slender rod. The system is supposed to be incapable of rotation about the axial line but it can assume any orientation in space, the latter being specified by two angles, say θ and φ. The motion of such a rotator is similar to that of a particle rotating over the surface of a sphere whose centre is fixed. Evidently, since the centre is fixed, no wave has to be ascribed to it. Further, since

the energy is wholly kinetic, the term V representing potential energy in Schrödinger's equation disappears and we can write

$$\frac{\nabla^2 \psi}{r^2} + \frac{8\pi^2 m}{h^2} E_{kin} \psi = 0$$

where r^2 is a function of the coordinates alone.

Assuming that $\nabla^2 \psi$ is independent of r^2, the solution by the method of "Spherical Harmonics" indicates that the proper values E_{kin} satisfying the above equation, is given by

68

$$r^2 . \frac{8\pi^2 m}{h^2} . \mathrm{E}_{kin} = n(n+1)$$

Since

$$\mathrm{E}_{kin} = \tfrac{1}{2} mv^2$$

$$r^2 . \frac{4\pi^2 m^2 v^2}{h^2} = n(n+1)$$

As the angular momentum $p_\varphi = \boldsymbol{mvr}$

$$\frac{4\pi^2}{h^2} p_\varphi{}^2 = n(n+1)$$

$$\therefore \qquad p_\varphi = \frac{h}{2\pi} \sqrt{n(n+1)}$$

These values of p_φ specify the quantum states of the simple rotator (n = 1, 2, 3, etc.).

According to the old quantum theory $p_\varphi = n. \, \boldsymbol{h/2\pi}$ as we shall see. But the application of wave mechanics shows that $\boldsymbol{n}$ should be replaced by $\sqrt{n(n+1)}$ and this is in close agreement with experimental data.

3. Particles in a potential box

Let us consider a stream of electrons, each of mass $\boldsymbol{m}$ and of total energy $\mathbf{E}$, moving parallel to the X-axis with velocity $\boldsymbol{v}$. Let the electrons be confined to the space between two points $x = 0$ and $x = 1$, within which region the potential $V = 0$ and beyond which the potential rise is such that it cannot be surmounted by the electrons.

The Schroedinger's wave equation for such system of particles is given by

$$\frac{\partial^2 \psi}{\partial x^2} + \frac{8\pi^2 m}{h^2} \, \mathrm{E} \, \psi = 0 \qquad \dots \qquad (1)$$

where E is a constant.

As ψ is a measure of the probability of finding the particle at a given point, we get the boundary conditions $\psi = 0$ at $x = 0$ and $x = l$, since the particle does not get out of the region defined by those limits. We have, therefore, a stationary wave, the wavelength of which is given by $\lambda = 2 \, l/n$, where $\boldsymbol{n}$ is an integer.

The general equation of motion in a stationary wave is

$$\frac{\partial^2 \psi}{\partial x^2} + \frac{4\pi^2}{\lambda^2} \, \psi = 0 \qquad \dots \, (2)$$

Comparing the coefficients of ψ in equations (1) and (2), we get

$$\frac{8\pi^2 m}{h^2} \mathrm{E} = \frac{4\pi^2}{\lambda^2} = \pi^2 \left(\frac{n}{2l} \right)^2$$

$$\therefore \qquad \mathrm{E} = \frac{n^2 h^2}{8ml^2} \qquad \dots \, (3)$$

This relation (3) shows that the energy of the electrons is restricted to a series of values which vary as the squares of the natural numbers. These values are energy levels and are characteristic of the box.

Paul Dirac (8 August 1902- 20 October 1984), England

4. Dirac's relativistic wave mechanics

The wave-mechanical theory described so far is a non-relativistic one suited to cases of small energy only. Dirac has succeeded to some extent to bring it into harmony with the principles of relativity by his relativistic wave equation of the electron. The line of argument in achieving this result may be briefly stated as follows:

The theory of relativity shows that for an adequate description of any event the three classical space-coordinates x, y and z alone are insufficient and a fourth time-coordinate *(ict)*, which stands on an equal footing with the others, is required. Now, according to the modern conception of the wave nature of matter, a moving electron is endowed with an associated wave which can be represented by an adequate wave equation as derived by Schrödinger. But

this wave equation of the electron is found not to satisfy the symmetry in the four-coordinates postulated by relativity, since it is of the *second order* in the *spatial* differential coefficients, while of the *first order* in the *time derivative*

Dirac succeeded in formulating a. wave equation for an electron moving in a potential field in such a way as to make it relativistically invariant, its form being of the first order in all the four variables. Although the rather complicated theory of Dirac cannot be given here in detail, it must be emphasized that

the solution of this new relativistic wave equation not only led to a complete interpretation of the fine structure of spectral lines, but also to a rational explanation of the spin and magnetic moment of the electron itself.

In Dirac's theory, in addition to the solutions corresponding to the normal electronic levels found experimentally, there were others which seemed to represent no observed facts.

These solutions predicted the existence of certain states, in which the electrons possessed a negative kinetic energy and hence did not correspond to particles in any usual sense. These states, however, could not be ignored, since transitions must theoretically occur between them and the normal states corresponding to positive kinetic energy.

Dirac suggested that such a difficulty might be avoided if it were supposed that

all the negative energy states are normally occupied and, further, that the totality of electrons in such states produces no external field and hence not physically observable.

For, in such a case, the process of an incident electron passing into the negative state, which is already filled normally, is impossible. Further, since there are an infinite number of states of negative energy even in a finite volume, there should be an infinite density of electrons in negative energy states in all space. If it now be assumed that the number of electrons existing in the universe is slightly greater than the number of available negative energy states, then some electrons, finding no place in these negative energy states will have to occupy states of positive energy; these are the ordinary electrons which are observed.

Let it be further supposed that from the sea of electrons occupying the negative energy states, electromagnetic fields, say in an act of absorption in the atom, can raise an electron and bring it into the region of positive energy states; it then takes its place as an ordinary observable electron. There now remains, however, an empty place in the midst of the negative energy states, called the *Dirac hole,* which has entirely the same character as a positive charge. Theoretical formulae show that this hole in the sea of negative states behaves as an exact counterpart of the electron, *i.e.,* a particle identical in mass and magnitude of charge with the ordinary electron, but with a positive sign for charge.

Since at the time when the theory was first proposed the only fundamental positively charged particle known was the proton, Dirac thought at first that the above-stated holes corresponded to protons, although he was quite aware of the difficulties arising from the difference in masses of the proton and the electron and from the very stable existence of the proton. But after the experimental discovery of the positron by Anderson in cosmic ray researches, this particle was identified with the Dirac hole.

Hence *in Dirac's theory the positron is considered as an unoccupied state* of *negative energy, a perfect image of the negative electron;*

its actual discovery argued to the essential correctness of the theory.

The most important consequence of Dirac's theory is the prediction of two novel processes, viz.,

(i) materialization of energy or pair production
(ii) annihilation of matter

In terms of the theory, the first of these consists in a photon of energy greater than $2m_0c^2$ (about 1MeV) raising an electron from a state of negative energy to a state of positive energy, which results in an *electron-positron pair production,* since the electron in the final positive energy state will be observed as an ordinary negative electron while the Dirac hole in the negative state produced in the process corresponds to a positron.

This process cannot take place in empty space, because energy and momentum cannot be conserved; but it can occur in the electric field in the neighborhood of a nucleus which can take up the extra momentum.

The second process, *viz., annihilation of matter,* corresponds to the reverse case of the. fall of an electron from a high positive state into a hole in the negative state. The law of conservation of momentum demands that if this process were to take place in empty space, two light quanta, each of energy $m_0c^2 = (1/2)$MeV would originate. Both the phenomena have been experimentally observed. Recent discoveries (1956) of the anti-proton and the anti-neutron, which bear the same relation to the proton and neutron respectively as the positron does to the electron, have added further support to the basic ideas of the theory. In spite of these successes,

Dirac's theory cannot be considered as perfect in every respect, on account of the several arbitrary and not readily understood assumptions involved, (such as the existence of an infinite density of electrons in space), as well as other very serious difficulties, with which it is confronted when the interaction of the negative energy electrons with the radiation field is considered.

PHYSICAL SIGNIFICANCE OF THE WAVE FUNCTION ψ

More than all these applications, the point which is of immediate interest to us here is the physical interpretation of ψ, the wave function, which appears in Schrödinger's equation. For, it is bound to throw light on the nature of matter waves and thereby enable us to solve the apparent contradiction involved in the concept of matter waves. This in turn might help us to reconcile the dualistic nature of both matter and radiation.

What, therefore, is the physical significance of ψ? Some authors like Millikan, for instance, are satisfied with the answer that ψ is merely an *"artifice of calculation",* i.e., an auxiliary mathematical quantity employed to facilitate computations relative to experimental results. This is true to a certain extent, but it is not quite in the spirit of theoretical physics to introduce an isolated mathematical function in a differential equation and rest content without enquiring into its physical significance.

A first simple interpretation of ψ was therefore attempted by Schrödinger himself in terms *of charge density.* Since a beam of electrons is diffracted like X-rays, one might use the optical analogy to arrive at the physical significance of ψ. In any electromagnetic wave system the energy density, *i.e.,* energy per unit volume, is equal to A^2, where A is the amplitude of the wave, and the number of photons per unit volume is equal to the energy density divided by $h\nu$, since the latter is the energy of the photon. Hence the photon density is equal to $A^2/h\nu$ which means that the photon density is proportional to A^2, since by $h\nu$ constant. By analogy, if ψ is the amplitude of the matter wave at any point in space, we may consider the particle density, *i.e.,* the number of particles per unit volume at that point, to be proportional to ψ^2. Hence

the square of the absolute value of ψ is a measure of the particle density.

If *e is* the electric charge of the particle, the charge density is obtained by multiplying the particle density by *e.* The quantity ψ^2 is therefore a measure of the charge density. On this view,

the amplitude of the electron wave ψ would represent the distribution of its electric charge

and the electron would have a diffuse structure; the electric charge being distributed continuously over a large volume, theoretically infinite, but actually limited to the wave packet.

This interpretation led to very satisfactory results when wave mechanics was applied to the stable states of the Bohr atom, to the emission of spectral lines, to the directional distribution of photoelectrons, to the intensity distribution in Compton scattering and to scattering processes of charged material particles as diffraction of electrons, scattering of a-particles by the nucleus, etc.

But the *main difficulties* raised against the simple pictorial interpretation of ψ are:

(i) The wave packet associated with the material particle must in course of time become dissipated so that it could not represent for long the particle concerned. This follows from the universal property of every kind of wave to spread through space. Observation, on the other hand, shows that material particles, such as electrons and protons, maintain their identity and preserve their charges intact. Even if one could devise a system of waves which would not scatter to any appreciable extent when the electron or proton was pursuing an undisturbed path through empty space, the waves must get scattered as soon as the particle interacts with matter; we have direct experimental evidence of this in the wave patterns they form on a photographic plate.

(ii) Consideration of the mutual action of the particles as a collision of the corresponding wave packets in ordinary three-dimensional space leads to very serious difficulties. For instance, when two electrons interact, it can be shown that their waves should be propagated in a six-dimensional space. In the same way; if a thousand electrons interact, their waves will be propagated in. a space of three thousand dimensions. Such a space can only be regarded as a purely abstract mathematical concept and as we cannot suppose waves to be more real than space through which they are propagated, the waves also must be considered to be of the same nature, viz., merely a mathematical artifice.

Against these objections we have a recent pronouncement of Schrödinger (1953) stating that

"particles are more or less temporary entities within the wave-field, whose form and general behavior are nevertheless so clearly and sharply determined by the laws of waves that many processes take place, as if these temporary entities were substantial permanent beings. The mass and charge of the particle defined with such precision must then be counted among the structural elements determined by the wave laws. The conservation of charge and mass in the large must be considered as a statistical effect based on the law of large numbers."

A second interpretation of ψ, generally accepted at present, was first suggested by Born in 1926 and then elaborated by Heisenberg and Bohr. According to this view, ψ^2 does not measure the particle density at any point but the *probability of finding the particle at that point at any given moment.* In general, a mechanical process has associated with it a wave process, the square of whose amplitude gives a measure of the mechanical process taking place at the point considered, If a material particle is considered as a wave packet, it becomes necessary to postulate the existence of a "pilot wave" for it, on account of the dissipating property of waves. The pilot wave is the one described by Schrödinger's equation. Physically the equation relates the amplitude ψ of the pilot-wave to the probability of finding the particle at that point. If the amplitude of the pilot wave at any point is zero, the probability of encountering the particle at that point is vanishingly small. The probability of finding the particle at a given point is proportional to the value of ψ^2 at that point. More exactly the probability of the particle being present in a volume $dxdydz$ is $|\psi|^2dxdydz$, the symbol $|\psi|^2$ being used instead of ψ^2, since according to the theory ψ is a complex number. ψ is then called the *probability amplitude*. Further, since the particle is *certainly* to be found somewhere in space

$$\iiint |\Psi|^2 \, dx.dy.dz = 1$$

the triple integral extending over all possible values of x, y, z. A ψ satisfying this relation is called a *normalized wave function*.

On this view *the wave is in no sense a physical phenomenon in, any region of space;* it is rather a mere symbolical representation of what we know about the corpuscle; it is *a wave of probability of non-physical character*. For instance, in the phenomenon of the diffraction of electrons, the electrons themselves behave as particles when they collide or interact with an object while the probability of their being at a point in question is governed by the associated waves. At a particular point the amplitude of the electron wave might correspond to one-fourth of an electron but this does not mean that an electron is divisible; if we try to detect an electron there, we shall find either a whole electron or none at all, the probability of finding it being 25% in a large number of observations.

The same probability interpretation can he applied to radiation also. When radiation interacts with matter it behaves as a particle and we must always deal with a whole photon or none at all; but if *we* try to find the photon at any given point we must have recourse to its wave aspect, the square of whose amplitude is the measure of the probability of finding the photon at that point. At a point where the wave amplitudes reinforce each other there is a high probability of finding the photon; at a point where they interfere the probability is infinitesimal. Such a statistical conception of light waves does away with the necessity of postulating an ether medium and is therefore considered a great asset in solving the riddle of radiation as electromagnetic waves without a medium.

It may be noted that
(i) so far as experimental evidence is concerned the two interpretations seem to be more or less justified;
(ii) in the charge density interpretation of ψ the electron wave has a real physical significance but the electron does not exist as a point charge since the charge is distributed over the large space of the wave packet, while in the probability interpretation the idea of the electron as a point charge is maintained, but the associated wave becomes purely symbolical, an analytical representation of certain probabilities and therefore not a physical phenomenon in the ordinary sense of the term. The physical interpretation of ψ remains therefore an extremely difficult problem, far from being fully solved.

The dualistic nature of both matter and radiation seems to receive the following further elucidation in the light of the foregoing interpretations of ψ. *The particle and wave aspects are strictly complementary,* in the sense that although the spatial distribution of particles can only be predicted by taking into consideration the concept of waves, yet when one tries to define precisely the wave aspect, the particle aspect recedes into the background and becomes indefinite being a density distribution according to the first interpretation or acquiring an uncertain statistical location according to the second.

This shows that reality, whether matter or radiation, is made up of a subtle and almost indefinable fusion of two antagonistic but complementary factors, the *continuous wave* and the *discontinuous particle;* it is a *discontinuous continuity* or a *continuous discontinuity* and hence not a simple but *complex unity.*

PRINCIPLE OF INDETERMINACY

From the probability interpretation of the wave function ψ, Heisenberg, in 1927, was able to draw a very interesting conclusion of far-reaching importance known as the *principle of indeterminacy* or the *uncertainty principle.*

According to wave mechanics, it is the wave associated with a particle that represents or symbolizes all that we know about the particle. Although the propagation of the associated wave obeys exact laws, yet from the probability interpretation of the wave function ψ it can only be said that the particle should lie in a region occupied by the wave and that its chance of being at a given point within this region is proportional to the wave amplitude at that point. This means that although the particle is somewhere within the wave packet moving with the group velocity, it is impossible to know where exactly the particle is and what is its exact velocity at any given moment. By making the wave packet very small *(i.e.,* practically zero except within a very small region) the position of the particle can be more or less fixed, but since it can be shown that the velocity spread of such a packet is large, the velocity of the particle becomes indeterminate. On the other hand, if we consider a large wave packet with many crests (*i.e.,* ψ having a value other than zero and constant over a large region), the velocity spread is so very small that the particle velocity can be fairly accurately determined, but the position of the particle becomes very uncertain.

Extending these results to the limiting cases of infinitely large and infinitely small wave packets it is readily seen that

certainty about velocity involves complete uncertainty about position and vice versa. Hence it is impossible to determine simultaneously both position and velocity of a particle with accuracy.

Experiment can never enable us to say precisely that a given particle occupies such and such a position in space and that its velocity has such and such a magnitude and direction. All that experiment can tell us is that the position and velocity of the particle lie in between certain limits; in other words, there is only a *calculable probability* that it has a certain position and velocity.

Heisenberg, who was the first to realize these consequences of wave mechanics, has expressed them mathematically by means of the following equations known as the uncertainty relation: $\Delta x.\Delta p \approx h$, where Δx and Δp are the respective uncertainties (or possible errors) involved in the simultaneous measurement of the position and momentum of a particle and **h** the Planck's constant, It states *that the product of the uncertainties in determining the position and momentum of a particle is approximately equal to Planck's constant.* According to it, the smaller the value of Δx, *i.e.,* the more exactly we determine the position, the larger the value of Δp, *i.e.,* the less exactly we determine the momentum. The converse is equally true.

The relation also brings out the fact that

it is the existence of the constant **h** which prevents us from knowing simultaneously both the position and motion of a particle with accuracy, while if **h** were zero such simultaneous accurate knowledge would be possible.

The constant **h** therefore represents an absolute limit to the accurate and simultaneous measurement of position and momentum, a limit which in the most favorable cases we may reach, but which we can never get beneath.

The essential point insisted upon in the uncertainty principle is *not the simultaneity of determination of the two quantities but the accuracy involved in the simultaneous determination.*

As a matter of fact, position and momentum can be simultaneously measured in the macroscopic bulk state as is done in ordinary physics, but the accuracy obtained is much less than that stated in the uncertainty relation. Since *h*, the ultimate limit of possible accuracy, is a small quantity (6.55×10^{-27}) the indeterminacy is completely masked by the experimental errors in the macroscopic state, while it becomes evident in the microscopic atomic state,

Although the principle of indeterminacy was first formulated in connection with wave mechanics, it was soon realized that it governs all Nature as a fundamental law. In every physical phenomenon there remains a margin of uncertainty which enshrouds it with a fundamental indefiniteness and discontinuity that cannot be overcome by any instrument of measurement, even of the highest precision. This generalized indeterminacy can be illustrated by one or two examples.

(i) **Determination of the position of a particle by a microscope.** Suppose we attempt to determine accurately the position of an electron using some instrument, such as a microscope of very high resolving power. Since the lower limit of resolution depends upon the wavelength of the light employed to illumine the particle, it follows that we must use radiation of the shortest wavelength, such as γ-rays, if we wish to determine the position as accurately as possible. But the employment of γ-rays involves the Compton effect so that the electron experiences a recoil. Thus, at the instant at which we locate the electron by irradiating it with γ-rays and observing the scattered γ-rays, its momentum undergoes a discontinuous change. Furthermore, there is indeterminateness about the magnitude of this change, for it will vary according to the direction in which the scattered γ-ray leaves the point of impact. We cannot limit closely the range of possible directions of the scattered γ-rays without a serious loss of resolving power which would involve less definition of the position of the electron.

Subjecting the case under study to a quantitative analysis, let the electron under the γ-ray microscope (Fig. 185) be irradiated in any direction by γ-rays of frequency ν.

From classical optical theory it can be shown that the resolving power of a microscope is given by $\Delta x = (\lambda/2)\sin\alpha$, where Δx represents the distance between two points which can just be resolved by the microscope and hence the possible error that is involved in the determination of the position of the electron, λ the wavelength of the light used and α the angular aperture of the microscope.

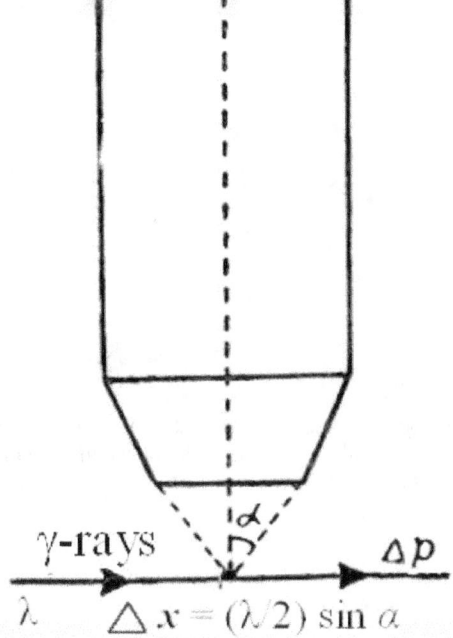

Now according to the corpuscular view, the minimum amount of light that could be used for irradiation is a single quantum $h\nu$. The electron can be actually seen only when it scatters this quantum into the microscope. But in this process the electron suffers a Compton recoil of the order of magnitude $h\nu/c$, the direction of which is indeterminate to the same extent as the direction of the scattered quantum. Since the scattered quantum can enter the microscope anywhere within the angle α, the uncertainty in direction of both the scattered quantum and recoil electron is given by α. Hence the component of momentum of the electron in a direction perpendicular to the axis of the microscope is uncertain to the amount Δp given by

$$\Delta p \approx 2\,\frac{h\nu}{c}\,\sin\alpha \approx \frac{2h}{\lambda}\,\sin\alpha$$

The product of the uncertainties in the simultaneous determination of the position and momentum is therefore

$$\Delta x \times \Delta p \approx \frac{\lambda}{2\sin\alpha} \times \frac{2h}{\lambda}\,\sin\alpha$$

i.e., $\Delta x.\Delta p \approx h$ (Heisenberg's uncertainty relation)

(ii) **Diffraction of a beam of electrons through a narrow slit.** From the corpuscular standpoint we have to regard the phenomenon of diffraction of electrons as occurring due to the deflection of the individual electrons at the slit, either upwards or downwards and the diffraction pattern recorded on a photographic plate as arising from the statistical superposition of the electrons in the beam. Every electron which is registered on the plate must have passed through the slit, but at what place in the slit is indefinite. If Δx is the width of the slit, the uncertainty in the specification of position of the electron perpendicular to the direction of flight is given by Δx (Fig 186).

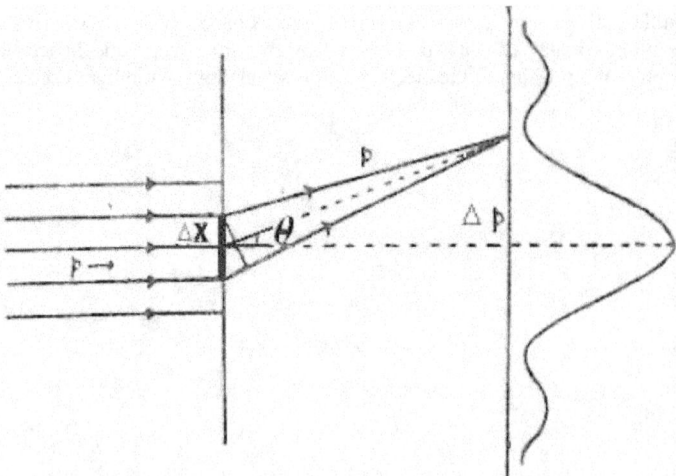

Fig 186

Since the electron is deflected at the slit it acquires an additional component momentum perpendicular to the original direction of flight. If p is the momentum of the electron, the component perpendicular to the initial direction is $p \sin \theta$, where θ is the mean angle of deflection. Since the electron may be anywhere within the diffraction pattern the uncertainty in the knowledge of the above-stated component is

$\Delta \mathbf{p} \approx 2\,p \sin \theta$

On the basis of the wave theory of diffraction

$\sin \theta = \lambda / 2\Delta \mathbf{x}$. Therefore, $\Delta \mathbf{x} = \lambda / 2 \sin \theta$

Hence the product of the uncertainties in the determination of the position and momentum of the electron in a given direction, i.e., perpendicular to its original direction of flight during its passage through the slit is

$$\triangle x \; . \; \triangle p \approx \frac{\lambda}{2 \sin \theta} \; . \; 2p \sin \theta \approx \lambda p$$

From De Broglie's theory $p = h/\lambda$, therefore, $\Delta \mathbf{x} . \Delta \mathbf{p} \approx h$

Thus once again we arrive at the uncertainty relation. These examples force upon us the conclusion that the fundamental indeterminacy that underlies every physical phenomenon cannot be overcome whatever attempt is made.

The existence of the quantum of action h introduces a finite disturbance which cannot be checked and which affects every pair of canonically conjugated magnitudes, such as position and velocity, energy and time, etc.

This new principle of indeterminacy is completely against the strict determinism inculcated by classical mechanics. According to the classical ideas the very definition of a particle implies that at any instant it occupies a *fixed position* in space and has a *fixed momentum*. The problem of classical mechanics is to predict exactly the position and momentum of the particle at any time *t* when its initial position and momentum are known; in other words, at every instant the particle has a position in space, which is perfectly determined and in course of time it describes a continuous curve known as its trajectory. But according to the principle of indeterminacy derived from wave mechanics, even the initial position and momentum of a particle cannot be known with accuracy; hence it is not possible to assign to any particle continuous trajectory and an instantaneous momentum. The term "particle" can no longer be used in the classical sense; it is something made up of a very deep and essential union of both the corpuscle and the wave, resulting in a certain unavoidable indeterminacy. Again, while the older mechanics claimed to apply exact and inexorable laws to every phenomenon, the new mechanics only gives us laws of probability and though these can be expressed in exact

76

formulae they still remain laws of probability. Hence probability takes the place of certainty or actuality in physical science. In the words of Bohr:

"the new quantum theory replaces the causal space-time coordination of. .classical physics by two apparently contradictory but actually complementary ideas of individuality and superposition"

CRITICISM OF THE THEORY OF MATTER WAVES

The great merit of the new theory of matter waves lies in the fact that it has solved all the problems for which it was designed, *viz.*, problems connected with the electronic structure, in the atom and spectral lines. The theory leads to equations whose solutions correspond to definite energy values, comparable to the energy terms of Bohr's atom model. Thus the quantum numbers of atomic structure and spectroscopy are inherent in the theory and are given in fundamental terms, not as arbitrary assumptions. The very discreteness of atomic processes appears as a natural outcome of the theory.

Wave mechanics also accounts for facts which are inexplicable by the older theories, such as half integral quantum numbers, zero point energy, etc. It has even led to predictions which later have been verified experimentally, as diffraction of electrons, resonance phenomena in spectroscopy, etc.

Furthermore, events impossible to the classical theory are found, when treated by wave mechanics, to have a very small but finite probability of taking place. On the basis of this prediction, natural radioactive process and artificial disintegration have received *a* very satisfactory explanation. Application of wave mechanics to these specifically nuclear phenomena has resulted in some understanding of the nucleus itself including its structure and energy content.

More than all these special problems, the general behavior of matter in the atomic state has been interpreted to a high degree of satisfaction by wave mechanics. It has shown that not only radiation but also matter exhibits sometimes wave-like properties and sometimes corpuscular properties; it has further led the way to treat the two concepts, wave and corpuscle, as complementary aspects of viewing one and the same objective process, which only in definite limiting cases admits of complete pictorial representation and is therefore nearly always subject to certain indefiniteness, given by the principle of uncertainty. Herein lies the great advance made by the new mechanics over the absolute determinism of classical mechanics.

But this final conclusion of wave mechanics, viz., the principle of uncertainty, *must be regarded not as a complete surrender of determinism* in physics, as many seem to think, *but as a further refinement in the right direction showing that the deterministic conception of Nature is neither complete nor universal.* For, the whole of the material universe, both in the macroscopic and microscopic states, appears to be subject simultaneously to determinism and indeterminacy. In other words, no physical phenomenon, whether large or small, is wholly deterministic or wholly indeterminate. It is true that in the large-scale phenomena dealt with in classical mechanics, as in astronomy, for instance, determinism predominates, while in the small-scale atomic phenomena indeterminacy is clearly perceived. But from this it does not follow that there is no determinism at all but only absolute indeterminacy in the microscopic state, while apparent determinism, *i.e.*, an admixture of determinism and negligible indeterminacy is restricted to the macroscopic state alone. Hence, we are inclined to believe that just as in macroscopic phenomena there is only apparent determinism, so also in microscopic phenomena the indeterminacy is only apparent. By that we mean the uncertainty lies in our ability to record events in timely fashion while the events themselves should proceed in certain deterministic manner governed by rigorous mathematical canons. It is only when rely on detecting a field in order to measure its interaction with matter that we obtain uncertain results due to the time lag between initiating the measurement and gathering the response. During such minuscule interval other events ensue that displace or speed the subatomic matter out of our field of measurement such that our harvested response represents unreal events within the accuracy of Planck's constant.

The following considerations seem to justify this opinion

(i) In general,

if **the doctrine of determinism** were to be rejected as completely wrong, there would be no order or regularity in physical phenomena

and no scientific knowledge of them could exist. But physics, even atomic physics, does exist: it is an actual fact and has proved its value by its progress and the number of its practical applications. We have a convincing proof of this in the modern art of artificial transmutation of elements, which is possible only if there exists a certain amount of determinism coupled with indeterminacy in material bodies, even in the atomic state.

(ii) **Planck's constant 'h'** which is considered to be the limiting barrier of determinism and a measure of indeterminacy, is really an enigma of Modern Physics; it still remains the unknown syllable in Nature's cross-word puzzle. Although its finite value shows that the microscopic state, where indeterminacy becomes apparent, is not merely a small scale representation of the greater macroscopic state of glaring determinism, yet the former tends asymptotically towards the latter. Such a tendency is possible only when both the states have common features, *viz.*, that they are partly deterministic and partly indeterminate.

Again, when one considers the strange way in which the constant h enters into Atomic Physics, viz., through the corpuscular aspect in the case of radiation, but through the wave aspect in the case of matter, it becomes possible to surmise that every physical reality, big or small, is complex, containing within itself, *i.e.,* independent from and prior to all experimental investigation, a double complementary aspect of determinism and indeterminacy. For, on this assumption alone one could understand why

an obviously wave form radiation acts as corpuscles and corpuscular matter behaves like a wave.

(iii) **The marked element of paradox** that appears in the wave mechanical conception of physical reality leading to the principle of indeterminacy, also seems to favor the opinion proposed. For,

waves of probability, pure abstractions as they are, are nevertheless propagated "in space" just as though they were continuous and homogeneous elastic waves.

Likewise, the particles whose existence is intermittent cannot be localized and defined with exactness. Thus the wave and particle of the new mechanics, are, each of them, a complex of indeterminacy and determinism. According to the principle of indeterminacy, both the position and velocity of a particle cannot be determined simultaneously with accuracy. This implies that there is an intrinsic limitation in the particle itself which prevents it from being defined with accuracy under all aspects at once.

(iv) **The very concept of probability,** which forms the basis of the principle of indeterminacy, implies the complementary double aspect of determinism and indeterminacy in every physical phenomenon. When the history of a single physical unit, whether electron or photon, is investigated, the associated wave symbolizes the probabilities of localization and the dynamic state of that unit; but when we are dealing with a very great number of identical units, for instance, a beam of electrons, then the wave represents the statistical distribution of the totality of units. Hence, in both cases statistical law must be used. Now a statistical law is one which defines the more or less constant mode of action of a great collection of similar things without stating anything about a particular individual of the collection, except in so far as it belongs to the collection. A statistical law is a law of averages and there is no finality or exactness about it when applied to particular individual cases. The reason for this is that when forming part of a system, a physical unit loses a large measure of its individuality, the latter tending to merge in the greater individuality of the system. To make **a** real individual of a physical unit belonging to a system, then, it is necessary to take this unit from out of the system, *i.e.,* to break the link which binds it to the system. The physical unit cannot be observed perfectly so long as it forms part of the system and the system is impaired once the individual unit has been identified. The concept of the physical unit thus becomes completely clear and properly defined, only if it is regarded as a unit completely independent of everything else.

Applying this general idea to a single particle we have a system formed of localization and dynamic state of the particle. The two lose their individualities to a large measure when they form part of the system. Hence, when we try to define exactly one of them, the other becomes most ill-defined. The same consideration can be applied to a system made up of a very great number of particles. Thus a close analysis of physical reality, as an individual unit or as a system of units, brings out clearly the intimate and unavoidable presence of the two aspects of determinism and indeterminacy in every physical phenomenon. Louis de Broglie, in his recent book *The Revolution of Physics* (1953) states that he is by no means convinced that the possibility has been extinguished of finding beneath the present statistical structure "a perfectly determinate reality".

CHAPTER 10

QUANTUM STATISTICS

Introduction

In studying the general behavior of a body which consists of a very large number of individual particles, the *statistical method* of investigation has to be employed. For instance, in order to arrive at a knowledge of the temperature and pressure of a gaseous mass, it is necessary to fix the position and velocity of each of the molecules contained in it.

The quantities, temperature and pressure, refer to the *macroscopic* or bulk state of the gas and can be measured by ordinary apparatus;

on the other hand,

the position and velocity of the individual molecules in the gas belong to the *microscopic* state and a complete knowledge of the continually changing motion of the molecules is not attainable.

Hence, in order to correlate the two states and thereby draw conclusions which can be subjected to practical tests, recourse is had to the calculus of probabilities and statistical mechanics. In this method the assigning of parameters to a large number of individual particles that constitute the assemblage is a very important factor and is known as *distribution function* or *statistics*.

Before the advent of the quantum theory, classical physics made use of a statistics, known as the *Maxwell-Boltzmann statistics*, with great success in the interpretation of many ordinarily observed phenomena such as temperature, pressure, energy, etc. But

the failure of the classical method to account adequately for several other experimentally observed facts, such as the black body radiation, photoelectric effect, specific heat at low temperature, etc., forced the issue in favor of the new quantum idea of discrete exchange of energy between systems,

and along with it a new quantum statistics was envisaged.

Further,

it was soon recognized that the new statistics should be subdivided into two categories, one holding good for *photons* and known as the *Bose-Einstein statistics* and the other referring to *elementary material particles*, such as electrons, and called the *Fermi-Dirac statistics*.

Our aim in this chapter is, after briefly stating the general principles involved in statistical mechanics, to deal with the classical statistics first and then take up the modifications required by the new quantum conception under the double aspect of *particle* and *wave* applicable to both radiation and matter. The classical method of statistics, besides its theoretical interest as a limiting case, serves as a guide for the quantum treatment. We shall then deal with several applications of the quantum statistics, which will bring out the utility of this new type of statistics in interpreting physical phenomena that remained unsolved or only partially solved by the classical method.

Principles of statistical mechanics

Statistical mechanics deals with the statistical features of the behavior of a system subjected to specified conditions, by a suitable combination of the laws of probability and dynamical principles. Hence, in this method of investigation we are not concerned with complete knowledge of the state of a dynamical system, but only with certain broad features of its behavior, which happen to be physically significant and are often of the nature of an average of some sort. Further, in any conclusion concerning probability which is drawn from statistical mechanics, the element of probability is introduced in describing the situation to which the conclusion refers and what is deduced from mechanical laws is only the relation between two probabilities.

Phase space

In the theory of mechanics it is shown that the motion of a dynamical system having n degrees of freedom can conveniently be described in terms of n coordinates,

$q_1 \, q_2 \, \ldots \ldots q_n$

together with n corresponding momenta,

$p_1 \, p_2 \, \ldots \ldots p_n$.

Each phase of any motion of the system is then represented by a set of values of the **n** q's and **n** p's ; and during **a** particular motion these values vary as definite functions of time. It is usual to regard the 2n quantities as the coordinates of a point in a space of 2n dimensions. Such a space is known as a *phase space* of the system. As the system executes a particular motion, its representative point traces out a certain trajectory in the phase space. Hence "a trajectory in the phase space" of the system is equivalent to a succession of sets of values of the q's and p's.

Distribution function

In applying the statistical method to a system made up of a very large number of individual particles, the position of each particle is fixed, but with a certain latitude and in the same way a velocity is assigned to each, but again with a certain latitude. (Thus probability is introduced into the problem). Then, the *distribution function*, which represents the number of particles assumed to be situated in each small element of volume and the number assumed to have momenta within a small but definite range, is usually expressed as

$$dn = f\left(x, y, z, p_x, p_y, p_z\right) dx\ dy\ dz\ dp_x\ dp_y\ dp_z$$

in the ordinary Cartesian coordinate system. In this relation **dn** is the number of particles having coordinates between x and $x + dx$, etc., and components of momentum between p_x and $p_x + dp_x$, etc. The product $dx.dy.dz$ is a volume element in coordinate space, and $dp_x.dp_y.dp_z$ a volume element in momentum space, and the product $dx.dy.dz.dp_x.dp_y.dp_z$ is defined as a volume element in phase space or *phase volume*, so that f defines the distribution in this phase space of six dimensions.

Calculus of probabilities and its laws

The *probability of an event* may be defined as the *ratio of the number of cases in which the event occurs to the total number.* Suppose we toss a coin a large number of times and count the number of times the "head" or *event occurs to the total number.* Suppose we toss a coin a large "tad" is uppermost. Experience shows that, provided the number of throws is sufficiently great, the "heads" and "tails" will be uppermost equal number of times, *i.e.,* the probability of each event or the ratio of the number of throws in which each event occurs to the total number of throws is 1/2.

> The *probability of a composite event is equal to the product of the probability of the individual component events,* provided these are independent.

Throwing of coin

Consider two coins say one silver and the other copper, which are tossed a large number of times and the number of times their "head" or "tails" arc uppermost is noted. Let a_1 and a_2 be the probabilities for the heads of the silver and copper coins to be uppermost respectively and b_1 and b_2 the probabilities for their tails to be uppermost. If we toss the silver coin, the probability ' a_1' of our getting the head is ½ ; if we toss the copper coin, the probability 'a_2' of getting the head is ½ . Hence in a *simultaneous single toss of the two coins,* the probability of getting the heads of both is $a_1 \times a_2 =$ ½ x ½ = ¼. Thus we see that the probability ($a_1 a_2$) of a composite event is the product of the probabilities (*i.e.,* of a_1 and a_2) of the individual component events.

If the two coins are *tossed a large* number *of times,* the following events are possible:

(1) heads of both coins up ---- $a_1 a_2$
(2) tails „ „ „ .... $b_1 b_2$
(3) head of silver coin and tail of copper coin up ... $a_1 b_2$
(4) tail of silver coin and head of copper coin up ... $a_2 b_1$

The possible events can therefore be represented by $(a_1 + b_1)(a_2 + b_2)$

If *the two coins* are *identical,* then $a_1 = a_2 = a$ and $b_1 = b_2 = b$. The possible combinations are now given by $a^2, b^2, 2\ ab$ [since events (3) and (4) cannot be distinguished from each other] i e., $(a + b)^2$. Here the chance of getting the heads or tails of both coins up is a^2 or $b^2 = $ ¼, while the chance of getting the head of the one and the tail of the other is $2ab =$ ½.

Extending the argument to 'n' identical coins, the possible combinations are given by the terms of the binomial expression $(a + b)^n$, *i.e.,* $a^n, a^{n-1}b, ... a^r b^s, ..., b^n$, where $a^r b^s$ means that in a throw r coins show heads and s show tails

$(r + s = n)$. This particular complexion $a^r b^s$ can be realized ^nC_r times, where ^nC_r is the coefficient of $a^r b^s$. By the binomial theorem,

$$^nC_r = \frac{n!}{r!\,(n-r)!} = \frac{n!}{r!\,s!}$$

In the binomial expression representing all possible combination, there are 2^n terms. The *a priori* probability of each complexion is therefore The probability W of the complexion $a^r b^s$ is given by

$$W = \frac{n!}{r!\,s!}\,2^{-n} \qquad\qquad \dots (1)$$

There are, of course, as many complexions of $a^r b^s$ as there are terms in the expansion of $(a + b)^n$, *i.e.*, $(n + 1)$. Among these, that which possesses maximum probability is given by the condition that ^nC_r is a maximum, which is the case when $r = n/2$, Hence

$$W_{max} = \frac{n!}{\left(\dfrac{n}{2}!\right)^2}\,2^{-n} \qquad\qquad \dots (2)$$

Throwing of dice

If we were to use dice instead of coins, we should have six possible states, not merely two as in the case of the coins. Each die has six sides which may be denoted by the letters a, b, c, d, e, f. Let a large number of n of dice be thrown which results in a particular complexion represented by

$$a^{n_1}\, b^{n_2}\, c^{n_3}\, d^{n_4}\, e^{n_5}\, f^{n_6}$$

This means that n_1 dice fall with the face 'a' up n_2 dice with 'b' up, etc.

By arguments similar to those used above, it can be shown that the possible number for such a complexion is given by

$$\frac{n!}{n_1!\,n_2!\,n_3!\,n_4!\,n_5!\,n_6!}$$

Since each aspect of a die has an *a priori* probability 1/6 on account of the six faces, the probability of the complexion under consideration is given by

$$W = \frac{n!}{n_1!\,n_2!\,n_3!\,n_4\,n_5!\,n_6!} \times 6^{-n} \qquad\qquad \dots (3)$$

General case

Consider a board with p cells arranged side by side and denoted by $a_1, a_2, a_3 \dots$ etc. Let a great number n of balls be thrown into the cells from a distance. Let us suppose than n_1 balls are found in the cell a_1, n_2 in a_2 .....etc.

The possible number of such a, complexion is given by

$$\frac{n!}{n_1!\,n_2!\,n_3!\dots\dots n_r!} = \frac{n!}{\xi\,n_r!}$$

Where

$$\prod_{k=1}^{k=s} n_k! = n_1! \times n_2! \times n_3! \times \ldots\ldots n_s!$$

Since the *a priori* probability for each complexion is $(1/p)$, the probability of the above complexion is

$$W = \frac{n!}{\prod\limits_{k=1}^{k=r} n_k!} \; p^{-n} \qquad\qquad \ldots (4)$$

When *n is very large,* this relation (4) can be reduced to a more simple and convenient form by the application of a famous theorem due to *Stirling,* hence known as *Stirling's theorem.* According to this theorem, we have

$$n! = (n/e)^n \quad \text{or} \quad \log n! = n \log n - n$$

Applying this

$$\log W = \log \left(\frac{n!}{\prod\limits_{k=1}^{k=r} n_k!} \; p^{-n} \right)$$

$$\log W \left(\prod_{k=1}^{k=r} n_k! \right) = n \log n - n - n \log p \qquad\qquad \ldots (5)$$

CLASSICAL STATISTICS

Classical statistics is rightly called *Maxwell-Boltzmann statistics,* since it takes its origin from Maxwell's law of distribution of molecular velocities and Boltzmann's theorem relating entropy and probability. Hence it is necessary to review briefly these two basic ideas.

Maxwellian distribution of molecular velocities
According to the kinetic theory of matter, a gas consists of a very large number of rigid and perfectly elastic material particles, called molecules, all of which have the same mass and move freely within the vessel containing the gas. Moving in all possible directions and colliding with each other these molecules acquire all possible velocities ranging from $-\infty$ to $+\infty$. Maxwell in 1859 conceived the idea that, at a given temperature when the gas is in a state of thermal equilibrium, there should be a law according to which the velocities of the molecules can be grouped, in spite of their apparently random and chaotic movements. He was able to derive that law of distribution of molecular velocities by the application of elementary ideas of probability as follows:-

Considering, for the sake of simplicity, a perfect, monatomic gas (where translational motion alone of the molecules is involved, rotational and vibrational motions being excluded) and assuming that the molecular density is unaffected by the molecular motions and collisions, probability considerations show that the number of molecules n_1 which have component velocities lying between
u and $\underline{u} + du$, v and $v + dv$ and w and $w + dw$,
along the three axes of the Cartesian coordinate system is given by

$$n_1 = n f (u, v, w) \, dv \, dv \, dw$$

where n is the number of molecules in unit volume. This is Maxwell's law of distribution of molecular velocities, which is usually expressed in the following two forms:-

$$(1) \qquad n_1 = n \, a \, e^{-bmc^2} \, d\tau$$

where

$$a = (b^3 m^3 / \pi^3)^{1/2}$$

$$c^2 = u^2 + v^2 + w^2 \text{ and}$$

$$d\tau = du \, dv \, dw.$$

b is a constant,
m the molecular mass,

$$(2) \qquad n_1 = n \left(\frac{m}{2\pi k \mathrm{T}} \right)^{3/2} e^{-mc^2/2k\mathrm{T}} \, d\tau$$

where k is Boltzmann's constant and T the absolute temperature of the gas. The identity of (1) and (2) is readily seen by putting $b = 1/2k\mathbf{T}$

Boltzmann's theorem on entropy and probability

A gas at a definite temperature and pressure is in a sensibly constant macroscopic state, but its constituent molecules are in incessant motion of a random and chaotic nature, so that the microscopic state of the gas is continually changing. The reason for this state of affairs must be naturally sought in the fact that

> of all the possible microscopic states the vast majority correspond to values of the quantities which make the macroscopic state practically constant.

If any microscopic state has a value very much different from that of the macroscopic state, the probability of its remaining in that state is small. When a substance is not in a state of macroscopic equilibrium, its state changes until equilibrium is established. This process involves naturally an increase in the number of possible microscopic states. Thus, it is evident that an *equilibrium macroscopic state is one for which the number of microscopic states is a maximum.*

On the other hand, according to the second law of thermodynamics in the form proposed by Cornot, the entropy of a system tends always towards a maximum and

> the maximum entropy corresponds to the maximum disorder, hence to the statistical condition of maximum probability.

From reasoning like this, Boltzmann concluded that there must be a relation between the thermodynamical entropy which has always a maximum value in cases of equilibrium and the maximum probability of the dynamical equilibrium state. If S is the entropy of an isolated system and W the number of possible microscopic states through which the system passes in a given macroscopic state, both S and W tend to increase to maximum values and according to Boltzmann's idea, S is a function of W:

$S = f(w)$

where W may be called the thermodynamic probability of the state of the system. In order to render this relation more explicit, let us consider two separate systems, having entropies S_1 and S_2 and thermodynamic probabilities W_1 and W_2. Then

$S_1 = f(W_1)$
and
$S_2 = f(W_2)$

The total entropy of the two systems is

$S_1 + S_2 = f(W_1) + f(W_2)$ .... (1)

But

the thermodynamic probability of the two systems taken together is $W_1 W_2$. Hence

$f(W_1 W_2) = S_1 + S_2 = f(W_1) + f(W_2)$ .... (2)

If this relation is to be satisfied, $f(W)$ must be a logarithmic function of W.

Therefore,

$f(W) = k \log W$

and

$S = k \log W$ ....(3)

This relation (3) can be established more rigorously as follows:

Considering unit volume of a perfect monatomic gas in thermal equilibrium, let there be n molecules in it. The velocity components u, v, w of each molecule may be represented by a velocity point. Dividing the unit volume into a very large number of elementary volumes assumed equal, these will contain n_1, n_2, n_3,......velocity points, such that

$n_1 + n_2 + n_3 + ... = n$

When the n_1 velocity points, contained in the first elementary volume are permuted among themselves, $n_1!$ complexions are obtained which are indistinguishable from one another, since they correspond to the same macroscopic state of the gas. Similarly, by permuting n_2 velocity points in the second volume element, $n_2!$ equivalent complexions are obtained and so on..

If, on the other hand, all the n velocity points are permuted in all possible ways, $n!$ distributions are obtained. These contain many equivalent complexions and we are interested only in those which remain distinct from one another for defining the macroscopic state of the gas. Now, the number of distinguishable complexions is given by

$$W = \frac{n!}{n_1! \, n_2! \, n_3! \dots}$$

On the assumption that n is very large, using Stirling's theorem,

$$\log W = n \log n - n - (n_1 \log n_1 - n_1 + n_2 \log n_2 - n_2$$
$$+ n_3 \log n_3 - n_3 + \dots)$$

$$= n \log n - n - (n_1 \log n_1 + n_2 \log n_2 + n_3 \log n_3 + \dots)$$
$$+ (n_1 + n_2 + n_3 + \dots)$$

$$= n \log n - (n_1 \log n_1 + n_2 \log n_2 + n_3 \log n_3 + \dots)$$

$$= - (n_1 \log n_1 + n_2 \log n_2 + n_3 \log n_3 + \dots)$$

$$= - \Sigma n_1 \log n_1$$

omitting the constant term ($n \log n$), as it does not effect the present consideration.

Replacing the summation by integration and using the relation

$$n_1 = n\,a\,e^{-bmc^2}\,d\tau,$$

We get

$$\log W = -\int n\,a\,e^{-bmc^2}\,d\tau\,\log(n\,a\,e^{-bmc^2}\,d\tau)$$
$$= -\int n\,a\,e^{-bmc^2}\,d\tau\,(\log n\,a + \log d\tau - bmc^2)$$
$$= -\int n\,a\,e^{-bmc^2}\,d\tau\,\log n\,a + bmc^2\int n\,a\,e^{-bmc^2}\,d\tau$$

since $(\log d\tau)$ can he omitted.

$$\log W = -n\log na + bmn\int c^2\,a\,e^{-bmc^2}\,d\tau$$

And since

$$\int c^2\,a\,e^{-bmc^2}\,d\tau = \overline{C}^2$$

Where $\overline{C}$ is the mean velocity.

$$\log W = -n\log(na) + bmn\,\overline{C^2}$$

Since

$$\overline{C}^2 = 3/2bm$$

Then

$$\log W = -n\log(na) + (3/2)\,n$$

Further, as $n \propto 1/V$ and $a \propto b^{3/2} \propto 1/T^{3/2}$, we get

$$\log W = \frac{N}{V}\log(T^{3/2}\,V) + \frac{G}{V}$$

where N is the Avogadro number and G a constant.

This result refers to unit volume. If V is the gram molecular volume,

$$\log W = N\log(T^{3/2}\,V) \qquad\qquad \cdots (4)$$

neglecting G.

To relate this result with the entropy, let us derive an expression for the entropy S of a perfect monatomic gas in terms of the temperature and the molecular volume.

Let dQ be the heat supplied to the gas which is utilized in doing external work $p\,dV$ and in raising the temperature by dT. Then

$$dQ = C_v\,dT + p\,dV$$

If dS is the change of entropy,

$$dS = dQ/T,$$

so that

$$TdS = C_v\,dT + p\,dV$$

Since pV = RT,

$$TdS = C_v\, dT + \frac{RT}{V}\, dV$$

Dividing throughout by T,

$$dS = C_v \cdot \frac{dT}{T} + R \cdot \frac{dV}{V}$$

Integrating,

$$S = C_v \log T + R \log V + \text{a constant.}$$

Omitting the additive constant,

$$S = C_v \log T + R \log V$$

Since, for a monatomic gas $C_v = (3/2)\, R$,

$$\begin{aligned} S &= (3/2)\, R \log T + R \log V \\ &= R\,(\log T^{3/2} + \log V) \\ &= R \log (T^{3/2}\, V) \end{aligned} \qquad \ldots (5)$$

Comparing this with relation (4), we get

$$S = (R/N) \log W = k \log W$$

Thus Boltzmann's theorem connecting entropy and probability is proved.

In classical thermodynamics the entropy is usually taken as the *difference* between the entropy in the actual state and the entropy in an arbitrarily chosen standard state. If the thermodynamic probability in the standard state is W_o, then

$$S = k \log W - k \log W_0 = k \log (W/W_0)$$

Further, the standard state is referred to that of absolute zero, since, according to Nernst's heat theorem, the entropy at absolute zero is zero so that W/W_o for that state is equal to unity. This means that

the standard state of absolute zero is one of perfect order, in which the molecules are either at rest or move with uniform velocity in parallel rows.

On this basis, Boltzmann's theorem of entropy and probability is written as

$$S = k \log N$$

where $N = W/W_o$ represents the number of probable ways in which a particular state (other than that of absolute zero) can be realized.

An expression for N may be derived by the use of statistical laws as follows:-

Let us suppose that the entire volume containing n molecules of a gas is divided into a number of separate elementary volumes or *cells,* in which the molecules are distributed. Let these cells be designed by $A_1, A_2, A_3, \ldots\ldots A_s$ and let them contain. $n_1, n_2, n_3, \ldots\ldots n_s$ molecules, each with energy $E_1, E_2, E_3, \ldots\ldots E_s$ respectively, subject, of course, to the following two conditions:

Total number of molecules: $n_1 + n_2 + n_3 + \ldots\ldots n_s = n$

Total energy: $n_1 E_1 + n_2 E_2 + n_3 E_3 + \ldots\ldots n_s E_s = E$

The possible number of ways this distribution can be realized is

$$\frac{n!}{n_1! \, n_2! \, n_3! \ldots \ldots n_s!} = \frac{n!}{\prod\limits_{k=1}^{k=s} n_k!}$$

To get the probability of this distribution we have still to consider the a *priori* probability of the distribution. For every one of the above possible ways there is no restriction on any one of the molecules occupying any one of the cells, since the molecules are considered *distinct* from one another, (each having a *recognizable individuality* of its own) and each cell can accommodate any number of molecules. Hence, considering the distribution of the n_s molecules in the A_s cells, any one of the n_s molecules can occupy any one of the A_s cells. This means that the probable number of ways in which the n_s molecules can be distributed in the A_s cells is $(A_s)^{n_s}$. In a similar manner, the molecules in the other cells can be grouped as $(A_1)^{n_1}$, $(A_2)^{n_2}$, $(A_3)^{n_3}$, etc. All such groupings can therefore be done in

$$\prod_{k=1}^{k=s} (A_k)^{n_k}$$

ways.

Hence the probable number of ways W, in which the desired distribution can be effected, is given by

$$W = \frac{n!}{\prod\limits_{k=1}^{k=s} n_k!} \times \prod_{k=1}^{k=s} (A_k)^{n_k} = n! \prod_{k=1}^{k=s} \left(\frac{A_k^{n_k}}{n_k!} \right)$$

Now, since $W_0 = n!$ is the probable number of ways in which the molecules can be arranged in the state of perfect order, therefore

$$\therefore \quad P = \frac{W}{W_0} = \prod_{k=1}^{k=s} \left(\frac{A_k^{n_k}}{n_k!} \right) \quad \ldots (6)$$

Equilibrium conditions and distribution law

The distribution law corresponding to the equilibrium state is governed by the *primary* condition that the state must possess *maximum* probability, as well as by the *two subsidiary* conditions concerning the total number of molecules and the total energy, stated above. For a gas in the steady state, *i.e.*, in thermal equilibrium,

$$S = k \log P = k \log \prod_{k=1}^{k=s} \left(\frac{A_k^{n_k}}{n_k!} \right)$$

$$\therefore \quad S/k = \Sigma \, [n_s \log A_s - \log n_s!]$$

By Stirling's theorem,

$$\log n_s! = n_s \log n_s - n_s$$

Hence

$$S/k = \Sigma \, [\, n_s \log A_s - n_s \log n_s + n_s \,]$$

Since, in the steady state S is constant, $dS = 0$ and $dS/k = 0$.

$$\therefore \quad \Sigma \, dn_s \, (\log A_s - \log n_s) - \Sigma n_s \, dn_s/n_s + \Sigma \, dn_s = 0.$$

Since $n = \sum n_s$ is a constant, $\sum dn_s = 0$ and also, $E = \sum E_s n_s =$ constant, $\sum E \, dn_s = 0$. Under these conditions, remembering also that A_s is not subject to variation, the above relation reduces to

$\log A_s - \log n_s = 0$

Using the method of *undetermined multipliers,* we obtain

$$\log A_s - \log n_s - \lambda - \beta E_s = 0$$

where λ and β are two undetermined constants, known as *Lagrangian multipliers.*

The above relation can be put in a more convenient form:

$$\log \frac{A_s}{n_s f} = \beta E_s \qquad \text{or} \qquad \frac{A_s}{n_s} = f \, e^{\beta E_s}$$

where β is a constant affecting the energy content of the n_s molecules distributed in the A_s cells and f is an unknown function whose value depends on the conditions of the case and is ordinarily called the *degeneracy parameter.*

The classical distribution law is therefore given by

$$n_s = (A_s/f) \, e^{-\beta E_s} \qquad \ldots (7)$$

From this relation, we see that

the number n_s corresponding to the cell A_s essentially involves the energy E_s belonging to this cell as well as the size A_s of the cell, and that in such a way that among cells of equal size, one with greater energy is not so well filled as one with smaller energy.

The fall in the value of n_s with increasing energy obeys an exponential law.

Maxwell's law of distribution of velocity can be deduced from the above relation by substituting the values of $A_s f$ and E_s, proper to the case; it can he shown also that $\beta = 1/kT$.

Satyendra Nath Bose (1 January 1894 – 4 February 1974), India

QUANTUM STATISTICS

The quantum statistics was first formulated in 1924 by Bose in deduction of Planck's radiation law by *purely statistical reasoning*, without having recourse to the roundabout method of employing absorbing and emitting oscillators, as was done by Planck. Certain fundamental assumptions were made in this theory, which were radically different from those of the classical statistics. Einstein in the same year utilized practically the same principles in evolving the kinetic theory of gases, as a substitute for the Boltzmann statistics. Thus a new quantum statistics, known as the *Bose-Einstein statistics* came to be accepted.

Enrico Fermi (29 September 1901 – 28 November 1954)

Two years later, in 1926, Fermi and Dirac found it necessary to modify the Bose-Einstein statistics in certain cases, on the basis of an additional principle, suggested first by Pauli in connection with the electronic structure in atoms and known as Pauli's exclusion principle. This led to the recognition of a second kind of quantum statistics, called the *Fermi-Dirac statistics.*

The essential difference between the three statistics, the classical and the two quantum ones, may be illustrated in the following simple manner: Let us suppose that there are only two members of a collection and only two cells to be occupied.

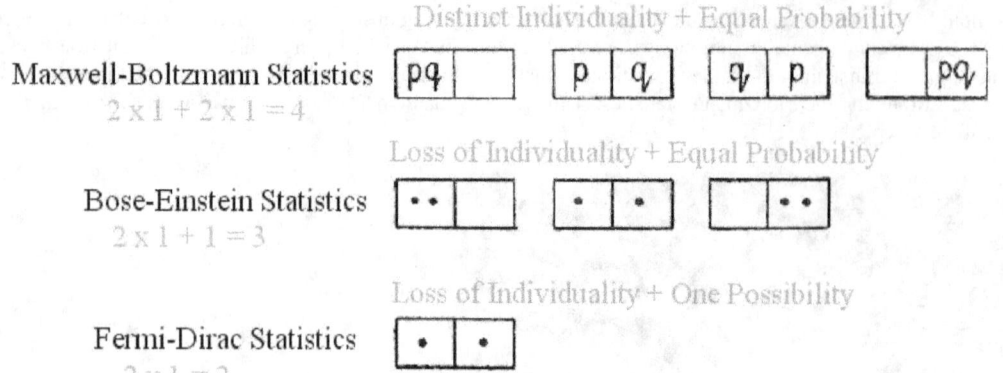

Fig. 187. Illustration of the three statistics.

The special feature of the classical *statistics* is that each member is assumed to have a *recognizable individuality,* hence *distinct* from the others, and each is as likely to be in one cell as in another. This means that, in the present case, each or both of the members can occupy any one of the two cells. We can have therefore *four possible arrangements,* each of which is to be counted in assigning a probability to the distribution, as shown in Fig. 187, where the two members are designated by two different letters *p* and *q*.

In the *Bose-Einstein statistics,* however, the members lose their individuality, hence are indistinguishable, and attention is concentrated on the cell rather than on the individual members. In the figure, therefore the members are represented by indistinguishable points. Each cell may have any number of points so that *three arrangements* are possible in the case considered.

In the *Fermi- Dirge statistics,* (based on Pauli's exclusion principle, according to which not more than one electron in a system can be in the same quantum state), although the idea of individualizing particles is given up and attention is concentrated on the cells as in the Bose-Einstein statistics, yet it is not possible to have more than one representative point in any one cell. A *cell can contain one point or none; each alternative is equally likely; but it is impossible that a cell should contain two or more.* Hence in the case under consideration, *only one arrangement is possible,* as shown in the figure.

When there are a large number of cells and comparatively few members to be arranged in them, there is no great difference between the three statistics, since the likelihood of any cell containing more than one member is very small. But when the number of members is comparable with the number of cells, the three statistics lead to divergent results. We shall now derive expressions for the distribution laws in the two quantum statistics, in a manner analogous to that employed in classical statistics, so that the divergence in results may be readily compared and studied.

A. Bose-Einstein statistics

Bose regards radiation in a temperature enclosure as composed of light-quanta or particles with energy $h\nu$ where ν is the frequency of the quantum. According to present views regarding matter and energy, a light quantum has a mass $h\nu/c^2$ and a momentum $h\nu/c$, where c is the velocity of light.

The light quanta in the enclosure *are indistinguishable* from one another,

i.e., no unchanging individuality can be assigned to them, since the walls are emitting and absorbing radiation. Further, even

the *number of light quanta is not necessarily unchangeable,*

unlike in classical statistics, where the number of particles is considered to he fixed. For, in every emission process in an atom, a new light quantum is formed and if a quantum of frequency ν is absorbed by the wall of the enclosure, it can be replaces by emission of several quanta of frequencies ν_1, ν_2 without any breach of the energy condition, provided $\nu = \nu_2 + \nu_1 + ..$. Einstein, dealing with the kinetic theory of a monatomic gas, assumes that

the molecules of the gas are like light quanta, indistinguishable from one another; but unlike in the light quantum case, *the number of molecules is conserved.*

In classical statistics, where the number of possible microscopic, states of a system is considered to be infinite, the phase space of the system can be divided into an infinite number of elementary regions of constant energy.

But in quantum statistics, the phase space is divided into *finite elementary regions of* constant energy content, which are sometimes denoted as *"sheets"* and each of these "sheets" is partitioned into still smaller *"cells" of finite size*, a cell representing the smallest phase-space that can contain any representative point at all. The finite number of sheets results from the quantum theory, according to which, only certain definite values of the energy of a system are permitted, so that

the number of possible microscopic states of the system is finite and the *phase-space can be divided only into finite regions of constant energy.*

A change in the energy of one of the systems is represented by its representative point jumping from one sheet to another without occupying positions intermediate between the sheets.

It is further legitimately assumed that *the elementary phase-cells have a finite size, of magnitude h^3, in accordance with Heisenberg's uncertainty principle.* For, this principle states that the position and momentum of a particle cannot be more exactly defined than is consistent with the relation $dp_x.dx \approx h$. Now, since a particle is defined by three pairs of conjugate coordinates, for each of which the above relation holds, it follows that

$$dp_x . dx . dp_y . dy . dp_z . dz = h^3 .$$

In other words, one cannot make a finer division of the phase-space than that given by the above relation, since it is impossible to decide by experiment in which of these cells a particle lies.

The problem to be solved may be stated as follows: Considering a sheet, say the s, in the phase-space let there be a definite number, A_s, of cells in it, whose energy content ranges between E_s and $E_s + dE_s$. A definite number of particles, say n_s are to be distributed among the A_s cells. The conditions to be observed are:

(i) the particles are indistinguishable from each other, so that there is no distinction between the different ways in which the n_s particles can be chosen;
(ii) each cell may contain 0, 1, 2 ... up to n_s identical particles;
(iii) the sum of the energies of all the particles in the different sheets taken together constitutes the total energy of the system.

The process involved in such a distribution of *identical* particles is known in Algebra as *combination with repetition*

which we shall employ here in a simple and inductive manlier, as follows:

Let the individual cells in the s sheet be denoted by c_1, c_2, c_3, etc., the number of them being, by definition, equal to A_s.
Let p_1, p_2, p_3, etc. represent provisionally the individual particles of the total number n, in the sheet.
We have to distribute these particles among the A_s cells and determine the number of distinguishable arrangements. To this end, let us consider an arbitrary arrangement such as

$$c_1 \; p_1 \; p_2 \; c_2 \; p_3 \; p_4 \; p_5 \; c_3 \; p_6 \; c_4 \; p_7 \; p_8 \; c_5 \; c_6 \; p_9 \cdots$$

In this distribution, p_1 and p_2 are supposed to be in the cell c_1,
p_3, p_4 and p_5 in cell c_2,
p_6 in c_3,
p_7 and p_8 in c_4,
no particle in c_5,
and so on.

This means that the first letter in the sequence should evidently be a "c". We therefore obtain all possible arrangements by writing down a "*c*" at the head of the series, which can be done in A_s different ways, and then writing down the remaining $(A_s - 1 + n_s)$ letters in arbitrary manner one after the other. The total number of these arrangements is, therefore,

$$A_s (A_s - 1 + n_s) ! \quad \ldots\ldots\ldots\ldots\ldots\ldots\ldots\ldots (1)$$

The cells among themselves and the particles among themselves can be permuted in
$A_s! \, n_s!$
ways. These, however, do not represent different states. Hence the number of distinguishable arrangements in the sth sheet is

$$\frac{A_s (A_s + n_s - 1)!}{A_s! \, n_s!} = \frac{(A_s + n_s - 1)!}{(A_s - 1)! \, n_s!} \quad \ldots(2)$$

Considering all the available sheets, with n_1 particles in the first, n_2 particles in the second and so on, the total number of distinguishable arrangements is given by

$$\prod_{k=1}^{k=s} \frac{(A_k + n_k - 1)!}{(A_k - 1)! \, n_k!} = \prod_{k=1}^{k=s} \frac{(A_k + n_k)!}{A_k! \, n_k!}$$

neglecting 1 in comparison with the large numbers of A_s and n_s. We call this the probability of distribution, according to Bose-Einstein statistics, so that we may write; on the analogy of the classical statistics,

$$P = \prod_{k=1}^{k=s} \frac{(A_k + n_k)!}{A_k! \, n_k!} \quad \ldots(3)$$

The distribution law of Bose-Einstein statistics can now be obtained by determining the *most probable distribution*, as in classical statistics.

Hence, substituting in the relation, $S = k \log P$, the value of P given by (3) we have

$$S = k \log \prod_{k=1}^{k=s} \frac{(A_k + n_k)!}{A_k! \, n_k!}$$

By Stirling's theorem, this becomes

$$S/k = \Sigma \, [(A_s + n_s) \log (A_s + n_s) - A_s - n_s$$
$$- A_s \log A_s + A_s - n_s \log n_s + n_s]$$
$$= \Sigma \, [(A_s + n_s) \log (A_s + n_s) - A_s \log A_s - n_s \log n_s]$$

Remembering that in the equilibrium state $dS/k = 0$ and that A_s is not subject to variation, we obtain, by differentiating the above relation and equating it to zero

$$\Sigma \, [dn_s \log (A_s + n_s) + (A_s + n_s) \, dn_s/(A_s + n_s)$$
$$- dn_s \log n_s - n_s dn_s/n_s] = 0$$

Of the two subsidiary condition,

Total number of molecules: $n_1 + n_2 + n_3 + \ldots\ldots n_s \qquad = n$ (constant)
Total energy: $\qquad n_1 E_1 + n_2 E_2 + n_3 E_3 + \ldots\ldots n_s E_s = E$ (constant)

in the case of light quanta the first condition drops out, while in the case of molecules of gas both conditions appear, as already indicated. Considering here the general case for which

$\sum dn_s = 0$

And

$\sum E_s \cdot dn_s = 0,$

And using the method of undetermined multipliers, relation (4) can be put as

$$\log (A_s + n_s) - \log n_s - \lambda - \beta E_s = 0,$$

which can be further reduced to the more convenient form

$$\log [(A_s + n_s) / n_s f] = \beta E_s$$

$$\text{or} \quad (A_s + n_s) / n_s = f e^{\beta E_s}$$

$$i.e., \quad (A_s / n_s) + 1 = f e^{\beta E_s}$$

$$\therefore \quad n_s = A_s / (f e^{\beta E_s} - 1) \quad\quad ...(5)$$

which represents the distribution law in the Bose-Einstein statistics.

Fermi-Dirac statistics

As already indicated, a special restriction, due to Pauli's principle, imposed on the Bose-Einstein statistics leads to the Fermi-Dirac statistics.

Let n be the number of particles in the gth sheet, distributed over the A_s cells in it. Among the A_s cells, n_s are occupied by one particle each and the remaining $(A_s - n_s)$ cells are empty, since, according to Pauli's principle, no cell can be occupied by more than one particle. The possible number of such a distribution is given by $A_s!$, corresponding to the permutations of the A_s cells. But on account of the indistinguishability of the particles, the same state is denoted by all distributions among $A_s!$, which only differ from one another by permutation of the n_s occupied cells or of the $(A_s - n_s)$ empty cells.

Hence the distinguishable arrangements in the gth sheet are

$$A_s ! / n_s ! (A_s - n_s) ! \quad\quad ...(6)$$

Considering all the sheets, we may write, as *in* the previous case

$$P = \prod_{k=1}^{k=s} [A_k ! / n_k ! (A_k - n_k) !] \quad\quad ...(7)$$

To find the most probable distribution, we take

$$S = k \log \prod_{k=1}^{k=s} [A_k ! / n_k ! (A_k - n_k) !]$$

Applying Stirling's theorem,

$$S / k = \Sigma [A_s \log A_s - A_s - n_s \log n_s + n_s$$
$$- (A_s - n_s) \log (A_s - n_s) + A_s - n_s]$$

$$= \Sigma [A_s \log A_s - (A_s - n_s) \log (A_s - n_s) - n_s \log n_s]$$

In the state of equilibrium, $dS = 0$. Hence

$$\Sigma \left[dn_s \log (A_s - n_s) + (A_s - n_s) \, dn_s \, / \, (A_s - n_s) \right.$$
$$\left. - dn_s \log n_s - n_s dn_s \, / \, n_s \right] = 0$$

remembering that A_s is not subject to variation.

Taking into account the two subsidiary conditions, viz.,
$\Sigma \, dn_s = 0$
And
$\Sigma E_s \, dn_s = 0$,
and applying the method of undetermined multipliers, we obtain

$$\log \left[(A_s - n_s) \, / \, n_s \right] - \lambda - \beta E_s = 0,$$

which can be put in the form

$$\log \left[(A_s - n_s) \, / \, n_s f \right] = \beta E_s$$

Or

$$(A_s - n_s) \, / \, n_s = f \, e^{\beta E_s}$$

$$\therefore \quad n_s = A_s \, / \, (f \, e^{\beta E_s} + 1) \qquad \qquad \ldots (8)$$

This is the distribution law of the Fermi-Dirac statistics, which, but for the + sign in the denominator, is the same as in the case of the Bose-Einstein statistics. The difference in algebraic sign, however, implies a fundamental difference between the two statistics, since the denominator in (8) is always greater than one, unlike in (5).

Comparison of the three statistics: the degeneracy parameter

The distribution laws according to the three statistics are:

$$\textbf{M-B} \quad f(E) = \frac{n_s}{A_s} = \frac{1}{A e^{E_s / kT}} \qquad \text{(Maxwell-Boltzmann)}$$

$$\textbf{B-E} \quad f(E) = \frac{n_s}{A_s} = \frac{1}{A e^{E_s / kT} - 1} \qquad \text{(Bose-Einstein)}$$

$$\textbf{F-D} \quad f(E) = \frac{n_s}{A_s} = \frac{1}{A e^{E_s / kT} + 1} \qquad \text{(Fermi-Dirac)}$$

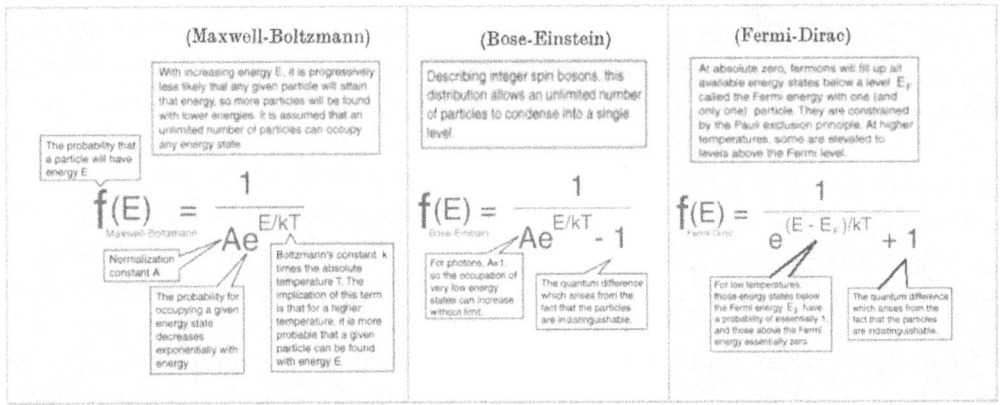

When A is very large compared with one (A >> 1), the distribution laws according to both B-E and F-D statistics reduce to that of the M-B statistics This is found to be the case for all ordinary gases at normal temperature and pressures. Hence, in the region in which the kinetic theory of gases is valid, there is practically no difference between the three statistics.

When A *is very small compared with one* (A << 1), the distribution will not be according to M-B statistics, but follow the B-E or F-D statistics. In such a case, the collection of the particles is said to be in a *state of degeneracy*, when nearly all the cells are filled, since the relations of the B-E and F-D statistics reduce to $n_s = A_s$ approximately. The determining factor of the degeneracy of a state is "A" which is therefore known as the "degeneracy parameter". We shall deal with this quantity in detail in the next section, where we shall study some important applications of the two kinds of quantum statistics.

APPLICATIONS OF THE QUANTUM STATISTICS

A. Bose-Einstein statistics
This has been applied with great success to two phenomena, viz., *black body radiation* and *as degeneration* which, in consequence, also form its experimental supports.

1. Black body radiation

Radiation from a black body at a temperature T absolute and in thermal equilibrium is supposed to consist of light quanta or photons of energy content hv, moving in all possible directions with the velocity of light c and therefore possessing the momentum $p = hv/c$. Each photon can be represented in a six-dimensional phase-diagram, defined by the space coordinates x, y, z and momentum coordinates p_x, p_y, p_z, such that

$$(p_x, p_y, p_z) = (hv_x, hv_y, hv_z) / c$$
$$(v_x, v_y, v_z) = v(\alpha, \beta, \gamma),$$

A, β, γ being the direction cosines of the motion of the photon. The components of the momentum vector p satisfy the condition:

$$p_x^2 + p_y^2 + p_z^2 = (hv/c)^2 \qquad \dots (1)$$

In order to find the distribution of energy in the spectrum of the black body radiation it is necessary to estimate the photons within the energy range hv and $h(v+dv)$ or frequency range v and $(v+dv)$ or the momentum range p and $p+dp$. For this purpose, we must, find the phase volume g described by these photons, which is given by

$$g = \iiint dx\,dy\,dz \iiint dp_x\,dp_y\,dp_z$$

Now

$$\iiint dx\, dy\, dz = V,$$

where V is the volume of the enclosure containing the radiation, and

$$\iiint dp_x\, dp_y\, dp_z = \int_{p}^{p+dp} d\left(\frac{4}{3}\,\pi\,p^3\right)$$

since the integration is to be carried out over the volume of a spherical shell of radii p and (p+ dp).

Changing the variable from p to v, the above expression becomes

$$\frac{4\pi}{3}\int_{v}^{v+dv} d\,(h v\,/\,c)^3, \quad \text{since } p = h v\,/\,c.$$

On integrating, we get

$$4\pi\,/\,3 \cdot (h^3/c^3) \cdot 3v^2 \cdot dv = 4\,\pi\,(h^3\,/\,c^3) \cdot v^2 \cdot dv$$

$$\therefore \qquad g = V \cdot \frac{4\pi v^2 \cdot dv \cdot h^3}{c^3} \qquad\qquad (2)$$

Introducing the quantum idea that each elementary phase-cell has the volume h^3, the number of cell per *unit volume* in the shell bounded by the frequencies v and v+dv is given by

$$\frac{g}{Vh^3} = \frac{4\pi v^2 \cdot dv}{c^3}$$

Taking account of the doubling of states due to polarization of the photons, the number of cells per unit volume available for the photons in the frequency range v and v+dv is given by

$$A_s = 2\,\frac{4\pi v^2 \cdot dv}{c^3} = \frac{8\pi v^2 \cdot dv}{c^3} \qquad\qquad \ldots (3)$$

If the number of such photons, denoted by n_s, are identical, *i.e.*, indistinguishable from one another, B-E statistics can be employed to determine the distribution of the n_s photons in the A_s cells. Remembering that

In the equilibrium state $dS = 0$.

subject to the *single* subsidiary condition
$\Sigma E_s \cdot dn_s = 0,$
(hence only one undetermined multiplier is to be used) we get; as already shown

$$\log\,(A_s + n_s) - \log n_s - \beta E_s = 0$$

Or

$$n_s = A_s\,/\,e^{\beta E_s} - 1 \qquad\qquad \ldots (4)$$

The *energy density* of radiation of frequency between v and v+dv is given by

$$n_s \cdot h\nu = (A_s / e^{\beta E_s} - 1) \cdot h\nu$$

Substituting the values of A_s ($= 8\pi\nu^2 \, d\nu/c^3$), β ($= 1/kT$) and E_s ($= h\nu$) we get the *energy distribution law*

$$dE = n_s \cdot h\nu = (8\pi\nu^2 \, d\nu \, / \, c^3) \cdot (h\nu \, / \, e^{h\nu/kT} - 1)$$

which in terms of wavelength λ becomes, [since $\nu = c/\lambda$ and $d\nu = - c \, (d\lambda/\lambda^2)$],

$$dE = - \frac{8\pi h}{c^3} \cdot \frac{c^3}{\lambda^3} \cdot \frac{cd\lambda}{\lambda^2} \cdot \frac{1}{e^{hc/\lambda kT} - 1}$$

$$= \frac{8\pi hc}{\lambda^5} \cdot \frac{d\lambda}{e^{hc/\lambda kT} - 1} \qquad \dots (5)$$

This is Planck's formula. Hence the Bose-Einstein statistics, while confirming the validity of Planck's law, has the merit of using only photons and no other hypothetical resonators in the deduction of the law governing the black body radiation.

2. Gas degeneration

Considering a volume V of a monatomic gas, let us suppose that the molecules in it are distributed in equi-energy cells lying between E_s and $E_s + dE_s$, E_s being the energy of each molecule. Each of these molecules can be represented in a six-dimensional phase-diagram defined by the space coordinates x, y, z and momentum coordinates p_x, p_y, p_z such that

$$(p_x, p_y, p_z) = m \, (v_x, v_y, v_z)$$

where m is the mass of the molecule, v_x, v_y, v_z are the three components of the velocity vector v of the molecule. This follows from the fact that the momentum p of the molecule is equal to mv.

Further, since the energy $E_s = (1/2) \, mv^2$ or $2mE_s = m^2v^2 = p^2$ the three components of the momentum vector p satisfy the condition:

$$p_x^2 + p_y^2 + p_z^2 = 2mE_s \qquad \dots (1)$$

The phase volume g_s at the disposal of each of the molecules considered is given by

$$g_s = \int \int \int dx \cdot dy \cdot dz \int \int \int dp_x \cdot dp_y \cdot dp_z$$

The first integral

$$\int \int \int dx \cdot dy \cdot dz = V.$$

The second integral refers to the volume included within the spherical shell of radii p and $p + dp$. Hence

$$\int \int \int dp_x \cdot dp_y \, , dp_z = \int_p^{p + dp} d\left(\frac{4}{3} \pi p^3\right)$$

Changing the variable from p to E_s, we get

$$= \frac{4\pi}{3} \int_{E_s}^{E_s + dE_s} d\,(2mE_s)^{3/2} \; [\text{since } p = (2mE_s)^{1/2}]$$

$$= \frac{4\pi}{3} \int_{E_s}^{E_s + dE_s} (2m)^{3/2} \cdot (3/2) \cdot E_s^{1/2} \cdot dE_s$$

$$= 2\pi\,(2m)^{3/2}\,E_s^{1/2}\,dE_s$$

$$\therefore \quad g_s = V \cdot 2\pi\,(2m)^{3/2}\,E_s^{1/2}\,dE_s \qquad \qquad \cdots (2)$$

As the volume of each elementary phase-cell, according to the quantum theory, is h^3, the number of cells, A_s, within the energy range E_s and $E_s + dE_s$, is given by

$$A_s = \frac{g_s}{h^3} = \frac{2\pi V\,(2m)^{3/2}E_s^{1/2}\,dE_s}{h^3} \qquad \qquad \cdots (3)$$

Supposing that n_s is the number of molecules distributed among these A_s cells and the molecules are indistinguishable from one another, B-E statistics can be employed.

In the equilibrium state $dS = 0$, subject to the two subsidiary conditions, viz. $\sum dn_s = 0$ and $\sum E_s\, dn_s = 0$. This leads to the relation

$$\log\,(A_s + n_s) - \log n_s - \lambda - \beta E_s = 0$$

or $\qquad n_s = \dfrac{A_s}{f\,e^{\beta E_s} - 1} \qquad \qquad \cdots (4)$

Substituting the value of A_s given by (3) and that of $\beta = 1/kT$,

$$n_s = \frac{2\pi V}{h^3} \cdot (2m)^{3/2} \cdot \frac{E_s^{1/2}\,dE_s}{f e^{E_s/kT} - 1} \qquad \qquad \cdots (5)$$

Since
$\sum n_s = n$ and $\sum n_s E_s = E$, where n is the total number of molecules and E the total energy,

$$n = \sum \frac{2\pi V}{h^3}\,(2m)^{3/2}\,\frac{E_s^{1/2}dE_s}{f e^{E_s/kT} - 1} \qquad\qquad E = \sum \frac{2\pi V}{h^3}\,(2m)^{3/2}\,\frac{E_s^{3/2}dE_s}{f e^{E_s/kT} - 1}$$

Changing the summation into integration and putting $E_s/kT = x$, so that $dE_s = kTdx$

$$n = \int_0^\infty \frac{2\pi V}{h^3} \, (2m)^{3/2} \, \frac{(kTx)^{1/2} \cdot kT \cdot dx}{fe^x - 1}$$

$$= \frac{V}{h^3} \, (2\pi mkT)^{3/2} \cdot \frac{2}{\sqrt{\pi}} \int_0^\infty \frac{x^{1/2} \, dx}{fe^x - 1}$$

$$= \frac{V}{h^3} \, (2\pi mkT)^{3/2} \cdot A \, (f) \qquad \qquad \ldots (6)$$

where

$$A(f) = \frac{2}{\sqrt{\pi}} \int_0^\infty \frac{x^{1/2} \, dx}{fe^x - 1}$$

$$E = \int_0^\infty \frac{2\pi V}{h^3} (2m)^{3/2} \frac{(kTx)^{3/2} \cdot kT \cdot dx}{fe^x - 1}$$

$$= \frac{V}{h^3} (2\pi mkT)^{3/2} \cdot kT \cdot \frac{2}{\sqrt{\pi}} \int_0^\infty \frac{x^{3/2} \, dx}{fe^x - 1}$$

$$= \frac{3}{2} \frac{V}{h^3} (2\pi mkT)^{3/2} \cdot kT \cdot B(f) \qquad \ldots (7)$$

Where

$$B(f) = \frac{4}{3\sqrt{\pi}} \int_0^\infty \frac{x^{3/2} \, dx}{fe^x - 1}$$

When f is greater than one, the integral functions, A(f) hand B(f), can be evaluated by expanding in series so that we obtain

$$A(f) = \frac{1}{f} + \frac{1}{2^{3/2} f^2} + \frac{1}{3^{3/2} f^3} + \ldots\ldots$$

$$B(f) = \frac{1}{f} + \frac{1}{2^{5/2} f^2} + \frac{1}{3^{5/2} f^3} + \ldots\ldots$$

When *f* is very large (*f* >> *1*), we get, to a first approximation, A(f) = B(f) = *1/f*, so that equations (6) and (7) reduce to

$$n = \frac{V}{h^3} (2\pi mkT)^{3/2} \cdot 1/f \qquad \ldots (8)$$

$$E = \frac{3}{2} \cdot \frac{V}{h^3} (2\pi mkT)^{3/2} \cdot kT \cdot 1/f \qquad \ldots (9)$$

$E/n = (3/2) kT$

Therefore,

$E = (3/2) nkT = (3/2) RT$ .... (10)

Thus we obtain the same result as in the classical statistics.

Gas degeneration

This sets in, as already indicated, *when f is very small compared to* one. In order to study the conditions under which the gas becomes degenerate, let us first obtain an expression for "f".
Taking the approximate relation (8),

$$f = \frac{V}{nh^3} (2\pi mkT)^{3/2}$$

If unit volume of the gas is considered, $V = 1$, so that

$$f = \frac{(2\pi mkT)^{3/2}}{nh^3} \qquad \text{... (11)}$$

If f is very small compared with unity, then, the gas, even if considered perfect, because the finite size of the molecules and the inter-molecular forces are negligible, would still show a departure from the laws of ideal gas due to the fact that the molecular velocities are subject not to the classical statistics but to the quantum statistics.

The gas in this condition is *degenerate*.

As the expression for 'f' contains three variables, viz., m the mass of the molecule, n the number of molecules per c.c. and T the absolute temperature of the gas, the *criterion of degeneracy* will be based on the magnitude of $(mT)^{3/2}/n$, according to the relation (11)

Hence the gas can become degenerate in three different ways:

(a) When the temperature 'T' is very low

Plotting the energy E of the gas as a function of its temperature T, a curve of the type shown in Fig. 188 is obtained.

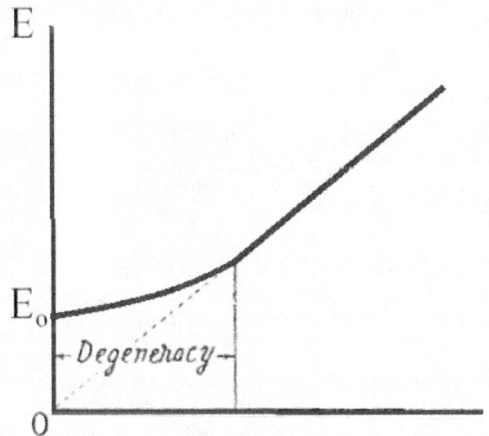

Below a certain value of T the full line shows the relation between E and T of a degenerate gas, while the dotted line through the origin that of a non-degenerate gas. E_0 is termed the *zero-point energy*, which will be understood when we deal with the application of the F-D statistics.

The temperature at which degeneracy begins to appear varies as $n^{2/3}/m$. It is about 5°K for helium gas.

It is evident that it is very difficult to test the condition of degeneracy at such low temperatures, not only on account of the technical difficulties, such as very high pressures involved, but also due to the fact that in that region, even according to classical statistics, gases no longer behave as ideal gases. At any rate,

gases are never degenerate at ordinary temperatures and pressures.

For example, considering hydrogen gas at room temperature (T= 300°K) the value of 'f' is found to be of the order of 10^5; for heavier gases f becomes still greater.

(b) When the number 'n' per c.c. is very large

This condition is realized in certain stars (e.g., white dwarfs) where the density attains high values of the order of 10^5, so that the gases in those stars are in a state of degeneracy, in spite of their very high temperature.

(c) When the mass 'm' is very small

This is the case for the so-called *electron gas* which we shall discuss below on the basis of the F-D statistics.

B. Fermi-Dirac statistics

This has many *applications* which have been carefully worked out by Sommerfeld, *but the most fundamental* among them is the *quantum electron theory of metals.* Hence we shall first consider this basic application and then pass on to the others.

1. Electron theory of metals

The various properties of metals, such as electrical and thermal conductivities, thermoelectricity, thermionic and photoelectric effects, etc., can be explained, as we have already seen, on the assumption that metals contain free electrons which constitute a perfect gas known as the *electron gas.* Riecke, Drude, Lorentz and others, since 1900, applying the classical statistics to the electron gas succeeded to a certain extent in explaining many of the phenomena dependent on the motion of the free electrons in metals. But, over and above the inadequate nature of such explanations,

very serious difficulties were encountered in the use of classical statistics, the chief among them being the specific heat of metals.

For these reasons the theory of the "electron gas" was discredited to some extent.

Sommerfeld, in 1928, however, revived the electron theory of metals on the basis of the new quantum statistics. According to him, the electrons in metals are not completely free but only partially so, in the sense that though they are not bound to any particular atomic system, yet are bound to the metal as a whole.

The interior of the metal is to be conceived as a region of uniform potential, positive relative to free space, so that work is required to extract an electron from the metal. The electrons in metals cannot therefore be compared to the free molecules of a gas obeying the classical statistics.

Moreover, owing to their light mass (nearly 1/2000 that of hydrogen) and dense packing,

the electrons in the metal should be assimilated to the molecules of a gas under very great compression (the calculated pressure for electron gas being of the order of 10^5 atmospheres), hence to a degenerate gas.

Further, since these electrons are assumed to be governed by Pauli's exclusion principle they should obey the Fermi-Dirac statistics.

To obtain the law of distribution for the electron gas according to F-P statistics we proceed exactly in the same way as in the application of B-E statistics to gas degeneration, given above, and obtain almost identical results, except for the following two modifications:

(i) *A positive sign appears before 1* in the denominator of the formula for n_s, viz.,

$$A_s / [f \exp(\beta E_s) + 1]$$

(ii) Owing to a special property of the electron, known as *spin,* of which we shall speak later at length, each cell of the momentum-space can *contain, not one electron alone* (*as* assumed above in the general development of the F-D statistics) *but two,* corresponding to the two directions of spin. This means that the quantity A_s appearing in the numerator of the formula must be multiplied by two.

Introducing these changes, appropriate to the case, we get

$$n = \Sigma \frac{4\pi V}{h^3} (2m)^{3/2} \frac{E_s^{1/2} dE_s}{fe^{E_s/kT} + 1} \qquad \ldots (1)$$

$$E = \Sigma \frac{4\pi V}{h^3} (2m)^{3/2} \frac{E_s^{3/2} dE_s}{fe^{E_s/kT} + 1} \qquad \ldots (2)$$

Let us leave aside, for the present, the spin factor. As before, changing the summation into integration and putting E_s/kT = x, we get

$$n = \frac{V}{h^3} (2\pi mkT)^{3/2} . A (f) \qquad \ldots (3)$$

$$E = \frac{3}{2} \frac{V}{h^3} (2mkT)^{3/2} . kT . B (f) \qquad \ldots (4)$$

Where

$$A (f) = \frac{2}{\sqrt{\pi}} \int_0^\infty \frac{x^{1/2} dx}{fe^x + 1}$$

And

$$B (f) = \frac{4}{3\sqrt{\pi}} \int_0^\infty \frac{x^{3/2} dx}{fe^x + 1}$$

(i) When $f \gg 1$, we can obtain $A(f)$ and $B(f)$ in the form of a series, as

$$A (f) = \frac{1}{f} - \frac{1}{2^{3/2} f^2} + \frac{1}{3^{3/2} f^3} - \ldots\ldots$$

$$B (f) = \frac{1}{f} - \frac{1}{2^{5/2} f^2} + \frac{1}{3^{5/2} f^3} - \ldots\ldots$$

so that to a first approximation $A(f)=B(f) = 1/f$, and equations (3) and (4) reduce to

$$n = \frac{V}{h^3} (2\pi mkT)^{3/2} . 1/f \qquad \ldots (5)$$

$$E = \frac{3}{2} \frac{V}{h^3} . (2\pi mkT)^{3/2} . kT . 1/f \qquad \ldots(6)$$

$$= (3/2) nkT \text{ (as in classical statistics).}$$

From equation (5), if n refers to unit volume

$$f = (2\pi mkT)^{3/2}/nh^3 \qquad \qquad \qquad ...(7)$$

Substituting the known values of m, n, k and h in relation (7),

the temperature, at which degeneracy sets in, can be estimated and it is found to be very high, about 36,000°K. Hence the electron gas will be almost completely degenerate even at ordinary temperatures (~300° K),

which justifies the assumption of Sommerfeld. From this it follows that we should carefully analyze the case proper to a degenerate gas, viz. $f << 1$.

(ii) When f <<1, both A(f) and B(f) assume asymptotic forms, *viz.*,

$$A\,(f) = \frac{4}{3\sqrt{\pi}}\,\alpha^{3/2}\left(1 + \frac{\pi^2}{8\alpha^2} + \,......\right)$$

$$B\,(f) = \frac{8}{15\sqrt{\pi}}\,\alpha^{5/2}\left(1 + \frac{5\pi^2}{8\alpha^2} + \,......\right)$$

where $\alpha = \log f$ or $f = e^{-\alpha}$

Substituting the value of A (f) in equation (3), we get

$$n = \frac{V}{h^3} \cdot \frac{4\pi}{3}\,(2mkT)^{3/2}\,\alpha^{3/2}\,(1 + \pi^2/8\alpha^2 + ...) \qquad ...(8)$$

This relation, to a first approximation, gives

$$\alpha = \left(\frac{3n}{4\pi}\right)^{2/3}\frac{h^2}{2mkT} \qquad \qquad \qquad ...(9)$$

where n refers to unit volume, i.e., V = 1 c. c.

To a second approximation, we have

$$\alpha = \left(\frac{3n}{4\pi}\right)^{2/3}\frac{h^2}{2mkT}\left[1 - \frac{1}{3}\left(\frac{4\pi mkT}{h^2}\right)^2\left(\frac{\pi}{6n}\right)^{4/3} + ...\right] ...(10)$$

Substituting the value of B (f) in equation (4), we have

$$E = \frac{3}{2} \cdot \frac{V}{h^3}\,(2\pi mkT)^{3/2} \cdot kT \cdot \alpha^{5/2}\left(1 + \frac{5\pi^2}{8\alpha^2} + ...\right) \qquad ...(11)$$

Replacing α by its value given by equation (10) and referring n again to unit volume (V = 1), we get

$$E = \frac{3}{40}\frac{h^2}{m}\left(\frac{6n}{\pi}\right)^{2/3}\left[1 + \frac{5}{3}\left(\frac{4\pi mkT}{h^2}\right)^2\left(\frac{\pi}{6n}\right)^{4/3} + ...\right] \qquad ...(12)$$

We are now in a position to study the important characteristics of the Fermi-Dirac distribution of the electrons in metals.

(i) Zero-point energy

Relation (12) shows that even at T=0, the system has a mean energy $\bar{E}_o$ given by

$$\bar{E}_0 = \frac{3}{40} \cdot \frac{h^2}{m} \left(\frac{6n}{\pi} \right)^{2/3} \qquad \qquad ...(13)$$

This property is closely, connected with the Pauli's exclusion principle involved in the F-D statistics.

In the classical theory, the absolute zero is characterized by the fact that the mean kinetic energy of the gas ($3kT/2$) disappears at that temperature, and accordingly the energy of every individual molecule also disappears; in other words, at absolute zero the gas molecules are at rest.

The case is different in the F-D statistics; here each cell can have only a single particle, in the state of lowest energy, all the cells of small energy are filled, and the limit of "filling up" of the system of cells is given by the number of electrons. This limit is characterized by the momentum p_0 of that cell, up to which the filling up reaches.

The total phase volume occupied by n particles at the absolute zero taking into account the spin factor = $(1/2) nh^3$.

This is equal to the volume of the momentum sphere $(4/3) \pi p_0^3$.

Hence $p_0 = (3nh^3/8\pi)^{1/3}$.

The limiting energy E_0 is then given by

$$E_0 = \frac{p_0^2}{2m} = \frac{1}{2m} \left(\frac{3nh^3}{8\pi} \right)^{2/3} = \frac{h^2}{2m} \left(\frac{3n}{8\pi} \right)^{2/3} \qquad ...(14)$$

If the spin factor is taken into account it can be shown that

$$\bar{E}_0 = \frac{3}{40} \cdot \frac{h^2}{m} \cdot \left(\frac{3n}{\pi} \right)^{2/3} = \frac{3}{5} E_0$$

(ii) Energy distribution

According to the classical theory the distribution of energy at temperatures other than the absolute zero will follow the familiar Maxwellian distribution curve, as shown in Fig. 189 for temperatures 300°K and 1500°K.

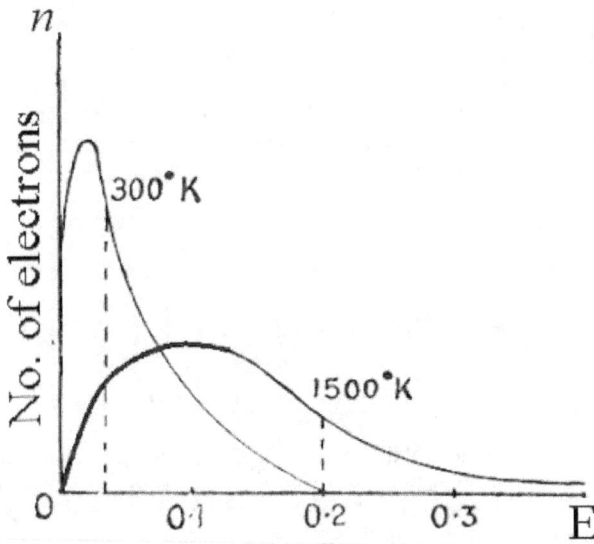

Fig. 189. Energy distribution classical statistics.

A few electrons have high energy but the majority of them are grouped round the mean value $3kT/2$ which at 300°K is equal to 0.04 electron volts and at 1500°K to 0.2 electron volts. Thus there is a change in energy of the electrons with change in. temperature.

The energy distribution according to the F-D statistics is quite different. Using relations (1) and (12), the energy distribution curves can be plotted at different temperatures of the electron gas. The curves thus obtained are as shown in Fig. 190 for temperatures 0°K and 1500°K.

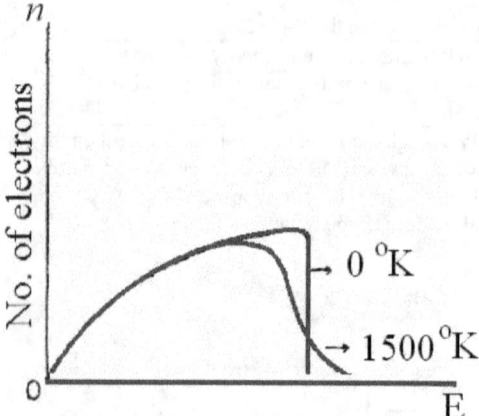

Fig. 190. Energy distribution F-D statistics.

Unlike in the classical theory, here, *not all electrons have zero energy at absolute zero ;* instead the curve for 0°K starts from the origin and rises as parabolic curve, convex upward, until a maximum value of E_0 is reached. The curve then suddenly drops to zero, indicating thereby that no electrons possess energy greater than E_0, or in other words, that up to file energy value E_0, the cells are completely filled, while the cells with greater energy values are empty. In this case, the quantity α, as the approximate relation (9) shows, varying as 1/T, becomes infinitely great and hence $f = e^{-\circ'}$ infinitely small. Comparison of relations (9) and (14) shows that we can put approximately $\alpha = E_0/kT$. The distribution function which is approximately valid for large values of α, *i.e.,* for low temperatures, is, then, using equation (1),

$$ n = \frac{4\pi V}{h^3} (2m)^{3/2} \cdot \frac{E^{1/2}}{e^{(E - E_0)/kT} + 1} \qquad \dots (15) $$

When T is zero, the exponential term in equation (15) is either infinity or zero, according as the variable E is greater or less than E_0. Thus we see that the density of the electrons in the phase space is zero for all values of the .energy greater than E_0. The corresponding velocity of the fastest electrons is obtained from

$$ E_0 = \frac{1}{2} m v_{max}^2 $$

and therefore is given by the relation

$$ v_{max} = \frac{h}{m} \left(\frac{3n}{8\pi} \right)^{1/3} \qquad \dots (16) $$

Now, if it be assumed that the number of free electrons per unit volume is of the same order of magnitude as the number of atoms per unit volume, the number of electrons *n* per unit volume in platinum is 6.6×10^{22}. Using relation (14) E_0 can be estimated. It is found to be 6 eV, so that the mean value E_0 is 3.6 eV, which is many times greater than the mean energy given by the classical theory even at as high a temperature as 1500°K, *viz.,* 0.2 eV. Thus we arrive at the astonishing result that

even at absolute zero the free electrons in a metal have enormous energies.

(iii) *Solution of the difficulty arising from the specific heat*
Experiments show that

metals obey the Dulong-Petit law, viz., that their specific heat referred to 1 mole is 6 calories per degree.

Now,

according to the classical theory, if there is equipartition of energy between the free electrons and the atoms in metals, the specific heat referred to 1 mole should be 9 calories per degree, which certainly is not warranted by experiment.

This is the chief difficulty against classical statistics, expressed here in a concrete manner. The F-D statistics solves it in a very elegant manner as follows:

Referring to F-D distribution curves (Fig. 190) the two curves at 0°K and 1500°K coincide over a good portion with each other. This means that

the *total energy of the electrons changes very little with rise of temperature.*

Hence the electrons will contribute very little to the change in energy of the metal as its temperature rises, i.e., to specific heat. Thus the difficulty regarding specific heat disappears in the new statistics. If we differentiate the expression for E given by equation (4) with respect to T, we get the specific heat at constant volume C_v, per electron, as

$$C_v = \frac{\pi^2 m k^2}{h^2} \left(\frac{8\pi}{3n} \right)^{2/3} T \qquad \cdots (17)$$

Hence the specific heat is proportional to the absolute temperature and vanishes at the absolute zero, in accordance with Nernst's heat theorem. On substituting numerical values in the above relation, it is found

that the contribution of the electrons at ordinary temperatures is of the order of 1/100th of that of the atoms. Hence we should not expect to detect it in specific heat measurements at normal temperatures.

It is only when very high temperatures, of the order of 10,000°C, are reached, that the tight packing of the electrons gradually becomes loosened and the electrons make a noticeable contribution to the specific heat.

(iv) *Asymptotic approach to Maxwellian distribution*

The sharp discontinuity that occurs for 0°K curve at E_0 becomes "rounded off" more and more as the temperature rises, as seen in the 1500°K curve. This means that as the temperature increases the electrons are gradually raised to higher states of energy, but the change in the electronic distribution begins to take effect only at the place where the Fermi function falls away. In this region, the curve approaches the energy axis asymptotically, where it can be shown that the distribution follows approximately Maxwell's law. Hence

there is no maximum energy limit for the electrons in metals, except at absolute zero; a few of them have all energies.

2. Thermal and electrical conductivities

These two closely allied phenomena were explained by Sommerfeld on the basis of the F-D statistics. His theoretical explanation depends, as already noted, on the legitimate assumption that the electron gas, responsible for the conduction properties, is in a state of complete degeneracy at ordinary temperatures, due to the light mass and dense packing of the electrons. Hence the statistical distribution of the electrons is not of the classical, but of the F-D type.

The function $F(v_x, v_y, v_z)$, giving the distribution depending on the three velocity components, is symmetrical when there is no preferential flow of electrons in any particular direction. In such a case, F depends only on $v = (v_x^2 + v_y^2 + v_z^2)^{1/2}$ and not on the three velocity components taken separately. If, however, there is an electric field gradient in the direction of the x-axis along which the electric current flows (electrical conduction), or if there is a temperature gradient in the x-direction, along which heat flows (thermal conduction), then the distribution of the electrons in unit volume is no longer symmetrical, and the new distribution law can be expressed in the form

$$F = F_0 + v_x \theta \qquad \cdots (1)$$

Where

$$\theta = -\, l/v \left(\frac{eX}{m} \cdot \frac{1}{v} \cdot \frac{\partial F_0}{\partial v} + \frac{\partial F_0}{\partial x} \right)$$

This expression for θ was deduced by Lorentz by analysis of the paths traced by electrons of velocity v and mean free path l under the action of the electric intensity X and of collisions with atoms which were assumed to be stationary.

The most important step in the theory is to determine F according to F-D statistics. Once this is done, the rest is purely a mathematical development, which we need not enter into, but only observe that in the case of pure electrical conduction $\partial F_o/\partial x = 0$ and $\partial T/\partial x = 0$, since it is assumed that there is no temperature gradient along the conductor. In pure thermal conduction, the current i is zero.

Considering first *electrical conduction,* for the solution of the problem we have to evaluate the expression for the intensity i of the electric current given by

$$i = en_x = e \int v_x \, F \, A_x \, dxdy \quad \dots \quad (2)$$

where e is the electronic charge, v_x the velocity component in the x-direction and A_x the number of phase cells within the range v_x and v_x+dv_x. Evidently $\int v_x \, F \, A_x \, dxdy$ gives the total number of electrons n_x passing along the x-direction.

Without proceeding further in the mathematical calculations, we give the final expression for the electrical conductivity $\sigma = i/X$ obtained by Sommerfeld:

$$\sigma = \frac{4}{3} \frac{e^2 \, l \, n}{\sqrt{2\pi m k T}} \qquad \text{for ideal gas} \qquad \dots \quad (3)$$

$$\sigma = \frac{8\pi}{3} \left(\frac{e^2 l}{h}\right) \left(\frac{3n}{8\pi}\right)^{2/3} \qquad \text{for degenerate gas} \qquad \dots \quad (4)$$

In the case of *thermal conduction,* the quantity of heat flowing per unit area per sec., is given by

$$Q = \tfrac{1}{2} m \frac{\int v_x \, v^2 \, F \, A_x \, dxdy}{\int dxdy} \qquad \dots \quad (5)$$

Remembering that $Q = - K (\partial T/\partial x)$ and $X = 0$, the thermal conductivity K is found to be

$$K = \frac{8}{3} \cdot \frac{ln k^{3/2} \, T^{1/2}}{(2\pi m)^{1/2}} \qquad \text{for ideal gas} \qquad \dots \quad (6)$$

$$K = \frac{8\pi^3}{9} \cdot \frac{l k^2 \, T}{h} \left(\frac{3n}{8\pi}\right)^{2/3} \qquad \text{for degenerate gas} \qquad \dots \quad (7)$$

For the ratio of thermal to electrical conductivities the following relations are obtained

$$K/\sigma = 2(k/e)^2 \, T \qquad \text{for ideal gas} \qquad \dots \quad (8)$$

$$K/\sigma = (\pi^2/3) \, (k/e)^2 \, T \qquad \text{for degenerate gas} \qquad \dots \quad (9)$$

From relations (3) and (4), we see that,

according to Sommerfeld's theory, $\sigma \propto 1/T^{1/2}$ for ideal gas, while σ is not explicitly a function of the temperature T for degenerate gas.

Experiments show that σ is inversely proportional to T which is consistent with the expression of the classical theory $\sigma = n \, e^2 l \, v / 4\alpha \, T$. But all the three relations involve n and l which are dependent upon the temperature T.

Relations (6) and (7) show that $K \propto 1/T^{1/2}$ for ideal gas and $K \propto T$ for degenerate gas. Experiments indicate that K is independent of temperature, in agreement with the classical theory expression $K = (1/3) \, n \, l \, v \, \alpha$. But, like the relations for σ, here also, all the three relations for K involve the more or less disposable constants n and l.

Equations (8) and (9), which refer to the Wiedmann-Franz law, agree with that of the old statistics [$K/\sigma = (4/e)^2 \, (\alpha/3) = 3 \, (k/e)^2 \, T$] as regards the dependence of K/σ on T alone.

Thus the Wiedmann-Franz law, which has been the cornerstone of researches on the two conductivities, is confirmed by the new statistics also.

The sole difference between the new and old statistics concerning K/σ lies in the value of the numerical constant, which is equal to 3 according to the classical theory and $\pi^2/3 = 3.3$ in the new.

Experimental values seem to agree better with the new theory than with the old.

From this comparative study of the two conductivities according to the old and new statistics, one realizes that although Sommerfeld's explanation is certainly are improvement over the classical interpretation, yet it is indecisive on several points and has therefore to be refined further. The refinements of the theory were attempted chiefly in the following two directions.

(a) The theory must be elaborated in such a way that it would be possible to *deduce the number of electrons taking part in the process of conduction and the change in this number with the temperature* from the properties of the atoms of the substance.

This, however, involves a very complicated problem in quantum mechanics, since an electron is not in this case bound to a definite atom but to the totality of the atomic structure which forms a regular crystal lattice.

Contrary to Sommerfeld's assumption, the potential of the totality of atomic system is not constant, but a space-periodic function. Hence the problem is to solve the Schrödinger's wave equation for a *periodic potential field* of the kind stated above. This can be achieved by various approximate methods which lead to the existence of discrete quantum energy states, separated by *gaps or potential hollows* that become narrower for higher energies,

as represented in Fig. 191.

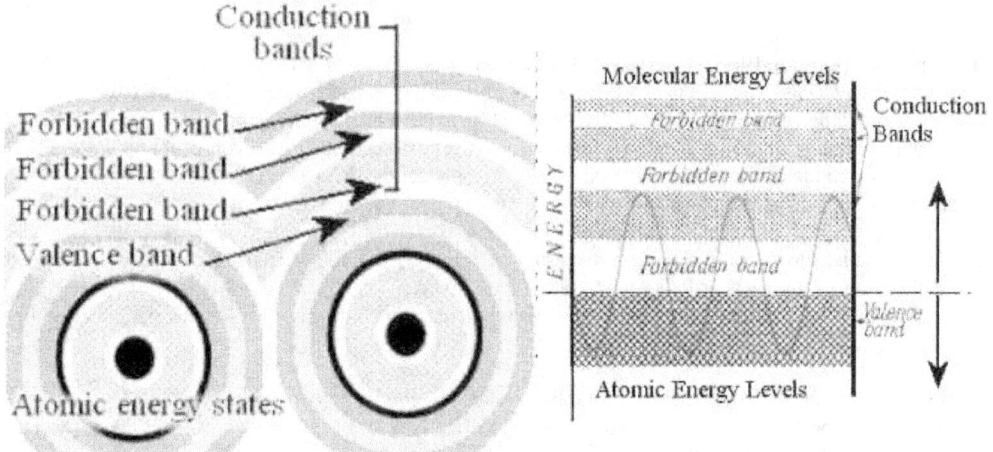

Fig. 191, Space-periodic potential field and discrete quantum energy states.

The number of these states depends upon the total number of electrons, while the width of the gaps on the nature of the substance. At absolute zero all the available energy states up to a certain value W are occupied by the electrons. At higher temperatures only a small fraction of the total number of electrons moves up to higher unoccupied states, but the rest remain below W, in conformity with F-D distribution law.

109

In the case of *metallic conductors,* the gaps are so narrow that the adjacent energy strips overlap. Now, if these are not yet completely occupied, the electron in them will move freely, *without doing work,* which means that the material will be a perfect conductor of electricity offering no resistance to the movement of the electrons. It is to be noted, however, that, since the energy states below W are practically occupied, electrical conduction mostly takes place in the unoccupied states above W which act as conduction channel.

If a P.D. is established between two points in the conductor, the energy states acquire a gradient in the direction of the applied field and an electron from an occupied level can move to a higher empty level without any change in its total energy. But the kinetic energy increases with a consequent decrease in potential energy. The electron now moves towards the positive point of applied potential constituting an electric current.

In **non-conducting insulators**, the distribution of electrons is entirely different. Energy states are grouped together closely but with wide gaps. At the absolute zero the occupied states cease at a gap. When the temperature is increased, the upper states still remain empty at moderate temperatures, since the gaps still cannot be crossed. At very high temperatures a number of electrons acquire sufficient energy to cross the wide gaps and enter the higher states. On establishing a P.D. the movement of electrons into unoccupied states does not occur, since the wide gap cannot be crossed. Hence there is no flow of electricity.

Semi-conductors have more or less the same distribution of energy states as insulators, but the semi-conducting properties are due to *localized impurities* in them, which give rise to occupied electronic states in the gap between the energy states and act therefore as a bridge. At moderately high temperatures electrons can move from the impurity states to higher unoccupied states, causing conduction as in metals. But as the electronic states in the gap due to localized impurities are discontinuous, conduction cannot take place at very low temperatures. With increase of temperature, as more and more electrons pass to higher states, the effect due to the thermal agitation of molecules is reduced, so that the electrical resistance of semi-conductors diminishes with rise in temperature, unlike conductors, as is actually found to be the case. In this manner, all gradations of conductivity and the dependence of conductivity on temperature can be explained qualitatively.

(b) The second line of improvement of-the theory is to determine *the free paths of the electrons on a wave mechanical basis.*

This has been attempted by Houston, Bloch, Peierls, Nordheim and others by replacing the classical encounters between electrons and atoms by a scattering of the electron waves as they traverse the crystal lattice of the metal. No scattering will occur if the ions of the lattice are fixed, since the electrons will then have indefinitely large free paths. If, on the other hand, the ions are in a state of thermal agitation, the electron waves are scattered and become more attenuated with rise of temperature. From the corpuscular stand point, this corresponds to a decrease in the mean free path (diminished probability of unobstructed passage) and the consequent increase in electrical resistance.

This theory leads to very satisfactory results. Houston, for instance, established that the electrical conductivity σ varied inversely as the absolute temperature T at ordinary temperatures and inversely as T^2 at very low temperatures, in agreement with experimental observations, although the calculated increase in conductivity at low temperatures is not as great as that actually found.

Nordheim was able to account for the experimentally established fact that conductivity of an alloy is not always a monotonic function of the proportion in which the two components are mixed, but is generally lower for the alloy than for either of the pure metals. The reason adduced for this state of affairs is that in alloys the ionic lattice is irregular and this irregularity results in increased scattering of the electron waves and, in consequence, the resistance is greater or the conductivity lower. Hence it appears that the interference of the electron waves plays an important role in the wave mechanical investigation of the phenomenon. It is to be mentioned, however, that the theory proposed fails to account for *superconductivity which* still remains unsolved to a great extent.

3. Thermoelectricity

According to classical statistics, the *Peltier coefficient* π is given by

$$\pi = \frac{2}{3} \cdot \frac{\alpha T}{e} \log \frac{n_1}{n_2} = \frac{kT}{e} \log \frac{n_1}{n_2}$$

Sommerfeld, using the F-D statistics obtained the expression

$$\pi = \frac{2\pi^2}{3} \cdot \frac{m(kT)^2}{eh^2} \left\{ \left(\frac{8\pi}{3n_1}\right)^{2/3} - \left(\frac{8\pi}{3n_2}\right)^{2/3} \right\}$$

Both theories give the same order of magnitude for the numerical value of π, which is a fraction of a millivolt as observed experimentally. But, while the classical statistics leads to a small value of the intrinsic potential difference, the new statistics gives, in general, a large value for that quantity, so that π turns out to be a smaller fraction of the intrinsic P.D. than in the former case.

Sommerfeld's expression for the *Thomson coefficient* is

$$\sigma = \frac{2\pi^2}{3} \cdot \frac{mk^2}{eh^2} \left(\frac{4\pi}{3n}\right)^{2/3} T$$

Experiment shows that $\alpha \propto T$, as required by the new statistics. The numerical value of σ obtained from the above expression is found to be only about one per cent of that given by the classical theory. But since the experimentally found value is much smaller than that of the classical statistics, the new theory seems to lead to better results.

Stern has pointed out that the discrepancy between theory and experiment very probably arises due to the illegitimate mixing up of two types of phenomena, *viz., equilibrium* and *transport*, associated with electricity in metals.

4. Thermionic effect

The phenomenon of thermionic emission has been investigated on the basis of the F-D statistics as follows:

The free electrons in the metal have an energy distribution even at absolute zero, varying from zero to a value $E_0 = (h^2/2m)\,(3n/8\pi)^{2/3}$. At this stage, however, no electron can leave the surface of the metal, since even with the maximum kinetic energy E_0 available the electron is unable to overcome the surface potential barrier. But as the temperature is raised, a finite but very small fraction of the electrons acquire energy greater than E_0, and these can escape overcoming the surface forces. Eventually a measurable thermionic current results, which increases rapidly with further increase in temperature. It is assumed that the thermionic current is produced by all those electrons which strike the surface of the metal with velocities, such that the components normal to the surface are greater than x_0, where x_0 is given by the relation: $(1/2)\,mx_0^2 = \omega$, where ω is the energy which an electron requires to escape from the surface of the metal.

The calculation of the thermionic current density for a metal at temperature T is made in Sommerfeld's theory in exactly the same way as in the classical theory of Richardson, but with the modification that the F-D distribution law for the degenerate electron gas must be substituted in the place of the M-B distribution.

If dn represents the number of electrons in unit volume having components of velocity between x and $x + dx$, y and $y + dy$, z and $z + dz$,

$$dn = F(x, y, z)\, dx\, dy\, dz,$$

where F represents ale F-D distribution function.

In the present case, we have to determine the number of electrons striking normally unit area of the surface of the metal per sec. Assuming the surface to be normal to the x component of velocity, the number of electrons having a velocity component normal to the surface between x and $x + dx$ is

$$dx \int_{-\infty}^{+\infty} \int_{-\infty}^{+\infty} F\, dy\, dz$$

Hence the number dN of electrons striking normally unit area of the surface per sec., with the velocity component lying between x and $x+ dx$ is given by

$$dN = x\, dx \int_{-\infty}^{+\infty} \int_{-\infty}^{+\infty} F\, dy\, dz$$

Since of the total number of electrons that strike the surface normally only those, whose velocity exceeds the value x_0 given by $\omega = (1/2)\, mx_0^2$, can escape through the potential barrier, the lower limit of integration for the x-direction should be x_0.

Therefore, the total number of electrons N emitted per sec. from unit area of the surface of the metal is given by

$$N = \int_{x_0}^{\infty} \int_{-\infty}^{\infty} \int_{-\infty}^{\infty} x\, dx\, F\, dy\, dz$$

The thermionic current I is given by

$$I = Ne = e \int_{x_0}^{\infty} \int_{-\infty}^{\infty} \int_{-\infty}^{\infty} x\, dx\, F\, dy\, dz$$

Now, the F-D distribution function F is given by

$$F = 2 \left(\frac{m}{h} \right)^3 \frac{1}{e^{-(w - E_0)/kT} + 1}$$

Where

$$E_0 = \frac{h^2}{2m} \left(\frac{3n}{8\pi} \right)^{2/3}$$

Substituting this value for F in the expression for I and integrating we get, for the case of complete degeneracy,

$$I = \frac{4\pi e m}{h^3} (kT)^2 \, e^{-(w - E_0)/kT}$$

$$= A'' \, T^2 \, e^{-b''/T}$$

where A" and b'' are constants, A" being equal to $4\pi e m k^2/h^3$, hence independent of the nature of the metal and $b'' = (\omega - E_0)/k$, dependent on the nature of the emitting surface, hence varying from metal to metal.

Comparing the above relation with the classical $T^{1/2}$ and T^2 formulae of Richardson, viz.

$I = A\,T^{1/2}\,e^{-b/T}$,

where
$A = ne\,(k/2\pi m)^{1/2}$ and
$b = eV/k = \omega/k$, and

$I = A'\,T^2\,e^{-b'/T}$,
where
$A' = Ce(k/2\pi m)^{1/2}$ and
$b' = \omega/k$,

we see that the new statistics decides in favor of the T^2 formula and not of the $T^{1/2}$ formula. But, as regards the coefficient of T^2 in the exponent, the new and the old formulae differ, being equal to ω/k in the classical ease and $(\omega - E_0)/k$ in the new theory.

To decide between the two formulae, when log I is plotted against 1/T, a straight line is obtained whose slope gives b. We can therefore determine $(\omega - E_0) = kb$ experimentally for different metals, and hence calculate ω by substituting for E_0 the value given by theory, viz., 0 for the classical and $3.63 \times 10^{-15} \times n^{2/3}$ for the new theory. By comparing the values thus found for ω with other measurements, it has been definitely settled that the quantum statistical formula is the correct one and not the classical one.

As a matter of fact, the quantity ω can be measured with sufficient accuracy by means of *diffraction of slow electrons by metal crystals*. Thus, for instance, Davisson and Germer found that $\omega \approx 17$ eV in the case of nickel. On the other hand, measurements on the thermionic effect gives values of kb in the region of 5 eV, in disagreement with the classical theory according to which ω should be equal to kb, *i.e.*, to 17 eV. The quantum theory, however, gives $E_0 \approx 12$ eV, if we assume that in nickel two electrons per atom are free, since nickel is divalent. This gives for $kb = \omega - E_0$, a value of about 5 eV in good agreement with the results of measurement of thermionic emission. It appears, therefore, that not ω but $\omega - E_0$ represents the thermionic work function $e\varphi$.

113

Kariamanickam Srinivasa Krishnan (4 December 1898 – 14 June 1961) India

A *new method of evaluation of the thermionic constants*. A further confirmation of the validity of the quantum statistical formula is obtained from the evaluation of the thermionic constants by a recent elegant method devised by K. S. Krishnan and S. C. Jain (1952) at the National Physical Laboratory of India.

The method is based on the measurement of saturated vapor pressure of the electron gas in equilibrium with the emitter at different temperatures. The saturation vapor pressure is determined by finding the rate of effusion into vacuum of electrons through a small hole in a thin wall of a chamber made of the given substance (Knudsen's effusion method). The chamber is in the form of a tube, and can be heated to any desired high temperature by sending a suitable heavy electric current through it. In order to eliminate the electrons emitted by the surface adjoining the effusion hole, the surface is covered by a thin mica sheet with a small hole punctured in it: this hole is arranged just in front of a bigger one in the wall of the chamber and serves as the effective effusion hole. The temperature of the chamber is readily determined with an optical pyrometer. The area of the effusion hole is measured indirectly by finding the rate of loss of a suitably chosen substance like naphthalene whose vapor pressure is known, kept in the chamber, in vacuo, the loss being due to the sublimation of the substance and the consequent effusion through the hole. The saturation current corresponding to the effusion of electrons through the hole, is determined with the usual Faraday cylinder technique.

Using the well-known thermodynamic relation of Clausius and Clapeyron, the saturation vapor pressure p of the electron gas in equilibrium with a metal at temperature T is given by p

$$p = \Delta T^{3/2} e^{-\phi/kT} \qquad \ldots (1)$$

where Δ is the vapor pressure constant, equal to $2 (2\pi m)^{3/2} k^{5/2}/h^3$ for a monovalent metal and φ the work function of the metal.

Considering the electrons that effuse out of a small hole of area s, their number n effusing out per sec., in all directions, *i.e.,* over the whole of the semi-solid angle 2π, is given by

$$n = s\, p \,/\, \sqrt{2\pi m k T} \qquad \ldots (2)$$

The saturation current i, as usually defined and extrapolated to zero applied field, corresponding to this effusion, per unit area of the effusion will be given by

$$i = ne \mathbin{/} s = AT^2\, e^{-\phi/kT} \qquad \dots (3)$$

Where

$$A = 4\pi emk^2/h^3$$

This equation (3) is very similar to the T^2 formula, and the constant A has the same expression as in the quantum theory formula. Representing the constants of this formula by A' and φ', for an ideally pure surface φ = φ' and A' = A(1— ρ), where ρ is the reflection coefficient of the emitting surface for electrons incident on the surface and . having sufficient energy to cross the barrier. Thus effusion from an aperture in a thin wall is analogous to the emission of electrons from an equal area of pure surface of metal kept at the same temperature, except for the transmission coefficient $(1 - \rho)$ that appears in the latter case.

The emission per unit solid angle varies with the cosine of the angle between the direction considered and the normal to the aperture. In the actual experiment, the number of electrons that effuse out within a well defined cone whose axis is along the normal to the hole is measured at different known temperatures, from which the saturation current i corresponding to effusion per unit area of the hole over the whole solid angle 2π can be readily calculated. From the experimental data, when **log $(i./T^2)$** is plotted against 1/T, a straight line is obtained, showing that A and φ are nearly independent of temperature (Fig. 192). The slope of the straight line gives — φ/k and the value of **log (i/T^2)** extrapolated to l/T → 0 gives the value of A.

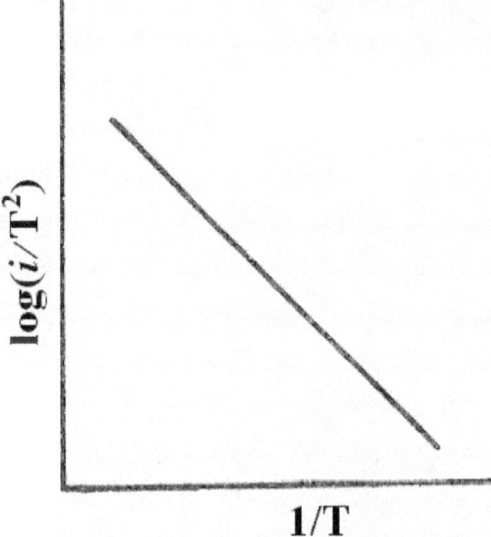

Fig. 192

The authors, using a graphite chamber for the emitter, obtained the values

φ = 4.62 ± 0.02 volts and
A = 60 ± 2 amp. cm^{-2} deg^{-2}

The method, though originally designed for graphite, can be used also for different metals by suitably coating the surface of the graphite chamber with the given metal under study to a sufficient thickness. The values obtained for A by this method decide in favor of the quantum theory formula.

This method of determining the thermionic constants, and in particular the coefficient A in the thermionic equation has the following advantages over the other methods:

(i) insensitiveness of effusion to the contamination of the surface of the chamber by absorption,
(ii) a knowledge of the effective area of the emitting surface or of the reflection coefficient of the electrons at the surface is not required.

Before we conclude the study of the thermionic emission in the light of the new quantum statistics, it is of interest to note that the distribution of energy round the highest values at temperatures greater than absolute zero is found to be similar to that given by classical statistics and this accounts for the experimental fact observed by Richardson that the electrons leaving the emitter have a Maxwellian distribution of velocity.

5. Photoelectric effect

There is a close similarity between the photoelectric and thermionic phenomena. In both cases electrons are emitted from metallic surfaces; the work function in the two cases has the same value, according to careful experimental researches. The main difference between the two lies in the manner in which the electron acquires the additional energy to enable it to escape through the surface of the emitter: in the *thermionic case,* the electron gets this energy due to the high temperature of the emitter, while in the *photoelectric case* by the absorption of radiation incident on the emitter. It can be, therefore, legitimately surmised that the thermionic electrons and photoelectrons have a common origin and are subject to similar laws.

According to Sommerfeld's theory, the ordinary photoelectrons, like the thermions, originate in the electron gas of the metal, obeying the F-D statistics. Hence the photoelectrons come from the effective top of the F-D distribution curve. These electrons even at absolute zero have an energy distribution varying from 0 to E_o. Hence at this temperature, if an electron with the maximum energy E_o absorbs a photon of energy hv, its energy becomes $hv + E_o$. When this energy equals the energy ω required for the electron to escape from the surface, the electron will just emerge with zero velocity. And for the electrons which are emitted with a velocity v, the relation is

$(1/2)\, mv^2 = hv + E_o, -\omega$

Comparing this with Einstein's equation

$(1/2)\, mv^2 = hv - \omega_o$

we see that ω_o of Einstein's relation is given by

$\omega_o = \omega - E_o,$

In the case of thermions, we have seen that $e\,\varphi = \omega - E_o$. Therefore, $\omega_o = e\,\varphi,$
i.e.,

at absolute zero the photoelectric work function is equal to the thermionic work function
in accordance with experimental results.

But, as we have already noted, the new theory predicts that the work function is much greater than that given by the classical theory.

Hughes and Dubridge have suggested the following graphical method of representing the photoelectric phenomenon according to the new statistics:

Let ABC represent the distribution of energy of the electrons in the metal according to the F-D statistics at some temperature T (Fig. 193). Let light of frequency v be incident on the metal surface and let each of a large number of electrons absorb a light quantum of energy hv. The new distribution in energy of these electrons is represented by A'B'C', on the assumption that the probability of absorption of energy by an electron is independent of its velocity. If ω_o is the energy required to penetrate the surface, only those electrons in the shaded area will escape and they will have energies ranging from zero to $(E_o + hv - \omega_o)$ or a little greater, depending on the extent to which the "foot" of the curve C' extends beyond $(E_o + hv)$. This velocity range is in qualitative agreement with experiment.

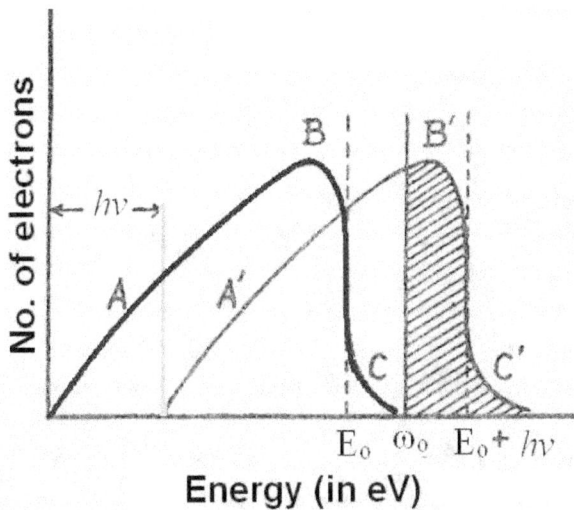

Fig. 193. Graphical representation of the photoelectric effect.

Since the energy distribution in the F-D statistics is almost, though not quite independent of temperature, the *photoelectric effect should be almost but not quite independent of temperature.*

Likewise, since the maximum energy of the Sommerfeld-.Fermi electrons increases with temperature, the *threshold frequency should move towards higher frequencies with increasing temperature.*

These two points have been confirmed by experiment. Coming back to the graphical representation, when hv has such a value that $(E_o + hv)$ almost coincides with ω_o, it is readily seen that the shaded area will comprise mainly of the foot of the curve C' which varies considerably with temperature, thus, accounting for the variation of threshold frequency and photoelectric current with temperature. The conclusions of the new statistics are, therefore, in much better agreement with experimental facts than those of the classical theory.

6. Contact potential difference (Volta effect)

The contact potential difference or the so called Volta effect is the name given to the phenomenon which consists in the appearance of an electric field in the dielectric surrounding two metallic conductors in contact.

As already pointed out, the contact P.D. is not to be confused with the intrinsic P.D. which gives rise to the thermoelectric effect at the junction of the two metals; but it is intimately related to the thermionic work function φ.

In order to apply the F-D statistics to the phenomenon, let us consider two pieces of different metals joined together at one end.

The number of electrons per unit volume is not the same in the two metals and it is assumed that the process of joining does not alter the concentration of the electrons in either metal at points remote from the junction.

Considering two such points, A in one metal and B in the other, let n_A and n_B be the electron densities at the two points respectively. According to the F-D distribution law, the maximum energy possessed by the electrons at A is given by

$$E_0^A = \frac{h^2}{2m} \left(\frac{3n_A}{8\pi} \right)^{2/3}$$

and the maximum energy possessed by the electron at B by

$$E_0^{B} = \frac{h^2}{2m}\left(\frac{3n_B}{8\pi}\right)^{2/3}$$

These maxima are not absolute, but they are values which are exceeded only by a very small proportion of the electrons.

When an electron is removed from the point A to the point B, the change in energy is given by

$$\frac{h^2}{2m}\left[\left(\frac{3n_A}{8\pi}\right)^{2/3} - \left(\frac{3n_B}{8\pi}\right)^{2/3}\right]$$

Equating this to the electrical work done, viz., $e\,(V_A - V_B)$ where e is the electronic charge, V_A and V_B the potentials at A and B,

$$e\,(V_A - V_B) = \frac{h^2}{2m}\left[\left(\frac{3n_A}{8\pi}\right)^{2/3} - \left(\frac{3n_B}{8\pi}\right)^{2/3}\right]$$

Or

$$V_A - V_B = \frac{1}{e}\left(E_0^{A} - E_0^{B}\right) \qquad \ldots \text{(1)}$$

The classical statistics also predicts an internal, P.D., but the value is different from that obtained in the new statistics. For example, in the case of silver and potassium, assuming that in each unit volume of these metals there are as many electrons as atoms, when the two are joined together, the P.D. developed across the junction is 4.2 volts with potassium negative, according to the new statistics, while the classical theory gives a considerably lower value, viz., 0.04 volt. This intrinsic P.D. is not the Volta effect.

An expression for the Volta effect may be obtained as follows:

Let C and D be points just outside the surfaces of the two metals.
The work W_1 required to remove an electron from the point A in the first metal to the point C just outside its surface is given by $W_1 = \omega_1 - E_o^{A}$, where ω_1 is the thermionic work function at the surface of the first metal.

The work W_2 required to remove an electron from A *via* the junction and the surface of the second metal to the point D just outside this second metal surface is given by

$$W_2 = w_2 - E_0^{A} + E_0^{A} - E_0^{B},$$

where ω_2 is the thermionic work function at the surface of the second metal.

The contact P.D. between the two metals $(V_1 - V_2)$ is, by definition, P.D. between C and D.

$$\therefore \quad V_1 - V_2 = \frac{1}{e}\,(W_1 - W_2)$$

$$= \frac{1}{e}\left[(w_1 - w_2) - \left(E_0^{A} - E_0^{B}\right)\right] \ldots \text{(2)}$$

This is quite different from equation (1) which gives the internal potential difference. Equation (2) can be expressed as

$$V_1 - V_2 = \triangle V = \phi_1 - \phi_2 + (T/11600)\,\log\,(A_2/A_1) \qquad \ldots \text{(3)}$$

where φ_1 and φ_2 are the work functions of the two surfaces, expressed in volts, T the temperature, A_1 and A_2 are the values of the constant A of the thermionic equation.

The experimental determination of the contact P.D. between two metallic surfaces is beset with the difficulty of keeping the surfaces free from contaminating films. This difficulty may be overcome to a certain extent by using filament electrodes which can be easily cleaned by heating. Even then the experimental values do not agree quite well with the values of ΔV calculated theoretically. This discrepancy is probably due to the fact that the thermionic constants are determined at high temperatures, while the contact P.D. is measured at room temperature. Experiments performed in high vacuum lead to results in better agreement with theory.

7. Paramagnetism of the alkali metals

We have already remarked that several metals, chiefly the alkalis, have very small paramagnetic susceptibility which is also practically independent of temperature and that the Curie-Weiss law, based on classical statistics, is quite incapable of explaining this experimental finding.

Pauli, in 1927, has been able to solve, to a certain extent, this difficulty by the application of the F-D statistics to the free electrons in the metal.

The valency electrons in the metal are assumed to be free and constitute the electron gas obeying the F-D distribution law and hence completely "degenerate" at the ordinary temperatures as at absolute zero. The electrons possess on account of their spin a magnetic moment. At absolute zero, two electrons occupy every phase-cell, with their spins in opposite directions,

so that their magnetic moments exactly balance each other. If an external field H is applied, the electrons will tend to direct their spins parallel to the field. Since, however,

there cannot be more than one electron in each cell with its spin parallel to the field, magnetization, *i.e.*, turning of the spin moment parallel to the field necessarily involves the removal of electrons from the lowest doubly occupied cells to cells of higher energy.

This means increase of kinetic energy, which goes on until it is compensated by the decrease of potential energy due to the orientation in the field. As only a few electrons jump to cells of higher energy, the paramagnetism will be much smaller than for systems not obeying Pauli's exclusion principle.

When the temperature rises, the uppermost sheets of the F-D distribution begin to be loosened, individual electrons being lifted out of the doubly occupied cells, so that now there are cells which are only singly occupied.

But this gives, as can be shown, only a second order effect. Hence the paramagnetism of an electron gas varies little with temperature.

On this basis, Pauli derived an expression for the susceptibility of the electron gas as follows:

Let dE be the difference in energy of electrons in successive cells in the region of maximum momentum; let μ be the magnetic moment of the electron; let x be the number of electrons removed from the *initially* highest x cells to the x *successive* higher ones.

For the x electron, the change in magnetic (potential) energy is equal to the change in kinetic energy caused by the jump. Hence we may write

$$2x \, dE = 2\mu H \qquad \qquad \dots (1)$$

The total magnetic moment is $2x \mu$, and the volume susceptibility χ_v is given by

$$\chi_v = \frac{2x\mu}{VH} \qquad \qquad \dots (2)$$

Substituting for H from (1), viz., $x \, dE/\mu$,

$$\chi_v = \frac{2\mu^2}{V dE}$$

dE can be shown to be equal to $4E_0/3N$, where $E_0 = (h^2/2m)(3n/8\pi)^{2/3}$

119

$$\chi_v = \frac{2\mu^2}{V} \times \frac{3N}{4} \times \frac{1}{\dfrac{h^2}{2m}\left(\dfrac{3n}{8\pi}\right)^{2/3}}$$

$$= \frac{3\mu^2 n}{2} \times \frac{2m\,(8\pi)^{2/3}}{h^2\,(3n)^{2/3}}$$

$$= 12\,\mu^2\left(\frac{\pi}{3}\right)^{2/3} n^{1/3}\,\frac{m}{h^2} \qquad \dots (3)$$

Since $\mu = eh/4\pi\,mc$, as will be proved later, substituting for μ and inserting numerical values, we get

$$\chi_v = 2\cdot209 \times 10^{-14}\,n^{1/3} \qquad \dots (4)$$

The theory indicates therefore that the paramagnetism due to free electrons will be small and that it will be independent of temperature if n is constant.

Pauli used this relation (4) to calculate the susceptibility of the electron gas in the alkali metals on the assumption that there is one free electron per atom. In the case of sodium and potassium, the theoretical values agree fairly well with the experimental values, but in the case of rubidium and caesium there is a wide divergence between the experimental and calculated values. This discrepancy may be due to the following imperfections of the theory:

(i) The assumption made as to the number of free electrons, *viz.,* one per atom. In order to bring the theoretical value into agreement with the experimental one, it is necessary to suppose that n is actually about 1/3 its present accepted value.

(ii) The theory neglects two important factors, viz., the diamagnetic effect of the atom cores and a possible diamagnetic induction effect of the free electrons.

(iii) The assumption that the electrons are entirely free is incorrect. In a metal, the so-called free electrons must interact to some extent with the ions and in a complete theory this interaction has to be taken into account.

In spite of these drawbacks, Pauli's theory is a valuable contribution towards the understanding of some of the puzzling features of the susceptibilities of metals.

CRITICISM OF THE QUANTUM STATISTICS

Based on a few simple postulates, such as the *indistinguishability of the particles,* the quantum nature of energy changes with the consequent finite *size of phase-cells* and the *exclusion principle of Pauli,* the quantum statistics has certainly made a great advance over the classical theory in the interpretation of phenomena, where elementary particles like photons and electrons play a prominent part. Further, its extension to nuclear particles has led to numerous and important consequences in the structure of nuclei, as we shall see later. Its chief merit lies in the fact that it provides the basis for dividing all the elementary particles of nuclear physics into two neat categories:
— the *bosons* after Bose and
— the *fermions* after Fermi.

At the same time, it must be noted that the new Statistics is not perfect in every respect. On the *physical side*, it appears to labor chiefly under two defects, *viz.*,

(i) the assumed *absolute freedom of the particles* and
(ii) the supposed *constant potential of the container of the particles.*

We have seen how the first assumption makes theoretical results very approximate in the cases of thermionic and photoelectric emissions and of paramagnetic susceptibilities of alkali metals. Sommerfeld, for similar reasons, has been obliged to attribute to the electrons in metals only a partial freedom, *i.e.*, they are bound to the metal as a whole, though not to any particular atomic system.

In order to overcome the second defect, several authors have attempted to refine the theory by a wave mechanical treatment of a space periodic potential field, as already indicated in connection with the thermal and electrical conductivities of metals.

From the *mathematical point of view* also, the quantum statistics is not above criticism in the use of Stirling's theorem, quietly assuming that in practice the particles are much more numerous than the cells, which is not true to facts.

Darwin and Fowler have shown that this rather illegitimate use of Stirling's theorem can be avoided entirely by considering the *average* rather than the "most probable" state of a system, as the analogue of the state of thermodynamic equilibrium. They have therefore developed a new method of working out statistical averages based on principles of cantour integration. This line of approach, however, is fairly complicated, and, in practice, leads to no new results.

CHAPTER 11

THE ATOM

General Introduction

Modern Physics deals chiefly with the atomic state of matter and it is by means of a detailed study of that ultramicroscopic state that remarkable results have been obtained, not only as regards the wonderful practical appliances of great utility, but also concerning the two basic aspects of physical phenomena, particle and wave, which are of utmost importance in the understanding of the mysterious universe around us. This fact has already been brought home to us by the matter treated in the first two parts. We shall now consider the life-history of the atom at close range and study the physics proper of the atom itself, with a view to appreciate more fully all that has been said so far.

From very early times the *atomic structure* of matter was suspected. It was surmised that all material bodies, whether elements or compounds, are fundamentally *granular* in structure, *i.e.*, made up of discrete particles separated by interspaces. These ultimate constituents of matter were called *'atoms'*, because they were supposed to be the smallest portion of an element which cannot be cut into still smaller portions. (The term *'atom'* comes from two Greek words meaning "not" and "cut"). Already in the fifth century B.C. Democritus, the Greek, had postulated that all matter was made up of atoms and that different substances were formed from the common primordial atoms which differed, however, in shape and size. In the first century B.C. Lucretius, the Roman, proposed the same atomic theory in his work, entitled *De Natura Reruns.*

This ancient concept of atomism, in spite of its attractive simplicity, involved some serious difficulties. First of all, if the different elements were constituted with the common fundamental units, the atoms, while the differences between elements were due merely to **size and shape** of the otherwise essentially same entities, the permanent existence of different elements in nature could not be adequately accounted for; nor could the idea of transmutation of elements have any great significance. Besides, it could not be easily understood how the specific nature or essential individuality of an element could be altered by a change in the mere size or shape of the atoms.

A second serious difficulty arose from the fact that the atoms were *not infinitely small*, since they had a size and shape proper to each element. Atoms were, therefore, extended bodies, however small their extension might be. In such a case it should always be possible to divide them into something smaller still, which goes against the very definition of the atoms, viz., the ultimate indivisible units.

On account of these objections the atomic theory of matter did not gain universal acceptance for a very long period of the Christian era, right up to the nineteenth century. Other theories which have had their origin also in a Greek school of thought, directed by Aristotle, found many supporters throughout the Middle Ages. These denied atomicity of matter and supposed that matter was *continuous in structure* and *infinitely divisible.* The physical universe was considered to be made up of "primary" matter, a fundamental, common and continuous substratum, which became ordinary matter when it received the impress of a definite form. The forms in the different elements were specifically different, but it was thought possible to infuse the right form into the common "primary" matter in order to transform one element into another, chiefly baser elements into gold. Hence, there followed the search for the Philosopher's Stone and the heroic but futile attempts of the alchemists, which would effect such changes.

Modern scientific researches, dating from the beginning of the nineteenth century, have, however, decided in favor of the atomic theory of matter, though the ancient concept of the atoms, as ultimate, indivisible units of matter, has to be substantially modified. The solid arguments adduced in favor of atomism are:

(i) *Compressibility of matter* which is clearly perceived in the gaseous state. If a gas, which is evidently a material body, is assumed to be continuous in structure, it is extremely difficult to understand how it can be compressed to a very small fraction of its original volume; if, on the other hand, it consists of discrete atoms with interspace, it is readily seen that they can be made to crowd more closely together. Further, it has become possible to give a satisfactory quantitative explanation of the behavior of gases with the help of the additional supposition that atoms are in rapid motion in a random fashion, as shown in the kinetic theory of gases. Matter in the liquid and solid states are compressible also, though to a less extent.

(ii) *The phenomena of diffusion, osmosis, Brownian movement, solubility*, etc., clearly prove the existence of discrete minute particles with spaces between, moving continuously, even in the smallest quantity of matter, in all the three states, solid, liquid and gaseous:

(iii) *The regular forms of crystals governed by definite simple laws* also argue to the arrangement of discrete atoms in a well-designed space-lattice. Most of the substances in the material universe have been shown to be crystalline in structure.

(iv) *The law of multiple proportions.* In the beginning of the nineteenth century, Dalton, the celebrated chemist, discovered a law governing chemical combinations, which placed the atomic theory on a quantitative basis and secured for him the title of "Father of the atomic theory". According to it,

when the same two elements combine to form different compounds, the different higher proportions of one element which combine with the same constant amount of the other are simple. integral multiples

(*i.e.,* not a fraction) of the lower. This law could be explained in an easy manner only on the assumption that each element consisted of identical but discrete atoms. The law of multiple proportions offered also a means for determining the relative weights of different atoms. This was undertaken during the early part of the nineteenth century, with Berzelius as the pioneer worker.

(v) *The periodic law among elements.* The discovery of a periodic law among elements arranged according to their atomic weights not only confirmed the atomic structure of the elements, but also indicated that atoms must be built up according to some system, which allowed certain repetition of properties periodically. This fact overshot the mark aimed at by the ancient atomists, who held that the atoms were the ultimate indivisible units that cannot be built up with other smaller units. As a matter of fact, further investigations showed that the atom could be broken up into parts and hence not indivisible.

Before we come to the consideration of the internal structure of the atom, it can be shown briefly how it is possible, with the findings of modern researches in Physical Science, to reconcile the two opposite views about the ultimate structure of matter, held firmly for such long periods of time by equally great thinkers.

The ancient atomists were right in so far as they maintained that material bodies which look apparently continuous in the bulk state, have, in reality, a discontinuous internal structure, built up of minute and discrete atoms; but they were wrong in thinking that differences between elements arose solely from the different sizes or shapes of the otherwise same entities, that the atoms were indivisible so that they could not be broken up into further smaller parts and finally that all the atoms of a given element were identical in all respects. For, the permanency of the different elements in nature, the phenomenon of transmutation of elements and the existence of different isotopes in one and the same element go directly against such an hypothesis. The defenders of the atomic theory could not, however, be greatly blamed, as they had to use the meager and even incorrect data then available.

A more serious charge could be made against them when they failed to explain how the atoms, having a definite size and shape, hence being extended bodies, could be considered as absolutely indivisible. For, anything extended must be capable of being divided further, while an absolutely indivisible entity, like a mathematical point, can only be a pure abstraction of the mind, which does not correspond to any reality. Their position can, however, still be defended by saying that the atom, the ultimate representative of an element, cannot be physically divided further without its ceasing to represent the element. Hence, from a concrete, physical point of view, the indivisibility of the atom can be maintained, though it is not completely true.

It is precisely on this point that the defenders of the continuous structure of matter scored a victory, since, logically and absolutely speaking, atoms which have a size and shape must be capable of being divided further and further "*ad infinitum*". Evidently they were wrong in denying an atomic discontinuous structure to matter, as experimental investigations have established beyond doubt. But they were quite correct when they maintained that the ultimate particle of matter, whatever it be, whether atom or something smaller still, must be infinitely divisible, since all material bodies, however minute they may be, are not mere mathematical points but possess the essential property of extension. Thus it appears that both the age-long theories defending discontinuity and continuity in the ultimate structure of matter are true but only partially, since *reality is a discontinuous continuity.* The reconciliation and synthesis of the two views are possible, thanks to the findings of Modern Physics concerning the ineffable admixture of the two aspects, the continuous and the discontinuous, in the material universe, valid in all scales, macroscopic and microscopic.

As regards the **internal structure of the atom** with which we are directly concerned here, progressive researches, chiefly those conducted by Rutherford on the scattering of α-particles by matter, showed that the atom could be split further into

(i) **a central positive nucleus** wherein the individuality of the atom resides and practically the whole of the mass of the atom is concentrated and

(ii) **configurations of electrons,** enveloping the nucleus at distances relatively great, which are responsible for the observed chemical and physical properties of the element concerned.

The aim in this Part III is to give a brief but clear account of the attempts made, both theoretical and experimental, to unravel the mysteries involved in the internal structure of the ultramicroscopic atom. We shall first deal with the various theories proposed to interpret the extranuclear electronic structure in the light of experimental observations and then pass on to the more interesting but difficult consideration of the innermost core of the atom, the nucleus, and gather the few important but incomplete data so far obtained concerning its structure, stability and activity.

SPECIAL UNITS USED IN ATOMIC PHYSICS

1. Length. For measuring distances of atomic order of magnitude very small submultiples of the ordinary unit of length, the centimetre, are used. Thus to express the *wavelength of ordinary light,* the unit used is

Angstrom = 10^{-8} cm.,

while in the case of X-rays and γ-rays, the unit is still smaller,

X.U. = 10^{-11} cm.

There are also two other special units:

(*a*) *The Bohr radius, i.e.,* the radius of the first orbit of the hydrogen atom according to Bohr's theory. It is given by the expression

$h^2/4 \pi^2 e^2 m$ and is equal to 0.53×10^{-8} cm.

(b) *The Compton wavelength* which is the shift in wavelength due to the Compton scattering of radiation, when the scattering angle is 90°. It is given by

h/mc and is equal to 24.17 X.U.

2. Mass. The most important unit in Atomic Physics is the *electron mass* or the rest-mass of the electron.

$m_o = 9.028 \times 10^{-28}$ grm.

Atomic masses are usually expressed in terms of the *mass unit*

(m.u.) which is equal to 1/16 of the mass of O^{16} and weighs 1.646×10^{-24} grm.

3. Electric charge. The fundamental unit is the

charge of the electron, *e*, which is equal to 4.767×10^{-10} e.s.u. or 1.59×10^{-20} e.m.u.

4. Energy. While the ordinary unit of energy is the **erg**, other units are used in Atomic Physics.

In *spectroscopy*, the natural unit is the *Rydberg* which is the ionization energy of the hydrogen atom.

In Modern Physics, the most common and arbitrarily chosen unit is the *electron-volt* (*e*V) which is defined as the energy possessed by an electron with its electrostatic charge and mechanical mass when it is accelerated by an electric field of P.D. of 1 volt. For instance, in the cathode ray stream of a discharge tube with P.D. of 50,000 volts, each electron possesses the energy of 50,000 electron volts.

The energy in ergs corresponding to one electron volt may be evaluated as follows:

$1\ e\mathrm{V} = 4.767 \times 10^{-10} \times 1/300 = 1.589 \times 10^{-12}$ erg (where *e* and V are expressed in *e.s.u.*)

Since the energy corresponding to 1 eV is very small, energies are often expressed in kilo-electron-volts and million-electron-volts:

$1\ \mathrm{K\ eV} = 10^{3}\ e V$ and
$1\ \mathrm{M\ eV} = 10^{6}$ eV.

The natural unit of energy is the self-energy of the electron, i.e., the energy equivalent to the rest-mass m_0 of the electron according to Einstein's mass-energy relation; its value can be found as follows:

$\mathrm{W} = m_0\ c^2 = 9.028 \times 10^{-28} \times (3 \times 10^{10})^2$ ergs
$= 9.028 \times 10^{-28} \times (3 \times 10^{10})^2\ [300/(4.767 \times 10^{-10})]\ /10^{6}$ Mev $= 0.5107$ Mev

It can be shown that this energy is equal to the energy corresponding to the Compton wavelength, according to the quantum theory:

Self-energy of the electron $=$ energy of the photon $= 5.107 \times 10^{5}$ eV.

Using Einstein's law of equivalence, the wavelength of the photon is given by

$$\lambda = \frac{12345}{V} \text{ A.U. (V being expressed in volts)}$$

$$= \frac{12345}{5 \cdot 107 \times 10^5} = 2 \cdot 416 \times 10^{-10} \text{ cm.}$$

$$= 24 \cdot 16 \text{ X.U.} = \text{Compton wavelength.}$$

The energies corresponding to 1 gram of mass and one mass unit can be evaluated as follows:—

Using Einstein's mass-energy relation, mc^2,

(a) the energy of
$$1 \text{ grm. of mass} = 1 \times (3 \times 10^{10})^2 \text{ ergs.}$$
$$= (300/e) \times 9 \times 10^{20}\ eV$$
$$\approx 6 \times 10^{26} \text{ MeV}$$

(b) the energy of
$$\text{one mass unit} = 1 \cdot 646 \times 10^{-24} \times (3 \times 10^{10})^2 \text{ erg}$$
$$= 1 \cdot 646 \times 10^{-24} \times (300/e) \times 9 \times 10^{20}\ eV$$
$$\approx 931 \text{ MeV}$$

Hence 1 gram of matter is a veritable storehouse of energy.

THE ELECTRONIC STRUCTURE OF THE ATOM

The extranuclear electronic structure of the atom has been chiefly investigated by the spectral properties of the atom. To account for the experimentally observed spectroscopic data of elements, several hypotheses have been successively proposed, which may be classified under the title **"the atom models"**. Thus we have

- the Thomson atom model,
- the Bohr atom model,
- the Sommerfeld relativistic atom model,
- the vector atom model and finally
- the wave mechanical atom model.

These different models mark the progress made in getting a more and more satisfactory interpretation of experimental data, until an almost perfect and complete understanding of the peripheral electronic structure of the atom is reached in the wave mechanical Model.

THE ATOM MODELS

Introduction

The first real foundation of the modern conception of the atom was laid by Faraday who discovered that in electrolysis, each atom, irrespective of the nature of the element, gave up or received a fixed quantity of positive or negative charge equal in magnitude to 1.59×10^{-19} coulomb.

A more definite idea of the intrinsic nature of the atom came into existence with J.J. Thomson's discovery of the electron and his measurement of its mass and charge, that led to the result of the electrical nature of matter, according to which the following two facts were clearly established:

(a) electrons enter into the constitution of all atoms,
(b) since the atom as a whole is electrically neutral the quantity of positive and negative charges in it must be the same.

THOMSON ATOM MODEL

J.J. Thomson, with the above two conclusions to guide him, give the first picture of the structure of the atom. The important points he had to tackle were:

(i) the total number of electrons in an atom, and
(ii) the way in which they were distributed along with the positive charges in the atom.

Using the phenomenon of X-ray scattering, he was able to determine the number of electrons in an atom as follows:

When a beam of X-rays passes through matter, it should be scattered, the scattering coefficient σ, according to the classical theory, being given by $\sigma = 8 \pi e^2 n/3m^2c^4$, where e and m are the charge and mass of the electron, n the number of electrons per unit volume and c the velocity of light.

Now σ can be experimentally determined, so that n can be estimated from the above relation. From the value of n thus obtained, the number of electrons per atom can be readily computed. He found that *this number was proportional to the atomic weight of the element.*

element.

The second point, regarding the arrangement of the electrons and the positive charges in the atom, was not so easy to solve. Thomson had no experimental data to help him. But, since the atom as a whole was stable, he argued that the electrons must be held by the positive charges by electrostatic forces.

He further made the assumption *that the positive charges were uniformly distributed in a sphere of atomic dimensions,* a conception which seemed to him most suited to mathematical treatment, while *the electrons were so arranged inside the positive sphere* that their mutual repulsions were exactly balanced by the force of attraction towards the centre of the sphere.

He then showed that in an atom with a single electron, like the hydrogen atom, the electron must be situated at the centre of the positive sphere.

In an atom with two electrons, like the helium atom, the two electrons must be symmetrically situated on opposite sides of the centre at a distance equal to half the radius of the positive sphere.

In the three-electron system the electrons should be at the corners of a symmetrically placed equilateral triangle, the side of which was equal to the radius of the sphere.

Proceeding in this manner, Thomson detailed the arrangement of electrons ranging from 1 to 100 inside the positive sphere.

With this model he attempted to account for the observed spectra of elements. He argued that since the electron was a charged particle, if it vibrated about its position of equilibrium, it should radiate energy according to the electromagnetic theory and the frequency of the emitted spectral line should be the same as that of the electron. Considering the case of the simplest hydrogen atom, he found that the above assumed mechanism should give rise to a spectral line in the whereabouts of 1400 A° which corresponds roughly to a line in the observed hydrogen spectrum.

Thus the Thomson atom model not only satisfied the requirements of the stability of the atom and the demands of the electromagnetic theory, but was also able to explain to a certain extent the origin of spectral lines. According to it, however, hydrogen can give rise only to a single spectral line, contrary to observed facts of several series and of several lines in each series. Evidently Thomson's model was defective somewhere. Soon the experimental conclusion arrived at by Lord Rutherford from the study of large angle scattering of α-particles definitely proved that the assumption of Thomson about the uniform distribution of positive charges in a sphere of atomic dimensions was wrong, since observation forced the conclusion that the positive charges in the atom should be concentrated in a very small region at the centre of the atom. This gave rise to a very different conception known as the nuclear atom model.

RUTHERFORD NUCLEAR ATOM MODEL

The nuclear atomic model resulted from the researches of Rutherford and his collaborators on the scattering of α-particles by thin sheets of matter. The scattering of α-particles by matter is evidently caused by the electrostatic repulsive forces governed by Coulomb's law that come into play between the α-particles and the positive charges of the atoms in the scattering material. The electrons of the scatterer on account of their extremely small mass as compared with that of the α-particles, will not effect to any appreciable extent the α-particles during their passage through the scatterer. Hence the deflection of an α-particle from its original path when it comes out of the scatterer is due to a great number of small deflections produced by the action of the positive charges of a large number of atoms. This is called 'multiple' or 'compound' scattering.

Experiments on the scattering of α-particles by thin foils of matter showed that although most of the α-particles suffered only a small deflection due to multiple scattering, yet there were a certain number that were scattered through much larger angles. For instance, it was found that 1 in 8000 alpha particles was deflected through more than 90° by a thin film of platinum.

127

This experimental finding could not be accounted for on the basis of the Thomson atom model. For, it can be readily shown that the structure of the atom assumed by Thomson would not result in a large deflection due to any single encounter. When an α-particle enters the positive sphere, the charge in the shell outside the path of the α-particle will exert no deflecting force on it. Hence, the farther the path of the α-particle from the centre of the atom, the greater will be the deflecting force. As the α-particle approaches the centre of the sphere, the shell of ineffective charge increases with the result that the force of repulsion becomes less and less (Fig. 194).

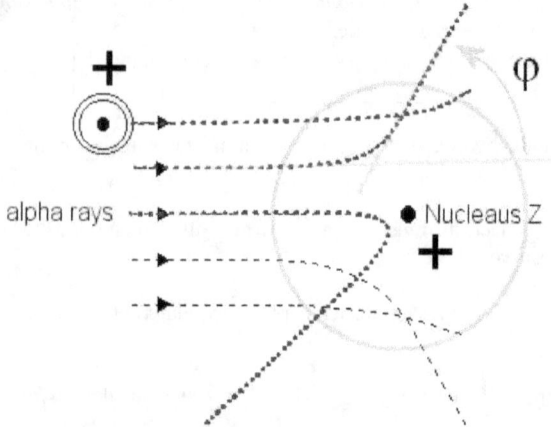

Fig. 194. α-particle scattering in Thomson atom model.

This means that there would be only a small deflection due to a single encounter. Calculating the probable angle of scattering of α-particles as a function of the atomic charges, it is found that

$$\mathbf{N_\varphi \ = \ N_0 \ e^{-\left(\varphi/\varphi_m\right)^2}}$$

where N_φ is the number of particles scattered at an angle φ, N_0 the total number of incident particles and φ_m the most probable angle of scattering. It is evident from this relation that the probability of large angle scattering is necessarily very small, since as φ increases N_φ will decrease very rapidly. For instance, when $\varphi = 30°$, the probability is of the order of 10^{-13}. Hence Thomson's picture of the atom was quite unable to account for the experimentally observed large angle scattering.

Rutherford therefore proposed, in 1911, a new type of atom model, capable of giving to an α-particle a large deflection due to a single encounter. Since it was necessary that the α-particle at some point in its path through the atom must experience much larger repulsive forces than that permitted in Thomson's model, he argued that such a thing was possible only if the whole of the positive charges of the atom, instead of being uniformly distributed throughout a sphere of atomic dimensions, as Thomson imagined was concentrated in a very small region at the centre of the atom. The central core of the atom in which the entire positive charge of the atom was concentrated was called the **nucleus.**

The electrons in the atom were, in consequence, assumed to be situated outside the nucleus in some sort of configuration. Thus Rutherford arrived at the conception of the nuclear atom model. Applying the same laws of probability to this atom model it was shown that the proportion of α-particles deflected through large angles was much greater than in the case of the Thomson's atom. The nuclear atom model was experimentally tested and confirmed by Rutherford and his co-workers.

EXPERIMENTS ON THE SCATTERING OF ALPHA PARTICLES

Theory
In order to simplify the somewhat complicated theory of the scattering of α-particles by matter, the following assumptions are made:

(i) The problem is treated from the classical point of view, *i.e.*, the scattering is .regarded as due to elastic impact of two particles, the α-particle and the nucleus, their wave aspect being neglected.

(ii) The nucleus and the α-particle are considered as point charges, mere centers of Coulomb electrostatic force ; thus the dimensions of the interacting particles are not taken into account.

(iii) The nucleus is considered to be so heavy that its motion during the impact may be disregarded.

Let an α-particle moving along PO approach the relatively heavy nucleus stationary at N (Fig. 195). Since both are positively charged, there is a force of repulsion between them, which being governed by the Coulomb's law of inverse squares, will increase enormously as the α-particle gets closer to the nucleus.

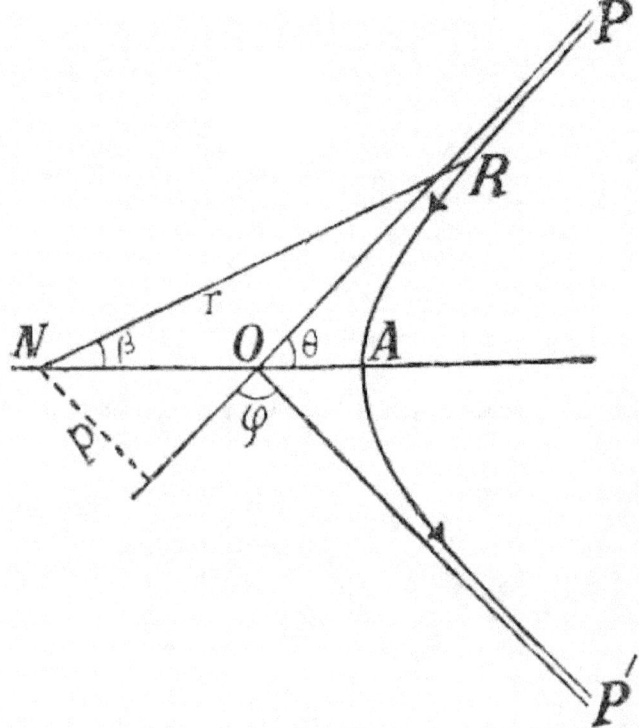

Applying the properties of motion in central orbits to the α-particle thus repelled by the nucleus, it can be shown that the path of the α-particle will, in general, change from a straight line to a hyperbola PAP' one of whose foci is N and whose asymptotes PO and PO give the initial and final directions of the α-particle.

Let *p* be the perpendicular distance of PO from N, *i.e.*, the shortest distance from the nucleus to the initial direction, which is called the *impact parameter.*

Let *m* be the mass of the α-particle, 2*e* its charge and v_o its initial velocity.

Let Z be the atomic number of the element that scatters the α-particle, so that Z*e* is the charge on the nucleus. The angle of deviation or scattering of the α-particle is evidently φ as shown in the figure.

Let R be the position of the α-particle at any instant *t*. Referring R to polar coordinates with N as the origin and NOA as the axis of reference, let NR = *r*, and RNA = β. Since the α-particle moves in a field of force, equal to +2Ze^2/r^2 (+ ve as the force is one of repulsion, directed away from N), we have to deal here with motion in a central orbit, the area velocity of which is given by (1/2) r^2 (*dβ/dt*) = h/2 (a constant).

Applying the principle of conservation of momentum, we have

$$m \, v_0 \, p = m \, . \, r \, . \, (d\beta/dt) \, . \, r = m \, r^2 \, (d\beta/dt) = mh$$

[$r \, (d\beta/dt)$ being the transverse velocity]

$$\therefore \qquad\qquad v_0 \, p = h \qquad\qquad \ldots (1)$$

Similarly applying the principle α-particle of conservation of energy, if v be the velocity of the A-particle at R,

$$\tfrac{1}{2} m \, v_0^2 = \tfrac{1}{2} m \, v^2 + 2 \, Ze^2/r$$

or $\qquad v^2 = v_0^2 - 4Ze^2/mr \qquad\qquad \ldots \qquad\qquad \ldots (2)$

since

$$v^2 = (\text{radial velocity})^2 + (\text{transverse velocity})^2$$
$$= (dr/dt)^2 + r^2 \, (d\beta/dt)^2$$
$$= (d\beta/dt)^2 \, [(dr/d\beta)^2 + r^2]$$
$$= h^2/r^4 \, [(dr/d\beta)^2 + r^2]$$

$$\therefore \qquad \frac{h^2}{r^4} \left[\left(\frac{dr}{d\beta} \right)^2 + r^2 \right] = v_0^2 - \frac{4Ze^2}{mr} \qquad \ldots (3)$$

Here we have a relation between r and β as a differential equation which will, when solved, give the nature of the path of the α-particle.

Putting

$$r = 1/u, \quad dr/d\beta = - \frac{1}{u^2} . \frac{du}{d\beta},$$

we get

$$\frac{h^2}{r^4} \left[\left(\frac{dr}{d\beta} \right)^2 + r^2 \right] = h^2 u^4 \left[\frac{1}{u^4} \left(\frac{du}{d\beta} \right)^2 + \frac{1}{u^2} \right]$$

$$= h^2 \left[\left(\frac{du}{d\beta} \right)^2 + u^2 \right]$$

Hence relation (3) becomes

$$h^2 \left[\left(\frac{dr}{d\beta} \right)^2 + u^2 \right] = v_0^2 - \frac{4Ze^2}{m} . u$$

Or

$$\left(\frac{du}{d\beta} \right)^2 + u^2 = \frac{1}{h^2} \left(v_0^2 - \frac{4Ze^2}{m} . u \right)$$

Replacing $1/h2$ by $1/v_0^2 p^2$ from (1),

130

$$\left(\frac{du}{d\beta}\right)^2 + u^2 = 1/p^2 - \frac{4Ze^2}{mv_0^2 p^2} \cdot u \qquad \qquad \dots (4)$$

Putting $(mv_0^2 p^2/2Ze^2) = l$.

$$\left(\frac{du}{d\beta}\right)^2 = \frac{1}{p^2} - \frac{2u}{l} - u^2$$

Or

$$\frac{(du)^2}{(1/p^2 - 2u/l - u^2)} = (d\beta)^2$$

$$\therefore \qquad \frac{du}{[(1/p^2 + 1/l^2) - (u^2 + 2u/l + 1/l^2)]^{1/2}} = d\beta$$

Or

$$\frac{du}{[(1/p^2 + 1/l^2) - (u + 1/l)^2]^{1/2}} = d\beta$$

Let

$$a = (1/p^2 + 1/l^2)^{1/2} \text{ and } z = (u + 1/l)$$

Then

$$d\beta = \frac{dz}{\sqrt{a^2 - z^2}}$$

$$\int d\beta = \int \frac{dz}{\sqrt{a^2 - z^2}} = -\cos^{-1}(z/a)$$

$$\therefore \qquad \beta = -\cos^{-1}(z/a) \quad \text{or} \quad \cos(-\beta) = z/a$$

Hence

$$\cos \beta = \frac{u + 1/l}{(1/p^2 + 1/l^2)^{1/2}},$$

Or

$$\cos \beta \left(\frac{1}{p^2} + \frac{1}{l^2}\right)^{1/2} = \frac{ul + 1}{l}$$

Or

$$ul + 1 = l (1/p^2 + 1/l^2)^{1/2} \cos \beta$$

$$= (l^2/p^2 + 1)^{1/2} \cos \beta$$

$$= \varepsilon \cos \beta, \qquad \text{where } \varepsilon = (l^2/p^2 + 1)^{1/2}$$

$$\therefore \qquad l/r = \varepsilon \cos \beta - 1$$

Or

$$r = \frac{l}{\varepsilon \cos \beta - 1} \qquad \qquad \dots (5)$$

This is the equation to a *hyperbola,* one of whose foci is occupied by the centre of force. The angle between the two asymptotes is equal to 2θ, where $\varepsilon \cos \theta = 1$ and $\theta = (\pi - \varphi)/2$, φ being the angle of scattering.

The magnitude of φ *will* depend on several factors, such as the charge of the nucleus Ze, the charge of the α-particle $2e$, its mass m, its velocity v_0 and the impact parameter p.

To derive an expression for φ let us consider the case of a central impact, *i.e.,* when the α-particle is directed straight towards N so that $p = 0$. On account of the repulsive force, the α-particle will be stopped at a certain distance b from the nucleus N and made to retrace its path, in which case φ is equal to 180° (Fig. 196).

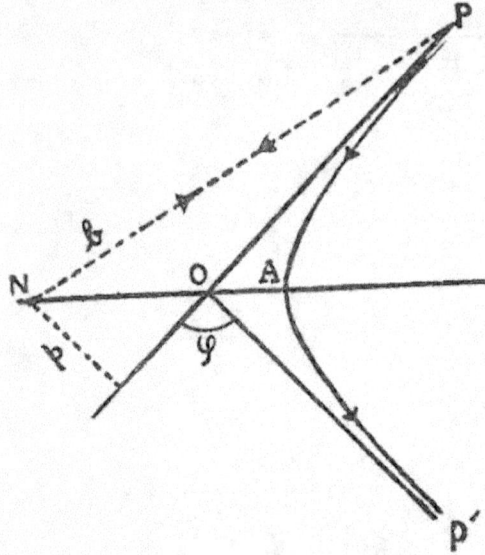

This distance of closest approach b can be determined by using the principle of conservation of energy. The electrostatic potential at a distance b due to the nucleus $= Ze/b$. This acts on a charge $2e$ of the α-particle. Hence the potential energy of the α-particle when it is at the distance b from the nucleus $= (Ze/b) 2e = 2Ze^2/b$.

Since the α-particle is momentarily stopped at the distance b, there its initial kinetic energy is completely changed into potential energy. Neglecting the very small amount of energy lost by the α-particle in its interaction with the peripheral electrons,

$$\frac{1}{2} mv_0^2 = \frac{2Ze^2}{b} \quad \text{or} \quad b = \frac{4Ze^2}{mv_0^2} \qquad \qquad \dots (6)$$

As it is not possible, in practice, to direct the α-particle exactly towards the nucleus, we must consider the case when $p \neq 0$. In such a case, the α-particle will be deflected through an angle φ which is less than 180° and will travel along the hyperbolic path PAP'.

Let V be the velocity of the α-particle at the vertex A. Using the principles of conservation of energy and of momentum,

$$\frac{1}{2} mv_0^2 = \frac{1}{2} mV^2 + \frac{2Ze^2}{NA} \qquad \qquad \dots (7)$$

$$m v_0 . p = m V . NA \qquad \qquad \dots (8)$$

Substituting from (6) for $2Ze^2$, viz., $(1/2) bmv_0^2$, equation (7) becomes

$$\tfrac{1}{2} m v_0^2 = \tfrac{1}{2} mV^2 + \tfrac{1}{2} (bmv_0^2/NA)$$

$$\therefore \qquad V^2 = v_0^2 \left(1 - \frac{b}{NA} \right) \text{ or } \frac{V^2}{v_0^2} = \left(1 - \frac{b}{NA} \right)$$

From equation (8) we get

$$p^2 = \frac{V^2}{v_0^2} \cdot NA^2 = NA^2 \left(1 - \frac{b}{NA} \right) = NA(NA - b)$$

Using the properties of the hyperbola, viz.

$$\varepsilon = 1/\cos\theta, \text{ where } \theta = (\pi - \varphi)/2, \text{ and } NO = \varepsilon \cdot OA,$$

$$NA = NO + OA = NO(1 + 1/\varepsilon) = NO(1 + \cos\theta)$$
$$= (p/\sin\theta)(1 + \cos\theta) = p \left(\frac{1 + 2\cos^2\theta/2 - 1}{2\sin\theta/2\,\cos\theta/2} \right)$$
$$= p \cot\theta/2$$

$$\therefore \quad p^2 = p\cot\theta/2\,(p\cot\theta/2 - b) \text{ or } p = p\cot^2\theta/2 - b\cot\theta/2$$

$$\therefore \quad b = p\,\frac{(\cot^2\theta/2 - 1)}{\cot\theta/2} = 2p\cot\theta = 2p\cot\left(\frac{\pi - \varphi}{2}\right)$$

Or

$$b = 2p\tan\varphi/2$$

Substituting for b from equation (6),

$$\tan\frac{\varphi}{2} = \frac{b}{2p} = \frac{2Ze^2}{pmv_0^2} \qquad \qquad \ldots (9)$$

This relation shows that Z; m and v_0, being kept constant, as the impact parameter p decreases from a relatively high value up to the limit zero, φ increases from 0 to 180°. This means that when the α-particle passes far away from the nucleus (*i.e.*, p great) the angle of scattering is very small. As the distance p diminishes, *i.e.*, as the α-particle passes closer and closer to the nucleus, the angle of scattering will be greater and greater and in the limiting case of the central impact ($p = 0$), $\varphi = 180°$, *i.e.*, the α-particle will be forced to retrace its path after approaching the nucleus up to a distance b.

We shall now deal with the case realized in the actual experiment where a narrow beam of α-particles is incident normally on a thin foil of the scatterer, and the scattered particles are detected by means of scintillations produced by them on a fluorescent screen normal to the direction of view.

Let n be the number of atoms per unit volume of the scatterer of thickness t.
Let Q be the total number of α-particles that strike unit area of the scatterer.
From simple probability considerations the number of α-particles (N) that are scattered through an angle φ and strike unit area of the fluorescent screen at a distance r (Fig. 197) can be estimated as follows:

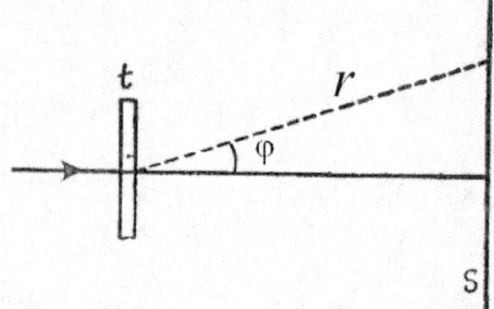

Assuming the atoms of the scatterer to be distributed at random like the molecules of a gas and hence utilizing the principles employed in the kinetic theory for calculating the mean free path, it can be shown that the probable number of α-particles coming within the distance of an impact parameter p of a nucleus is given by

$$\pi \ p^2 \ Q \ t \ n \hspace{3cm} \dots (10)$$

Hence, the number of α-particles having an impact parameter between p and $p + dp$ is

$$d \ (\pi \ p^2 \ Q \ t \ n) = 2\pi \ p \ Q \ n \ t \ dp$$

After impact, these particles will be deflected through an angle between φ and φ +dφ. Thus the number of α-particles scattered between angles φ and φ +dφ is

$$2\pi \ Q \ n \ t \ p \ dp \hspace{3cm} \dots (11)$$

From equation (9),

$$p = -\frac{1}{2} \ b \ \cot \ \varphi/2 \ \text{ and } \ dp = \frac{1}{2} b \times - \frac{1}{2} \text{cosec}^2 \ \varphi/2 \ . \ d\varphi$$

Substituting these values in equation (11) the number of α-particles scattered between φ and φ +dφ is

$$= 2\pi \ Q \ n \ t \times \frac{1}{2} \ b \ \cot \ \varphi \times - \frac{1}{4} \ b \ \text{cosec}^2 \ \varphi/2 \ . \ d\varphi$$

$$= -\frac{1}{4} \ \pi \ Q \ n \ t \ b^2 \ \cot \ \varphi/2 \ \text{cosec}^2 \ \varphi/2 \ . \ d\varphi$$

All these α-particles will strike the screen within a circular annulus of area

$2\pi r \sin \ \varphi \times r \ d\varphi \ = 2\pi r^2 \sin \ \varphi \ d\varphi.$

Hence the number N of α-particles striking unit area of the screen at an inclination φ from the incident direction is given by

$$N = \frac{\frac{1}{4}\pi \ Q \ n \ t \ b^2 \ \cot \ \varphi/2 \ \text{cosec}^2 \ \varphi/2 \ d\varphi}{2\pi r^2 \ \sin \ \varphi \ d\varphi}$$

$$= \frac{Q \ n \ t \ b^2 \ \cot \ \varphi/2 \ \text{cosec}^2 \ \varphi/2}{8r^2 \ . \ 2 \ \sin \ \varphi/2 \ \cos \ \varphi/2}$$

$$= \frac{Q \ n \ t \ b^2}{16 \ r^2 \ \sin^4 \ \varphi/2}$$

134

Substituting the value of b from eqn. (6),

$$N = \frac{Q\, n\, t\, .\, 16\, Z^2\, e^4}{16\, r^2\, m^2\, v_0^4\, \sin^4 \varphi/2}$$

$$= \frac{Q\, n\, t\, (Ze)^2\, e^2}{r^2\, m^2\, v_0^4\, \sin^4 \varphi/2}$$

This is known as the *Rutherford scattering formula* which states that, if Rutherford's conception of the nuclear atom is correct, the number of the α-particles N striking unit area of a fluorescent screen at a distance r from the point of scattering must be proportional to

(1) $1/\sin^4 \varphi/2$,
(2) the thickness t of the scatterer,
(3) the square of the nuclear charge $(Ze)^2$ and
(4) $1/(mv_0^2)^2$, i.e., inversely to the square of the initial kinetic energy or to the fourth power of the initial velocity.

Experimental verification
These theoretical predictions were tested and verified by a series of experiments conducted by Geiger and Marsden in 1913 and by Chadwick in 1920.

The experimental arrangement used by Geiger and Marsden consisted essentially of a strong cylindrical metal box B fixed to a graduated circular platform A which could be rotated in the air-tight joint C. The box was closed airtight at the top by a ground glass plate P and evacuated through the tube T (Fig . 198).

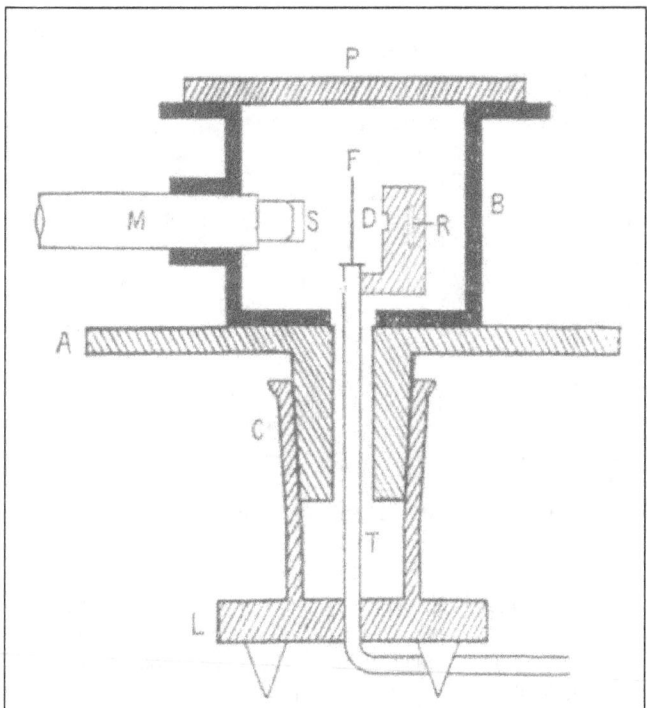

Fig. 198. Apparatus of Geiger and Marsden for the study of α-particle scattering.

Inside the box were arranged the source of α-particles R (a tube filled with radon) and the scattering foil F (gold, silver, or platinum). These were mounted independently of the box. A low power microscope M to which a zinc sulphide screen S was rigidly attached was fixed to the box, facing the scattering foil. On account of the independent mounting of the foil and source, the microscope with the fluorescent screen could be rotated along with the box, while the scatterer and source remained fixed. This device enabled them to detect the α-particles scattered over a wide range of angles, from 5° to 150°.

The experimental procedure and the results obtained may be summarized as follows:

(i) *Angular distribution of the scattered particles*
The number of scattered particles N at different angles ; *i.e.,* for different values of φ were counted by means of the scintillations produced on the screen S and observed through the microscope M with the given source and scatterer. A graph was drawn with log ($1/sin^4$ φ/2) against log N and it was found to be a straight line inclined at 45° to the axes, as was to be expected, since the two quantities plotted were proportional **to** each other, according to the theory.

(ii) *Dependence of scattering on the thickness of the foil*
The thickness *t* of the foil was varied and the number N of α-particles scattered in a given direction (*i.e.,* φ constant) was counted for each thickness. It was found that N/*t* was constant as predicted by the theory. On doubling the thickness, for instance, the number of scattered particles was doubled also, provided the thickness was small. When the thickness of the foil was so great as to cause an appreciable decrease in velocity of the α-particles, the number of scattered particles increased rather more rapidly than required by the theory. This increase was due evidently to the decrease in velocity and when corrections were made for this disturbing effect, the results agreed with the theory.

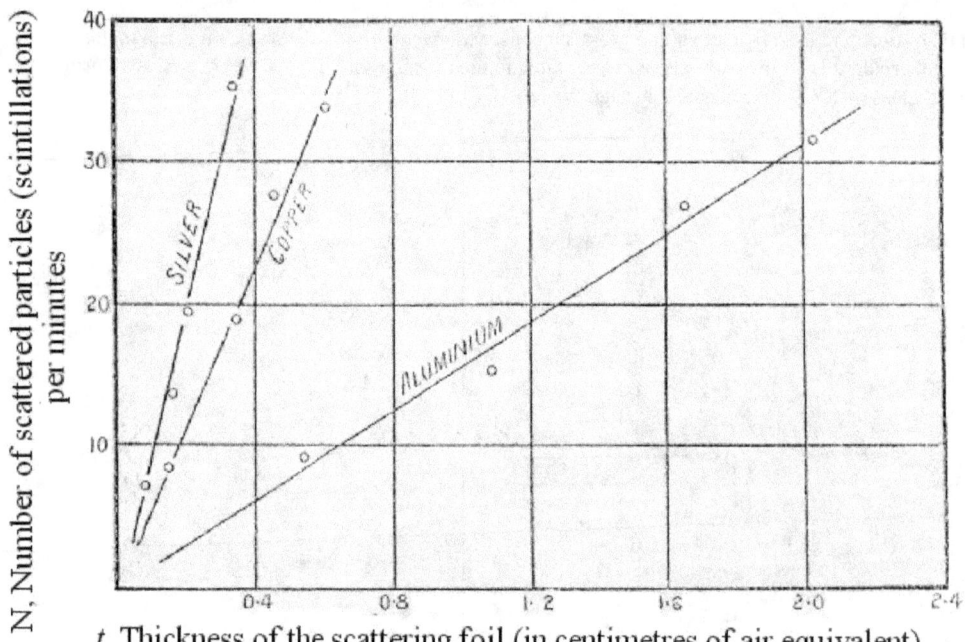

(iii) *Variation of scattering with the nuclear charge*
Using different elements for the scatterer, the number of α-particles N scattered at a given angle φ was counted in each case. The initial kinetic energy of the α-particles was determined by an independent magnetic deflection experiment. The total number Q of the α-particles was obtained by counting the scintillations on the fluorescent screen in the absence of the scatterer. Using Rutherford's formula, the quantity $(Ze)^2$ was evaluated. Assuming the value of *e,* the values of Z for different scatterers were obtained, which agreed with the atomic numbers of the elements known from other sources.

Accurate experiments of Chadwick gave for Z in the case of copper, silver and platinum the values 29.3, 46.3 and 77.4 respectively, while according to the periodic table they are 29, 47 and 78. This excellent agreement confirms the validity of Rutherford's nuclear theory of the atom.

(iv) *The effect of the velocity of α-particles on the scattering*

The initial velocity v_o of the incident α-particles was varied by placing absorbing screens of mica between the source and the scatterer, and its value obtained by finding the range R of the α-particles and applying Geiger's law : $R = av_o^3$. Geiger and Marsden showed Nv_o^4. to be a constant over a range of velocities, such that the number of scattered particles varied as 1 : 10, which, therefore, verified the fourth conclusion of the theory.

These experiments on the scattering of α-particles by thin metallic foils confirmed Rutherford's theory so well that the nuclear atom model was at once universally adopted. Further, the experimental data enabled the value of b, the distance of closest approach of the α-particle to the nucleus, to be determined. It was found to be of the order of 10^{-12} cm. for gold and smaller for lighter elements.

The distance of closest approach of the α-particle to the nucleus is found to be of the order of 10^{-12} cm for gold and smaller for lighter elements, the order of magnitude of the linear dimensions of the nucleus.

Considering the process of scattering as a kind of collision between the α-particle and the nucleus, the experimental value of b, *viz*. 10^{-12} cm., may be taken as giving the order of magnitude of the linear dimensions of the nucleus. This shows that the nucleus occupies an extremely small part of the atom, since the atomic diameter is of the order of 10^{-8} cm. In the space between the nucleus and the outer confines of the atom there are a few electrons, whose masses are very small and dimensions more or less of the same order of magnitude as that of the nucleus, but whose charges exactly balance the positive charges concentrated in the nucleus. Hence it has been remarked that

the atom consists chiefly of empty space.

Drawbacks of the Rutherford nuclear atom

Rutherford's atom model, though unanimously accepted, was not free from limitations, the chief of which arose from considerations of the stability of the atom as a whole. For, it became obvious that in

the nuclear atom *equilibrium could not be secured by the operation of electrostatic forces alone* between the positively charged nucleus and the negative electrons outside the nucleus.

For instance, considering the case of an atom with two electrons, the nuclear charge $2e'$. If the electrons are symmetrically placed at a distance r from the nucleus, then

The force of attraction between the nucleus and each of the electrons is $2e^2/r^2$
The force of repulsion between the electrons is $e^2/4r^2$.

Since

the force of attraction is eight times greater than that of repulsion, the condition of stability is not satisfied and the electrons will fall into the nucleus, thus destroying the stable structure of the atom.

In order to overcome this difficulty,

Rutherford suggested that the electrons might be assumed to revolve round the nucleus, like the planets round the sun at such a speed that the mechanical centrifugal force would just balance the net excess of electrostatic attraction

and in consequence stability could be secured.

But such an assumption brought, in its wake, another very serious difficulty from the point of view of the electromagnetic theory, according to which,

a revolving electron should radiate energy continuously.

Now this energy can only come from the atomic system, which will therefore steadily lose energy. As a result, the electron will approach the nucleus by a spiral path, giving out a radiation of constantly increasing frequency and finally fall into the nucleus. Thus the orbital motion of the electrons destroys the very purpose for which it was postulated, viz. the stability of the atom.

Further, emission of radiation of constantly increasing frequency has no experimental basis, since atoms are actually found to emit discrete spectral lines of definite frequency.

One is therefore forced to conclude that either the Rutherford atom model with revolving electrons is defective or the classical electromagnetic theory fails in the present case.

The dilemma was solved in 1913 by Niels Bohr, a Danish physicist, who, admitting the failure of the classical theory, applied with remarkable success the quantum theory to the Rutherford nuclear atom with revolving electrons. This leads us to the consideration of the Bohr atom model.

BOHR ATOM MODEL

Bohr's celebrated theory of atomic structure is an application of Planck's theory of quanta to the Rutherford nuclear atom in an attempt, extraordinarily fruitful, to define the nature of the orbits in which the electrons might revolve round the nucleus and to explain the origin of spectral lines of elements.

According to the quantum theory, the energy of a system or exchange of energy between different systems occurs not in a continuous fashion permitting all possible values as demanded by the classical theory, but in a discrete quantified form as integral multiples of an elementary *quantum* of energy *hv*.

Bohr built his theory of atomic structure on the following *two fundamental postulates* conformable to the quantum theory.

(i) The first postulate, referring to *the electronic structure,* states that the electrons cannot revolve in all possible orbits as suggested by the classical theory, but only in certain definite orbits satisfying quantum conditions. These orbits may therefore be considered as *privileged orbits;* they must also be treated as *stable* and *stationary,* since the motion of the electron in them, though governed by the ordinary laws of mechanics and electrostatics, is not subject to the electromagnetic theory demanding continuous radiation of energy. In other words, the privileged quantum orbits are the *non-radiating paths* of the electron. Thus the difficulty with regard to the stability of the atom is overcome.
(ii) The second assumption concerning the *origin of spectral lines* states that radiation of energy takes place only when an electron jumps from one permitted orbit to another. The energy thus radiated, which is equal to the difference in the energies of the two orbits involved, must be a quantum of energy *hv*.

Bohr took, for the application of these ideas, hydrogen, the simplest of all elements, which had already been investigated extensively and was known to contain only one electron. According to the Rutherford nuclear model, the hydrogen atom should consist of a single charged positive nucleus (proton) and a single electron outside the nucleus. Assuming that the laws of mechanics and electrostatics hold good for the orbital motion of the electron round the nucleus, the path of the electron can be taken, as a first approximation, to be a circle with the nucleus at the centre. Bohr tackled the two problems connected with the hydrogen atom, viz.,

-    the electronic structure and
-    the origin of its spectral lines

ELECTRONIC STRUCTURE

Nature of the privileged quantum orbits
Considering the general case of a linear simple harmonic oscillator, its displacement x at any instant t is given by

$$x = A \sin 2\pi \nu t \qquad \qquad \dots (1)$$

where A is the amplitude and ν the frequency. As the total energy of the oscillator changes from all kinetic, at the equilibrium position, to all potential at the maximum displacement, it can be determined by computing the kinetic energy at the equilibrium position.

The kinetic energy of the oscillator at the instant t

$$= \frac{1}{2} m (dx/dt)^2$$

where m *is* the mass of the oscillator and *dx/dt* its linear velocity at the instant considered.

138

From eqn. (I),

$dx/dt = 2 \pi \nu A \cos 2\pi\nu t$

and at the equilibrium position dx/dt is maximum;

$(dx/dt)_{max.} = 2\pi\nu t,$

since the maximum value of $\cos 2\pi\nu t = 1$.

$$\therefore \quad \text{The total energy} = \frac{1}{2} m [(dx/dt)_{max}]^2$$

$$= \frac{1}{2} m (2\pi\nu A)^2$$

$$= 2\pi^2\nu^2 A^2 m$$

According to the quantum theory, this energy should be an integral multiple of $h\nu$.

$$\therefore \qquad\qquad nh\nu = 2\pi^2\nu^2 A^2 m, \quad \text{where } n \text{ is an integer,}$$

$$\text{or} \qquad\qquad nh = 2\pi^2 A^2 \nu m \qquad\qquad\qquad \dots (2)$$

The momentum p_x of the oscillator at the instant t is given by

$$p_x = m (dx/dt) = m.2\nu\pi A \cos 2\pi\nu t$$

$$\text{Putting} \quad m . 2\pi\nu A = B, \quad p_x = B \cos 2\pi\nu t$$

$$\therefore \qquad\qquad p_x / B = \cos 2\pi\nu t$$

$$\text{From eqn. (1)} \quad x/A = \sin 2\pi\nu t$$

$$\therefore \quad x^2/A^2 + p_x^2/B^2 = 1$$

This means that the relation between p_x and x is given by an ellipse. If we draw a graph with x as abscissae and p_x as ordinates, it will be an ellipse whose semi-major and semi-minor axes are A and B respectively.

Considering such an ellipse (Fig. 199),
Let dx' be the width of an element at a distance x from the origin.
Let p_x be the value of the ordinate corresponding to x.
Then the area of the element considered is $p_x . dx$.
The area of the ellipse is obtained by the integration of $p_x. dx$ over a complete cycle, which is known as the "phase integral", represented by

$$\therefore \qquad \oint p_x . dx = \text{area of the ellipse}$$

$$= \pi \times A \times B$$

$$= 2\pi^2 A^2 \nu m$$

$$= nh \qquad\qquad \text{from eqn. (2)}$$

Thus the phase integral of a linear oscillator is an integral multiple of h, the Planck's constant. Applying this result to the electron in uniform circular motion, which is equivalent to a harmonic oscillator, and replacing the linear momentum p_x by the angular momentum p_φ and the linear element dx by the angular element $d\varphi$ appropriate to the case, we can write

$$\oint p_\varphi . d\varphi = nh$$

But $p_\varphi = I\,\omega$, where I is the moment of inertia of the electron and ω the angular velocity. The angular velocity being assumed constant, p_φ is also constant.

$$\therefore \qquad \oint p_\varphi \cdot d\varphi = p_\varphi \cdot \oint d\varphi = p_\varphi \int_0^{2\pi} d\varphi = p_\varphi \cdot 2\pi = nh$$

$$i.e. \qquad\qquad p_\varphi = n\,(h/2\pi)$$

Hence; according to the quantum theory, the angular momentum of the electron moving in a circular orbit can have only those values which are integral multiples of $h/2\pi$, or conversely the *permitted circular orbits of the electron are those in which the angular momenta are integral multiples of $h/2\pi$.*

Radius, frequency and energy of the permitted orbits

In order to derive expressions for these quantities that define more precisely the permitted orbits of the electron in the hydrogen atom, let **M** and **E** be the mass and charge of the nucleus; *m*, *e* and *v* the mass, charge and linear velocity respectively of the electron which is assumed to revolve in a circular orbit of radius *a*. In general, the nuclear charge **E** = **Z**e, where Z is the atomic number of the element; for hydrogen $Z = 1$ and $E = e$. The nuclear mass **M** is so large compared to the electronic mass *m* that, for the present, the nucleus is assumed to remain at rest.

The electrostatic force of attraction between the nucleus and the electron is Ee/a^2.
The centrifugal force of repulsion between the two resulting from the circular motion of the electron $= mv^2/a$. The system will be stable, if

$$Ee/a^2 = mv^2/a$$
$$\therefore \qquad v^2 = Ee/am \qquad\qquad \dots \qquad\qquad \dots\;(1)$$

Introducing the quantum condition for the orbit,

$$p_\varphi = I\omega = n\,(h/2\pi)$$
$$I\omega = ma\omega^2 = ma^2 \cdot v/a = mav$$
$$mav = nh/2\pi$$
$$v = nh/2\pi ma \qquad\qquad \dots \qquad\qquad \dots\;(2)$$

Dividing (1) by (2), we get

$$v = \frac{Ee}{am} \cdot \frac{2\pi ma}{nh} = \frac{2\pi Ee}{nh} \qquad \dots \qquad\qquad \dots\;(3)$$

Again from (2) $a = nh/2\,\pi\,mv$
Substituting the value of v from (3),

$$a = \frac{nh}{2\pi m} \times \frac{nh}{2\pi Ee} = \frac{n^2 h^2}{4\pi^2 Eem} \qquad\qquad \dots\;(4)$$

Thus,

the **radius** *a* of the permitted orbit is directly proportional to n^2, since all the other quantities are constant. This means that the radii of successive permitted orbits are proportional to the squares of the integers 1, 2, 3,,,,,, These integers are called the *quantum numbers* of the respective orbits.

The radius of the first smallest orbit in the hydrogen atom can be readily calculated from the above relation by using the known values of *h*, $E = e$, and *m* and putting it $n = 1$. It turns out to be equal to 0.53×10^{-8} cm., and is known as the *Bohr radius*. The diameter of the first orbit is, therefore, of the order of 10^{-8} cm., which agrees with the values of the diameters of atoms computed by various other methods.

Orbital frequency

If f be the orbital frequency,

$$f = \frac{\omega}{2} = \frac{v}{2\pi a} = \frac{2\pi \mathrm{E}e}{nh} \cdot \frac{1}{2\pi a} = \frac{\mathrm{E}e}{nha}$$

$$= \frac{\mathrm{E}e}{nh} \cdot \frac{4\pi^2 \mathrm{E}em}{n^2 h^2} = \frac{4\pi^2 \mathrm{E}^2 e^2 m}{n^3 h^3} \quad \ldots \qquad \ldots \ (5)$$

According to the classical theory, this orbital frequency is equal to the frequency of the spectral line emitted by the atom. But .we shall presently see that it is not so according to Bohr's theory.

Orbital energy

The total energy W of the electronic system is equal to the sum of the kinetic and potential energies.

The kinetic energy, from (1), is $(1/2)\,mv^2 = \mathrm{E}e/2a$
The potential energy $= -\ \mathbf{E}e/a$

$$\therefore \qquad\qquad \mathrm{W} = \frac{\mathrm{E}e}{2a} - \frac{\mathrm{E}e}{a} = -\frac{\mathrm{E}e}{2a}$$

Substituting the value of a from (4),

$$\mathrm{W} = \mathrm{W}_n = -\frac{\mathrm{E}e}{2} \cdot \frac{4\pi^2 \mathrm{E}em}{n^2 h^2} = -\frac{2\pi^2 m \mathrm{E}^2 e^2}{n^2 h^2} \quad \ldots \qquad \ldots \ (6)$$

W_n being the energy of the electron when it is in the nth orbit, or the energy corresponding to the nth orbit.

In this relation, since all the quantities except n are constants, the orbital energy is inversely proportional to the square of the quantum number of the orbit. Evidently

for any one particular orbit the energy is constant, which means that as long as the electron remains in that orbit it cannot lose energy by radiation,

in contradiction to the classical electromagnetic theory.

The interpretation of the negative sign associated with the expression for the orbital energy is important. As $'n'$ increases, the absolute numerical value of the energy decreases, but on account of the negative sign, the actual energy will increase. This means that

the outer orbits have greater energy than the inner ones.

In the case of the hydrogen atom, taking the relation

$$\mathrm{W}_n = -\frac{2\pi^2 m \mathrm{E}^2 e^2}{n^2 h^2}$$

Since

$$\mathrm{E} = e, \qquad \mathrm{W}_n = -2m \left(\frac{\pi e^2}{h} \right)^2 \cdot \frac{1}{n^2}$$

Energy of the nth orbit is thus:

$$W_1 = -\frac{2 \times 9 \times 10^{-28} \cdot \pi^2 (4 \cdot 77 \times 10^{-10})^4}{(6 \cdot 55 \times 10^{-27}) \times 1^2} = -2 \cdot 155 \times 10^{-11} \text{ erg.}$$

$$W_2 = -\frac{2 \cdot 155 \times 10^{-11}}{2^2} \qquad\qquad = -0 \cdot 535 \times 10^{-11} \text{ erg.}$$

$$W_3 = -\frac{2 \cdot 155 \times 10^{-11}}{3^2} \qquad\qquad = -0 \cdot 238 \times 10^{-11} \text{ erg.,}$$

and so on.

Since the first orbit has the least energy it is the most stable and is the one which the electron occupies in the normal unexcited atom.

Expressing the orbital energy in terms of the more convenient unit, viz. volts,

Energy in ergs = $eV/300$,

where V is in volts and e in e.s.u. (4.77 x 10^{-10})

$$V_n = -\frac{2m(\pi e^2)^2}{n^2 h^2} \times \frac{300}{e} = -\frac{13 \cdot 6}{n^2} \text{ volts,}$$

when the known values of m, e and h are substituted.

The negative sign leads to another important conception regarding atomic structure. It means that the electron is bound to the nucleus by attractive forces so that energy must be supplied to the electron in order to separate it completely from the nucleus.

In this sense the orbital energy is known as the *binding energy* or *work function,* which is then considered as a positive quantity. In the case of the hydrogen atom the binding energy of the electron in the nth orbit is given by $13.6/n^2$ volts. Energy required to extricate the electron from inner orbits is therefore greater than from outer orbits. If energy equal to that obtained by an electron subjected to an electrostatic .field whose P.D. is 13.6 volts be supplied to the electron in the hydrogen atom, the electron will be completely expelled and the positively charged nucleus alone will be left over. The atom, in such a case, is said to be *ionized.*

ORIGIN OF SPECTRAL LINES

Summary of the experimental researches made on spectral lines before the time of Bohr

In order to appreciate the importance of Bohr's theory of spectral lines, it is necessary to make a brief survey of earlier researches on the spectra of elements and of the results obtained.

A scientific study of the spectra of elements became possible only when the wavelengths of the spectral lines could be accurately measured. Direct measurement of wavelengths is based on the phenomenon of interference.

Young, in 1801, was the first to make such estimates of wavelengths from data obtained with Newton's rings.

Twenty years later, Fraunhofer developed the diffraction grating as a. method of measuring wavelengths.

The next important advance was made by Angstrom who, in 1868, published an elaborate table of wavelengths in the solar spectrum, measured with three carefully ruled gratings.

Measurements with gratings approaching modern accuracy were made by Rowland, who, about 1885, introduced the concave reflection grating and greatly improved the technique of ruling the grating and using them.

Michaelson's introduction of the interferometer and his use of that instrument in 1895 for measuring the wavelength of the red cadmium line marks the latest important step in precision measurement of wavelengths.

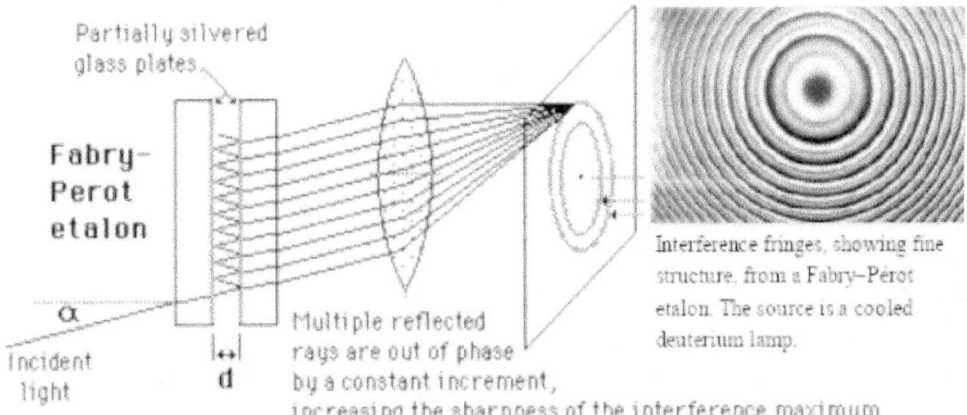

Interference fringes, showing fine structure, from a Fabry–Perot etalon. The source is a cooled deuterium lamp.

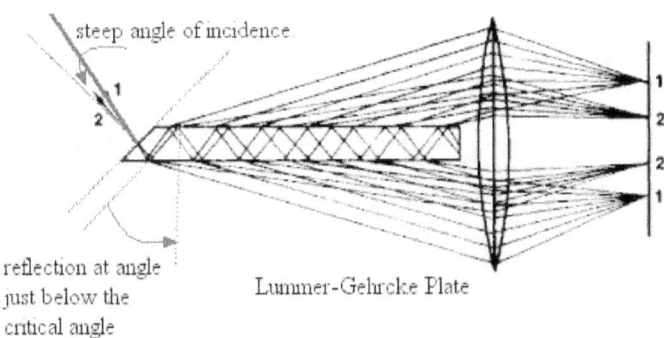

With the subsequent arrival of other types of highly perfected interferometers, such as Febry-Perot interferometer, Lummer Gehrcke plate, etc. measurement's of wavelengths are now possible with an absolute accuracy of one part in a million, the relative accuracy being considerably higher.

Dependable wavelength measurements thus having been made available, careful analysis of the great number of lines constituting the spectrum of an element was undertaken, as it was early realized that the study of the spectral radiations emitted or absorbed by the elements would greatly help in determining the structure of the atom. Most of the atomic spectra are very complex, and the lines appear at first sight to be distributed at random. But by the use of various kinds of evidence such as

(a) the physical appearance of the lines— for instance, "sharp" and "diffuse",

(b) a physical change in the element affecting certain lines in a similar manner-for instance, compressing the emitting gas, if one line is broadened, the companion lines are broadened at the same time,

(c) the behavior of the lines when the emitting atom is subjected to a magnetic field, etc., it was shown that the spectral lines of an element could be grouped into several *series.*

Of such series for any one element, the most intense are four, called the *principal, sharp, diffuse* and *fundamental.* Similarities among these series in the spectra of several elements pointed that not only there were some real underlying factors common to these series but also there existed some fundamental mechanism common to all elements as the origin of characteristic line spectra. Physical similarities in the spectra of elements having thus been noted, several

143

investigators sought for numerical relations in the lines of a series as well as between the different series of a given element and of different elements.

Balmer's empirical relation

Balmer, in 1885, succeeded in obtaining a simple relationship for the wavelengths of the lines in the hydrogen spectrum which can be represented by the formula

$$\frac{1}{\lambda} = \bar{\nu} = R\left(\frac{1}{2^2} - \frac{1}{n^2}\right)$$

Where
R is a constant, n = 3, 4, 5, etc,
λ the wavelength and

$\bar{\nu}$ the wavenumber, i.e., the reciprocal of the wavelength expressed in cms, a quantity by which it is usual to designate spectral lines.

Energy of the third orbit W3 - 92 = — 0'535 x 10 -11 erg. 2155 x 10 -11 572 PHYSICS OF THE ATOM

By substituting for *n* in the formula the successive values 3, 4, 5, 6, we obtain the wave numbers of the four lines H_α, H_β, H_γ and H_δ in the visible region, forming part of a series known as the *Balmer series* which actually contains as many as thirty lines so far measured and extends well into the invisible ultra-violet region, the lines becoming progressively closer and fainter until they merge.

The chief merit of this formula lies in the fact that it was not only later justified by Bohr's theory but also correctly predicted that no lines of longer wavelength than H_α would be found and that the series would "converge" as *n* assumes very large values.

Rydberg's formula

Rydberg, as a result of detailed researches, established in 1889 that all the series in optical spectra could be arranged according to a general relation of the form

$$\bar{\nu} = \bar{\nu}_\infty - R / (n + \mu)^2,$$

where

R was found to be a universal constant for all series, now called the *Rydberg constant,*
n an integer,
μ a fraction less than unity which is practically constant for all the lines of a series,

$\bar{\nu}_\infty$ the limiting or convergent wave number in the series, corresponding to n = ∞ .

He proved also that the Balmer's formula is a special case of the above general relation and estimated the value of the constant R as 109,720 cm^{-1}.

The great merit of Rydberg's formula consists in the fact that wave number of any line can be expressed as the

difference of two terms, one fixed, represented by $\bar{\nu}_\infty$ and the other variable which is obtained by giving different integral values to *n*. Rydberg's formula can therefore be written in a simpler form as

$$\overline{\nu} = R\left[\frac{1}{m^2} - \frac{1}{n^2}\right]$$

where *m* is fixed and *n* variable.

Ritz combination principle

It occurred to Rydberg that combinations of terms other than those giving the four chief series might correspond to spectral lines observed to be present in spectra but not belonging to the series. Ritz, in 1908, generalized this idea of Rydberg into a principle which achieved remarkable results in the classification of spectral lines. It may be stated thus:

By a combination of the terms that occur in the Rydberg or Balmer formula, other relations can be obtained holding good for new lines and new series.

For instance, series other than that of Balmer in the hydrogen spectrum were predicted even before they were actually discovered by Paschen and Brackett, thanks to this principle. Taking the first two lines, H_α and H_β of the Balmer series, we may represent them by

$$\overline{\nu}_\alpha = R\left[\frac{1}{2^2} - \frac{1}{3^2}\right] \text{ and } \nu_\beta = R\left[\frac{1}{2^2} - \frac{1}{4^2}\right]$$

Combining these two as

$$\overline{\nu}_\beta - \overline{\nu}_\alpha = R\left[\left(\frac{1}{2^2} - \frac{1}{4^2}\right) - \left(\frac{1}{2^2} - \frac{1}{3^2}\right)\right]$$
$$= R\left(\frac{1}{3^2} - \frac{1}{4^2}\right),$$

this represents a new line-indeed, the first line of a new series in the infra-red, discovered by Paschen. Similarly the second line of the same series can be obtained by forming the difference of H_γ and H_α and so on. In like manner, another series, also in the infra-red, discovered by Brackett, can be obtained with the combination principle.

Ritz combination principle has maintained itself in the whole realm of spectroscopy, both in the optical and X-ray regions, as an exact physical law with the degree of accuracy that characterizes spectroscopic measurements.

It gave Bohr the clue to interpret atomic spectra in terms of the quantum theory. For, according to classical idea, it was very difficult to imagine a mechanism which could emit spectra with the observed features. It was natural to assume that the higher members of a series were of the nature of overtones. Among acoustic vibrations many cases are met with, in which the frequencies of the overtones are not integral multiples of the fundamental frequency. But no cases are known in which the frequencies converge to an upper limit. And in particular, the Ritz combination principle is without analogy in the classical theory of vibrations. In characteristic atomic spectra, as in thermal radiation, the quantum theory succeeded where the classical theory failed, as proved by Bohr.

Bohr's theory of the origin of spectral lines

Since the electron revolving in any one of the permitted quantum orbits cannot radiate, according to Bohr's first postulate, the question naturally arises whether the Bohr atom could radiate at all. It is an experimental fact, however, that atoms do radiate. Bohr solved the difficulty dexterously by the use of his second postulate. It has been shown that the outer orbits have greater energy than the inner ones. Now, supposing that the electron jumps, due to some reason or other, from an outer to an inner orbit, there must be an excess of energy equal to the difference between the energies of the two orbits involved. On the principle of conservation of energy one may legitimately assume that the excess of energy has been radiated, How? That question one cannot answer. The assumption must be treated as an axiom based on observation. Introducing now the quantum condition of the second postulate, the excess of energy thus radiated must be a quantum of energy $h\nu$, so that the frequency of the emitted spectral line is equal to ν. If W_{n1} and W_{n2} be the energies of the inner and outer orbits of quantum numbers n_1 and n_2 between which electronic transition takes place, then

$h\nu = W_{n1} - W_{n2}$ or $\nu = (W_{n1} - W_{n2})/h$

This is known as *Bohr's frequency condition* which is one of the most important principles used in theoretical interpretation of spectra. It leads in the present case to a formula **of** the spectral series. Substituting the values for W_{n1} and W_{n2}, viz.

$$W_{n_1} = -\frac{2\pi^2 m E^2 e^2}{n_1^2 h^2}$$

$$W_{n_2} = -\frac{2\pi^2 m E^2 e^2}{n_2^2 h^2}$$

$$h\nu = W_{n_2} - W_{n_1} = \frac{2\pi^2 m E^2 e^2}{h^2}\left(\frac{1}{n_1^2} - \frac{1}{n_2^2}\right)$$

$$\therefore \qquad \nu = \frac{2\pi^2 m E^2 e^2}{h^3}\left(\frac{1}{n_1^2} - \frac{1}{n_2^2}\right)$$

It may be noted that the frequency ν of the emitted spectral line is not the same as the orbital frequency

$$f = 4\pi^2 m E^2 e^2 / n^3 h^3$$

contrary to the demands of the classical theory. In terms of the wave number $\bar{\nu}$, a spectral line is given by

$$\bar{\nu} = \frac{1}{\lambda} = \frac{\nu}{c} = \frac{2\pi^2 m E^2 e^2}{ch^3}\left(\frac{1}{n_1^2} - \frac{1}{n_2^2}\right) \qquad \ldots(7)$$

This relation suggests a very simple picture of the origin of spectral lines. A spectral line is emitted when an electron initially revolving in an orbit of quantum number n_2 drops to an inner orbit of quantum number n_1, the wave number being given by the above relation. A whole series of lines corresponds to the electronic transitions from various outer orbits to a given inner orbit.

The equation also offers a means of testing the validity of Bohr's theory.

In the first place, Bohr's formula which contains a constant quantity $2\pi^2 m E^2 e^2/ch^3$ and the difference between the two terms $1/n_1^2$ and $1/n_2^2$ is very similar to the empirical relation of Balmer and Rydberg.

Secondly, comparing Bohr's relations with the empirical formula of Balmer, which had been stated long before Bohr proposed his theory, we see that the quantity $2\pi^2 m E^2 e^2/ch^3$ corresponds to the Rydberg constant R. The empirical value of R obtained by Balmer from his data of the hydrogen spectrum is 109,677.7 cm^{-1}. Now evaluating the constant of the Bohr relation $2\pi^2 m e^4/ch^3$ (since $E = e$ for hydrogen) by substituting the known values of the quantities π, e, m, c and h, we get 109,740 cm^{-1}. The agreement between the two values is so close that it offers an excellent proof of the soundness of Bohr's theory.

The equation predicts also the different series actually observed in the hydrogen spectrum, giving even the numerical values of the wave number of any line in each series. Making $n_1 = 2$ and $n_2 = 3, 4, 5$, etc., the lines of the Balmer series are obtained. If we put $n_1 = 3$ and $n_2 = 4, 5, 6$, etc., we get the Paschen series in the infrared, discovered by Paschen in 1909.

If $n_1 = 1$ and $n_2 = 2, 3, 4$, etc., we have the Lyman series in the ultra-violet region, first observed by Lyman in 1914.

Brackett discovered in 1922 another series in the infra-red corresponding to $n_1 = 4$ and $n_2 = 5, 6, 7$, etc.

Diagrammatic representation of the series spectrum of the hydrogen atom in the light of Bohr's theory

The spectral lines of the different series may be represented diagrammatically in the following two ways:

1. *Orbit diagram* (Fig. 200) gives a pictorial representation of Bohr's privileged orbits and the transitions of the electron from outer orbits to inner orbits, as shown in the figure, giving rise to different series. The orbits corresponding to n_1 = 1, 2, 3, etc., are called K, L, M. etc. shells, a nomenclature used in the study of X-ray spectra. It should be emphasized that this mode of representation of the origin of spectral lines does not correspond to the actual make up of the atom. The picture of an atom with its several electrons revolving in the various privileged orbits, is as far from the real structure of the atom, as are, for instance, "lines of force" from the actual structure of a magnetic field. Though there are no such lines in a magnetic field, yet what they represent, *viz.*, the fact that a magnetic field involves energy, is real. In a similar ways though the concept of orbits may not be true to reality, yet we are reasonably sure that the energy associated with these various orbits is real.

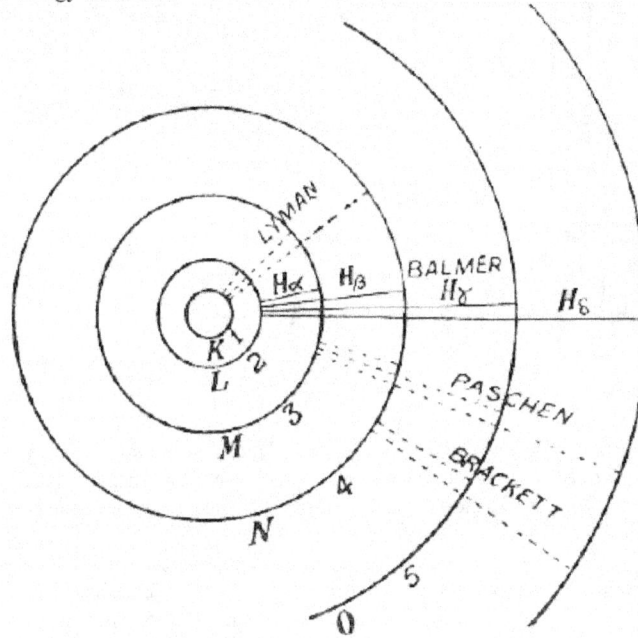

Fig. 200. Diagram showing the origin of the series spectra of the hydrogen atom.

2. *Energy level diagram*. Much more important than the above geometrical concept of orbits is what is known as the energy level diagram, where the different discrete energy states of the atom are represented by horizontal lines and the transition of the atom from one to the other of these states, giving rise to a spectral line, by a vertical line connecting

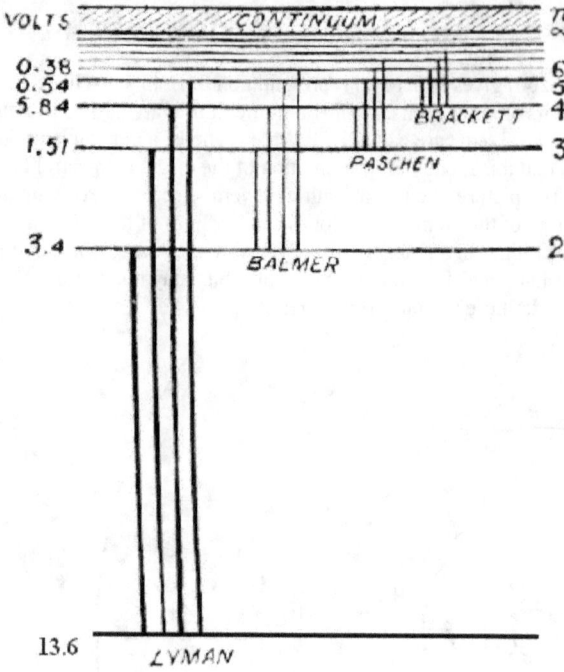

the two states involved (Fig. 201).

Fig. 201. Energy level diagram for the hydrogen atom.

The lowest energy level ($n = 1$) corresponds to the normal unexcited state of the atom. Transitions to this level from higher levels give rise to the Lyman series. The other series are similarly indicated in the figure. The energy levels crowd together when the value of n becomes great. The diagram clearly brings out the fact that each series ends at a particular energy level. It summarizes also beautifully the process of *excitation* which takes place in a discharge tube containing hydrogen gas subjected to electronic bombardment. Certain molecules of hydrogen are dissociated into atoms, and these either by collisions or through some other cause become "excited", that is to say, the electron is removed from its normal level to one of higher energy. After a brief interval, it returns to a level of lower energy, with the emission of a spectral line of frequency characteristic of the transition process, which is always given by Bohr's frequency condition

$$\nu = \frac{W_{n2} - W_{n1}}{h} = \frac{\triangle W}{h}$$

A single atom can, at a particular instant, emit only one spectral line of frequency corresponding to a particular transition taking place at that instant. But, in the actual production of the spectrum in the laboratory a very large number of atoms participate. In some, one kind of transition occurs, while in others, other kinds of transition, so that several series of lines appear. Further, the number of atoms involved is so large and there are so many transitions of a given kind occurring each second, that to an observer the lines appear to be produced continuously and to be of constant relative intensity.

Continuous spectrum beyond series limits

Regions o continuous spectral emission are observed on the short wavelength side of the limits of the Balmer and Lyman series, which we have represented by the portion marked "continuum" in the energy level diagram. Taking, for instance, the Balmer series, the first line, the 'head' of the series (H_a) corresponding to $n_2 = 3$, is 6563 A° and the 'limit' of the series ($n_2 = \infty$) is 3670 A°. Inside the series, there is a total of thirty-one lines as computed from the spectra of certain nebulae and of the solar corona at a total eclipse.

Beyond the limit of the series there is a region of continuous emission, which we are considering now. It is due to the existence of states of higher energy than that corresponding to an infinitely large quantum number.

In terms of the Bohr model,

these energy states are represented by electron moving in parabolic and hyperbolic orbits. Electrons with excess kinetic energy approach the nucleus, curve round it once and then fly off again. Since these motions are periodic they are not quantized, but form a continuous range of energy values.

If one such electron is captured by the ionized hydrogen atom in the encounter, its final state must be a stationary orbit, *i.e., it* must take up a quantized orbit. The excess energy involved in such a process is radiated as monochromatic light but since all energy states above that of quantum number $n_2 = \infty$ are possible, the observed spectrum is the additive effect of many such emissions, *i.e.,* a continuous spectrum extending towards shorter wavelengths from the limit of that series whose constant term corresponds to the quantum state in which electron in captured.

Bohr's correspondence principle

Bohr's theory gives only the frequencies or wave numbers of the spectral lines and says nothing about the nature (whether polarized or not) and intensity of the lines, whereas classical theory of light emission gives a deep insight into the forms of the vibrations and is able to draw conclusions about the relative intensities of the vibration components. To make up for this deficiency in his theory, Bohr discovered a principle which, establishing a correspondence between the classical and the quantum theories, enabled him to make use of the results of the classical theory about intensity and polarization in his theory also.

According to the classical theory, the frequency of the spectral line is the same as the orbital frequency of the electron, whereas it is not so in Bohr's theory.

But it can be shown that under certain circumstances, *viz.,* for transitions between states whose quantum numbers are relatively high (say 51 and 50) the frequency of the spectral line coincides very nearly with the orbital frequency.

$$\text{The orbital frequency } f = \frac{4\pi^2 m E^2 e^2}{n^3 h^3} = \frac{4\pi^2 m e^4}{n^3 h^3}.$$

$$\text{Since } R = 2\pi^2 m e^4 / c h^3, \quad f = \frac{2Rc}{n^3} \qquad \dots (1)$$

The frequency of the spectral line is given by

$$\nu = \frac{2\pi^2 m e^4}{h^3}\left(\frac{1}{n_1^2} - \frac{1}{n_2^2}\right)$$

$$= Rc\left(\frac{1}{n_1^2} - \frac{1}{n_2^2}\right)$$

Since n_1 and n_2 are relatively high quantum numbers, the above relation can be written as

$$\nu = Rc\left\{\frac{1}{n^2} - \frac{1}{(n+\Delta n)^2}\right\}$$

$$\approx Rc \cdot \frac{2\Delta n}{n^3} \qquad \dots (2)$$

as Δn is small compared with n (e.g., *1* compared with 50).

Now, if $\Delta n.=1$, $\nu' = 2Rc/n^3 = f.$

Thus,

the frequency given by the quantum theory for two very large quantum numbers and separated by unity becomes identical with the orbital frequency and hence with the classical frequency, referred to the *fundamental* mode of vibration.

If $\Delta n > 1$, *but still $<< n$, $v = f.\Delta n$,* which means that in the region of high quantum numbers in which the change is by several units, the quantum transitions correspond to *overtones* or harmonics of the classical frequency.

Hence, when the quantum number has sufficiently great values, there is *coincidence,* while for moderately great values there is *correspondence* between the results of the classical and quantum theories.

From this fact, Bohr drew a very important conclusion, *viz*

., for large quantum numbers, the behavior of the atom tends asymptotically to that which would be expected on *the basis of the classical theory, which* is the correspondence principle.

Applying this general principle to the question of intensity and polarization of spectral lines, the classically computed data about them holds *perfectly correctly* even to the lines obtained on the basis of the quantum theory in the case of very high quantum numbers and *approximately* correctly for moderately great quantum numbers.

The correspondence principle has been particularly useful in predicting with certainty which transitions are forbidden, i.e., it leads to certain *selection rules* as required by experiment, which may be stated in a general way as follows:

A spectral line is forbidden in emission if the corresponding harmonic component does not occur in the Fourier's series of the classical periodic process.

Thus the correspondence principle has been very fruitful, though only an approximate rule. The problem of direct calculations of intensity and polarization of the light emitted by an atom as well as that of selection rules have been decisively solved by wave mechanics. By describing the quantum states as wave states one is able to make quantitative statements about the intensity of spectral lines.

It is to be noted that the question of intensity is, in reality, a *statistical problem* which therefore does not directly come within

the compass of the quantum theory that deals with individual events in the atom and hence offers no measure for the frequency with which they occur.

It is this frequency of occurrence that is involved in all questions of intensity.

The classical theory of radiation, however, uses mechanics to derive from a given orbital curve the complex of vibrations contained in it together with their amplitudes. In contrast with this, the correspondence principle asserts that

the unknown statistics of individual quantum processes is actually given by the classical calculation; by calculating the amplitudes of the classical spectrum we obtain the correct numbers for the frequency of occurrence of the corresponding quantum processes.

An adequate cause for this assertion is obtained in wave mechanics which establishes a synthesis between the classical idea of continuity and the quantum concept of discontinuity.

The motion of the nucleus

In the simple theory developed above, it was assumed that the mass **M** of the nucleus was so great compared with the mass **m** of the electron that the nucleus remained fixed at the centre of the circular orbits. This is rigorously true only if the mass of the nucleus is infinite.

In reality, this is not the case. For instance, in the hydrogen atom the nuclear mass is only about 2,000 times that of the electron. On account of its finite mass the nucleus will also revolve. But the electron and the nucleus will evidently rotate about their common centre of gravity O (Fig. 202), the former in the larger circle of radius r_1, while the latter in the smaller one of *radius* r_2. A simple theorem on centre of gravity gives the relation $\mathbf{M}r_2 = \mathbf{m}.r_1$. As before, if **a** represents the distance between the nucleus and the electron, we get

$$r_1 = a - r_2 = a - mr_1/\mathrm{M}$$

$$(1 + m/\mathrm{M})\, r_1 = a ; \qquad \therefore\ r_1 = a\mathrm{M}\,/\,(\mathrm{M} + m) \qquad \ldots (1)$$

$$r_2 = mr_1/\mathrm{M} = am\,/\,(\mathrm{M} + m) \qquad \ldots (2)$$

Since the two masses are revolving in circular orbits, we shall have two equations as conditions of stability, viz.

$$Ee/a^2 = mv^2/r_1 \quad \text{for the electron}$$
$$Ee/a^2 = MV^2/r_2 \quad \text{for the nucleus,}$$

Where the velocity of the electron $v = \omega\, r_1$ and the velocity of the nucleus $V = \omega\, r_2$, ω being the common angular velocity of the system about O.

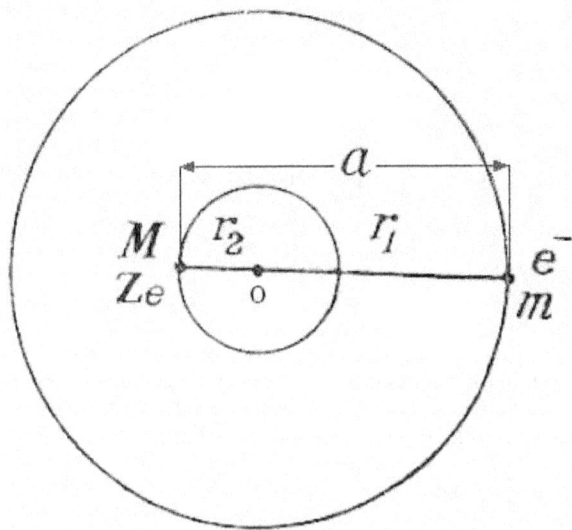

The total kinetic energy is made up of two parts, that of the electron and that of the nucleus. Therefore,

$$\text{Total K.E.} = \tfrac{1}{2}\, mv^2 + \tfrac{1}{2}\, MV^2$$
$$= \tfrac{1}{2}\, m\omega^2 r_1^2 + \tfrac{1}{2}\, M\omega^2 r_2^2$$

Substituting the values of r_1 and r_2 from equations (1) and (2) and simplifying we get

$$\text{Total K.E.} = \tfrac{1}{2}\, \{\, mM\,/\,(M+m)\,\}\,\omega^2 a^2$$
$$= \tfrac{1}{2}\, \mu\, \omega^2\, a^2$$
$$\mu = \frac{mM}{M+m} = \frac{m}{(1 + m/M)} \qquad \ldots (3)$$

usually called the "reduced mass".

The expression for the kinetic energy in the present case where the motion of the nucleus is taken into account differs from that in which the nucleus is supposed to be at rest in that the reduced mass μ replaces the mass *m* of the electron.

Without further following through the entire deduction write the final equation for the wave numbers of the spectral lines of the atom, in which the nuclear mass is not infinite:

$$\bar{\nu} = \frac{2\pi^2 m E^2 e^2}{ch^3} \cdot \frac{M}{M+m} \left(\frac{1}{n_1^2} - \frac{1}{n_2^2} \right) \qquad \ldots (4)$$

We see that the factor multiplying the quantity within the bracket, which factor gives the Rydberg constant, depends upon the ratio of the mass of the electrons to that of the nucleus. For, m is replaced my $mM/(M+m) = m/(1+m/M)$. If M is infinite this expression is reduced to m. But actually the ratio m/M is not the same for different atoms. Therefore,

the Rydberg constant will vary from element to element, though the variation will be small, since the correction factor, $1/(1 + m/M)$ is small.

Including the atomic number Z by the relation $E = Ze$, equation (4) becomes

$$\bar{\nu} = \frac{2\pi^2 me^4}{ch^3} \cdot Z^2 \cdot \frac{M}{(M + m)} \cdot \left(\frac{1}{n_1^2} - \frac{1}{n_2^2} \right) \quad \dots (5)$$

The Rydberg constant for any element is given by

$$R_z = \frac{2\pi^2 me^4}{ch^3} \cdot \frac{M_z}{M_z + m}$$

$$= R_\infty \cdot \frac{1}{1 + m/M_z} \quad \dots (6)$$

where $R_\infty = 2\pi^2 me^4/ch^3$ and M_z the mass of the nucleus of the element of atomic number Z.

If $M_z = \infty$, the expression reduces to $R_z = R_\infty = 2\pi^2 me^4/ch^3$.

The constant R_∞ can be computed by substituting the values of the quantities involved; but they are not known with spectroscopic precision. It is therefore preferably estimated from the spectroscopically observed value of the Rydberg constant ; for hydrogen $R_H=109,677.7$ cm.$^{-1}$ and the known value of $m/M_H = 1/1840$ using the relation

$$R_H = R_\infty \{ 1/(1 + m/M_H) \}$$

The value of R_∞ thus found is 109,737.4 cm^{-1}.

Equation (5) can be written as

$$\bar{\nu} = Z^2 \cdot R_z \left(\frac{1}{n_1^2} - \frac{1}{n_2^2} \right) \quad \dots (7)$$

which expresses the modification to be made when the motion of the nucleus is included, and holds good for all elements.

The above relation has been experimentally confirmed in the three following cases:

1. *Spectrum of the singly ionized helium*
The helium atom (Z = 2) has a mass four times that of hydrogen, a two-fold positive, charge and normally two peripheral electrons. If it is singly ionized by the removal of one of the electrons, the residual atom, usually represented by He$^+$, will have a hydrogen-like structure, consisting of a positive nucleus about which a single electron rotates. It is to be expected therefore that the spectrum of singly ionized helium should be similar to that of hydrogen, but there will be differences due to the nuclear charge of the former being two instead of one and its nuclear mass being nearly four times that of the latter.

Considering first the effect of the increased nuclear charge and neglecting the difference in nuclear mass, the ionized helium should give series of spectral lines represented by the equation

$$\bar{\nu} = 4R_H \left(\frac{1}{n_1^2} - \frac{1}{n_2^2} \right)$$

This is identical in form with that of hydrogen; but the wave numbers of the He^+ lines would be four times as large as those of the lines in the corresponding series of hydrogen, since $4R_H$ replaces R_H.

Such series have been actually observed in the *spark* spectrum of helium and are known as the Fowler and Pickering series, the former corresponding to $n_1 = 3$, while the latter to $n_1 = 4$. This not only verifies Bohr's theory but also provides an irrefutable proof for the nuclear charge of helium being double that of hydrogen.

Careful examination of the spectra of the two atoms, however, showed that the coincidence was not quite exact. Taking, for instance, the Pickering series, the second line in it is given by

$$\bar{\nu} = 4R_H (1/4^2 - 1/6^2) = R_H (1/2^2 - 1/3^2)$$

which is the wave number of the first (H_α) line of the Balmer series. Accurate spectroscopic measurements showed that the two lines do not coincide but the second Pickering line is shifted towards the shorter wavelength side of the H_α line.

Theory accounts for this discrepancy by considering the difference in nuclear masses in the two cases, which results in a slightly different value of the Rydberg constant for the two atoms, R_{He} being greater than R_H.

From the general expression for the Rydberg constant (equation 6) we get

$$R_H = R_\infty . \{ M_H / (M_H + m) \}$$
$$R_{He} = R_\infty . \{ M_{He} / (M_{He} + m) \}$$

For corresponding lines in the two spectra

$$\frac{R_{He}}{R_H} = \frac{M_{He}(M_H + m)}{M_H(M_{He} + m)}$$

Since the mass of the helium nucleus is nearly four times that of the hydrogen nucleus,

$$\frac{R_{He}}{R_H} = \frac{4M_H(M_H + m)}{M_H(4M_H + m)}$$

$$= \frac{4(M_H + m)}{4M_H + m}$$

Since the numerator is slightly greater than the denominator, $R_{He} > R_H$. In consequence, a line in the helium spectrum will have a slightly greater wave number and hence shorter wavelength than the corresponding line in the hydrogen spectrum, as experimentally observed.

The values of the Rydberg constant estimated from the hydrogen and helium spectra are

$$R_H = 109,677 \cdot 7 \ cm.^{-1} \ ;$$

$$R_{He} = 109,722 \cdot 4 \ cm.^{-1}$$

Other hydrogen-like ions whose spectra have been observed are Li++ and Be+++ and the values of the Rydberg constants obtained are

$R_{Li} = 109,728.9 \ cm^{-1}$
$R_{Be} = 109,730.8 \ cm^{-1}$

Hence the Rydberg, constant is least for hydrogen and increases with the increase of nuclear mass, approaching the universal limit

$R_\infty = 109,737.4 \ cm^{-1}$

which corresponds to a nucleus of infinite mass.

2. Determination of the ratio of the mass of the electron to that of the proton : m/M_H
Taking the relation

$$\frac{R_{He}}{R_H} = \frac{4(M_H + m)}{4M_H + m}$$

$$\therefore \quad \frac{R_{He} - R_H}{R_H - {}^1/_4 R_{He}} = \frac{4M_H + 4m - 4M_H - m}{4M_H + m - M_H - m}$$

$$= \frac{3m}{3M_H} = \frac{m}{M_H}$$

Substituting the values for R_{He} and R_H obtained from spectroscopic data, m/M_H can be calculated and is found to be 1/1840, which is in excellent agreement with the value obtained by other methods.

Further, determination of m/M_H is equivalent to measuring the specific charge e/m of the electron. For,

$$\frac{m}{M_H} = \frac{e/M_H}{e/m} \qquad \text{or} \qquad \frac{e}{m} = \frac{e/M_H}{m/M_H}$$

Now e/M_H is the charge carried by a gram ion of hydrogen, a constant which is known with great accuracy (96,494 coulombs). Hence, e/m is readily obtained which comes out to be 5.31×10^{17} e.s.u. or 1.77×10^7 e.m.u., a result much more accurate than can ever be obtained from experiments on the deflection of the cathode rays.

3. The discovery deuterium or heavy hydrogen of mass number two
According to theory, for atoms of the same value of Z there should be lines of slightly different wave numbers if their nuclei have different masses. Hence, in the case of hydrogen, if the isotope of mass number 2 existed, there should be isotopic components of the Balmer lines on the short wavelength side. Calculation showed that the separation between the isotopic components would be

1.793,1.326, 1.185 and 1.119 A° for the first four' lines
H_α, H_β, H_γ , and H_δ respectively.

We have already described the very interesting circumstances that led to the discovery of deuterium. We shall here summaries the researches of Prof. Urey and his collaborators, who succeeded, in 1932, to prove the existence of deuterium.

They used a twenty-one foot concave grating, capable of producing a dispersion of 1.3 A° per mm, and photographed the lines of the Balmer series. Using first ordinary hydrogen in the discharge tube, they found that H_α and H_β had weak satellites separated by exactly the calculated distance. In itself, the appearance of a very weak line near each Balmer line would not have been quite convincing; but Urey and his colleagues made their proof definite by preparing three samples of hydrogen in which the proportion of H^2, if it existed, had been raised by fractional distillation, and showing that the intensities of the lines in the positions calculated for H^2 were in agreement with the proportion of the heavy isotope believed to be in the sample.

Thus the third sample which was prepared so as to be considerably richer in H^2 than the others gave more intense lines at the required positions. With this last sample they obtained 1.791, 1.313, 1.170 and 1.088 A° for the separation of the isotopic components of the first four Balmer lines.

Their experiments also showed that H^2 was present in ordinary hydrogen in the proportion of 1 in 5,000.

Such were the triumphs of Bohr's theory, which, in their turn, established beyond doubt the essential correctness of the quantum atom model conceived by Bohr. The general principles used by Bohr were also successfully applied to a great number of phenomena, such as the excitation and ionization of atoms, complex spectra emitted by atoms with many electrons and in particular X-ray spectra. We shall now consider some of these applications.

APPLICATIONS OF BOHR'S THEORY

I. EXCITATION AND IONIZATION OF ATOMS

According to the theory of Bohr, in order that the atom may radiate, the electron must abandon its normal orbit and move temporarily to an outer orbit of greater energy. When this happens, the atom is said to be in an *excited* state. It is evident that the shifting of the electron to a higher orbit demands a supply of energy to the atom. The process of thus increasing the *internal* energy of the atom is called *excitation of the atom*. If the excitation is so intense that the electron is raised to the outermost permitted orbit ($n = \infty$), the atom is left with a net positive charge and is then said to be *ionized*. This extreme type of excitation is called *ionization of the atom*. It is clear that

the process of excitation and ionization is an *absorption phenomenon,*

since the atom absorbs energy sufficient to raise itself from the normal to the excited or ionized state.

The state of excitation being abnormal it lasts only for a very short time, of the order of 10^{-8} sec., after which the atom comes back to its normal state, the return being accompanied by the emission of radiation according to Bohr's frequency condition.

Critical potentials

It is to be expected that just as radiation of energy takes place in a discontinuous manner as quanta of energy, so also the absorption of energy involved in excitation and ionization of the atom takes place in discrete quanta. This supply of energy to the atom in the absorption process is most conveniently made by bombarding the atom with an outside electron accelerated by an electric field whose P.D. must have definite values to excite the atom to its different quantized states.

As long as the accelerating potential remains below a certain critical value, known as the *critical potential,* the colliding electron cannot excite the atom. Such collisions are known as *elastic collisions,* whose characteristic property is the conservation of both kinetic energy of translation and of momentum without any conversion of the external kinetic energy into internal *energy or vice versa.*

But

if the accelerating potential exceeds the critical value mentioned above, the colliding electron loses a considerable amount of its kinetic energy and the atom struck by the electron suffers internal changes of energy, which leads to its excitation or even ionization. Such collisions are called *inelastic collisions,* to which the laws of classical impact can no longer be applied.

It is usual to distinguish two kinds of critical potentials, *viz., excitation potential* and *ionization potential.*
The excitation potential is that accelerating potential which imparts to the impinging electron enough energy to make an electron of the impacted atom move from the normal to a higher orbit. It is also called *radiation potential,* because it causes the atom which has absorbed energy corresponding to that potential to emit radiations, when it returns from the excited to the normal state, or *resonance potential,* because it is possible to provoke the same excitation of the atom and the subsequent emission of the same radiation as is caused by a bombarding electron also by a light radiation whose energy content equals that of the impinging electron.

The ionization potential is that accelerating potential which makes the impinging electron acquires sufficient energy to knock an electron completely out of an atom and thereby ionize the atom.

Taking a concrete example, viz., the hydrogen atom, we have seen that the energy of the nth orbit expressed in electron volts, is given by the relation

$W = -13.6/n^2$

From this we get
— 13.6, — 3.4, — 1.51 ... 0 eV for the energies of the 1st, 2nd, 3rd ... orbits.

Hence the energy to be supplied to the atom to raise it:

to the first excited state is (13.6 - 3.4) = 10.2 eV,
to the second excited state is (13.6 — 1.51) =12.09 eV,
to the ionized state is (13.6 — 0) = 13.6 eV.

Thus, definite and discrete amounts of energy are required to excite the atom to its different quantized states. It can be readily seen that 10.2, 12-09 volts are resonance potentials, while 13.6 volts the ionization potential of hydrogen.

The number of resonance potentials is ordinarily great as compared with the ionization potentials. In the case of atoms with several electrons, there can be more than one ionization potential. They are called the first order ionization potential, second order ionization potential, etc., according as the atom is singly ionized, doubly ionized etc. But their number is limited to a few, since most of the electrons are interlocked in closed shells incapable of being affected by ordinary external agents, as we shall see.

Helium has two ionization potentials 24.5 and 78.6 volts;
Magnesium also two 7.6 and 15 volts.

Making the legitimate assumption that practically all the energy of the impinging electron is absorbed, by the struck atom in the inelastic collision, critical potentials enable us to determine the wavelengths of the radiations emitted by the excited atom as it returns to the normal state and vice *versa*. Thus, if V be the critical potential, and ν the frequency of the radiation emitted by the atom, then

$eV = h\nu = hc/\lambda$ or
$\lambda = hc/eV$

Expressing V in volts and λ in Angstroms we have already seen that

$\lambda = (12345/V)$ in A°

Examples
If the accelerating potential is such that the impinging electron raises an atom of hydrogen struck by it to the first excited state, i.e., the electron in it moves from the normal first orbit to the second, the radiation which the atom will emit when it returns to the normal will be the first line of the Lyman series, whose wavelength, as experimentally measured in 1216 A°. Since the accelerating potential in this case in 10.2 volts, the wavelength of the radiation according to the above relation is $\lambda = (12345/10.2)$ 1210 A° in excellent agreement with the experimental value.

The wavelength corresponding to the ionization potential of the hydrogen atom (13.6 volts) is equal to 12345/13.6 = 908 A° which is in the far ultra-violet.

In the case of mercury vapor, the first resonance line is experimentally found to have

$\lambda = 2536$ A.
From this we can get the value of its first resonance potential.

V = 12345/2536 = 4.87 volts.

It is to be noted that when an atom is excited to a sufficiently high state, in returning to its normal state, the electron might "drop" into several of the intervening permitted orbits and thereby generate the entire spectrum of the atom.

Methods of excitation of atoms

Atoms may be excited and even ionized chiefly by the following three methods:

A. Electronic bombardment

An extensively used and important method of excitation of atoms is by bombarding them with suitably accelerated electrons. As already stated, slow moving electrons whose velocity is below a certain critical value, cannot excite the atom on which they collide. It can be shown that in such *"elastic"* collisions the transfer of energy from the colliding electron to the atom is negligibly small and hence insufficient to excite the atom.

Let us consider the elastic collision of an electron of mass **m** and velocity **u** with an atom of mass M, which for the sake of simplicity is assumed to be initially at rest After collision let the electron travel with a velocity v in a direction inclined at φ to its initial direction and the atom acquire a velocity V and move in a direction making an angle θ with the same initial direction of the electron (Fig. 203).

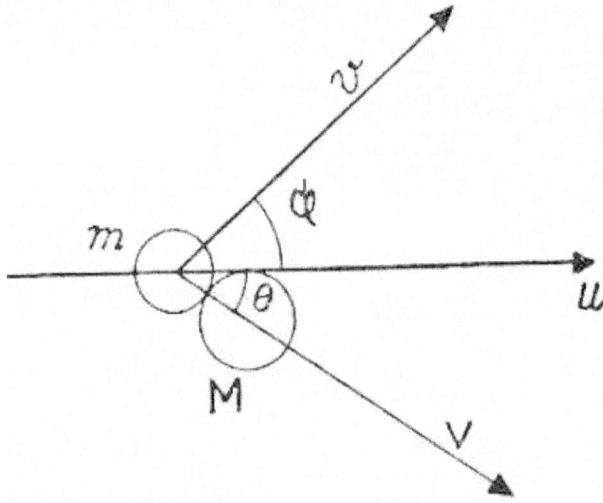

Applying the laws of conservation of energy and of momentum appropriate to the case, we get

$$\tfrac{1}{2}\,mu^2 = \tfrac{1}{2}\,mv^2 + \tfrac{1}{2}\,MV^2 \quad (1)$$
$$mu = mv\cos\varphi + MV\cos\theta \quad (2)$$
$$0 = mv\sin\varphi - MV\sin\theta \quad (3)$$

Rewriting (2) and (3) as

$$mv\cos\varphi = mu - MV\cos\theta$$
$$mv\sin\varphi = MV\sin\theta$$

and squaring and adding; we obtain

$$m^2v^2 = m^2u^2 - 2MmuV\cos\theta + M^2V^2$$

Substituting for m^2u^2 from (1),

$$m^2v^2 = m\,(mv^2 + MV^2) - 2MmuV\cos\theta + M^2V^2$$

or
$$0 = mMV^2 - 2MmuV\cos\theta + M^2V^2$$
$$= MV^2\,(m + M) - 2MmuV\cos\theta$$
$$= V\,(m + M) - 2mu\cos\theta$$

$\therefore$
$$V = \frac{2mu}{(M + m)}\cos\theta$$

The kinetic energy acquired by the atom is

$$\frac{1}{2}\,MV^2 = \frac{1}{2}\,M\cdot\frac{4m^2u^2}{(M + m)^2}\cdot\cos^2\theta$$
$$= \frac{1}{2}\,mu^2\cdot\frac{4Mm}{(M + m)^2}\cdot\cos^2\theta$$
$$= E\cdot\frac{4Mm}{(M + m)^2}\cdot\cos^2\theta$$

The maximum K.E., E_{max}, which the atom can acquire, is given, when $\cos^2\theta = 1$. Since E_{max} is also the maximum K.E. that the electron can lose in the process,

$$E_{max} = E\cdot\frac{4Mm}{(M + m)^2}$$

$\therefore$
$$\frac{E_{max}}{E} = \frac{4Mm}{(M + m)^2} \approx \frac{4m}{M}$$

since **m** is very small compared with M, even in the case of the lightest hydrogen atom, for which **m**/M = 1/1840, therefore,

$$\frac{E_{max}}{E} \approx \frac{4}{1840} \approx \cdot002$$

Thus the energy transferred from the electron to the atom in an elastic collision, even at its maximum value, is inappreciable. It is to be noted also that in an elastic collision if the colliding particles have energy other than that associated with their translational motion, for instance, rotational energy or internal energy, no conversion of this energy into kinetic energy of translation or *vice versa* takes place.

But, when the velocity of the colliding electron exceeds the critical value, *'inelastic'* collision occurs, in which the electron loses a much larger amount of energy than that computed above and the struck atom suffers internal change of energy leading to its excitation. For example. allowing a stream of electrons, accelerated by an electric field, whose P.D. can be varied, to traverse a vessel containing sodium vapor and observing the result with a spectroscope, nothing is seen until the accelerating P.D. reaches 2.09 volts. For voltages slightly above this value the well-known D lines appear and those only. According to the quantum picture of the origin of spectral lines, there must have been a change of state of the Na atom corresponding to an energy drop equivalent to 12345/5893 ≈ 2.09 volts, which is exactly the accelerating P.D. applied. Hence the presumption is very strong that the electrons have, by collisions, transferred to the Na atoms sufficient energy to raise the latter from the normal to the excited state of 2.09 volts higher, so that the atoms in returning from that excited state to the normal emit the D line doublet.

On further increasing the accelerating P.D., the D lines continue to appear, but no others, until the P.D. reaches 5.12 volts, beyond which, many more lines are produced, in fact a complete spectrum.

The convergence wavelength of the principal series of the Na atom, to which the D lines belong as the first, is known to be 2410 A° and is due to the electron falling from infinity to the lowest normal state. The energy change involved in the emission of this line is 12345/2410 ≈ 5.12 expressed in volts, which is exactly equal to the accelerating P.D. of the bombarding electron. Hence we may legitimately conclude that an electron accelerated by a P.D. of 5.12 volts colliding with a Na atom imparts to it sufficient energy to raise an electron in it to infinity, *i.e.*, to ionize the atom. When the atom returns to the normal state, the electron might drop into any of the intervening quantum levels and thereby generate the entire arc spectrum of sodium.

B. Collision of atoms at high temperatures
This method is known as *thermal excitation* or *thermoluminiscence*, since the excitation of the atom and the subsequent emission of radiation are due to a rise in temperature. At ordinary temperatures, it can be shown from kinetic theory principles that the probability of atoms which possess energies required for excitation by collision is negligible. But at high temperatures, with the increase in translational energy of the atoms, the proportion of atoms, which have sufficient energy to excite one another by collision, is large enough to produce a perceptible luminescence.

When a vapor begins to radiate due to its high temperature, the first spectral line observed corresponds to a transition from the lowest excited state to the normal. On further increase of temperature, the translational motion becomes increasingly violent and is able to produce higher excitation, so that new spectral lines appear.

For instance, just a few lines are observed at the temperature of the Bunsen flame (1870°C) in the electric arc (3500°C) many more lines appear, chiefly of the neutral atom; in the electric spark discharge the lines of the ionized atom are seen.

Inelastic collision of the second kind
There is ample evidence that a process, converse to the above, is also possible, viz. an excited atom may lose its excitation energy by collision with another unexcited atom, so that it returns to its normal state *without emitting radiation*. Such collisions are called *collisions of the second kind* in order to distinguish them from collisions, where the kinetic energy of translation is converted into excitation energy as in the previous case and which are known as *collisions of the first kind.*

Collisions of the second kind are subdivided into two categories according as the excitation energy lost appears:

A. *Non-radiative collision of the second kind* is where energy is lost purely as kinetic energy of translation of the atoms involved in the collision.

B. *Sensitized fluorescenc* where the excitation energy of the initially unexcited atom is subsequently emitted of its own characteristic frequencies (*sensitized fluorescenc*).

As an illustration of these phenomena the experiments of Franck and Cario on mixtures of two substances in the vapor state may be cited:

Sensitized fluorescence. Some mercury and thallium were placed in a quartz tube which was heated in an oven to 800°C, so that an appreciable fraction of the mixture was kept in the vapor state. Light from a quartz mercury are lamp was made to fall on the mixture. When the latter was examined with a spectroscope a number of lines characteristic of thallium, of wavelengths, 5351, 3776, 3319, 3230, 2918 and 2238 A° were seen. The mercury resonance line, 2536 A°, was also there, but it was much weaker in intensity than when no thallium was present. The observed phenomenon was explained as follows:

The Hg atoms are raised to the first excited state by resonance absorption of the 2536 line from the mercury arc lamp. Some of these excited Hg atoms by collisions of the second kind with unexcited Tl atoms transfer their excitation energy to the latter, which, in consequence, get excited and radiate their characteristic lines during their return to the normal state. That such a process actually takes place is proved by the following facts:

(a) thallium vapor alone is not excited to resonance by the 2536 mercury line.

(b) all the lines emitted by thallium except one (2238 A°) are of wavelengths greater than 2536 A°, which means that they correspond to energy transfers in the Tl atom less than the energy corresponding to the mercury 2536 A° line;

(c) the quenching or reduction in intensity of the resonance line of the mercury vapor indicates that the Hg atom has lost its excitation energy and returns to its normal state without emission of radiation. Hence it is concluded that it is a case of sensitized-fluorescence.

The anomalous thallium line, 2238° A, cannot evidently be excited by the transfer of the excitation energy of Hg atom (λ = 2536 A°) to the thallium atom, since the former is of shorter wavelength than the latter and in consequence corresponds to a greater energy than could be supplied by the excited Hg atom. The extra energy required must, therefore, have come from some other source. Since such anomalous lines appear only at high temperatures (800°C), it is reasonable to suppose that the required additional energy is provided solely by the kinetic energy of thermal motion of the atoms. This explanation has been confirmed by experiments with mixtures of Hg — Cd, Hg— Pb, Hg — Bi and Hg — Zn.

Evidence is thus furnished that

both the excitation energy of the Hg atom and the kinetic energy of thermal motion may combine in one elementary process to raise another atom, such as Tl, Cd, etc., to a high excited state

Non-radiative collision of the second kind has also been observed in the case of mercury-argon mixture. Cario showed that

the presence of argon in mercury vapor very materially reduced the intensity of the resonance radiation of the mercury vapor, without, however, exciting the characteristic radiations of argon, when the mixture was irradiated by the 2536 line.

Examination of the light transmitted through the mixture showed that there was no diminution in absorption by mercury due to the presence of argon. Hence the observed phenomenon can be explained only by assuming that nonradiative collisions of the second kind take place between the Hg atoms and the Ar atoms, in which the excitation energy of the former is transferred to the latter in the form of kinetic energy alone.

C. Irradiation of atoms with light

Atoms can be excited also by energy supplied in the form of light. For instance, irradiating hydrogen with ultra-violet light and supposing that the incident light contains one of the wavelengths of the Lyman series, say, that corresponding to the second line λ = 1026 A°, then the atoms which absorb this light will be raised to the second excited state, *i.e.*, the electron will move from the first to the third orbit. In returning to the normal, these excited atoms can do so either in a single step with the emission of light of the same wavelength (1026 A°) or in two successive steps, from the second to the first excited state and then to the normal, in which case two lines will be emitted, *viz.*, the first line of the Balmer series (6563 A°) and the first line of the Lyman series (1216 A°).

The phenomenon of emission of light by atoms which have been excited by absorption of light is called *fluorescence,* where the spectral lines emitted have always frequencies less than that of the incident light.

It was first discovered by Stokes, as early as 1852, but an adequate explanation could not be had on the basis of the classical theory of resonance. Bohr's quantum picture, on the other hand, offers a simple and satisfactory interpretation. When an atom is irradiated with light of frequency corresponding to one of its absorption lines, the return of the atom from the excited to the normal state may occur *in stages.* Thus the transition from the level of energy W_1 to that of W_2 may take place in many steps, as

$$W_1 — W_a, W_a — W_b, W_b —W_c .... W_n — W_2,$$

where $W_a, W_b, W_b, W_c .... W_n$ are energies of intermediate levels and correspondingly great number of frequencies are generated, all of which, however, are lower than the incident frequency given by $(W_1 — W_2)/h$.

Measurement of critical potentials

The critical potentials of excited atoms have been experimentally determined by the two following methods:

I. Electrical method. The general *principle* of this method consists in passing accelerated electron through gases and vapors and measuring the resulting current with a sensitive galvanometer. If the current thus measured be plotted against the steadily varied accelerating potential, the curve obtained presents marked discontinuities at potentials corresponding to the inelastic collisions resulting in the excitation of the atoms of the gas or vapor under study to the different quantum states. Hence the method uses *excitation by electronic bombardment.*

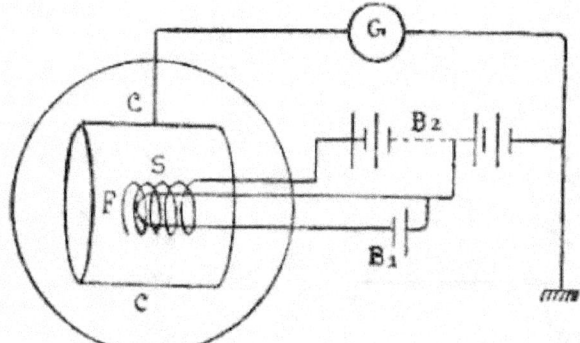

Fig 204. Experimental arrangement for measuring the critical potentials.

The *experimental arrangement* (Fig. 204) consists essentially of a filament F which becomes a source of thermionic electrons when heated by a battery B_1. Surrounding the filament is a spiral grid S which by means of a battery B_2 can be maintained at any desired positive potential with respect to the filament. Around the grid S at a distance relatively greater than that between S and F is a metallic cylinder CC which is connected to a sensitive galvanometer G. The filament, grid and cylinder are arranged inside a glass bulb containing the gas or vapor under study, maintained at any desired low pressure.

The thermionic electrons emitted by F are accelerated towards the grid S by a known positive potential V. When they enter the space between S and CC, they will have an energy equal to eV. If these electrons suffer no energy loss on collision with the atoms of the gas, they will travel to the cylinder and be recorded by the galvanometer. As the accelerating voltage is increased, the current to the cylinder is increased until the energy of the electrons is just the right amount to raise the atom of the gas from the normal to one of its excited states. When this happens there is a drop in the current, indicating that many electrons have given up their energy to the gas atoms causing the excitation of the latter. Plotting the current measured by the galvanometer against the steadily varied accelerating potential, a curve, as shown in Fig. 205, is obtained.

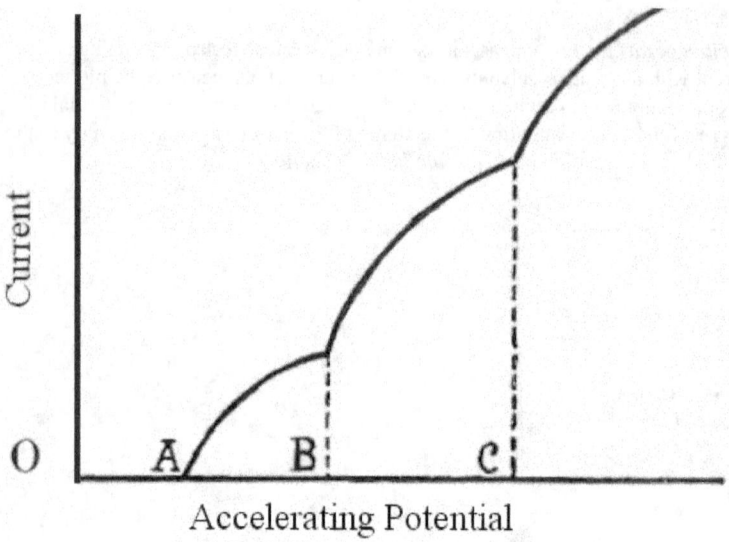

The discontinuities in the curve due to the drops in the current give the excitation potentials *e.g.*, OA is the first excitation potential, OB the second, OC the third, etc. As the voltage is increased a stage is reached at which ionization of the gas sets in resulting in a very large increase in current to the cylinder.

This is a *direct* method in which critical potentials, both resonance and ionization, are directly measured by the readily known accelerating potentials of the electrons which by collision with the atoms give them sufficient energy to rise to well defined excited states and finally to be ionized. Lenard was the first, in 1908, to measure ionization potentials of a number of gases, such as air, hydrogen, etc. by this method. Franck and Hertz, in 1913, utilized it to measure the first resonance potential of mercury vapor. The apparatus of these pioneer workers, however, did not lend itself to a clear differentiation between resonance and ionization potentials. Davis and Goucher, in 1917, perfected the apparatus in such a way as to render evident the discrimination sought for and measured the various resonance potentials as well as the ionization potential of mercury.

EXPERIMENT OF DAVIS AND GOUCHER

The modification introduced by Davis and Goucher in the original apparatus (Fig. 204) were:

(i) The galvanometer was replaced by a quadrant electrometer E with a grounding key K, so that a P.D. could be established between S and C.

(ii) The filament F was so connected to the battery B_2 that the P.D. between F and S was *less than* and in *the opposite direction to the* P.D. between S and C.

(iii) A coarse wire gauze S' was placed surrounding the grid S and close to the cylinder C. By means of a third battery B_3 this gauze could be maintained at a *small* positive or negative potential with respect to C.

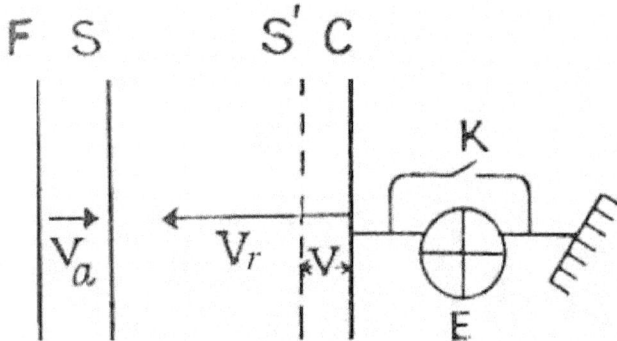

These modifications are schematically shown in Fig. 206.

Let V_a represent the P.D. between F and S, hence the accelerating potential of the electrons.
Let V_r represent the P.D. between C and S; it is known as the retarding potential, since the experimental condition requires that $V_r > V_a$ acts in a direction opposite to that of V_a. On account of this fact, the electrons accelerated froth F towards S cannot reach C. Hence, as long as no positive ions are produced in the space between S and C, the cylinder C can acquire no charge. But, if the accelerating potential exceeds a certain limit and in consequence if the energy of the electrons becomes great enough to ionize some of the gas molecules in the space between S and C by inelastic collisions, the positive ions produced are attracted towards C and give it their positive charge. This results in a so-called ionization current which is measurable by the rate at which the electrometer acquires the charge when the grounding key is open. Hence if V_a is gradually increased, keeping V_r always greater than and opposed to V_o, then the least value of V_o for which the cylinder. C gets any positive charge can be found.

This value of V_a certainly represents a critical potential, but it is not evident whether it is an ionization potential or merely a resonance potential. For, the charge received by C may not be necessarily due to positive ions; it may also be caused in the following manner:
When the energy of the bombarding electrons corresponds to the resonance potential, the molecules of the gas get simply excited without being ionized. These excited molecules returning to the normal state emit resonance radiations which proceeding in all directions fall on the inside walls of the cylinder C. If these radiations be of sufficiently short wavelength, active photoelectrically, they will expel photoelectrons from C, which in consequence will acquire a positive charge, and this also will be registered by-the electrometer. In other words, on account of the direction of the field between C and S the photoelectrons emitted by C will be directed towards S, which means a photoelectric current will flow in exactly the same direction as the true ionization current due to the ionization of the gas. In order to distinguish between the true ionization current and the false one, i.e., the photoelectric current, the wire gauze S' is used. Let V be the P.D. between C and S', which is weak and whose direction can be reversed. Now if the direction of V is such that *S' is positive with respect to C,* the photoelectrons ejected by C will reach S', while those given out by S' cannot reach C. In this case C will acquire a positive charge which will be registered by the electrometer. If V is reversed, so that *S' is negative with respect to C,* photoelectrons from C cannot reach S', while those from S' will reach C which therefore acquires a negative charge, that will deflect the electrometer needle in the opposite direction. Thus changing the direction of V reverses the direction of the photoelectric current. Supposing now that ionization really takes place in the gas, the positive ions produced are driven by the field through S' on to C and this will occur, whatever be the direction of V, since V is very small compared with V_r. Hence, the true ionization current cannot be reversed by reversing the direction of V and thus it can be differentiated from the false one. It is to be noted that although the photoelectric current might mask the real ionization effect, yet it enables one to determine the excitation potentials, since it will begin abruptly as soon as a resonance potential is reached.

163

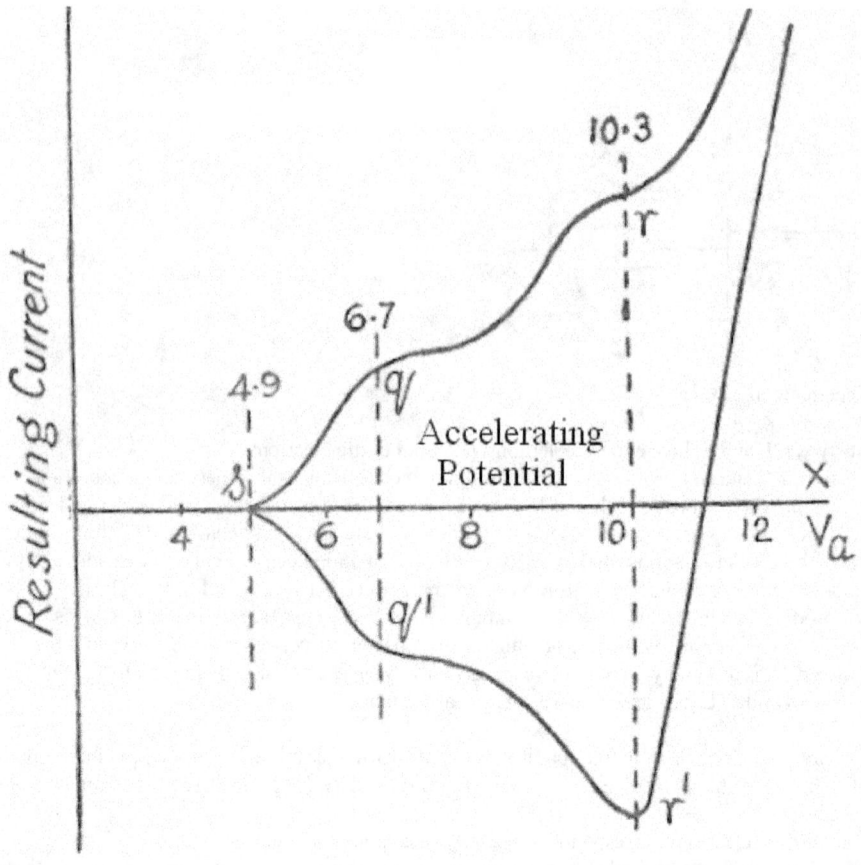

Davis and Goucher, using such a device, obtained for mercury vapor, curves of the kind shown in Fig. 207 by plotting the resulting currents to the cylinder C against the accelerating potentials V_a. The curve *s q r* above the X-axis is obtained when S' is positive, while the curve *s q' r'* below the X-axis when S' is negative with respect to C. No current is observed until V, reaches 4.9 volts. Beyond that value a current is recorded, which can be reversed by reversing V, indicating thereby that it is a photoelectric current. At a point r' on the lower curve, corresponding to $V_a = 10.3$ volts, a sudden reversal of the direction of the current takes place indicating that the cylinder is beginning to collect positive ions. The potential 10.3 volts, at which the true ionization current begins to flow, is the ionization potential of mercury. The beginning of the current at 4.9 volts, (*s*) is due to the photoelectric effect caused by the mercury resonance line 2536 emitted by the Hg atom excited to the first quantum state above the normal and returning to the normal. The experimental value is 12345/4.9 = 2520 A°. The agreement is as good as can be expected, since it is difficult to measure critical potentials to an accuracy greater than 0.1 volt. When the accelerating potential reaches 6.7 volts corresponding to *qq'* in the curve, there is a rapid increase in the photoelectric current, which is evidently due to the excitation of the actually existing mercury line 1849 A. Hence 4.9 and 6.7 volts are resonance potentials, while 10.3 volts the ionization potential of mercury.

The electrical method is subject to a systematic error on account of the initial velocity of emission of the electrons from the hot filament. The actual energy of the electrons will therefore be greater than that corresponding to the accelerating potential. The correction for this error is best determined by measuring known critical potentials with the apparatus used.

2. Spectral method. It consists in measuring the wavelengths of the spectral lines emitted by the excited atoms reverting to the normal state. Having obtained photographs of the spectra of the elements under test, the spectral lines corresponding to the different critical potentials are first identified, which is not easy in practice. Then their wavelengths are measured and using the relation $V = hc/e\lambda$ the critical potentials are readily calculated.

In these measurements one should remember that

the first line in the principal series of the *arc* spectrum corresponds to the *first resonance potential,* while the last, the so-called convergence line of the series, to the *first ionization potential,*

since the successive lines are supposed to correspond to states in which the optical electron is removed farther and farther from the nucleus, so that the last line, *i.e.,* the convergence limit should correspond to the case when the electron is entirely removed from the atom which is thereby ionized. The intervening lines in the spectrum will give the other resonance potentials.

In the *spark* spectrum one gets the spectral lines of the already ionized atom and the convergence limit gives the *second order ionization potential.*

For instance, in the *arc* spectrum of helium the singlet series of lines has a convergence wave number of 198,228 cm^{-1} or a wavelength of 504.2 A°, which gives the first ionization potential of 12345/504.2 = 24.46 volts.

In the *spark* spectrum, one gets the spectral lines of the ionized helium atom and the convergence of the first series is 438,800 cm^{-1} which gives the *second order ionization potential* of 54.14 volts.

This method, though an *indirect* one, is capable of giving more accurate results than the electrical method, since the measurement of wavelengths can be done with greater precision than that of accelerating potentials. It has been used with great success by Millikan, Bowen and others. The great difficulty in this method is to identify correctly the lines corresponding to the different resonance potentials, but it is best suited for determining the ionization potential, since the convergence line is easily identified and its wavelength accurately measured.

2. X-RAY SPECTRA

The emission spectrum of X-rays consists of two kinds, known as the *continuous X-ray spectrum* and *X-ray spectral lines,* while the absorption spectrum is characterized by *critical absorption limits.* Bohr's quantum picture of the atom gives a very satisfactory interpretation of these three phenomena met with in X-ray spectra.

A. CONTINUOUS X-RAY SPECTRUM

The existence of a continuous X-ray spectrum is shown by the fact that the intensity of the reflected X-ray beam from a Bragg spectrometer never falls to zero but only to a minimum value, as we have already remarked. The characteristic line spectrum is superposed on such a continuous background of varying intensity, which means that there is a series of uninterrupted wavelengths besides the definite wavelengths in an X-ray spectrum. The wavelengths of the continuous spectrum are found to be quite independent of the material of the target, which fact differentiates them from the characteristic X-rays, but are determined by the voltage across the X-ray tube.

Experimental study. Duane and Hunt were the first to study, in 1915, the main features of the continuous spectrum. Using a Coolidge tube with a given target and high excitation voltages of the order of several tens of thousands of volts and thus obtaining a heterogeneous X-ray beam, they measured the intensities of ionization corresponding to different wavelengths with a Bragg spectrometer for different excitation voltages. When the experimental data were plotted, *i.e.,* intensity of ionization I *vs.* wavelength λ, curves as shown in Fig. 208 were obtained.

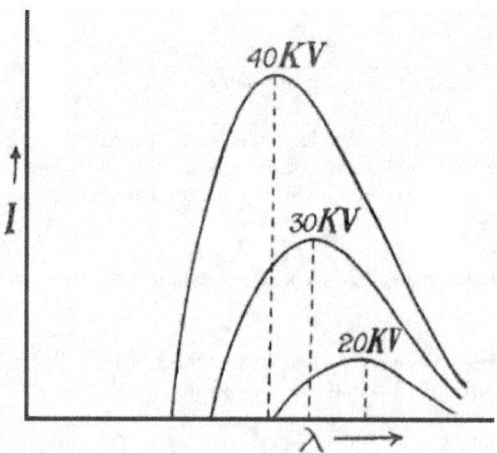

Fig. 208. Continuous X-ray spectrum.

It is seen that for each excitation voltage, starting at the long wavelength side the curve rises to a maximum and then drops rapidly to zero. This indicates that there is a well-defined minimum wavelength, no radiation occurring with a wavelength lower than that critical value. The greater the excitation voltage, the smaller is the value of the minimum wavelength limit. The intensity reaches a maximum at a definite wavelength and the position of the maximum depends on the excitation voltage, being displaced towards the short wavelength side as the voltage is increased.

Duane and Hunt discovered a very simple relation between the minimum wavelength limits and the excitation voltage, viz., λ_{min} is *inversely proportional to the applied voltage* V *or* v_{max} *is directly proportional to* V. If the limiting frequencies are plotted against the applied voltages, a straight line graph passing through the origin is obtained. This empirical *law of Duane and Hunt* is expressed analytically as

$$eV = h v_{max} = hc/\lambda_{min}$$

which is exactly like Einstein's law of equivalence. This law has been verified by different observers over a wide range of potential, from 5,000 to 100,000 volts.

It may be noted that the above relation affords one of the *most reliable and direct means of measuring Planck's constant 'h'*. From the experimental curves (Fig. 208) λ_{min} for a given excitation voltage V can be accurately measured. Then assuming the values of *e* and c, *h* can be calculated using the Duane and Hunt law.

The total intensity of the continuous spectrum given by the area enclosed by the experimental curve, corrected for several sources of error (such as absorption of the rays in the space between the target and ionization chamber, incomplete absorption of the rays in the ionization chamber, production of secondary electrons and characteristic radiations) is found to be very nearly proportional (i) to the square of the applied voltage for a given target and (ii) to the atomic number of the target when a constant potential is applied. The shift of the maximum intensity position towards the short wavelength side is analogous to the shift of maximum intensity in the black body radiation with rise of temperature.

Theoretical explanation. The continuous spectrum is believed to be emitted as a result of the deflection of the cathode ray electrons by the strong fields surrounding the nuclei of the atoms of the target, in contrast to the line spectrum which arises due to changes in the internal energy of the atoms of the target ionized by the cathode electrons. If an electron passes close to the positive nucleus, its path about the nucleus would be a hyperbola. The acceleration of such an electron, according to the classical theory, will cause the emission of a pulse of radiation.

The intensity distribution curve for such a pulse can be found by a Fourier analysis of the pulse in monochromatic wave trains. But the existence of a limiting minimum wavelength in the intensity distribution can hardly be explained by the classical theory.

The quantum theory has been more successful in the interpretation of this special sharp "cut off" feature of the continuous spectrum. For, considering the case as the inverse of the photoelectric effect, just as a radiation of frequency v causes the ejection of an electron of energy eV from the irradiated atom, so also an electron of energy eV hitting an atom excites a radiation of frequency v, both phenomena being governed by the same Einstein's quantum law of equivalence $hv_{max.} = eV$ which is the same as the empirical Duane- Hunt law. If the kinetic energy of the electron is increased, the limiting frequency also will be increased. In addition to this maximum frequency, we should also expect a whole spectrum of lower frequencies, because only very few of the electrons give up the whole of their energy in a single encounter with the nuclei. Most of them require repeated collisions before being brought to rest, leading thus to a continuous spectrum with, however, a sharp out off at a certain maximum frequency given by the relation $eV = hv_{max}$.

The phenomenon may be treated also as *a special case of Bohr's theory of spectral lines*. The collision of the cathode electron with the atom of the target may be considered as an initial state made up of a neutral atom *plus* a free electron with kinetic energy eV. If the atom has to remain neutral after the collision, the final state is a neutral atom plus a free electron with kinetic energy eV', where V' is less than V. Applying Bohr's frequency condition, radiation of any frequency may be emitted in the process, given by
$hv = e (V — V')$.
The maximum frequency will occur when $V' = 0$ in which case $hv_{max} = eV$ which is Duane-Hunt law.

The general relation also predicts continuous spectrum extending from $v = 0$ to $v_{max} = eV/h$ in complete accord with experiment.

B. CHARACTERISTIC X-RAY SPECTRA

When X-rays fall upon any material, other rays arise to which the name of *secondary radiation* has been given. This secondary radiation is complex in character and consists of three different types of rays, viz.,

(i) scattered X-rays,
(ii) corpuscular rays or a stream of fast moving electrons produced by photoelectric effect and
(iii) characteristic X-rays, which, as the name itself implies, depend on the nature of the substance from which they arise.

We have already dealt with the first two types and we are now concerned with the third.

There are **two methods of producing characteristic X-rays:**

(a) *Fluorescence or excitation by irradiation*:
When primary X-rays from a hard X-ray tube are made to fall on an element, the latter emits its characteristic X-rays, provided the primary are harder than the characteristic rays to be produced.

(b) *Excitation by electronic bombardment*:
The characteristic X-rays of a material can be excited also by using the material as the target in the X-ray tube and thus subjecting it to direct bombardment by the cathode electrons. In this case the velocity of the cathode ray must be sufficiently high to excite the particular characteristic X-rays of the material of the target.

Experimental study. The characteristic X-rays wore discovered and their properties first analyzed chiefly by means of their absorbability. Thus Barkla and Sadler, in 1908, examined the characteristic X-rays from different materials by measuring their mass absorption coefficients in aluminum. From the systematic variation of the mass absorption coefficients of the X-rays originating from the different substances, they were able not only to differentiate the characteristic X-rays from the two other types of rays, but also to establish their dependence on the atomic weights of the elements from which they arise, their hardness increasing with increasing atomic weight.

They also found that the characteristic X-rays from a substance fell of the element emitting into two, more or less homogeneous, groups which they designated as the K and L series.

With the advent of Bragg's X-ray spectrometer, not only the existence of series in characteristic X-rays was confirmed, but also more series than the two of Barkla were discovered. It was also found that each series consisted of a group of lines. Bragg, working with his ionization spectrometer and using different anticathodes, such as platinum, osmium, palladium, rhodium, etc., in the X-ray tube, established beyond doubt that the peaks in the ionization *vs.* glancing angle curves represented spectral lines, the wavelengths of which were characteristic of the target emitting the rays. He further showed that these observed wavelengths were components of the K and L series, so that Barkla's "homogeneous rays" were in reality a mixture of spectral lines forming a narrow band of wavelengths.

 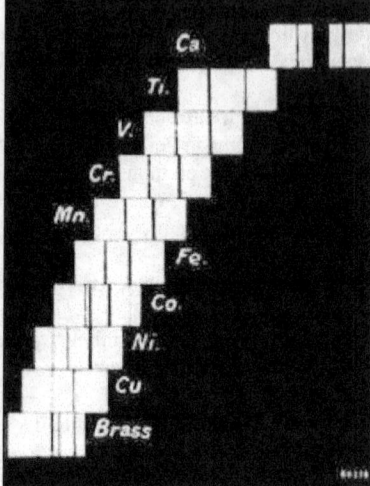

Photograph of lines of the K series from different elements (Moseley)

But a systematic and comprehensive study of the characteristic X-ray spectra was first made by Moseley in 1913-1914. He used the "rotating crystal photographic method" with a crystal of potassium ferrocyanide for diffracting the X-rays. The chamber containing the photographic plate was exhausted to reduce absorption of the soft X-rays in the air. His researches covered a range of wavelengths from 0.4 to 8A°, employing thirty-eight different elements, from aluminum to gold, as targets in the X-ray tube. The photographic plate when developed showed traces of dark lines of varying intensities, each one corresponding to a spectral line characteristic of the target. Moseley was able to draw the following important conclusions from the detailed analysis of the spectral lines thus obtained:

(a) The spectral lines could be classified into two distinct groups, one of short wavelengths which was identified with the K series by means of the value of absorption coefficient in aluminum and the other of comparatively longer wavelengths similarly identified with the L series.

(b) Unlike optical spectra, the X-ray characteristic spectra were much simpler consisting of comparatively fewer lines. They were of the same type for all elements except that the scale of wavelengths was changed, corresponding lines occurring, in general, at shorter wavelengths, the greater the atomic weight of the element in which the lines photo.

(c) A very simple relationship existed of a particular line and the atomic number Z pf the element emitting that line. On plotting the square root of the frequencies of a given line against the atomic numbers of the elements emitting that line, a straight line was obtained, as shown in Fig, 209.

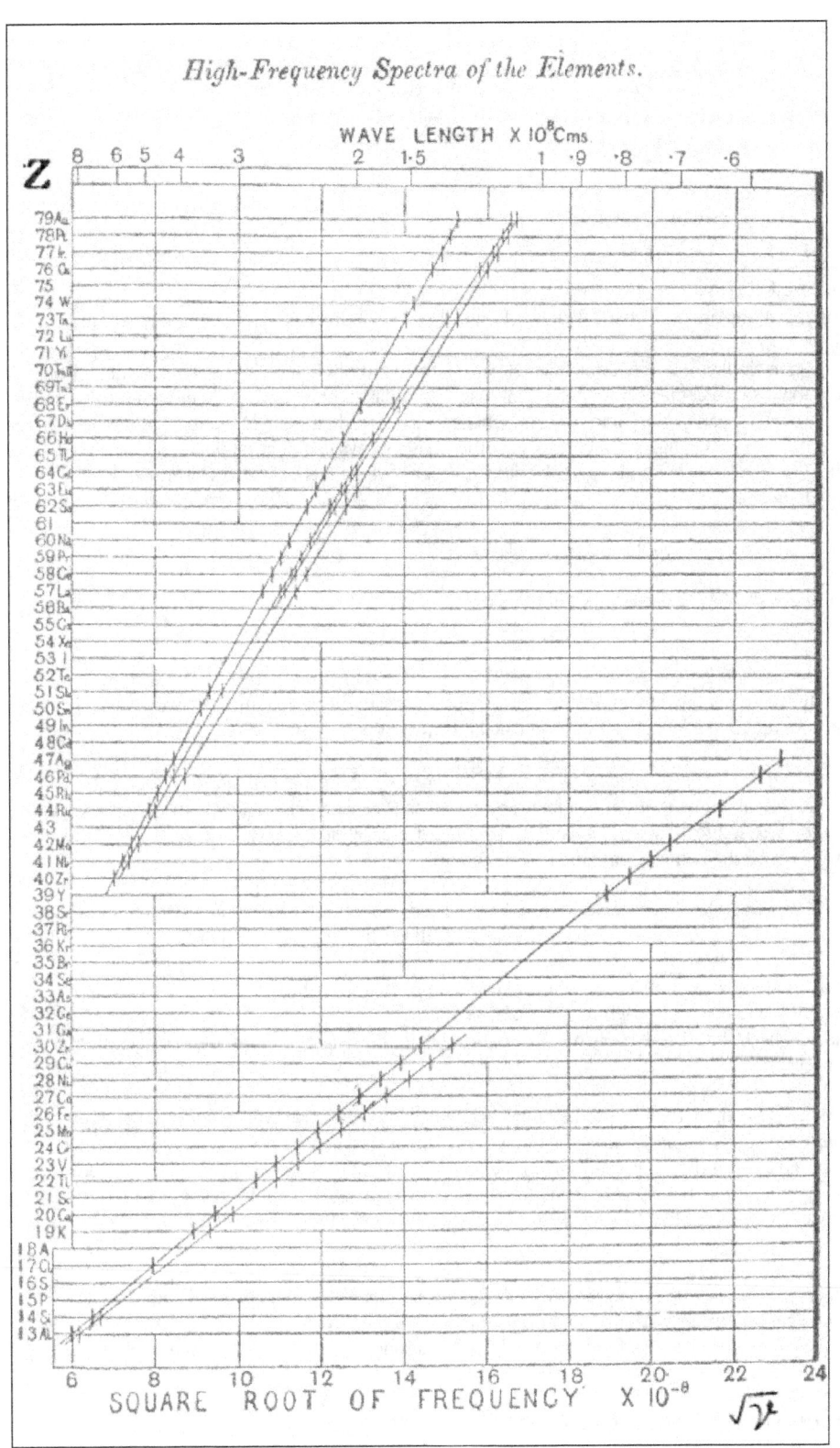

High-Frequency Spectra of the Elements.

Fig. 209. Moseley diagram.

The same linear relation was found to hold good for any line in any series.

The equation of any one of the straight lines on the Moseley diagram is given by

$$\nu = a\,(Z - b)^2$$

where a and b are constants, characteristic of the line under consideration.

Moseley's researches showed also that

the determining factor in the arrangement of elements in the Periodic table should be the atomic number and not the atomic weight,

thus removing the discrepancy in the order of certain elements from the point of view of their atomic weights. For example, we find that argon ($_{18}Ar^{40}$) comes before potassium ($_{19}K^{39}$), similarly cobalt ($_{27}Co^{58.9}$) before nickel ($_{28}Ni^{58.7}$) etc., thus in the reverse order as regards their atomic weights, but correct according to their chemical properties. On the basis of the atomic number, we see that the arrangement is justified. This shows that as regards the properties of the atom the atomic number is much more important than the atomic weight, which fact is further confirmed by the isotopic constitution of elements.

X-ray line spectrum analysis also perfected the Periodic table by

(i) *the discovery* of *new elements, e.g.,* hafnium (72), illinium (61), masurium (43), rhenium (75), etc., indicated by gaps in the regular progression in the character of X-ray spectra with increase in atomic number, and

(ii) *the determination of atomic numbers of rare-earths,* thereby fixing their positions in the table. In fact, the study of the characteristic X-ray spectra from the time of Moseley provided a comparatively rapid means for the analysis of substances which otherwise required a lengthy and complicated chemical process.

The number of lines present in the X-ray spectrum depends both on the element and the excitation voltage.

The heaviest elements, such as uranium, thorium, emit a pretty complete spectrum, where all the series K, L, M, N, are present. As we proceed towards lighter elements, the number of lines decreases and the series disappear one after another, until finally with very light elements only the K series remains.

As regards the excitation voltage, it has been found, as is to be expected, that higher voltages are required to excite the series of greater frequencies. Further, at a definite excitation voltage all the lines of the corresponding series or sub-series appear together. In this way it has been found that

there is only one excitation potential for the K-series, three excitation potentials for the three sub-groups of the L series, five corresponding to the five sub-groups of the M series, etc.

The excitation potential has been shown to be very nearly proportional to the square of the atomic number of the element emitting the spectrum.

The *X-ray spectra* have been found, to a first approximation, *to be independent of the isotopic constitution of the emitter as well as its state of chemical combination.* Several workers, notably Duane and Shimizu, Siegbahn and Stenstroem and Cooksey have carefully compared some of the emission lines of ordinary lead and of specimens of radioactive lead. They have not been able to observe any measurable difference, which indicates that the isotope constitution of the elements, chiefly the heavy ones like lead, does not influence the X-ray spectra, unlike in the case of optical spectra.

The X-ray spectrum of an element is, to a first approximation, independent of its state of chemical combination. There is evidence, however, of a certain influence of the chemical combination on the X-ray spectrum, manifested by a very small displacement of the spectrum, chiefly in the case of light elements and in particular with sulphur, phosphorus and chlorine.

SOME PECULIARITIES OF THE CHARACTERISTIC X-RAY SPECTRA

1. Fine structure. Later investigations with more refined apparatus than that used by Moseley have extended very much our knowledge of the characteristic X-ray spectra and at present it is believed that practically the entire spectra of most elements have been mapped out. In addition to the K and L series, others such as M, N and O have been found. Of these the K series alone is single, while the L series consists of three sub-groups, M five sub-groups and N seven.

The K series is composed of four lines, the L series more than thirty lines, etc. They are designated by the symbols $K_{\alpha 1}$, $K_{\alpha 2}$, $K_{\beta 1}$, $K_{\beta 2}$, $L_{\alpha 1}$, $L_{\alpha 2}$, $L_{\beta 1}$, $L_{\beta 2}$, etc. and are known as the fine structure of X-ray spectra.

2. Satellites. With improved technique in X-ray spectrography, many more lines, other than those mentioned above, were discovered. Most of them were rather faint and were usually found close to and on the short wavelength side of the more intense lines; hence they were called "satellites" or "second order" lines; as they could not be readily fitted in the conventional energy level diagram, unlike the "parent" lines, they were called also "non-diagram" lines.

Siegbahn and Stenstroem observed these satellites in the K spectra of elements from Cr (24) to Ge (32), while Coster, Thoraeus and Richtmyer in the L spectra of elements from Cu (29) to Sb (51) and Hjalmar, Hindberg.and Hirsch in the M series of elements from Yb (70) to U (92). Most first order lines are found with such satellite structure of varying intensity, which varies also from line to line as well as from element to element. The total number of the satellite lines now known far exceeds the total number of the parent lines. On account of the low intensity of the satellites, reliable information on their various characteristics, such as their wavelength positions, their intensity variation with the atomic number of the radiating material or the excitation potential etc., is difficult to obtain, Careful and patient researches by different workers have, however; furnished the following data:

(i) The excitation potential of certain satellites is definitely greater than that of the corresponding parent line, as was shown by the experiments conducted by Baccklin, Druyvesteyn and others.

(ii) Siegbahn has shown that as many as five satellites can be excited by the side of the normal $L_{\alpha 1}$ line of Mo by progressively raising the excitation potential. Their successive appearance might well correspond to the atom, once, twice, thrice, etc., ionized, since the intensity of ionization can be increased by increasing the excitation potential, in analogy with the higher orders of optical "spark" spectra. Hence the satellites are sometimes called the X-ray *spark lines.*

(iii) In the case of the spectrum of Fe it has been found possible to avoid the intense ionization of the atoms and thereby prevent the appearance of satellites by causing the emission of the lines by the fluorescence method.

(iv) While the intensity of the K satellites is found to decrease in a continuous manner with increasing atomic number of the radiating material, the intensity of the L_{α} satellites decreases rather abruptly as the atomic number increases from 47 to 50 and to increase again rather abruptly at about 75; between atomic numbers 50 and 75, the L_{α} satellites are practically unobservable.

3. Auger effect. One of the methods of producing the characteristic X-ray spectra of an element is to irradiate the element with primary X-rays that are harder than the characteristic rays to be produced. This is known as *fluorescence* which is preceded by the photoelectric process, *i.e.,* the ejection of electrons from the atoms of the elements.

Prof. **Pierre Victor Auger** (May 14, 1899 – December 24, 1993), France

Under certain circumstances, however, the photoelectric effect alone takes place and that in a complex manner, without the accompanying fluorescent radiations. This is known as *Auger effect,* since although the phenomenon was first observed by Wilson in 1923, yet it was Auger, who, in 1925, clearly demonstrated its existence. Auger took expansion photographs with a Wilson's cloud chamber filled with argon, using X-rays whose frequencies were much higher than necessary to eject the K-electron of argon.

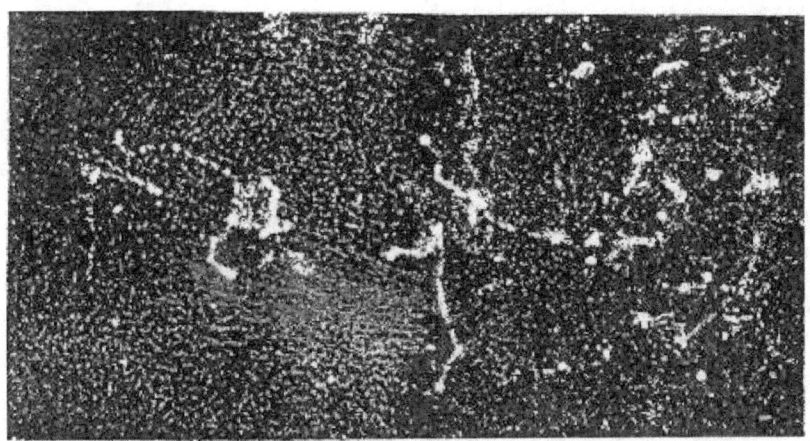

Auger effect (cloud chamber photograph)

In the photographs he found, along with the expected tracks of the K-photoelectrons, other, very short, ball-like tracks, occurring at the starting points of the photoelectron tracks as seen in the photo above. On the assumption that these short tracks were due to low speed electrons, he surmised that

there should be a *simultaneous ejection of two or more electrons from the same atom, i.e., a compound photoelectric effect,* different from the normal one caused by the incident X-ray beam at the various shells of the atom

Pursuing his researches on the short tracks, produced by, what we shall henceforth call as 'Auger electrons' in order to distinguish them from the normal photoelectrons, Auger was able to gather the following characteristics of the phenomenon:

(i) The photoelectron and the accompanying Auger electron arise at the same point.

(ii) The length of the Auger electron track is independent of the frequency of the incident X-rays, whereas the length of the photoelectron track increases with the frequency.

(iii) The direction of ejection of the Auger electron is independent of that of the photoelectron.

(iv) Not all photoelectron tracks show an Auger electron track at their source.

(v) The length of the Auger electron track increases as the atomic number of the irradiated gas increases.

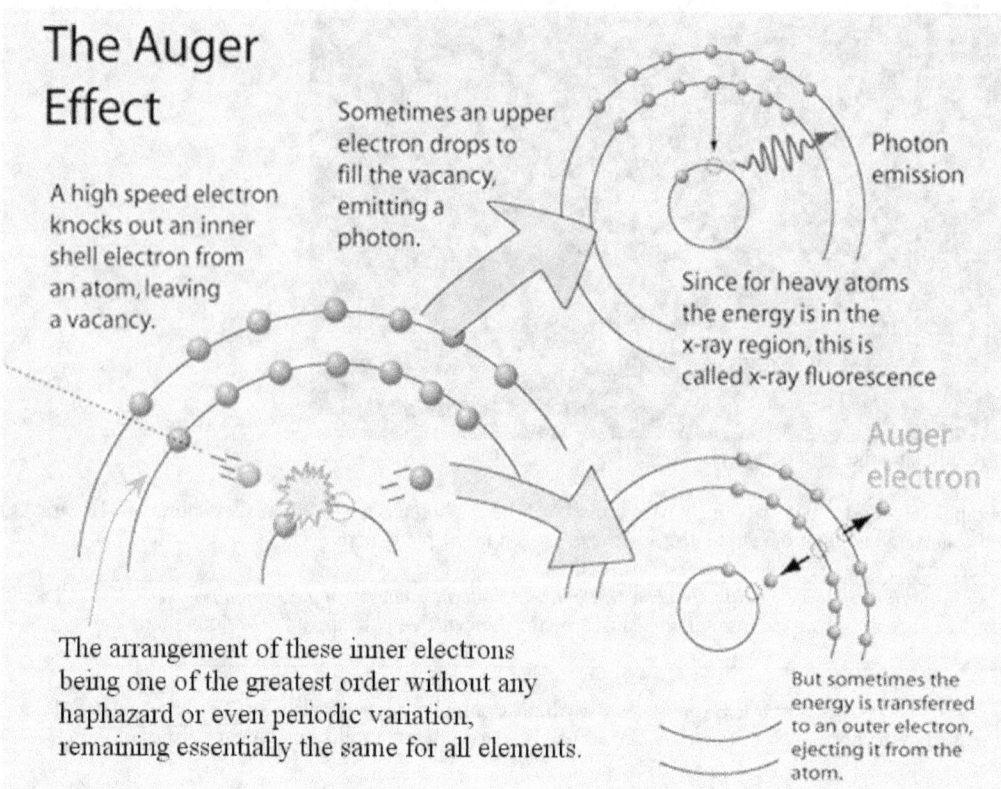

The Auger Effect

A high speed electron knocks out an inner shell electron from an atom, leaving a vacancy.

Sometimes an upper electron drops to fill the vacancy, emitting a photon.

Photon emission

Since for heavy atoms the energy is in the x-ray region, this is called x-ray fluorescence

Auger electron

The arrangement of these inner electrons being one of the greatest order without any haphazard or even periodic variation, remaining essentially the same for all elements.

But sometimes the energy is transferred to an outer electron, ejecting it from the atom.

Any plausible theory of the characteristic X-ray spectra must be able to account for all these experimentally observed peculiarities also.

Theoretical explanation. Bohr's theory offers a cogent explanation of the main features of the characteristic X-ray spectra.

Considering the atom of an element of atomic number Z, it is made up of a central positive nucleus of charge Ze and a swarm of Z electrons surrounding the nucleus and arranged in a definite manner, *i.e.*, different groups or shells. There are reasons to believe, as we shall see later, that the innermost shell corresponding to the first quantum orbit (n = 1) can contain only two electrons, the next (n = 2) 8 electrons, the third (n = 3) 18 electrons and so on. These successive shells have been designated by the letters K, L, M, etc.

In the normal state of the atom, the Z electrons are, in general, so arranged that the innermost shells are all full with their quota of electrons, while at the outer incompletely filled shells there are a few stray electrons.

The electrons in the innermost shell are very strongly bound to the nucleus so that extraction of these electrons will involve a large amount of work which will be greater the nearer the shell is to the nucleus—hence maximum for the K shell, less for the L shell, still less for the M shell, and so on. Further, the arrangement of the electrons in the innermost shells will be regular and essentially the same for all elements, except the very light ones. The only varying factor with the increase of atomic number is the increasing pull of the nucleus due to its charge Ze on these inner groups of electrons. The electrons in the partially filled outer shells are characterized, on the other hand, by a comparatively small amount of work of extraction; they are differently constituted in different elements; they are more exposed than the inner electrons to external disturbances, such as collisions, chemical force, etc., which can easily displace them to still outer virtual levels unoccupied in the normal state of the atom (excitation) or even remove them completely from the atom (ionization).

Since high quantum energies are required for the excitation of X-ray spectra, which are further found to be

174

independent of the state of chemical combination of the element, it can be inferred that the X-rays arise from the activity of the electrons in the innermost shells.

The arrangement of these electrons being one of the greatest order without any haphazard or even periodic variation, remaining essentially the same for all elements, unlike in the case of the electrons in the outermost shells, the comparative simplicity and similarity of the X-ray spectra of all elements, without any of the periodicity characteristic of optical spectra, are readily understood.

At the same time, the different stability of the inner groups of electrons arising from the increasing nuclear pull as the atomic number increases, explains the steady shift in the X-ray spectrum as we pass from one element to another. The same variation of nuclear pull accounts also for the experimentally observed increase of the excitation voltage as the square of the atomic number.

Supposing that an atom in the target of an X-ray tube is bombarded by a high velocity electron which possesses enough energy to penetrate into the atom and knock out one of the electrons in the innermost shells, say one of the two electrons in the K shell, the vacant space in the K shell will soon be filled by another electron from an outer shell, say for instance, by an electron from the next L shell. In such a case the atom will undergo a transition from the K to the L state, the excess of energy being emitted as a quantum of energy. The emission of K_a line is ascribed to an atomic process of this kind. The quantum $h\nu_a$ corresponding to the K_a line is given by Bohr's frequency condition:

$$h\nu_a = W_K — W_L$$

where W_K and W_L are the energies of the atom in the K and L shells respectively.

In a similar manner, the K_β line originates when an electron from the M shell goes to fill the vacancy in the K shell, and its frequency is given by

$$h\nu_\beta = W_K — W_M$$

The picture may be extended to account for or predict any one of the characteristic emission lines in any series. For instance, when an electron from the L shell is knocked out of the atom or gone over to occupy a vacant space in the K shell, the vacancy thus created in the L shell may be filled by an electron from the higher M, N, etc. shells, thus giving rise to the lines of the L series. In an actual target, many atoms are simultaneously involved, giving rise to the observed lines grouped into the different series. The same series of events will happen, even if the process is started by a primary X-ray of sufficiently high frequency falling on the emitting material.

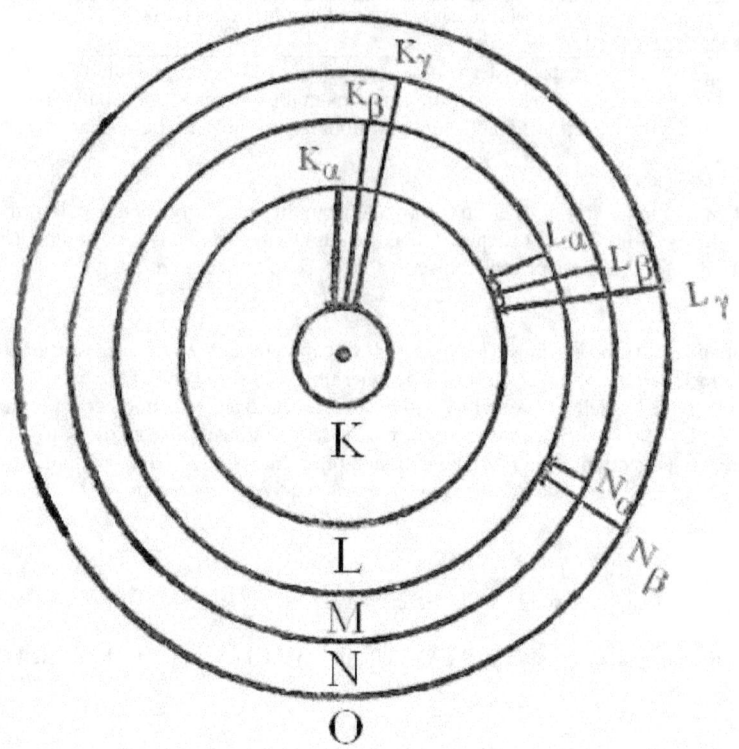

Fig. 210. Kossel's diagram.

Kossel's diagram. The origin of the X-ray lines may be represented schematically by a diagram due to Kossel (Fig. 210). It is seen that the lines of a given series, say K, are formed by the transition of an electron to the K shell from a number of other outer shells. It may be noted that the series in the optical spectra observed with the lightest elements (H, He) are merely the X-rays of these elements. In the case of hydrogen, the Lyman, Balmer and Paschen series can thus be assimilated to the K, L and M series respectively. In a general way the X-ray spectra of the light elements tend to resemble optical spectra by becoming more complex and losing their atomic character. For the heaviest elements the K and L series belong, strictly speaking, to the X-ray region, while the outer P, Q series to the ultra-violet region. As we proceed from light to heavy elements, the number of electrons increases and in consequence the number of shells complete with their quota of electrons increases also, which will make the number of lines in a series as well as that of the series increase, as actually observed.

Considering the high terms of a series, the departure levels of these will belong to the outer parts of the atom, subject to external influence. The lines corresponding to such transitions will share with the lines of optical spectra properties which are not possessed by the lines arising from the deeper levels that are practically independent of external influence.

X-ray energy level diagram. Like optical spectra X-ray spectra also can be conveniently interpreted in terms of energy levels, as shown in Fig. 211.

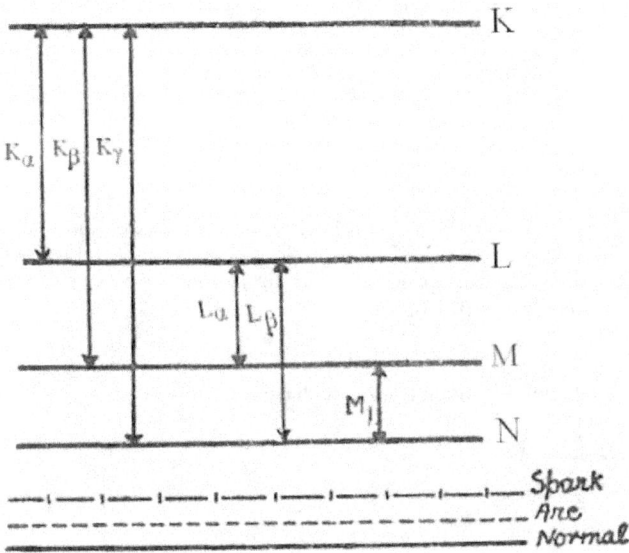

Fig. 211. X-ray energy level diagram.

Lines indicating transitions between X-ray levels on such diagrams are often drawn with an arrow-head at each end, to indicate that as an electron moves upward the atom moves downward. Thus in the emission of K_α line, an electron goes from the L to the K shell, while the atom drops from the higher K level to the lower L level; the K_β line arises from the transition of the atom from the K to the M level; the loss of energy in this case being greater, the K_β line is of higher frequency or shorter wavelength than the K_α line. The transition of the atom from K to the normal state, representing capture of a free electron of negligible kinetic energy from outside the atom into the vacancy of the K shell, would cause the emission of the series limit of the K lines. In a similar way the lines belonging to L, M, N series may occur.

It may be noted that just above the **normal level** lies a whole set of optical levels connected with the emission of lines in the *arc* **spectrum**, and above this set another belonging, like the X-ray levels, to the ionized atom and associated with the *spark* **spectrum** in the visible and ultra-violet regions.

Ordinary X-ray spectra, in a sense, constitute an extension of the spark spectrum of higher levels of energy, resulting from an electronic vacancy in the interior of the ionized atom rather than from the excitation of one of its outer electrons.

Quantitative confirmation of Bohr's theory. It has been found that there is even a fair quantitative agreement between Bohr's theory and the experimental findings in X-ray spectra, chiefly with reference to two points, viz., excitation potential and Moseley's law.

Excitation, potential
The necessary condition for the emission of an X.-ray line, say K_α, is that a vacancy must be created in the K shell by the removal of one of the two K electrons. This electron cannot stop in the L shell because the latter is already full, nor in any other full shell for the same reason; it has to go to the outer incomplete shell or be removed completely from the atom. This means that the energy involved in the ejection of the K electron must be greater than that corresponding to any of the completed shells, roughly equal to the energy required to ionize the atom with respect to the K electron. Since this energy is supplied by the excitation potential, it is readily understood why in the excitation of an X-ray line, the potential required must be greater than that which corresponds to the wavelength of the line, in fact, greater than that corresponds to any of the lines in a given series, as experimentally found.

Considering, for example, an X-ray tube with molybdenum target, it is experimentally found that the excitation potential for the K series is 20,000 volts, which we may regard, for reasons given above, as the ionization potential of the molybdenum atom for the K electron. Since this ionization potential measures the binding energy of the K electron, the latter is equal to 20,000 volts.

Now, according to Bohr's theory, the binding energy of the electron in a hydrogen-like atom, *i.e.,* consisting of a nucleus of charge Ze and a single electron, is given by

$V_n = 13.6 \times Z^2/n^2$ volts.

Hence the excitation potential is directly proportional to the square of the atomic number, as observed experimentally. Since for the K electron n = 1, and for molybdenum Z = 42, $V_K = 13.6 \times (42)^2 = 23,980$ volts, which agrees at least as regards the order of magnitude with the experimental value. The discrepancy of nearly 20% is due to the fact that the molybdenum atom containing 42 electrons is decidedly not hydrogen-like.

Moseley's law
Taking Bohr's expression for the frequency of a spectral line and putting E = Ze, we have

$$\nu = R \cdot Z^2 \left(\frac{1}{n_1{}^2} - \frac{1}{n_2{}^2} \right)$$

where R, the Rydberg constant, is expressed in frequency units.

Considering the simplest possible values $n_1 = 1$, $n_2 = 2$,

$\nu = (3/4) R \cdot Z^2$

Hence $\nu \propto Z^2$, as Moseley experimentally found.

For the K_α line of Cu, Moseley found the wavelength to be 1.54 A°. Hence the frequency of that line is

$\nu = c/\lambda = 3 \times 10^{10} / (1.54 \times 10^{-8}) = 1.95 \; 10^{18}$.

From Bohr's relation, substituting for R and Z,
(R =109,700 cm^{-1} = 109,700 x 3 x 10^{10} in frequency units, Z = 29)

$\nu = (3/4) \times 1.097 \times 10^5 \times 3 \times 10^{10} \times (29)^2 = 2.07 \times 10^{18}$

The agreement is surprisingly good, which shows that the K series of the X-ray spectrum corresponds to the Lyman series of hydrogen.

Following the analogy, if we put $n_1 = 2$, and $n_2 = 3$,

$\nu = (5/36) R \cdot Z^2$.

For the L_α line of platinum (Z =78), Moseley found the wavelength to be 1.32 A°, which gives for its frequency

$\nu = 3 \times 10^{10} / (1.32 \times 10^{-8}) = = 2\;27 \times 10^{18}$

From Bohr's relation,

$\nu = (5/36) \times 1.097 \times 10^5 \times 3 \times 10^{10} \times (78)^2 = 2.78 \times 10^{18}$

The agreement in this case is not quite so good, which is due to the fact that there is a correction to be made for the so-called "screening effect", i.e., the effect on the nuclear charge due to the presence of the electrons;

Bohr's theory is rigorously true only in the case of a single electron system. In Moseley's relation $\nu = a (Z -b)^2$, the correction for the screening effect makes its appearance in the constant b which he found to be equal to 1 for the K series and 7.4 for the L series.

In the case of atoms where there are several electrons, the electric force acting on the electron which comes to fill the vacancy in the K shell in the emission of a K line will not be Ze but $(Z-1)e$ on account of the single electron still remaining in the K shell.

The L shell being only the second nearest to the nucleus, there will be greater number of electrons occupying the central shells and, in consequence, the value of b for the L series will increase and the apparent nuclear charge will be still smaller than $(Z-1)e$.

Apart from this screening effect, the agreement is sufficiently near to recognize the similarity between the L series of X-rays and the Balmer series of hydrogen, in the numerical example considered above.

Moseley also found experimentally the valise of the constant a appearing in his relation to be 82,303 cm^{-1} for the K_α line. According to Bohr's theory, for the corresponding line

$$a = (3/4)\, R = (3/4) \times 1.097 \times 10^5 = 82{,}275 \text{ cm}^{-1},$$

a striking agreement which is a further confirmation of Bohr's theory.

Of the other peculiarities of the X-ray emission spectra, *viz.*, the fine structure, the satellites and the Auger effect, the first one can be explained adequately only on the basis of the more perfect vector atom model, while the other two can be interpreted fairly well with a slight modification of Bohr's theory. Hence we shall consider here the theoretical explanations offered for the satellites and Auger effect, postponing the fine structure to a later section.

Explanation of satellites. Wentzel, in 1921, and Druyvesteyn in 1927, have proposed a theory of satellites on the principle of *multiple ionization of the inner electrons.* It is assumed that the initial and final states of the atom, which give rise to the satellites, are doubly ionized. Such states, called the KK, KL, KM, LL, LM, etc. atomic states, are possible, since it is experimentally found that, for instance, in the case of the K satellites, the excitation energy is bound to be equal to that required to eject a K electron and in addition an L electron. Now an atom in that doubly ionized "KL state" may undergo a radiative transition into any one of a number of other possible states of double ionization *e.g.*, KL → KM (an electron falling from the M shell into the L shell), or KL→ LL (an electron falling from the L into the K shell). Estimates of atomic energy indicate that the loss of energy in KL → LL transition should be slightly greater than that in the normal transition K →.L which gives rise to the K lines; hence, the former transition should give rise to satellites close to and on the short wavelength side of the K_α lines. In a similar manner, the transition KL → LM should give rise to satellites on the short wavelength side of the K_α lines. Moreover, the vacancies may have different quantum characteristics (different *l* values) corresponding to the various sub-shells. Thus we may explain the multiplicity of the observed satellite structure.

Some satellites are believed to originate in transitions between states of *triply ionized atoms;* these constitute *satellites of second and higher orders.* Representation of the second or higher order lines in X-ray spectra by energy-level diagrams is extremely difficult and has not been achieved so far.

As far as it has been possible to make theoretical calculations, the predictions of wave mechanics lend support to this explanation for the origin of satellites. It must be noted, however, that the theory has hardly advanced beyond the qualitative stage and hence it is possible that at least some of the satellite lines originate due to same other atomic process also.

Explanation of Auger effect. When a K photoelectron is ejected from an atom, the vacancy in the K shell may be filled by one of the L electrons; the energy thus set free may be transformed into a light quantum and the K_α line emitted. But it is not necessary that this should happen. Instead, the quantum of energy corresponding to the K_α, line may be absorbed or converted by the L shell of the atom into energy necessary for the ejection of an L electron *plus* the kinetic energy of that electron. In such a case, two electrons are expelled simultaneously from the same atom in a single elementary action which may be termed *auto-ionization* or *nonradiative transition* for obvious reasons. One of the electrons is the normal K photoelectron and the other is the Auger electron from the **L** shell.

If the kinetic energy of the Auger electron is 2m.v 2 , neglecting relativity corrections,

$(1/2)\ mv^2 = h\nu_\alpha - (W_L' - W_L)$ --(1)

where

$h\nu_\alpha$ is the energy corresponding to the K_α line,
W_L' **is** the energy of the atom with two L electrons missing,
W_L is the energy of the atom in the L shell, so that
$(W_L' - W_L)$ is the work required to remove the second L electron.

Now,

$h\nu_\alpha = (W_K - W_L)$ --(2)

Therefore

$(1/2)mv^2 = (W_K - W_L') = (W_K - 2W_L)$ --(3)

if we assume that the energy required to remove the second L electron is nearly the same as that required to remove the first. This relation (3) shows

that the energy of the Auger electron is independent of the wavelength of the incident X-rays.

It is also possible that the quantum corresponding to the K_α line may eject an electron from the M shell, leaving the L shell undisturbed. In such a case, energy of the Auger electron would be

$(1/2)mv^2 = h\nu_\alpha - (W_{KM} - W_L) = (W_K - W_{LM})$ --(4)

where W_{LM} is the energy of the atom with one electron missing in the L shell and another in the M shell.

If we follow the process represented by relation (3) a step further, it may happen that the two vacancies produced in the L shell are filled by the transfer of two M electrons, and that the quanta thus liberated are absorbed in the M shell, producing two more Auger electrons from the same atom. These are called *secondary* Auger electrons whose energies will be less than that of the *primary* Auger electron. Auger observed such secondary electrons from Br_{35} and gases of higher atomic number.

In these heavy gases, even *tertiary* Auger electrons, ejected from the N shells by radiation resulting from the filling of vacancies in M shells, were occasionally observed, The analysis of the directions of ejection of these tertiaries establishes the independence of the direction of the Auger electron from that of the primary photoelectron.

The Auger effect implies that an atom in a state of single ionization may automatically become doubly ionized. Hence it offers a means of studying the intensity of the satellite lines which originate from transitions between states of double ionization in the atom. As **a** matter of fact, it has been possible to explain the anomaly met with satellites in connection with their variation of intensity with atomic numbers on the basis of the Auger effect.

The Auger effect has been used also in the analysis of *the fluorescence yield* of X-rays in the different shells of a number of elements of different atomic numbers. It is evident that the greater the number of non-radiative Auger transitions in an element the less will be the yield of fluorescent radiations from it.

Analysis of the Auger effect renders the discontinuous shell structure of the atomic edifice almost directly perceptible and even gives, by comparison of the tracks of the electrons produced, an estimate of the respective energies of the shells.

The Auger effect seems to fit in well with the theory of Klein and Rosseland who were able to establish the existence of a "quantum induction" which allows non-radiative transformations with the direct conversion of excitation energy into kinetic energy, analogous to collisions of the second kind.

It may be noted that non-radiative transition processes of the Auger type play a part of great significance in the interpretation of phenomena in such diverse fields as X-ray spectra, internal conversion of γ-rays, internal pair production, atomic and molecular spectra and the capture of mesons by nuclei.

C. X-RAY ABSORPTION SPECTRA

The study of the absorption of X-rays of different wavelengths in different elements has shown that the X-ray phenomena are atomic in character and that the atom has a discontinuous structure endowed with discrete energy levels, in beautiful confirmation of the Bohr atom model.

Experimental study. Investigation of the absorption spectra of X-rays has been made with the X-ray crystal spectrometer both by the ionization chamber method and by the photographic method.

For a particular setting of the crystal and for a monochromatic beam of a definite wavelength λ given by Bragg's law $n\lambda = 2d \sin \theta$, the ionization current is measured, first without the absorber, and then with the absorber of known density ρ and thickness x, placed in the path of the beam either before or after the crystal. These measurements give the quantities I_0 and I in the relation

$$I = I_0 \, e^{-(\mu/\rho)\rho x}$$

from which the mass absorption coefficient μ/ρ may be computed. By a careful choice of the excitation potential the second order reflections can be eliminated and thus it is possible to obtain values of μ/ρ for different wavelengths.

In the photographic method, the blackening on the plate is a measure of the intensity to be measured. The variation of μ/ρ for different substances can be determined by using them as absorbers for the same monochromatic beam.

If the values of μ/ρ in a *given absorber*, say Pb, be plotted against the wavelength λ for a range, say 0.1 to 1.2 A°, a complicated curve is obtained as shown in Fig. 212.

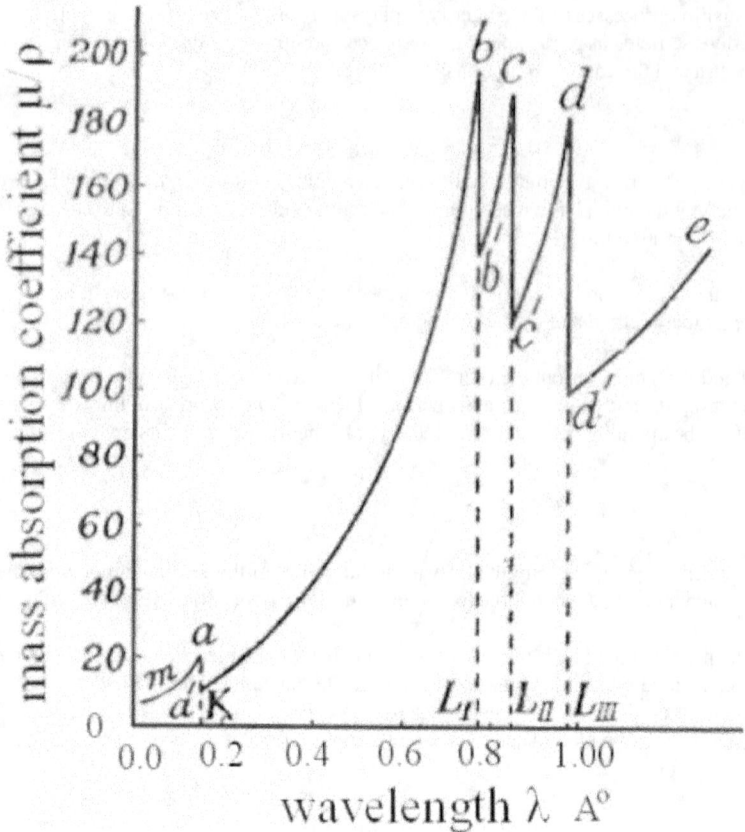

Fig. 212. Variation of absorption coefficient with wavelength

Beginning at a point m, μ/ρ increases - rapidly with increasing λ until the point a is reached, at which the value of μ/ρ suddenly drops to a'.

With further increase in wavelength, however, μ/ρ again increases rapidly until a point b is reached at which there occur three consecutive drops at close range, b to b', c to c' and d to d'.

With still longer wavelengths the value of μ/ρ again increases rapidly. These sudden drops in the absorption curve are known as *critical absorption limits* or *critical absorption wavelengths*. The first drop (aa') occurs at λ = 0.14 A° which is just smaller than the shortest wavelength of the K lines in the emission spectrum of the substance. The next three consecutive drops (bb', cc' and dd') occur at λ = 0.78, 0.813, 0.95 A°, each of them just shorter than the shortest wavelength in each of the three sub-groups of the L emission spectrum.

The first discontinuity is the K *absorption limit,* while the next three are the L *absorption limits.*

If the absorption carve could be followed beyond e, it will be found that in the region of 3.2 and 5A°, a group of five "breaks" occurs representing the five M absorption limits and still further at about λ = 14 A°, a group of seven N limits. The sharp discontinuities at certain well-defined wavelengths indicate that at each of these wavelengths a new absorption phenomenon occurs, characterized by the appearance of a new spectral series.

The shape of the absorption curve between the various discontinuities is such as could be approximately represented by the relation:

$$\mu/\rho = C\lambda^3 + \sigma/\rho,$$

where C is a constant which assumes different values along the various branches of the curve and σ/ρ the scattering coefficient which, if assumed constant, *i.e.,* independent of the wavelength, according to the classical theory, shows that the true photoelectric absorption coefficient τ/ρ varies as λ^3 . The general trend of the curve shows that absorption diminishes rapidly as λ decreases, which expresses the well-known fact that X-rays become more penetrating and less absorbed as their wavelength decreases.

If λ be kept constant and Z the atomic number varied by using different absorbers, the law governing the absorption of X-rays by different substances can be studied. When μ/ρ is plotted against Z from experimental data, curves very similar to those of μ/ρ *vs.* λ are obtained, which at once indicates that μ/ρ must be approximately symmetrical both in Z and λ. The discontinuities occur at longer wavelengths the lower the atomic number of the absorber.

The shape of the curve between the discontinuities for reasons given above, may be represented approximately by the relation

$$\mu/\rho = KZ^3 + \sigma/\rho$$

where K is a constant which remains fixed along a given branch, but suddenly changes in value at the discontinuities. Hence the true absorption coefficient τ/ρ varies as Z^3, which means that elements of higher atomic number absorb X-rays of a given wavelength to a very much greater extent than those of the lower atomic number, except at the discontinuities.

From these various experimental absorption data, Bragg and Pierce deduced the relation

$$\tau_a = CZ^4 \lambda^3,$$

which is known as *Bragg- Pierce law,* τ_a being the atomic absorption coefficient.

It is to be noted that, in contrast with the apparently chaotic region, absorption state of affairs in regard to the absorption of radiation in the optical region, we find a comparative simplicity in the empirical laws of the of the absorption of X-rays

Additional peculiarities of the X-ray absorption spectra

In the first place, unlike in the optical region, in the X-ray region there is *no selective absorption of spectral lines.* Instead of single lines of absorption which are exact reversals of particular emission lines, as found in optical spectra, we have here critical absorption discontinuities, hence absorption bands with definite edges, the absorption coefficient in regions of wavelengths shorter than that of the discontinuity being a monotonic function of wavelength with no unusual features.

Secondly, like the emission spectra, the absorption spectra of X-rays, when not considered in great detail, are *independent of chemical combination.* The mass absorption coefficient of an element, for X-rays of a given wavelength, has been found, in the early experiments, to be approximately the same whether the element is in combination with other elements or not, which is consonant with the atomic nature of X-ray absorption. More recent researches by Berengren and Lindh, however, appear to show that chemical combination has an influence on the absorption of X-rays, chiefly in the case of light elements, and in particular for sulphur, phosphorus and chlorine. The position of the absorption discontinuities has been shown to vary by some 10 to 15 X.U. according to the chemical compound used to absorb the rays, the shift being towards shorter wavelengths and the amount of displacement greater, the larger the valency of the compound.

Finally, it has been found that *fringes appear in the immediate neighborhood of the absorption discontinuities on the short wavelength side.* They have a structure more or less complex and the absorption edges show a *finite width,* when observed with a high resolving power instrument. Stenstroem was the first to discover them which were later confirmed by Hertz, Fricke, De Broglie and others. They manifest themselves chiefly with elements whose energy levels are relatively weak.

Any correct theoretical interpretation of the phenomenon of X-ray absorption must he able to account for these peculiarities also.

Theoretical explanation. Absorption, in general, is the reverse of emission. Hence, just as a photon of K_α radiation is emitted when an electron drops from an L shell into the K shell, so also it should be possible in the reverse process for a photon of this frequency to be absorbed by the atom while one of the K electrons goes up into an L shell, provided, of course, there is a vacancy in the L shell into which it can go.

But in any atom with atomic number greater than ten, the L shell normally contains as many electrons as can get into it. In such cases, the K electron cannot occupy the next outer shell, but, in general, may have to leave the atom completely. Hence it is that the K_α lines are not actually observed as single absorption lines, *i.e.,* exact reversals of particular emission lines, unlike the selective absorption of spectral lines in the optical region.

If in the incident beam there is a photon which has enough energy to remove one of the electrons in the inner complete shells from the atom entirely, that photon is absorbed, its energy being spent in the ionization of the atom with respect to that particular electron that has been ejected. This process of complete removal of an electron from the atom is the photoelectric effect.

If the incident photon has more energy than required to extract the electron, the extra amount may be communicated to the ejected electron in the final of kinetic energy. As a result of this process, the incident beam will lose a photon which means that it will diminish in intensity by a true absorption phenomenon. This absorption is also called *fluorescent absorption,* since the energy thus absorbed reappears under the form of a fluorescent radiation at the moment of the electronic reorganization in the atom, which follows the ionization of the atom, conformably to the ideas of Bohr.

If W_K represents the energy required to remove a K electron from the atom, then the photon of frequency v can eject a K electron, provided v is equal to or greater than v_k where $hv_k = W_K$. The explanation of the absorption curve with its discontinuities now becomes clear. If the wavelength is progressively increased, which is the same as decreasing the frequency v from a high value, at a frequency $v = v_k$, the absorption suddenly drops, i.e., absorption in the K shell ceases. The curve representing the absorption coefficient plotted against wavelength, will therefore show a sudden drop towards the side of longer wavelength at $\lambda = \lambda_k = c/v_k$, the K absorption limit. As the wavelength is further increased, similar discontinuities will occur every time the value of λ just equals that required for the ejection of an electron from the three sub-groups of the L shell, the five sub-groups of the M shell, etc. Since the energy W_K required to remove the K electron completely from the atom is somewhat greater than the energy change involved in the production of the emission lines of the K series when the atom undergoes transitions between its K state and the other quantum states such as L, M, etc., it is to be expected that the critical absorption wavelength will be just shorter than that of any of the lines in a given series, as is actually found.

As the frequency of each absorption limit is a measure of the energy required to remove an electron from the corresponding X-ray level to the periphery of the atom, the difference in the frequency of two absorption limits gives the frequency difference between the two X-ray levels involved, and by Bohr's frequency condition also the frequency of the resulting emission line. Thus, from data of X-ray absorption limits the whole structure of X-ray atomic levels can be built up which serves to represent a wide range of facts in a compact and clear form.

It is to be noted that the absorption limits represent the different effective electronic shells in the atom; the number of the former gives the number of the latter and the frequencies of the former are a measure of the energies of the latter, which thus brings out the discontinuous structure of the atom with discrete energy states as well as the atomic character of X-ray phenomena.

The *influence of chemical combination, on the absorption spectra as well as the existence of fringes in the neighborhood of the discontinuities* show, however, that there are exceptions to the general interpretation given above.

In the optical region the effect of chemical combination and the reversal of emission lines prove that the electronic transitions involved refer to the peripheral levels of feeble energy.

We have supposed above that in the X-ray region it is the inner levels, hardly affected by external influence, that are involved. If, however, one of the levels, even in this case, is sufficiently superficial, say, if an electron from the K shell jumps to one of the outer shells of feeble energy, containing many vacant spaces, without being completely ejected from the atom, it becomes evident how the K absorption can be affected by the chemical state of the absorber and also how absorption lines can occur in the neighborhood of the K discontinuity.

Such a state of affairs is simply a *prolongation of optical phenomenon in the X-ray region.* This explanation was first suggested by Kossel. In certain cases, however, the fringes extend too far from the head of the absorption discontinuity to be satisfactorily accounted for by the above-stated line of reasoning.

Hence, some authors have suggested that such fringes would correspond to multiple ionization of the atom, viz., a given quantum of primary radiation ejects, for instance, not only a K electron but also an M electron at the same time, to some superficial levels, similar to the mechanism of production of the satellite lines in the emission spectra.

Wave mechanical considerations have led to better results. With an electron removed from an inner shell, the residual ion will be surrounded by a field, due to the nucleus and the remaining electrons, and will approximate to a hydrogen nucleus. For an electron moving in such a field, there will exist a discrete set of quantum states, with associated wave functions. The field (can be considered as the extension outside the atom of the equivalent central field that is introduced in the zero-order stage of perturbation theory. The electrons remaining in the atom can be regarded as occupying the innermost quantum states in this field, leaving all the rest of them unoccupied. It should, therefore, be possible for the electron which is removed from an inner shell, instead of proceeding to an indefinite distance, to stop in one of these unoccupied outer states. According to this theory, we are led to expect that, instead of a single atomic level, such as the K shell, there would be closely spaced sequence of energy levels. The uppermost of these represents removal of the electron into a state of rest at infinity and is thus a level belonging to the ionized atom. The frequency corresponding to such removal from the K shell gives the K absorption limit. The lower-lying levels, corresponding to states in which the electron removed from the K shell remains attached to the atom in one of the outer unoccupied quantum levels, belong to the neutral atom, and are sometimes called resonance levels.

Hence, a graph of the absorption coefficient should show, over and above the K absorption edge, a series of absorption lines on the long wavelength side, each arising from atomic transitions into one of those lower lying atomic levels of the neutral atom. The absorption edge would constitute the series limit for these lines. The lines should be very closely spaced, since the energy differences between the levels should be of the same order of magnitude as the differences between the ordinary optical levels.

SOMMERFELD RELATIVISTIC ATOM MODEL

Bohr's simple theory of circular orbits, in spite of its many successes, was found inadequate to explain certain details in the spectrum of hydrogen. Close observation with instruments of high resolving power had already shown that the individual spectral lines were not really single but consisted of several very fine lines packed together, which were called the "fine structure". For instance, the H_a line of the Balmer series was found to consist of several components. This fine structure of spectral lines could not be explained by Bohr's theory which assumed that there was only one orbit for each quantum number n, whereas the observed fine structure suggested that for any given quantum number n there might be several orbits of slightly different energies.

Sommerfeld, in 1015, guided by the above suggestion, modified Bohr's theory, to make it fit with the additional experimental data, by introducing the ideas of motion of the electron in elliptical orbits and of the consequent relativistic variation of the mass of the electron. The Bohr atom thus improved upon, is known as the relativistic atom model, the main features of which may be briefly stated as follows

Elliptical orbits

Sommerfeld argued that, since the electron is moving around and under the influence of a massive nucleus, like a planet round the central massive sun, it might describe elliptical orbits as well. Considering therefore the electron (e^-) moving in an elliptical orbit, its position at any instant can be fixed in terms of polar coordinates r and φ, where r is the radius vector, *i.e.,* the distance of the electron from the nucleus (E^+) at one of the foci of the ellipse and φ the vectorial angle, *i.e.,* the angle which the radius vector makes with the major axis of the ellipse (Fig. 213).

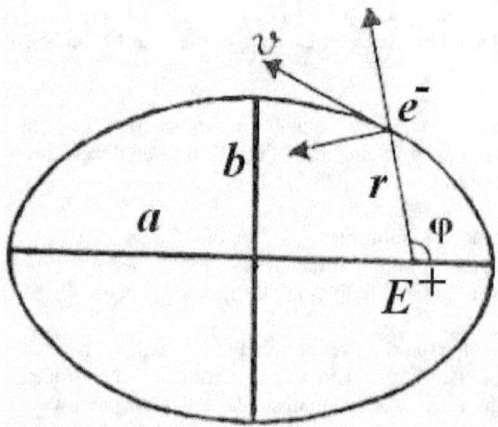

Fig. 213.

The tangential velocity v of the electron at the instant considered can be resolved into two components, one *radial, i.e.,* along the radius vector equal to dr/dt and the other *transverse,* the right angles to the radius vector equal to $r(d\varphi/dt)$. Corresponding to these two components we have a *radial momentum* p_r equal to $m(dr/dt)$ and an *angular* or *azimuthal momentum* p_φ equal to $mr^2 (d\varphi/dt)$, where m is the mass of the electron,.

Sommerfeld assumed that, since the elliptical orbits should satisfy the quantum. conditions just as the circular orbits, the circle being only a special case of the ellipse, to each of these two momenta the phase integral of the quantum theory might be applied, so that

$$\oint p_r . dr = n_r . h$$

and

$$\oint p_\varphi . d\varphi = n_\varphi . h \qquad \dots (1)$$

Thus two new quantum numbers n_r and n_φ have appeared replacing the single one n of Bohr's theory.

The three quantum numbers are related by the equation:

$$n = n_r + n_\varphi \dots\dots\dots\dots\dots\dots (2)$$

where
n is called the total quantum number and is identical with the quantum number of Behr's circular orbit,
n_r the radial quantum number and
n_φ the angular or azimuthal quantum number.

The total energy W of the system is partly potential and partly kinetic, the latter being further subdivided into radial and angular. Hence

$$W = P.E. + radial\ K.E. + angular\ K.E.$$

$$= -\frac{Ee}{r} + \tfrac{1}{2} m \left(\frac{dr}{dt} \right)^2 + \tfrac{1}{2} mr^2 \left(\frac{d\varphi}{dt} \right)^2 \qquad \dots (3)$$

From these three fundamental equations, which define the motion of the electron in the elliptical orbit satisfying the quantum conditions, it can be shown that

$$1 - \epsilon^2 = \frac{b^2}{a^2} = \frac{n_\varphi^2}{(n_\varphi + n_r)^2} \qquad \dots (4)$$

where ϵ is the eccentricity of the ellipse whose semi-major and semi-minor axes are a and b respectively and

$$W = -\frac{2\pi^2 m E^2 e^2}{h^2}\left(\frac{1}{n_\varphi + n_r}\right)^2 \qquad \dots (5)$$

From equation (4) we get

$$(1 - \epsilon^2)^{1/2} = \frac{b}{a} = \frac{n_\varphi}{n_\varphi + n_r} = \frac{n_\varphi}{n}$$

This shows that for a given value of the total quantum number there can be only a limited number of elliptical orbits that are permitted, having the same value for the major axis but different eccentricities. Further n_φ cannot be zero, since the ellipse would then degenerate into a straight line passing through the nucleus. Also n_φ cannot be greater than n, since b is always less than a. When n_φ is equal to n, the path becomes circular, since then $b = a$ and the eccentricity ϵ is equal to zero.

Hence for a given value of n, n_φ can assume only n different values, which means there can be only n elliptical orbits of different eccentricities, the orbits becoming less eccentric the higher the value of n_φ, until finally circular with zero eccentricity when $n_\varphi = n$. For instance, when $n = 4$, four different ellipses are permitted, usually designated by 4_1 4_2, 4_3, 4_4, of which the last one is evidently a circle.

It may be noted that the orbits need not be concentric it is enough that they are confocal, as shown in Fig. 214; nor is it necessary that they should be in the same plane or have a common major axis.

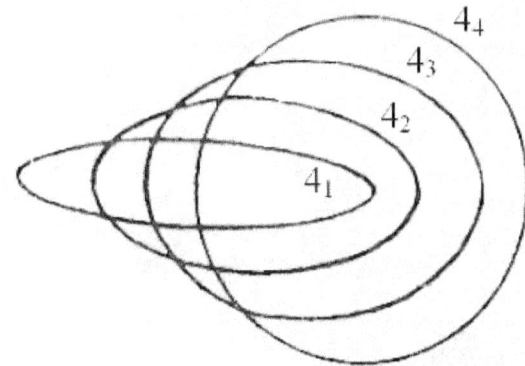

Equation (5) which gives the total energy of an elliptical orbit defined by the quantum numbers n, n_φ and n_r, viz.

$$W = -\frac{2\pi^2 m E^2 e^2}{h^2}\left(\frac{1}{n_\varphi + n_r}\right)^2 = -\frac{2\pi^2 m E^2 e^2}{h^2 n^2}$$

shows
(i) that the energy is the same for all the ellipses having the same total quantum number, i.e.- elliptical orbits which have the same value for n, though of different eccentricities, have the same energy and hence are indistinguishable from this point of view, e.g., the orbits 4_4, 4_3, 4_2, 4_1 have all the same energy;

(ii) that all orbits have the same value of the semi-major axis *a* possess the same energy, since the length for the semi-major axis is determined solely by the total quantum number $n = n_r + n_\varphi$. Hence also, the energy of any of these permitted elliptical orbits is identical with that of a circular Bohr orbit, whose radius is equal to the semi-major axis of the ellipse. This, in turn, means that the theory of elliptical orbits, in spite of the two new quantizing conditions involved, introduces no new energy levels other than those given by Bohr's theory of circular orbits. No new spectral lines, which would explain the fine structure, are therefore predicted. The extension to elliptical orbits is not, however, without effect, since it enabled Sommerfeld to proceed further and find a solution to the problem of the fine structure of spectral lines on the basis of the variation of the mass of the electron with velocity.

Relativistic variation of electronic mass

The velocity of the electron moving in an elliptical orbit varies considerably at different parts of the orbit, being a maximum when the electron is nearest to the nucleus. According to the theory of relativity, this variation of velocity means variation of the mass of the electron. Although this relativistic effect becomes appreciable only at very high velocities, yet it is sufficient, in the present case, to make the energy of the electron in a more eccentric orbit greater than in a less eccentric one, even if the major axis remains unchanged.

When allowance is made for the relativistic variation of electronic mass, as Sommerfeld did, the path of the electron is found to be no longer a simple ellipse; it is, indeed, no longer a closed figure, but is transformed into a complicated curve, known as a *rosette*, (Fig. 215)- a *precessing ellipse or a wobbling or gyrating ellipse.*

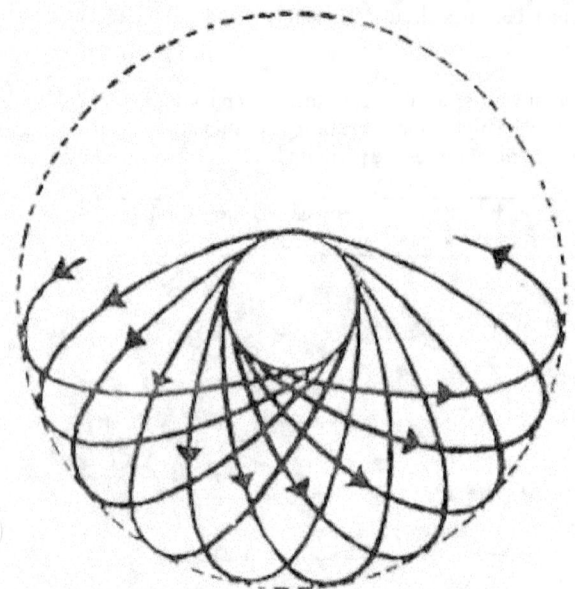

As the electron traverses the elliptical path, the major axis revolves slowly in the plane of the ellipse with a constant angular velocity around the nucleus situated at the focus of the ellipse. Such an orbit is *doubly periodic*: besides the original period of revolution there is also the period of the precessional motion.

The total energy W of the system, corrected for the relativistic variation of the mass of the electron, can be shown to be

$$W = -\frac{2\pi^2 m \mathrm{E}^2 e^2}{h^2} \left[\frac{1}{n^2} + \frac{4\pi^2 \mathrm{E}^2 e^2}{c^2 h^2} \left(\frac{n}{n_\varphi} - \frac{3}{4} \right) \frac{1}{n^4} + \cdots \right] \cdots (6)$$

This expression is only approximate as it is limited to the first two terms of an expansion in series, but the other terms are unimportant. It differs from the previous relation for W (eqn. 5) in that it contains an additional term representing the effect of the relativity correction, *viz.*

188

$$- \frac{2\pi^2 m e^4}{h^2} \frac{Z^4 \alpha^2}{n^4} \left(\frac{n}{n_\varphi} - \frac{3}{4} \right) \qquad \qquad \dots (7)$$

Where

$$E = Ze \quad \text{and} \quad \alpha = \frac{2\pi e^2}{ch}$$

This shows that energy depends not simply on the total quantum number n, but also upon the azimuthal quantum number n_φ, on account of the presence of the ratio n/n_φ. For a given total quantum number n, different energies are obtained for the different orbits corresponding to the different values of n_φ.

The correction term (eqn. 7) is a very small quantity as it involves α^2, whose value, readily estimated from the known values of e, c and h is found to be only 5.3×10^{-5}. α is called the *fine structure constant* and is equal to 7.28×10^{-3}.

The relativity correction, therefore, results in splitting up a given energy level W_n into n levels differing slightly from one another in energy. This splitting up of each energy level naturally gives rise to a fine structure of single spectral lines, on application of the usual Bohr's frequency condition.

A final important remark must be made in connection with the explanation of the fine structure, viz. not all the theoretically possible transitions of the electron from one elliptical orbit to another actually occur. According to a principle known as the *selection rule,* of which we shall speak more in detail later, transitions can take place only between orbits for which the azimuthal quantum number changes by +1 or -1, *i.e.,* $\Delta n_\varphi = \pm 1$.

Application to the fine structure of the H$_\alpha$ line

Let us now see how the theory proposed can be applied to interpret the fine structure of the H$_\alpha$ line. This line, being the first in the Balmer series corresponds to the change in the total quantum number: $n_2 = 3$ to $n_1 = 2$. Since there are three permitted orbits of the former, viz., 3_3, 3_2 and 3_1 and two of the latter, viz., 2_2 and 2_1 six different transitions are theoretically possible:

$$3_3 \to 2_2, \; 3_3 \to 2_1, \; 3_2 \to 2_2, \; 3_2 \to 2_1, \; 3_1 \to 2_2, \; 3_1 \to 2_1$$

To compute the frequency corresponding to any one of these transitions, it is convenient to determine first the value of a constant, known as the *hydrogen doublet constant* which is simply the difference in energy between the two orbits 2_2 and 2_1, expressed in wave number units. Taking the relation for W corrected for relativity, we may express it as:

$$W_{n, n_\varphi} = - RZ^2 c h \left\{ \frac{1}{n^2} + \frac{\alpha^2 Z^2}{n^4} \left(\frac{n}{n_\varphi} - \frac{3}{4} \right) \right\}$$

where $W_{n, n\varphi}$ is the energy corresponding to an orbit of total quantum number 'n' and azimuthal quantum number 'n_φ' and $R = 2\pi^2 m e^4 / ch^3$. Using the above expression,

Energy of the 2_2 orbit is

$$W_{2, 2} = - RZ^2 c h \left\{ \frac{1}{2^2} + \frac{\alpha^2 Z^2}{2^4} \left(\frac{2}{2} - \frac{3}{4} \right) \right\}$$

Energy of the 2_1 orbit is

$$W_{2,\,1} = -\,R\dot{Z}^2\,c\,h\left\{\frac{1}{2^2} + \frac{\alpha^2 Z^2}{\cdot 2^4}\left(\frac{2}{1} - \frac{3}{4}\right)\right\}$$

Therefore, the energy difference between the orbits 2_2 and 2_1 is

$$W_{2,\,2} - W_{2,\,1} = RZ^2\,c\,h\left\{\frac{\alpha^2 Z^2}{2^4}\left(\frac{2}{1} - \frac{3}{4}\right) - \left(\frac{2}{2} - \frac{3}{4}\right)\right\}$$

$$= \frac{RZ^4\alpha^2\,c\,h}{2^4}$$

Putting $\Delta W = W_{2.2} - W_{2.1}$ and $Z = 1$, the corresponding frequency $\Delta v = \Delta W/h$ and expressed in wave number units,

$$\Delta\bar{v} = \Delta v/c = \Delta W/ch = R\alpha^2/16.$$

This quantity $R\alpha^2/16$ is the hydrogen doublet constant which we shall use as a unit in the sequel, representing it by "d". Its numerical value can be readily calculated, substituting the values of where
$R = 109,700\ cm^{-1}$ and
$\alpha^2 = -5.3 \times 10^{-5}$,
and is found to be $0.364\ cm^{-1}$

Again, expressing the general relation for energy as

$$W_{n,n_\varphi} = -\,\frac{RZ^2 ch}{n^2} - \frac{RZ^4\alpha^2 ch}{n^4}\left(\frac{n}{n_\varphi} - \frac{3}{4}\right)$$

we see that the *first term* on the right-hand side represents the *uncorrected energy* while the *second* gives the *relativity correction*. Replacing Z by 1 and expressing the energy in wave number units, we get:

$$\overline{W}_{n,n_\varphi} = -\,\overline{W}_n - \frac{R\alpha^2}{n^4}\left(\frac{n}{n_\varphi} - \frac{3}{4}\right)$$

where $\overline{W}_n$ corresponds to the uncorrected energy. From this relation, we readily obtain the energies of the five

different orbits involved in the present case:

$$n = 3 \begin{cases} \overline{W}_{3,3} = -\,\overline{W}_3 - \dfrac{16d}{3^4}\left(\dfrac{3}{3} - \dfrac{3}{4}\right) = -\,\overline{W}_3 - \dfrac{4d}{81} \\[2mm] \overline{W}_{3,2} = -\,\overline{W}_3 - \dfrac{16d}{3^4}\left(\dfrac{3}{2} - \dfrac{3}{4}\right) = -\,\overline{W}_3 - \dfrac{12d}{81} \\[2mm] \overline{W}_{3,1} = -\,\overline{W}_3 - \dfrac{16d}{3^4}\left(\dfrac{3}{1} - \dfrac{3}{4}\right) = -\,\overline{W}_3 - \dfrac{36d}{81} \end{cases}$$

$$n = 2 \begin{cases} \overline{W}_{2,2} = -\overline{W}_2 - d\left(\dfrac{2}{2} - \dfrac{3}{4}\right) = -\overline{W}_2 - \dfrac{d}{4} \\[2mm] \overline{W}_{2,1} = -\overline{W}_2 - d\left(\dfrac{2}{1} - \dfrac{3}{4}\right) = -\overline{W}_2 - \dfrac{5d}{4} \end{cases}$$

Now, the wave number of the single transition to be expected, if there were no splitting due to relativity correction, is given by $\overline{W}_3 - \overline{W}_2$. Representing it by $\overline{v}$ and the wave numbers of the six different possible transitions by

$$\overline{v}_a, \overline{v}_b \; \overline{v}_c, \overline{v}_d, \overline{v}_e \text{ and } \overline{v}_f,$$

$$(3_3 \to 2_2), \quad \overline{v}_a = \overline{v} - \frac{4d}{81} + \frac{d}{4} = \overline{v} + \frac{65}{324}d$$

$$(3_3 \to 2_1), \quad \overline{v}_b = \overline{v} - \frac{4d}{81} + \frac{5d}{4} = \overline{v} + \frac{389}{324}d$$

$$(3_2 \to 2_2), \quad \overline{v}_c = \overline{v} - \frac{12d}{81} + \frac{d}{4} = \overline{v} + \frac{33}{324}d$$

$$(3_2 \to 2_1), \quad \overline{v}_d = \overline{v} - \frac{12d}{81} + \frac{5d}{4} = \overline{v} + \frac{357}{324}d$$

$$(3_1 \to 2_2), \quad \overline{v}_e = \overline{v} - \frac{36d}{81} + \frac{d}{4} = \overline{v} - \frac{63}{324}d$$

$$(3_1 \to 2_1), \quad \overline{v}_f = \overline{v} - \frac{36d}{81} + \frac{5d}{4} = \overline{v} + \frac{261}{324}d$$

Applying the selection rule, viz., only those transitions for which $\Delta n_\varphi = \pm 1$ are allowed, $(3_3 \to 2_1)$, $(3_2 \to 2_2)$ and $(3_1 \to 2_1)$ are forbidden, so that $\overline{v}_b$, $\overline{v}_c$ and $\overline{v}_f$ will not be observed. Hence according to the theory, the fine structure of H_α line should consist only of three lines, whose wave numbers are:

$$\bar{\nu}_a = \bar{\nu} + \frac{65}{324}\,d$$

$$\bar{\nu}_d = \bar{\nu} + \frac{357}{324}\,d$$

$$\bar{\nu}_e = \bar{\nu} + \frac{63}{324}\,d$$

Energy level diagram

The above result may be represented by the usual energy level diagram as follows:

On the vertical energy scale the positions of $\bar{W}_3$ and $\bar{W}_2$ are arbitrarily chosen (Fig. 216).

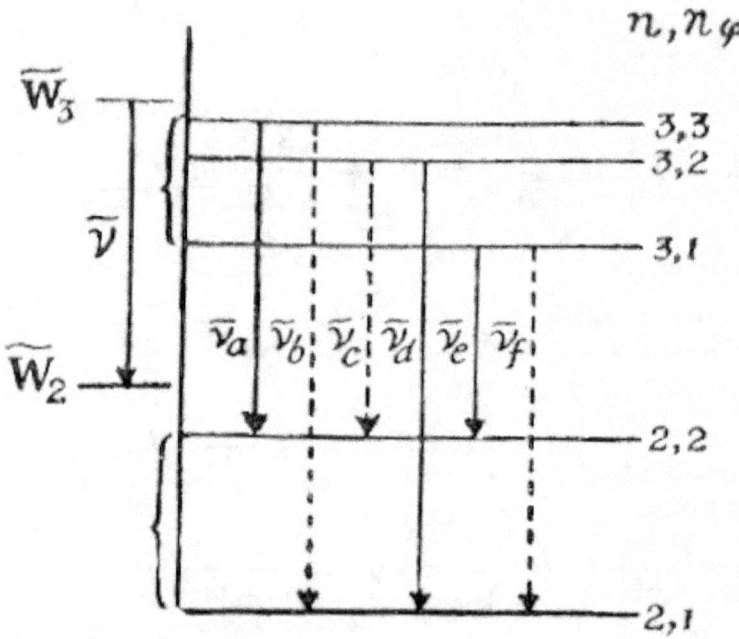

Fig. 216. Fine structure of H_a line.

The horizontal lines representing the different energy levels can be drawn by taking an arbitrary scale for d and calculating the correction factors in terms of d. The transitions allowed by the selection rule are shown by continuous lines, while the forbidden ones by dotted lines.

Comparison of theory with experimental data

Experimentally, he lines of the Balmer series appear as very close *doublets* and the microphotometer record taken by Kent, Taylor and Pearson give the doublet separation for H_a as 0.318 cm⁻¹. More accurate measurements by G. Hansen who used specially prepared Luminer-Gehrcke plates for resolution gave the following values for the doublet separation of the lines of the Balmer series:

$$\mathrm{H}_\alpha \qquad \mathrm{H}_\beta \qquad \mathrm{H}_\gamma \qquad \mathrm{H}_\delta \qquad \mathrm{H}_\varepsilon$$
$$0{\cdot}316 \qquad 0{\cdot}317 \qquad 0{\cdot}328 \qquad 0{\cdot}322 \qquad 0{\cdot}324 \text{ cm.}^{-1}$$

Direct observation is rendered difficult on account of the blurred character of the H lines, which is due to several causes, such as the heat motion of the emitting H atom with the resulting Doppler effect, Stark effect, etc. But there is an. indirect method of observation, employed by Paschen, which enables the doublet separation to be deduced from the structure of the lines of the singly ionized helium atom which gives much sharper lines on account of the fact that, its mass being four times as great as hydrogen, the Doppler effect is less and its nuclear charge being twice as great, the Stark effect is also less. The origin of the observed doublet separation is readily understood on the basis of the theory proposed:

$$\text{The separation between } \bar{v}_a \text{ and } \bar{v}_d = \frac{292}{324}\, d$$

$$\text{,,} \qquad \text{,,} \qquad \text{,,} \quad \bar{v}_a \text{ and } \bar{v}_e = \frac{128}{324}\, d$$

$$\text{,,} \qquad \text{,,} \qquad \text{,,} \quad \bar{v}_d \text{ and } \bar{v}_e = \frac{420}{324}\, d$$

Hence the components $\bar{v}_a$ and $\bar{v}_e$ being very close together are not easily resolved in practice. Thus only a doublet

separation of $\bar{v}_d$ and $\bar{v}_a\ (\approx \bar{v}_e)$ could be observed. The theoretical value of the separation between $\bar{v}_a$ and $\bar{v}_d$ agrees

pretty well with the experimental value. Further, since the relativity correction term is proportional to $1/n^4$, the component levels for $n = 4$ will be crowded more closely than those for $n=3$. Since H_β originates in transitions from $n=4$ to as $n = 2$ the expected doublet separation for H_β should be greater than for H_α; the separation for H_γ should be still greater. The slight increase in value of the separation with higher members is also confirmed, at least qualitatively, by experimental data.

Though Sommerfeld's theory is thus fairly well verified, yet it leads only to three components for the fine structure of H_α line, while there should be really five, as we shall see. The relativistic atom model, therefore, has met with a partial success alone, not through any fault of the selection rule chosen, but due to the omission of a new factor in the problem, viz., the spin of the electron,
which the next atom- model will incorporate and thereby account for all-the five components.

It must be said that Sommerfeld's theory, in spite of its limited success, was a further progress, in the right direction, in the understanding of the origin of the complex structure of spectral lines, in so far as it introduced new quantum numbers and a selection rule that represent more adequately the actually observed facts than Bohr's theory.

It has the additional merit of being one of the first cases in which the theory of relativity was applied to the microscopic atomic state. The following points are worthy of note in connection with Sommerfeld's theory:

(*a*) *Evaluation of the universal constants e, m and h from spectroscopic data.* Utilizing the Bohr-Sommerfeld theories, it is possible, starting from spectroscopic measurements, to work back and determine the numerical values of the above-mentioned important constants. The calculation is all the more interesting because in recent years wavelength measurements have reached a very high precision and hence the values of the constants can be fixed with great exactness.

The first step in the calculation is to evaluate *e/m* using the relation

$$e/m = \frac{e/M_H}{m/M_H};$$

Where
e/M_H is the Faraday, known from electrolysis ;
m/M_H is obtained from the expression:

$$m/M_H = \frac{R_H - R_H}{R_H - \frac{1}{4}R_{He}}$$

where R_{He}, and R_H are known from the analysis of the respective spectra.

Next, the electronic charge e is found as follows:

$$\alpha = \frac{2\pi e^2}{ch} \quad \text{(fine structure constant)}$$

$$R_\infty = \frac{2\pi^2 m e^4}{ch^3} \quad \text{(Rydberg constant)}$$

$$\therefore \quad \alpha^3/R_\infty = \frac{8\pi^3 e^6}{c^3 h^3} \times \frac{ch^3}{2\pi^2 m e^4}$$

$$= \frac{4\pi e(e/m)}{c^2}$$

$$\therefore \quad e = \frac{\alpha^3 c^2}{4\pi R_\infty (e/m)}$$

In this relation, e/m is already determined;
α^3 can be obtained from the expression for the hydrogen doublet constant
$d = R\alpha^2/16 = 0.364$,
where R stands for R_H ;
R_α is also got using the relation
$R_H = R_\alpha \{1 / (1 + m/M_H)\}$;
c^2 is known, c being the velocity of light (3×10^{10} cm./sec.).

Knowing e/m and e, m, is also found.

To find h, the relation $\alpha = 2\pi e^2 /ch$ is used; $h = 2\pi e^2 /c\alpha$.
From the known values of α, c and e, h is evaluated.

(b) Spectroscopic confirmation of the theory of relativity.
According to the theory of relativity, adequately represented by

$$m = m_0/\sqrt{1 - \beta^2} = m_0 (1 + \tfrac{1}{2} \beta^2 + \ldots\ldots),$$

the value of the hydrogen doublet constant is found to be 0.364 cm.[-1]. The older "absolute" theory, assuming that space was absolute and that the electron was spherical in shape, led to a different relation:

$$m = \frac{3}{4} \cdot \frac{m_0}{\beta^2}\left(\frac{1+\beta^2}{2\beta} \log \frac{1+\beta}{1-\beta} - 1 \right)$$
$$= m_0 \left(1 + \frac{2}{5}\beta^2 + \ldots\ldots \right)$$

Using this, it can be shown that the value of the hydrogen doublet constant is only
$(4/5)\ (R\alpha^2/2^4) = 4/5 \times 0.364 = 0.29$ cm[-1]

This value, however, is incompatible with Paschen's result obtained from the measurements of He[+] lines. The same may be said of the doublet constant in the X-ray region. Taking the results all together, one may draw the conclusion that the "absolute" theory comes to grief on the spectroscopic data which, on the other hand, offer an indirect confirmation of the theory of relativity.

VECTOR ATOM MODEL

Introduction

The vector atom model is an extension of the Rutherford-Bohr-Sommerfeld atom model (described in the preceding sections) on new lines, devised to cover new fields of experimental research.

We have seen that Bohr's theory could solve the spectroscopic problem of only the simplest atom, hydrogen or hydrogen-like. In the case of atoms having a system of 30, 40 or even more (up to 92) electrons whirling round the nucleus, it seems impossible to calculate the energy of the system and from this the frequency of the radiation emitted in the passage between two states W_{n1} and W_{n2}. As a matter of fact, after laborious attempts to perform the calculations, the search for the solution in such cases has bean abandoned. The reason for this lies in the fact that

the general problem of determining the orbits of three or more bodies, attracting or repelling one another in accordance with the inverse square law, is beset with insurmountable difficulties.

A large number of calculations have been made for He, Li and the alkali metals, but without any conclusive result. The impossibility of solving the problem of more complex atoms is not merely a mathematical one. It has a more fundamental origin in the basic conception itself. It is due to the fact that the atom is not subject to purely classical mechanics. Further, the original simple theory of Bohr was absolutely incapable of explaining the fine structure of spectral lines even in the simplest one electron systems, such as H and He[+] atoms.

Sommerfeld's modification, though giving a theoretical justification of the Splitting of individual spectral lines of hydrogen into fine structure components, met with only a partial success, as it could not predict the correct number of the fine structure lines. Moreover, it gave no information about the relative intensities of the lines whose frequencies alone it predicted.

Hence these older atom models were quite inadequate to tackle more complex atoms. Soon, other new experimental data; such as the anomalous Zeeman effect, Stark effect, etc., brought out this insufficiency of the older ideas with still greater force. Another drawback of the Bohr atom model was that it did not throw any light on the distribution and arrangement of electrons in atoms. The most fundamental objection raised against the Bohr atom model was that it involved the use of two theories which are essentially opposed to each other:

the quantum theory was invoked to account for the existence of stationary orbits and for the frequencies of the radiation emitted, while the motion of the electron in its orbit obeyed the law of classical mechanics.

New ideas were therefore introduced partly by analogy, partly by empirical methods, in the interpretation of more complex spectral phenomena and their relation to atomic structure, which finally resulted in what is known as the *vector atom model.* Among the physicists who have contributed substantially to the accomplishment of this really monumental work, mention must be made of Bohr, Sommerfeld, Uhlenbeck, Goudsmit, Pauli, Landé, Stern and Gerlach.

The two essential elements that characterize the vector atom model and differentiate it from other models are:

195

(i) the conception of spatial quantization or quantization of direction, and
(ii) the spinning electron hypothesis.

Spatial quantization

We have seen that the Bohr -Summerfield orbits are all quantized as regards their *magnitude, i.e.,* their size and form. But quantum theory demands more than this, viz., quantization also of *direction* or orientation of the orbits in space, *i.e.,* it selects from the continuous manifold of all possible positions of the orbits in space, permitted by classical ideas, a discrete number conformable to quantum conditions. The introduction of such a spatial quantization makes the orbits *vector* quantities.

To quantize spatially, we need, of course, a certain preferred direction with respect to which the orbits may receive their orientation. Such a favored direction may be obtained by an external field of force. But, in such a case, we have no longer, even for the hydrogen atom, pure elliptical orbits, since they would be deformed by the external field. To overcome this difficulty, we must pass on to the limiting case when the external force tends to zero since thus the disturbance of the orbits is reduced to the minimum, while at the same time the possibility of their orientation with respect to the direction of the field remains. To determine the permitted orientations relative to the field direction, we are guided by the fact that *the projections of the quantized orbits on the field direction must themselves be quantized.*

This new idea of spatial quantization was first proposed and worked out on the basis of the quantum theory by Sommerfeld with the interpretation of the splitting of spectral lines under the influence of an external magnetic field, *i.e.,* the Zeeman effect,

Spinning electron

The Bohr atom model even with Sommerfeld's modifications could not completely explain the complex fine structure of even the single line in the spectrum of the simplest atom, as we have seen. Naturally then, so it could not tackle the problem of fine structure of more complicated systems. Thus, for instance, it has been known for many years that the yellow sodium line is actually a doublet. With a spectroscope of sufficiently high resolving power it is found that the yellow D line splits into two components, D_1 and D_2, with a measurable separation of 6 A_o. Close triplets are likewise found for the lines in the spectrum of magnesium.

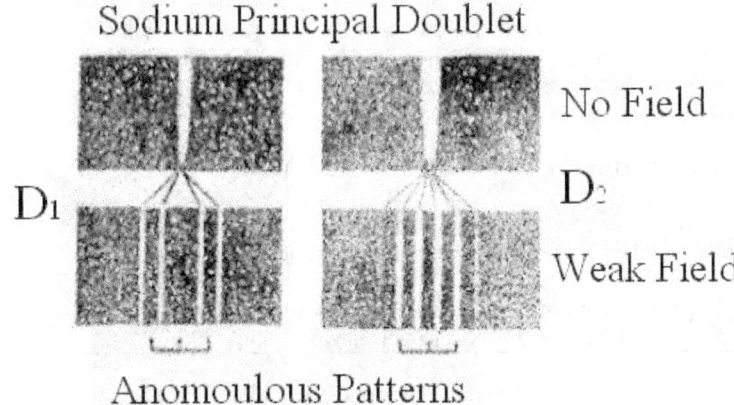

Sodium Principal Doublet

No Field

D₁   D₂

Weak Field

Anomoulous Patterns

Zeeman pattern of the sodium **D** lines in a weak field (H. E. White)

In other cases the splitting becomes much more complex. The heavier the atom the greater is the separation of the components : *e.g.,* while the doublet of the sodium D line is separated only by 6 A°, the mercury triplet extends over 1400 A°. Hence the name "fine structure" is not appropriate in the case of heavier atoms; *"multiplet structure"* is to be preferred.

Again, when an external magnetic field of *moderate* strength is applied to the source which produces the spectrum of sodium, say a discharge tube containing sodium vapor, each of the doublets mentioned above splits up still further, the separation being proportional to the field strength. Thus the D_1 line splits into 4 components, while the D_2 line into 6, as seen in the photo on the previous page. Such complex magnetic splitting of spectral lines is called the anomalous Zeeman effect, as distinguished from the normal triplet Zeeman effect.

In order to explain satisfactorily some of these intricate spectral phenomena, two Dutch physicists, Uhlenbeck and Goudsmit, put forward, in 1925, the hypothesis of the spinning electron, according to which the electron revolves not only in an orbit round the nucleus but also about an axis of its own, somewhat like a planet in the solar system. In other words, the electron is endowed with a spin motion over -and above the orbital motion.

The "spinning electron" thus introduced brought in profound modifications in the atom model. In general, a body rotating about an axis gives rise to a mechanical angular momentum; further, if the rotating body carries an electric charge, the latter will revolve also along with the body. But a revolving charge is equivalent to a circuital current, giving rise to a magnetic moment.

> Hence the rotation of a charged body about an axis produces a mechanical momentum as well as a magnetic moment.

In the older atom model, the electron was supposed to have only orbital motion round the nucleus, which would produce a mechanical momentum and a magnetic moment. Now including the "spinning electron" idea, the electron will be endowed with two angular momenta and two magnetic moments, one due to orbital motion and the other due to spin. In consequence, the total angular momentum of the atom will be no longer due simply to the orbital motion of the electron, but also due to the spin of the electron. Likewise, the total magnetic moment of the atom is due to both the orbital and spin magnetic moments. In other words, the atomic magnet is the resultant of two magnets, one arising from the orbital motion and the other from the spin motion.

According to the quantum theory, the spin motion is to be quantized, like the orbital motion, which will therefore introduce a new quantum number, known as the spin quantum number, in addition to the orbital quantum number. Further, as both the orbital and spin motions are to be quantized not only in magnitude but also in direction according to the idea of spatial quantization, they are to be considered as *quantized vectors.* It is this special feature which is a valuable addition to the atom model. Since the different component parts that determine the state of the atom, such as orbital and spin motions, are all quantized vectors, the atom model built on such considerations is aptly called the *vector atom model,* to which vector laws apply.

Experimental study of the behavior of atoms in non-homogeneous magnetic fields, initiated by Stern and Gerlach in 1921, gave independent evidence for the reality of spatial quantization and spinning electron, so that the two fundamental postulates of the vector model of the atom may be considered to rest on sure experimental basis.

QUANTUM NUMBERS ASSOCIATED WITH THE VECTOR MODEL OF THE ATOM

In the vector atom model, to each of the component parts is assigned a quantum number, the numerical value of which may conveniently be thought of as the length of the vector which represents the angular momentum of that component part.

In vector analysis, angular momentum is represented by a straight line whose direction is parallel to the axis of rotation and whose length is proportional to the magnitude of the momentum.

The quantum numbers associated with each of the electrons of a given atom are the following:

(*a*) *A total quantum number n*, identical with the one used in the Bohr-Sommerfeld theory. It can take only integral values 1, 2, 3, etc.

(*b*) *An orbital quantum number l* which may take any integral value between 0 and (n-1) inclusively. Thus if n =4, then *l* can have values 0, 1, 2, 3. This quantum number is similar to the azimuthal quantum number n_φ of Sommerfeld's theory, which later came to be expressed by the letter k ($n_\varphi = k$). The two are related by
$l = k - 1$,
since k or n_φ of the older theory cannot assume the value 0, while *l* of the present theory can be equal to zero. The orbital angular momentum p_l of the electron is given by
$p_l = l \cdot h/2\pi$,
as it is quantized. It may be noted that according to wave mechanics the orbital angular momentum has the value

$$\sqrt{l(l+1)} \cdot h/2\pi$$

By convention, an electron for which $l = 0$ is called *s* electron; $l = 1$ is called *p* electron; $l = 2$ is called *d* electron; $l = 3$ is called *f* electron, etc

(*c*) *A spin quantum number s*, the magnitude of which is always 1/2. At first sight, this half integer value, as it contradicts the integral multiple rule of quantization of angular momenta, appeared to raise a serious difficulty against the spinning electron theory. But an answer was readily found in the following facts:

The use of half integers for *'s'* consistently leads to results which are in complete agreement with experimental facts. Further, the wave mechanical treatment of the spinning electron in the form given to it by Dirac leads automatically to this half-integral property and thus we possess a theoretical justification also. The spin angular momentum $p_s = s \cdot h/2\pi$, where $s = 1/2$; but according to wave mechanics

$$p_s = \sqrt{s(s+1)} \cdot h/2\pi$$

(*d*) *A total angular quantum number j,* which refers to the resultant angular momentum of the electron due to both orbital and spin motions, is the numerical value of the vector sum of *l* and *s.* Hence, $j = l + s$, and the value of *j* is naturally a. half integer, since one of its components *s* is always equal to 1/2 . It is usually expressed as $j = l \pm 1/2$, *plus* sign when *s* is *parallel* to the *l* vector and *minus* sign when *s* is *antiparallel.* The total angular momentum of the electron p_j is given by $p_j = j \cdot h/2\pi$ and more correctly by

$$p_j = \sqrt{j(j+1)} \cdot h/2\pi$$

according to wave mechanics.

If the atom is placed in a magnetic field so that some preferential axis in space is thereby determined, due to the fact that the field exerts a directive influence on both the orbital and spin motions which constitute elementary magnets, three more quantum numbers will be associated with the electron.

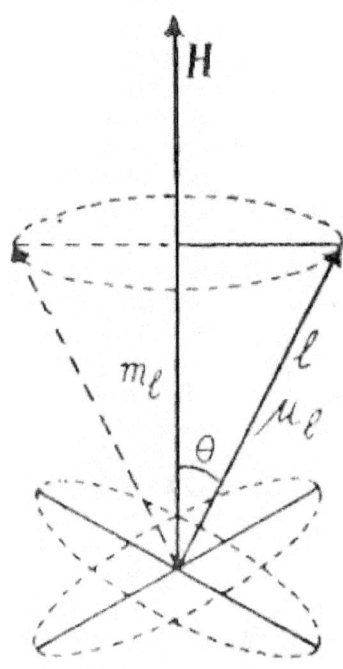

Fig. 217

(e) *A magnetic orbital quantum number m*, which is the numerical value of the projection of the orbital vector *l* on the field direction. m_l is an integer and may have only any of the $(2l+1)$ values from $-l$ to $+l$ including zero. For, according to spatial quantization, the projection of *l* in the field direction must itself be quantized. Hence *l* can be inclined to the field direction only at such discrete angles that its projection m_l may also be an integer. For instance, if the vector *l* is inclined to the field direction at an angle θ (Fig. 217), its projection $m_l = l \cos \theta$. *Now* since m_l has to be an integer and $\cos \theta$ can never exceed unity, the permitted values of m_l are from $+l$ to $-l$ at unit intervals, *viz.*, l, $(l-1)$, $(l-2)$,...1. 0, -1,... $(l-1)$, $-l$. Using the laws of arithmetical progression, it is seen that the total number of possible values of m_l is $(2l+1)$.

Conversely, the permitted orientations of the *l* vector relative to the field direction is also $(2l+1)$. Thus, for instance, if *l* = *1*, the permitted orientations of *l* are 3, for which m_l has the values + 1, 0, -1 . If *l* = 2, the permitted orientations are 5, for which m_l =+ 2, + 1, 0, **— 1, —** 2, as shown in Fig. 218.

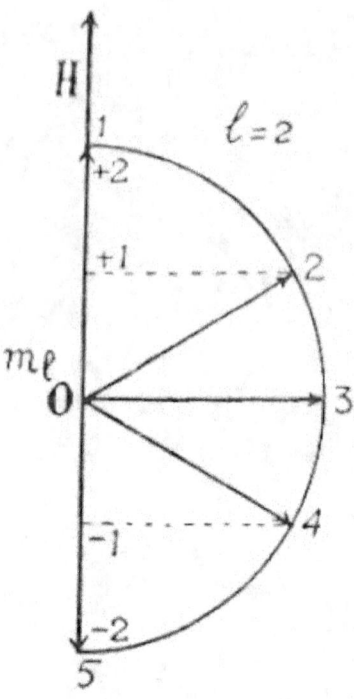

Fig. 218

(*f*) *A magnetic spin quantum number* m_s which is the numerical value of the projection of the spin vector *s* on the field direction. By analogy with the orbital vector *l*, the vector *s* can have also only $(2s + 1)$ permitted orientations with respect to the field direction according to spatial quantization and in consequence m_s can have any of the $(2s + 1)$ values from -*s* to + *s*, without however, including zero, since *s is* always equal to 1/2 and never zero. This means that m_s can have only two values +1/2 or -1/2 .

(*g*) *A magnetic total angular momentum quantum number* m_j *which* is the numerical value of the projection of the total angular momentum vector *j* on the field direction. Since we are dealing with a single electron, *j* can have only half integral values, since $j = l \pm 1/2$ and in consequence m_j must have only half integral values. The permitted orientations of *j* with respect to the field direction are $(2j+1)$ and hence m_j can have only $(2j + 1)$ values, from +*j* to - *j*, zero excluded.

Note.
The quantum numbers m_l and m_s come into play only in very strong magnetic fields where the coupling between *l* and *s* is broken, while m_j is effective in ordinary fields.

Having thus defined the state of each of the electrons in a given atom, we shall now turn our attention to the state of the atom as a whole.

The resultant quantized vectors representing the atom as a whole can be obtained by
(*i*) suitably combining the vectors of the individual electrons in the atom and
(*ii*) the application of the rule of spatial quantization.

It is to be noted that while *small* letters as *l, s, j* and *s*, p, d, f, etc., are used to designate the state of the *electron, capital* letters as L, S, and S, P, D, F, etc., are used to describe the state of the *atom as a whole.*

Coupling schemes

There are several ways in which the different vectors of the electrons may combine to give the vectors representing the atom as a whole. The method of combination depends on the interaction or "coupling" between the component vectors, since the orbital and spin motions of the electron producing magnetic fields result in mutual perturbation.

It is usual to distinguish two types of combination, know as the *Russell-Saunders* or *L-S coupling* and j-j *coupling.*

1. The L-S coupling is the one which occurs most frequently and hence is known also as the *normal coupling.* In this type the several spin vectors s of the electrons combine to form a resultant vector S; the several orbital vectors l of the electrons likewise combine to form a resultant vector L ; and then the S and L combine to make the vector J which represents the total angular momentum of the atom. The process may be symbolically represented as

$$(s_1 + s_2 + s_3 + \dots) + (l_1 + l_2 + l_3 + \dots) = S + L = J$$

This sort of coupling is the most natural one when the interactions between the individual spins on the one hand and the individual orbital momenta on the other are very strong.

The guiding principles in this type of coupling are:

(i) All the three vectors L, S and J are quantized.
(ii) L is always an integer including zero, *i.e.,* L may be 0, 1, 2, 3, etc.
(iii) S is an integer or half integer depending on the number of electrons involved and on the direction of the spin vectors. Thus, for instance, for an atom containing a *single* electron, S can have only the value 1/2 ; for a . two-electron system S may have either 1 or 0 depending on whether the spin vectors are parallel or antiparallel. For a three-electron system, S is 3/2 or 1/2 and for a four-electron system, S may be 2, 1 or 0, due to the same reason, as readily seen from Fig. 219.

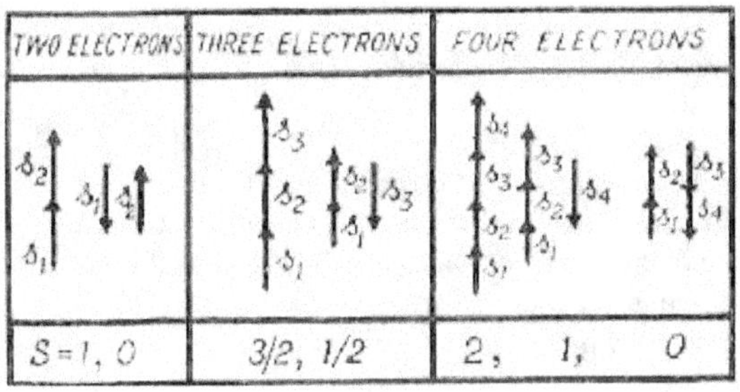

TWO ELECTRONS	THREE ELECTRONS	FOUR ELECTRONS
$S = 1, 0$	3/2, 1/2	2, 1, 0

Fig. 219

It follows that

S *is an integer for an even number of electrons, and half integer for an odd number of electrons.*

(iv) Hence J, the vector sum of L and S must be an *integer* (0, 1, 2, 3, etc.) if S is an integer, i.e , for even *electron systems* and *half integer* (1/2, 3/2, 5/2, etc.) if S is an half integer, *i e.,* for odd *electron systems.* It can be proved that, in general, the possible number of values, which J can assume, are

(2S + 1) if L > S and (2L + 1) if L < S.

The quantity (2S + 1) is known as the *multiplicity (r)* of the L state. It is simply the permitted values of J for a given value of L. r =(2S + 1). In particular, if L = 0, J can have only one value, viz., J =S.

J must always be positive, never *negative,* since it represents the total angular momentum of the atom.

2. The j-j coupling. Under certain circumstances, the interaction between the spin and orbital vectors in each electron may be stronger than that between either the spin vectors or the orbital vectors of the different electrons. If this is the

201

case, each electron may be considered separately and its contribution to the total angular momentum of the atom may be obtained by combining first its individual spin and orbital vectors by the relation
$$j = l + s.$$
The vector sum of all the individual j vectors of the electrons gives the total angular momentum J of the atom. The sort of coupling may be symbolically represented by

$$[(s_1 + l_1) + (s_2 + l_2) + (s_3 + l_3) + \ldots] = (j_1 + j_2 + . j_3 + \ldots) = J$$

These two types of "coupling" are limiting cases, between which a whole range of intermediate types may occur, which makes the problem very difficult of treatment. *For most known cases, however, the L-S coupling is effective.*

Application of spatial quantization

According to the quantum theory, the resultant vectors representing the atom, obtained by the coupling schemes described above, are quantized not only in magnitude but also in direction. Hence the number of permitted orientations of L, S and J with respect to a preferential field-direction are $(2L + 1)$, $(2S + 1)$ and $(2J + 1)$ respectively. The corresponding magnetic quantum numbers

$$m_L = \sum m_l ,$$
$$m_S = \sum m_s ,$$
$$m_J = \sum m_j ,$$

can only have $(2L + 1)$. $(2S + 1)$ and $(2J + 1)$ values respectively. It is to be noted that in the one-electron system, i.e., an *atom with a single effective electron, the state of the atom as a whole is identical with the state of the electron,* so that L = l, S = s and J = j.

OTHER IMPORTANT PRINCIPLES USED IN CONJUNCTION WITH THE VECTOR MODEL OF THE ATOM

In order to be able to use the somewhat artificial structure of the vector model of the atom in explaining atomic and spectral phenomena, the following important principles are required.

(i) Pauli's exclusion principle, sometimes termed also *equivalence principle,* states that every completely defined quantum state in an atom can be occupied by only one electron. In other words, it is impossible for two electrons in an atom to be identical as regards all their quantum numbers, *i.e.,* one of the two in such a case will be excluded from entering into the constitution of the atom. The principle may be stated in yet another way;

two systems of quantum numbers which are deducible from each other by interchange of two electrons represent only one state.

Thus understood, the principle enunciates the *indistinguishability* of electrons which have identical quantum numbers. Hence the name "equivalence principle".

It is clear that the principle defines a certain minimum individuality of the electron in the atom. Among the several quantum numbers associated with the electron, four, viz., n, l, and m_l or m_s or n, l, j and m_j can be shown to be strictly required to specify completely the state of any particular electron, the former set being used in the presence of a strong magnetic field which breaks the coupling between l and s in the electron and the latter in the rest of the cases.

The principle was first introduced by Pauli to explain certain experimental facts, such as

(a) the occurrence or non-occurrence of spectral lines in optical and X-ray spectra—it was found that for the missing energy states of the atoms, all the four quantum numbers of the electrons agreed and

(b) the complete scheme of the successive formation of the atoms as arranged in the periodic table, discovered by Bohr, Stoner and others—such as arrangement required the association of the above-stated four quantum numbers with each electron.

Naturally then, the principle finds its chief use in the elucidation of electronic structure and atomic spectra. It has also been very helpful in defining the special quantum property of closed shells, as well as accounting for certain limitations of terms multiplicity actually observed. It has been realized later that the principle is much more universal and fundamental, holding good for the totality of electrons in any molecule, nay, even for the more comprehensive system of conduction electrons that belong to a metal in the bulk state on the one hand and for the constituent particles of the nucleus in the ultramicroscopic state on the other.

The principle, though justified by the new quantum mechanics, has not yet received a rigorous theoretical proof and for the present it must be regarded as something *empirical added* to and regulating the vector model of the atom.

(ii) The selection rules. It is experimentally found that all the possible combinations of permitted energy states of an atom do not actually appear as spectral lines. Selection rules are certain principles, which give the reason for such a state of affairs.

The first selection rule devised in connection with the older atom model referred to the azimuthal quantum number n_φ or k and it was $\Delta k = \pm 1$, as we have seen. This was formulated on an empirical basis, but was later justified theoretically by Rubinowicz by applying the theorem of conservation of momentum to the process of emission of an electromagnetic radiation by an atom. This simple rule, however, was soon found inadequate.

For the vector atom model, three selection rules have been devised, one for L, another for J and a third for S.

(a) *The selection rule for L.*
Most of the observed spectral lines are due to transitions between states, in which a single electron jumps from one orbit to another, and in such cases, the selection rule is $\Delta L = \pm 1$, *i.e.,* only those lines are observed for which the value of L changes by ± 1.

(b) *The selection rule for J.*
Spectral lines arise only when transitions take place between states for which $\Delta J = \pm 1$ or 0, $0 \rightarrow 0$ being, however, excluded.

(c) *The selection rule for S.*
This is given by $\Delta S = 0$, which means that states with different S (hence different multiplicities) do not combine with one another. Theory and experiment, however, show that this selection rule is adhered to less and less strictly as the atomic number increases. Hence it is only an approximate rule holding good in the case of light atoms.

Note.
In the presence of a magnetic field, the orbital magnetic quantum number, m_L either does not change or changes by ± 1, *i.e.,* $\Delta m_L = 0$ or ± 1. The spin magnetic quantum number m_S remains unchanged, *i.e.,* $\Delta m_S = 0$. In consequence

$\Delta m_S = 0$ or ± 1

These selection rules were first introduced on a purely empirical basis in the study of optical and X-ray spectra and Zeeman effect. Later, they were obtained by a rigorously deductive method in wave mechanics and hence they now rest on a permanent theoretical basis.

These rules furnish invaluable help in the allocation of observed spectral series to the proper quantum number. With their aid, energy level diagrams can be constructed both for the natural complex multiplet lines and for the Zeeman effect of such lines.

Transitions which contravene these rules are sometimes observed, but the intensities of such *"forbidden lines"*, as they are called, are usually weak in comparison with the normal lines.

(iii) The intensity rules have been devised to supplement the selection rules, in order to predict also the intensity of the lines that occur. These were originally postulated on an empirical basis in the study of optical and X-ray spectra. Later, a full theoretical derivation was given on a wave mechanical basis.

The intensity rules are:

(*a*) *Those transitions are strong, giving rise to intense lines, in which L and J change in the same sense; the transitions are the weaker, the more the change in direction of L and J is different.*

(*b*) *A transition in the decreasing sense* ($L \rightarrow L$ -1) *is, ceteris paribus* (with all other factors or things remaining the same), *stronger than a transition in the increasing sense* ($L \rightarrow L$ +1).

(*c*) *The case of oppositely directed transitions does not occur,* in general, either in X-ray spectra or in doublet spectra; because it would lead to a final state, in which (J — L) would be two units greater than in the initial state, which is forbidden.

Hence we may distinguish the following cases:
$\Delta L = -1$, $\Delta J = -1$; most intense line (a)
$\Delta L = -1$, $\Delta J = 0$; less intense (a)

$\Delta L = +1$, $\Delta J = +1$; weaker (b)
$\Delta L = +1$, $\Delta J = 0$; weakest (a and *b*)

$\Delta L = -1$, $\Delta J = +1$ no line (c)
$\Delta L = +1$, $\Delta J = -1$ no line (c)

Note.
It can be shown both theoretically and experimentally that the total intensity of all the lines coming to or starting from any given J term of a multiplet is proportional to (2J + 1).

(iv) The interval rule. Landé discovered a rule regarding the interval in frequency between the different levels constituting a multiplet. It is called the *Landé interval rule* and states that the frequency interval between two levels with total angular momenta (J + 1) and J respectively is proportional to (J +1**).**

There are, however many deviations from this interval rule, chiefly in cases when the coupling scheme follows neither the L-S type, nor the *j-j* type, but is intermediate between these extreme cases.

(v) The Landé splitting factor 'g' is simply a factor which affects the ratio of the magnetic moment to the mechanical angular momentum of atomic particles. It characterizes therefore atomic and sub-atomic magnets, such as the orbital and spin magnets.

In order to determine this 'g' factor we shall first of all derive an expression for the magnetic moment corresponding to the orbital motion of the electron.

MAGNETIC MOMENT DUE TO ORBITAL MOTION

BOHR MAGNETON

Considering the simple case of an electron moving in a central orbit, such as an ellipse, the current *i* due to the motion of the electron round the orbit is given by $i = e/c\mathbf{T}$ e.m.u., where *e* is the charge in e.s.u., **c** the velocity of light and **T** the periodic time. Applying Ampere's theorem, this current gives rise to a magnetic moment

$$\mu_l = \frac{e}{c\mathrm{T}} \cdot \mathrm{A} \qquad \ldots (1)$$

where A is the area enclosed by the orbit. Since the areal velocity of motion in a central orbit is $(1/2)\, r^2\, (d\theta/\, dt)$,

$$\mathrm{A} = \int_0^{\mathrm{T}} {}^1/_2\, r^2 \left(\frac{d\theta}{dt} \right)\, dt$$

Now the angular momentum $p_1 = mr^2 (d\theta/dt)$, or

$$\tfrac{1}{2} r^2 \frac{d\theta}{dt} = \frac{p_l}{2m}$$

$$\therefore \qquad A = \int_0^T \left(\frac{p_l}{2m}\right) dt$$

Since p_l and m are constants,

$$A = \frac{p_l T}{2m} \qquad\qquad \dots (2)$$

Substituting this value of A in eqn. (1),

$$\mu_l = \frac{e}{cT} \cdot \frac{p_l T}{2m} = \frac{e}{2mc} \cdot p_l \qquad\qquad \dots (3)$$

As p_l, according to the quantum theory, is equal to $l.(h/2\pi)$,

$$\mu_l = \frac{eh}{4\pi mc} \cdot l \qquad\qquad \dots (4)$$

Hence the orbital magnetic moment μ_l is directly proportional to the orbital quantum number l.

If $l = 0$, $\mu_l = 0$;
If $l = 1$, $\mu_l = eh/4\pi mc$, etc.

In general, the orbital magnetic moment is an integral multiple of the quantity $eh/4\pi mc$ which represents the *smallest constant unit* in terms of which the magnitude of atomic and sub-atomic magnetic moments are measured. It is known as the **Bohr magneton,** usually represented by the symbol M_n. Its value can be readily computed from the known values of *h*, *e/m* and c.

$M_n = eh/4\pi mc = 9.21 \times 10^{-21}$ erg. oersted^{-1}

As the Landè splitting factor g is a factor that affects the ratio μ_l/p_l, let us find this ratio.

$$\frac{\mu_l}{p_l} = l \, \frac{eh}{4\pi mc} \times \frac{1}{l \, \dfrac{h}{2\pi}} = \frac{e}{2mc}$$

The same ratio can be written as

$$\frac{\mu_l}{p_l} = g \cdot \frac{e}{2mc}, \text{ where } g = 1$$

Thus we have introduced 'g'. In the case of the orbital motion of the electron, the value of g is equal to 1.

This way of putting things, at first sight, might appear futile, but it is not really so, since it implies that for other categories of atomic and sub-atomic magnets the ratio of the magnetic moment to the mechanical momentum is not always equal to *e/2mc*. For instance, in the case of the spin of the electron we are compelled, for various reasons, both theoretical and experimental, to assign for the corresponding ratio μ_s/p_s a value *e/mc*, so that the value of g in this case is two, not one. The value of g varies, in general, between 1/2 and 2.

MAGNETIC MOMENT DUE TO ELECTRON SPIN

On this legitimate assumption that the value of g for the spinning electron is 2, the magnetic moment corresponding to the electron spin is readily estimated as follows:

$\mu_s / p_s = 2e/2mc,$
Therefore,
$\mu_s = e\, p_s /mc = e\, s\, h /2\pi\, mc$

Since $s = 1/2$, $\mu_s = eh/4\pi\, mc = M_n$. *The magnetic moment due to electron spin is therefore equal to a Bohr magneton.*

The Landé splitting factor was originally introduced by Landé, on an empirical basis, to explain the anomalous Zeeman effect ; it was later theoretically justified by wave mechanics. It is called the *splitting factor* because it is the determining cause of the anomalous splitting of spectral lines in an external magnetic field.

APPLICATION OF THE VECTOR ATOM MODEL TO ATOMIC AND SPECTRAL PHENOMENA

The great advance made by the vector model of the atom lies in the fact that it offers an adequate and rational explanation of several complex atomic and spectral phenomena which cannot be satisfactorily explained on the basis of the older models. To substantiate this statement four important and interesting cases will now be considered; *viz.*

(i) the electronic structure in atoms,
(ii) the fine or multiplet structure of spectral lines,
(iii) *the* Zeeman effect with anomalous characteristics and
(iv) the Stark effect.

ELECTRONIC STRUCTURE IN ATOMS

Admitting, that the number of electrons in the *normal* atom increases steadily as one passes from light to heavy elements, one is interested to know whether the several electrons in a given atom are distributed at random or in a regular manner. Since, in general, a system is most stable when its potential energy is least and as the potential energy of the atom is smaller the lower the value of the total quantum number n designating the electronic orbit, it would seem that all the electrons in the atom should tend to crowd into the lowest orbit. But investigations based on the X-ray spectra and periodic classification of elements have proved the contrary, *viz.*, that

> the electrons do not all crowd into the lowest orbit of minimum energy but are arranged with a certain regularity in different orbits or shells.

The **X-ray spectra** of elements strongly suggests that the *electrons in the atom are arranged in different groups or shells* round the nucleus, all the electrons having the same total quantum number n forming a shell. Thus, the innermost shell, known as the K shell, consists of electrons whose total quantum number $n = 1$; the next shell, called the L shell, contains electrons, for which $n = 2$ and so on. The complete scheme, as far as it is required in building up the electronic structure in the atoms of all the elements, is:

Shell →	K	L	M	N	O	P	Q
n →	1	2	3	4	5	6	7

For the conception and nomenclature of these electronic shells, we are indebted to Lewis, Langmuir, Bohr and Stoner.

The **periodic table** is an arrangement of different elements that exist in Nature in a scheme based on their chemical properties and atomic weights, first drawn by Mendeleeff in 1871 and then gradually perfected, chiefly by the study of the X-ray spectra of elements.

Moseley's work on the characteristic X-ray lines, for instance, settled in an unambiguous manner that the determining factor in the arrangement is not the atomic weight but the atomic number which assigns the numerical position of the elements in the classification as 1, 2, 3, etc., up to 92.

This table of elements presents a complex scheme, with *seven horizontal rows, called* **periods,** in which the chemical and physical properties of the elements vary gradually as a periodic function of the atomic number, and *eight vertical columns, called* **groups,** each of which contains elements manifesting similar properties.

The number of elements in the different periods is not the same, *e.g.,* the first period consists only of two elements, the-second and third eight elements each, the fourth and fifth eighteen elements each, the sixth thirty-two and the seventh sixteen elements.

As the atomic number gives also the number of electrons in the atom, it readily follows that *the atoms of successive elements in the periodic table are formed by the addition of one more electron at each step.*

Such a classification of elements, with its clear manifestation of periodic variations in the properties of the elements, confirms, in its turn, the grouping of electrons in different shells. For, it readily suggests the idea of the formation, at regular intervals, of a *new shell* which, as it develops in the course of the period, passes through the same characteristic phases, determined by the number and distribution of its component electrons.

There is also a further implication that a new shell which makes the beginning of a new period, will, in general, be formed only when the inner ones are completed with their quota of electrons. It must, however, be noted that there occur cases where a new outer shell is started before the inner one is completed.

To determine the number of electrons required to complete each one of the shells, we have a clue in the periodic occurrence of monatomic elements, chemically inactive, hence known as inert gases, viz., helium, neon, argon, krypton, xenon and radon. Since their atomic numbers are *2.* 10, 18, 36, 54, and 86 respectively, the number of electrons in their respective atoms are also 2, 10, 18, 36, 54 and 86.

THE PERIODIC TABLE OF ELEMENTS

Period	Gr. I	Gr. II	Gr. III	Gr. IV	Gr. V	Gr. VI	Gr. VII	Gr. VIII			Gr. 0	Electronic shells
I	1 H 1·008										2 He 4·003	K
II	3 Li 6·94	4 Be 9·02	5 B 10·82	6 C 12·01	7 N 14·008	8 O 16	9 F 19				10 Ne 20·183	K, L
III	11 Na 22·997	12 Mg 24·32	13 Al 26·97	14 Si 28·06	16 P 30·98	16 S 32·07	17 Cl 35·46				18 A 39·944	K, L, M
·IV	19 K 39·10	20 Ca 40·08	21 Sc 45·10	22 Ti 47·90	23 V 50·95	24 Cr 52·01	25 Mn 54·93	26 Fe, 27 Co, 28 Ni 55·84, 58·94, 58·68				K, L, M, N
	29 Cu 63·54	30 Zn 65·38	31 Ga 69·72	32 Ge 72·60	33 As 74·91	34 Se 78·96	35 Br 79·92			36 Kr 83·7		
V	37 Rb 85·48	38 Sr 87·63	39 Y 88·92	40 Zr 91·22	41 Cb 92·91	42 Mo 95·95	43 Tc 99	44 Ru, 45 Rh, 46 Pd 101·7, 102·9, 106·7				K, L, M, N, O
	47 Ag 107·88	48 Cd 112·41	49 In 114·8	50 Sn 118·7	51 Sb 121·77	52 Te 127·61	53 I 126·92			51 Xe 131·3		
VI	55 Cs 132·91	56 Ba 137·37	57-71 Rare earths	72 Hf 178·6	73 Ta 180·88	74 W 183·92	75 Re 186·31	76 Os, 77 Ir, 78 Pt 190·2, 193·1, 195·2				K, L, M, N, O, P
	79 Au 197·2	80 Hg 200·61	81 Tl 204·4	82 Pt 207·2	83 Bi 209	84 Po 210	85 At 211			86 Rn 222		
VII	87 Fr 223	88 Ra 226·05	89-103	Actinide	series							K, L, M, N, O, P, Q

RARE EARTHS									
VI 57-71	57 La 138·9	58 Ce 140·1	59 Pr 140·9	60 Nd 144·3	61 Pm 147·0	62 Sm 150·4	63 Eu 152·0	64 Gd 156·9	65 Tb 159·2	66 Dy 162·5	67 Ho 164·9
	68 Er 167·2	69 Tm 169·4	70 Yb 173·04	71 Lu 174·99						

ACTINIDE SERIES									
VII 89-103	89 Ac 227	90 Th 232·12	91 Pa 231·1	92 U 238·07	93 Np 237	94 Pu 242	95 Am 243	96 Cm 245	97 Bk 249	98 Cf 249	99 E 254
	100 Fm 255	101 Mv 256	102 No 256	103						

Rydberg made a suggestion that these numbers could be arranged in a simple numerical series as follows:

$$2 \times (1^2 \quad + \quad 2^2 \quad + \quad 2^2 \quad + \quad 3^2 \quad + \quad 3^2 \quad + \quad 4^2 \quad + \ldots)$$

$$\leftarrow \text{He} \rightarrow$$
$$\longleftarrow\!\!\!-\!\!\text{Ne}\!-\!\!\!\longrightarrow$$
$$\longleftarrow\!\!\!-\!\!\!-\!\!\text{A}\!-\!\!\!-\!\!\!\longrightarrow$$
$$\longleftarrow\!\!\!-\!\!\!-\!\!\text{Kr}\!-\!\!\!-\!\!\!-\!\!\!\longrightarrow$$
$$\longleftarrow\!\!\!-\!\!\!-\!\!\!-\!\!\text{Xe}\!-\!\!\!-\!\!\!-\!\!\!-\!\!\!\longrightarrow$$
$$\longleftarrow\!\!\!-\!\!\!-\!\!\!-\!\!\!-\!\!\text{Rn}\!-\!\!\!-\!\!\!-\!\!\!-\!\!\!-\!\!\!\longrightarrow$$

Such a series suggests that the *electrons in these inert gases can be arranged in shells:*
Helium atom, for instance, has a single shell of two electrons;
Neon has two shells, the first containing two electrons and the second eight electrons;
Argon has three shells, the first one containing two electrons, the second eight, the third eight also;
Krypton has four shells, the first three the same as argon, but the fourth eighteen electrons, and so on.

The factor 2 outside the bracket suggests *some sort of symmetry,* a *pairing of electrons.*

Further investigations showed that Rydberg's suggestion had to be modified as regards the order of the numbers within the bracket.

Thus, in the case of Kr, the arrangement should be
$2 (1^2+2^2+3^2 +2^2)$,
the outermost shell containing eight electrons like argon, but the third one eighteen electrons.

Similarly for Xe the order should be
$2 (1^2+2^2+3^2 +3^2+2^2)$

and for Rn
$2 (1^2+2^2+3^2+4^2+3^2+2^2)$.

This would indicate that the number of electrons in the shells increases up to a certain limit and then decreases. The chemical inactivity and the consequent stable structure of these elements suggest the general law that *2, 10, 18, 36, etc., electrons tend to make a very stable combination in the atom.*

On the other hand, the next element *beyond* any one of these inert gases has one more electron than necessary to form a stable structure, e.g. Li (3), Na (11), K (19), Rb (37), Cs (55), the metals of the alkali group, whose properties are strikingly similar.

The second element beyond an inert gas has two superfluous electrons, *e.g.,* Mg (12), Ca (20), Sr (38), Ba (56), Ra (88), the alkaline earths, also with similar properties.

Going in the other direction the next element *before* an inert gas lacks one electron to make a stable structure, *e.g.,* F (9), Cl (17), Br (35), I (53), forming the halogen group, again with similar properties.

The second next before the inert gases, viz., O (8), S (16), Se (34), Te (52), are likewise similar.

Considerations of this kind indicate a **very fundamental principle in electronic structure, viz., the chemical and physical properties of an atom are determined more by the number and arrangement of the electrons in the outermost shell than by the total number of electrons in the atom.**

Conversely, by studying the chemical and physical properties of atoms, one may obtain information as to the number of electrons in the outermost shell. Then paying attention to the position of the corresponding element with respect to the nearest inert gas, it is possible to determine the whole arrangement of electrons in the different shells. Thus, for instance, in the case of Na, the arrangement would be two electrons in the first K shell, 8 electrons in the second L shell and one in the third M shell, as yet incomplete, represented ordinarily by the abbreviation (2, 8, 1). Similarly the arrangement in Mg is (2, 8, 2), in F (2, 7), in O (2, 6). Similar reasoning determines the number of electrons required to complete .different shells. The K shell is complete or closed with two electrons, the L shell with 8 electrons, the M shell with 18 electrons, the N shell with 32 electrons, the O shell with 18, the P shell with 12, and the Q shell with 2.

Thus the shell character of the electronic structure in atoms, *i.e.,* the property which finds its expression in the periodicity of elements with respect to their physical and chemical properties was discovered by reasoning based on empirical data.

The **vector model gives a rational explanation of such an empirically derived shell property of atoms,** showing why the electrons are grouped in shells, a certain fixed number of these being required to complete a shell, and why the number and distribution of electrons outside closed shells are mainly responsible for the chemical and physical properties of elements, thus leading up to a logical construction of the periodic table.

Considering, for instance, the *K shell,* for the electrons in it $n = 1$, $l = 0$; hence $m_l = 0$. Thus three of the four quantum numbers that characterize the electron are fixed. The fourth, viz,, m_s can assume only the two values $+1/2$ and $-1/2$. It is readily seen that the maximum number of electrons in the K shell can be *only two,* since a greater number would mean that two or more should necessarily be identical as regards all their four quantum numbers, which is against Pauli's exclusion principle. *The shell is therefore completed or closed with 2 electrons.*

For the electron in the next L *shell,* $n = 2$, $l = 0$ or 1. In the case $n = 2$, $l = 0$, $m_l = 0$, and $m_s = \pm 1/2$; hence there can be only two electrons according to Pauli's principle. In the other case $n = 2$, $l = 1$, m_l can have $(2l + 1)$ values, *i.e.,* three values 1, 0, - 1, and with each of these may be associated the two values of m_s, *i.e.,* $\pm 1/2$, so that there will be $2(2l+1)$ or 6 possible sets of values for the four quantum numbers characterizing the electrons. Hence in this case six electrons can coexist. The *L shell* with two sub-groups ($n = 2$, $l = 0$, and $n = 2$, $l = 1$) *is therefore completed when it contains 2 + 6 = 8 electrons.*

The electrons in the third *M shell* have $n = 3$; while $l = 0, 1, 2$. Three sub-groups can therefore be distinguished in this shell, first $n = 3$, $l = 0$, second $n = 3$, $l = 1$, third $n = 3$, $l = 2$. The first and second sub-groups are completed by two and six electrons respectively for reasons given above. The third is completed with $2(2l + 1)$ electrons, *i.e.,* 10 electrons since $l = 2$. Hence the *total number of electrons required to complete the M shell is 18.*

Arguing in a similar manner it can be shown that the N *shell* has four sub-groups, since $n = 4$, while $l = 0, 1, 2, 3$ and that the first three are completed by 2, 6, 10 electrons respectively, while the fourth by $2 (2 \times 3 + 1) = 14$ electrons. This brings up the total to 32 electrons to complete this shell.

Thus the vector model in conjunction with Pauli's principle supplies the reason why the electrons of an atom do not all occupy the most stable shell closest to the nucleus but dispose themselves in different shells. From what has been said above the following important conclusions can be drawn:

(i) *In electronic configuration of an atom there can be only $2n^2$ electrons with the same total quantum number n.*

(ii) *In the nth shell there are n sub-groups and in each sub-group whose orbital quantum number is l, there can be only $2(2l + 1)$ electrons which are known as "equivalent electrons".*

A direct consequence of Pauli's Principle is that every closed shell or sub-shell is balanced with respect to angular momenta both as regards orbital and spin motions **and hence contributes nothing to the total angular momentum** of the atom as a whole.

The proof of this is essentially contained in the statement that a closed shell of orbital quantum number l is built up of $2(2l + 1)$ electrons. Since the $(2l + 1)$ different values of m_l and the two different values of m_s are just used up in this way by the $2(2l + 1)$ electrons, there is nothing more of m_l or m_s which can contribute to the atom as a whole.

In other words, for each closed sub-group as well as for the whole closed shell

209

$$m_L = \sum m_l = 0$$
$$m_S = \sum m_s = 0$$

Thus, for instance, taking the L shell, for the first closed sub-group ($n = 2$, $l = 0$), $m_l = 0$ and m_s is equal to $+1/2$ for one and $-1/2$ for the other, so that $\sum m_s = 0$. For the second sub-group ($n = 2$, $l = 1$) the vector sum of the three m_l values, viz., $+1$, 0, -1 is evidently zero, so that $\sum m_l = 0$; similarly the vector sum of the two m_s values, viz., $+1/2$, $-1/2$ is also zero.

If, therefore, m_L and m_S are zero for a closed shell, $L = 0$, $S = 0$ and $J = L + S = 0$, i.e., the total angular momentum of a closed shell is zero. Hence

a closed shell can contribute nothing to the atom as a whole. From this it follows that the total angular - momentum of the atom as a whole is determined only by the electrons outside closed shells. *This fact explains why the chemical and physical properties of elements depend only on the electronic configuration outside closed shells,*

i.e., on the electrons in the outermost incomplete shells, which are, in consequence, known as time *valency or optical* electrons.

Having thus fixed the maximum number of electrons in each shell and sub-shell, it is possible to build up elements genetically and show that they can be arranged in a periodic system, starting from hydrogen and ending at uranium. The general procedure is to picture the nucleus of atomic number Z, initially stripped of all its electrons, as capturing one by one the Z electrons necessary to make a neutral normal atom. The quantum numbers chosen for each added electron are such as to place this electron in a vacant permissible shell, which makes *the energy of the system a minimum,* the condition required for maximum stability In deciding which orbit the next captured electron will permanently occupy, spectroscopic and critical potential data arc used.

When normal atoms of different elements are built like this, it is found that their chemical and spectroscopic behavior are determined by the electron configuration of the outermost shell, as is expected from Pauli's principle. Elements with outermost shells of similar electronic structure possess in a large measure equivalent chemical and optical properties. This explains the occurrence of periods in the classification of elements. Thus the alkalis, to which belong Li, Na, K, Rb and Cs, are characterized by a single electron in the outermost shell; the alkaline earths Mg, Ca, Sr, Ba and Ra have two electrons outside closed shells; the halogens F, Cl, Br and I lack one electron to make up a closed shell; the inert gases Ne, A, Kr, Xe and Rn have the outermost shell closed with eight electrons.

The vector model leads therefore to a periodic system agreeing closely with the one built on empirical data. It further presents an *ideal* system, much simpler than the real one of the chemists, and going deeper into the internal structure of the atom, since the real system gives only the peripheral properties of the atom, which are of interest to the chemist and the spectroscopist in the visible region, while the ideal system describing the interior of the atom is of interest to the X-ray spectroscopist.

It is to be noted that *the ideal periodic system given by the vector model does not, however, completely agree with the real one.* The two coincide up to argon, but then with potassium and calcium, the N shell begins to form before the completion of the preceding M shell ; likewise the O shell is initiated irregularly with Rb (37), while it has to begin only with Hf (72). Likewise, in the real system the O shell has only 18 electrons although 30 electrons would be required according to the ideal system to complete it, the P shell only 12 electrons, while 72 would be necessary to fill it entirely and the Q shell 2 electrons when 98 would be its maximum quota of electrons.

These discrepancies are accounted for as follows: The minimum potential energy condition may not always be such that one shell must be completed before an electron settles in the next shell. Further, since the deviations of the real system from the ideal one concern only the less tightly bound electrons of the outermost shells, if we regard as representatives of the elements not only the normal neutral atoms, but also the ionized ones, the theoretical rule is found to be exactly verified.

SOME EXAMPLES OF ELECTRON CONFIGURATIONS WITH THEIR MODERN SYMBOLIC REPRESENTATIONS

In order to illustrate the theory proposed above, the electronic configurations in some elements will now be considered.

Hydrogen (Z = 1)

This has only one electron which will naturally occupy the first K shell for which $n=1$, in order to give least energy to the system. According to Pauli's principle there can be only two electrons in the K shell, or the single electron can exist in two different states for $m_s = +1/2$ and -1/2. The symbolic representation is: 1s, where 1 referring to the total quantum number of the shell and s the type of electron, since $l = 0$ in this case. On account of the single electron which cannot complete the K shell, one may expect the atomic hydrogen to be very active chemically, combining its single electron with another hydrogen atom to form a helium-like pair, the molecule of hydrogen which is known to be comparatively inactive.

Helium (Z = 2)

This has two electrons, both of which, according to Pauli's principle, can occupy the first K shell. Both spectroscopic and ionization potential data show that to be really the case. Further, the two electrons complete the shell, which renders the helium atom an *inert* monatomic gas, completing the first period of the periodic system. The normal helium atom is therefore represented by $1s^2$, *i.e.*, two electrons, for which $n = 1$ and $l = 0$, the total number being affixed at the top of s. The rectangular enclosure indicates the electrons interlocked in closed shells.

Lithium (Z = 3)

This has three electrons, two of which occupy the first K shell and the third the next L shell, in order to keep the potential energy a minimum. The third electron in the incomplete L shell is the *valance* electron. Hence the neutral atom of Li is represented by $1s^2 2s$. Optical spectrum of Li shows that the lowest orbit is such that this third electron is really $2s$ and not $2p$ or any other. The valence $2s$ electron is much more free, i.e., much more loosely bound than the $1s^2$ electrons, as is seen from the fact that Li is strongly monovalent and electropositive. The *arc spectrum* of Li is predominantly that to be expected from a single electron system of the alkali type. The *spark spectrum*, on the other hand, which is ascribed to Li^+ is similar to the arc spectrum of He. This is what we should expect from theory as a singly ionized Li atom contains the same number of electrons as the normal He atom. The frequencies of corresponding lines are, however, much higher than in He, because the stronger nuclear charge in Li causes all energy levels to lie much lower.

Sodium (Z = 11)

This has a total number of 11 electrons. According to Pauli's principle, 10 of them will be interlocked in the first two K and L shells, completed by 2 and 8 electrons respectively, so that the remaining single electron will have to occupy the third M shell. The two electrons in the K shell are represented by $1s^2$. Of the eight electrons in the L shell two will be in the first sub-group ($l = 0$) and hence represented by $2s^2$; the remaining six in the second sub-group ($l = 1$) and hence by $2p^6$. The last electron in the M shell is the valence electron which is found to be an s electron from energy considerations as well as from spectroscopic data. Hence the electron configuration of the normal sodium atom is represented by

$$\boxed{1s^2\, 2s^2\, 2p^6} \quad 3s$$

The arrangement receives confirmation from the fact that sodium is *monovalent* and *electropositive*.

Magnesium (Z = 12)

Of the 12 electrons, 10 occupy the first two shells while the remaining 2 will have to go to the third M shell. These 2 electrons in the outermost incomplete M shell are the valence electrons making Mg *divalent*. A study of the arc and spark spectra shows clearly that the two valence electrons are s electrons. Hence the normal Mg atom is represented by

$$\boxed{1s^2\, 2s^2\, 2p^6} \quad 3s^2$$

Aluminum (Z = 13)

Leaving aside the 10 electrons occupying the first two closed shells, the remaining three go to the M shell. Hence Al is *trivalent*. Spectroscopic study and energy considerations show that the 11th and 12th electrons are s electrons, while the 13th is a p electron. The symbolic representation of Al atom is

$$\boxed{1s^2\, 2s^2\, 2p^6} \quad 3s\, 3p$$

Copper (Z = 29)

Of the 29 electrons, 2 will go to the first shell, 8 into the two sub-groups of the second L shell and 18 to the third M shell, distributed as 2, 6 and 10 in the three sub-groups. What is left over is only one electron, the valence electron, which occupies the fourth N shell. Hence copper should be *monovalent* and its electronic configuration is

$$\boxed{1s^2\ 2s^2\ 2p^6\ 3s^2\ 3p^6\ 3d^{10}}\qquad 4s$$

The arc spectrum of copper should therefore resemble that of the alkalis like Li, Na, etc. It is to be noted, however, that the last of the electrons which completes the 3d sub-shell is rather lightly bound and comes off easily so that copper frequently exhibits a valency of two as an alternative to the expected valency of one, as clearly seen from the existence of both cupric oxide (CuO) and cuprous oxide (Cu_2O).

Thallium (Z = 81)

The total number of electrons is therefore 81. The normal atom of Tl is represented, applying Pauli's principle and paying due attention to energy considerations, by

$$\underset{\text{K}}{1s^2}\quad \underset{\text{L}}{2s^2\,2p^6}\ \underset{\text{M}}{3s^2\,3p^6\,3d^{10}}\quad \underset{\text{N}}{4s^2\,4p^6\,4d^{10}\,4f^{14}}\quad \underset{\text{O}}{5s^2\,5p^6\,5d^{10}}\ \underset{\text{P}}{6s^2\,6p}$$

Hence it has the first four shells complete. Before the fifth O shell is completed, the sixth P shell is started. Of the three electrons in the last shell, two are *s* electrons and one *p* electron. Hence Tl should be *trivalent*.

These few examples will serve to demonstrate the methods and some of the fundamental principles employed in ascertaining the distribution of electrons in atoms. The same procedure is followed in the case of all the other elements in the periodic table.

Note 1. Electron configuration and valency

The chemical valency is the governing factor in the formation of chemical compounds from elements. From what has been said above concerning electronic structure, it is easily seen that the number of electrons in the outermost incomplete shell determines the valency of an element, due attention being paid to its position with respect to the nearest inert gas. Thus the elements in the same group of the periodic table have the same valency, the first group monovalent, the second group divalent, the third trivalent, etc., while in one and the same period the valency first increases and then decreases by unity at each step. Further,

> since an element occurring between two successive inert gases may be considered to be either *after the first* or *before the second* inert gas, it may have both positive and negative valencies,

e.g.,

Sulphur being six steps after neon is a positive sextavalent; as in SO_3; it is also two steps before argon and hence a negative divalent, as in H_2S.

Carbon is four steps after helium and four steps before neon, occupying an intermediate place between the two inert gases ; it has therefore a valency four, both positive and negative.

This rule, however, cannot be generalized on account of the preference shown by elements to valency of one sign over that of the other, *e.g.,* chlorine nearly always exercises its negative valency as in HCl, AgCl, NaCl, $AlCl_3$, in preference to its positive valency **as in** Cl_2O_7. The partiality of carbon towards the negative valency seems to be at the basis of great number of organic compounds which it is able to form. Further investigations have brought out the following points about the cause and nature of valency:

(a) Since the electrons in the outermost shell, being at the boundaries of the atom, will be involved when two atoms become united, it is readily understood why valency is due to these outer electrons.

(b) A more fundamental reason for the existence of valency is the tendency of the electron systems in atoms to form stable groupings like those of the inert gases whose outermost shell always contains eight electrons. Hence *the formation of an octet by an atom appears to be the real cause of its valency.*

When an atom lacks the number to complete the octet in its outermost shell, it will readily combine with another atom which is in excess of eight electrons in its outermost shell by exactly the number required by the first. In the exchange process, the two atoms would be ionized, the one that loses its electrons becoming a positive ion, while the other that gains the electrons becoming a negative ion. The oppositely charged ions thus formed attract one another strongly and give rise to what is known as *electrovalent* compound, occurring practically in all inorganic salts.

There is yet another way of formation of the octet stable grouping, viz., two atoms may mutually complete octets by a *common, sharing of a pair of electrons.*

This form of union gives rise to what is called a *covalent* compound, occurring in all organic and many inorganic substances.

(c) While in the first two periods valency is largely based on the tendency towards octet formation, this relationship becomes more complicated in the other periods containing elements of higher atomic number, probably due to the fact that electrons from inner shells also may influence valency transactions.

Note 2

Semiconductors and transistors
The electron configuration and valency of atoms throw light on the electrical behavior of semiconductors in general and of their special version, known as transistors, that are fast growing in importance in electronics as able competitors to thermionic valves.

Semiconductors are solid crystalline substances having electrical properties which may be regarded as intermediate between those of metallic conductors and of insulators. As already stated, at low temperatures the bands of energy levels in semiconductors are either completely filled with electrons or completely empty, with gaps separating the filled and unfilled bands. But due to localized impurities, the above mentioned gaps are much narrower in a semiconductor than in an insulator, so that electrons in the lower filled band have a reasonable probability of gaining sufficient energy from the thermal vibrations of crystal lattice to enter the higher, nearly empty band. Hence

In semiconductors the number of free charge carriers, which can cause current to flow when an e.m.f. is applied, is very limited, and is a function of temperature or of impurities present. This means also that if the impurities are completely removed from the substance, the latter will behave as an insulator.

Copper oxide, silicon, selenium, germanium, etc. are some of the semiconductors that are of special importance in electronics.

Let us now consider the inner mechanism that determines their electrical behavior. Taking a concrete example, say **silicon**, this belongs to the fourth group of the periodic table and hence is tetravalent; it occurs in the diamond type of crystal lattice where each silicon atom is surrounded by four nearest neighbors at the corners of a tetrahedron, the binding being by electron-pair or covalent bonds, i.e., each of the four valency electrons of a particular atom forming a pair with one of similar electrons of the four nearest atoms. Hence in absolutely pure silicon, since all the electrons are held in the valence bonds, no mobile carriers of electricity are available and hence such a material cannot conduct electricity at low temperatures. If pure silicon is to become a conductor, the valence bonds must be broken and the electrons set free, which can be done only by the expenditure of energy corresponding to the gap between the filled and unfilled bands. This energy can be supplied in several ways. In the thermal process the crystal is raised to a moderately high temperature.

With increase of temperature, as the valence bands are broken and more and more of the valence electrons cross the gap, the effect due to the thermal agitation of molecules is reduced. Hence the electrical resistance of the substance diminishes with rise in temperature.

The specific resistance ρ of a pure or intrinsic semiconductor varies with temperature according to the relation
$\rho = A \exp(E_G/2kT)$
where
A is a constant of the material,
E_G the energy gap between filled and unfilled bands,
k the Boltzmann constant and
T the absolute temperature.

It is readily seen that as T increases ρ will decrease. The values of E_G for some of the important semiconductors are:

Germanium	—	0·76	eV
Silicon	—	1·20	,,
Cuprous oxide	—	1·40	,,
Diamond	—	7·00	,,

It is also possible to dislodge the electrons from the valence bonds and make them move from the filled to the unfilled energy bands oven at ordinary temperatures by the action of light, provided the quantum energy by of the light exceeds the energy gap E_G. Thus in the case of pure silicon, when the quantum energy exceeds the critical value of 1.2 eV, the valence electrons that have been set free and have crossed over the gap can move freely in the crystal lattice. Under these circumstances, if an external field is applied, a steady drift of the free electrons towards the positive terminal will take place, constituting a photoelectric current. This means that pure silicon has now become a conductor of electricity.

The empty space left behind when an electron escapes from the valence bond is called a *"hole"* with the consequent creation of an *electron-hole pair*. Since the electrons in the atoms of a crystal always tend to arrange themselves in valence bonds, a hole in an atom will be soon filled by another electron from an adjacent atom, thereby shifting the hole to that adjacent atom. This hole in turn will be filled up by an electron of the next nearest atom and so on. Thus the holes also move about in the crystal in the same way as the free electrons, with a steady drift towards the negative terminal of the applied external field. The holes therefore behave as though they were identical with the electrons but with positive charge. Conduction by holes is essentially the same as that by electrons except for the difference in sign of the "effective" mobile charge and for the different mobilities, the electrons usually moving twice as fast as the holes.

Impurity in semiconductors
Semiconductors, in general, owe their electrical behavior to the presence of minute amount of impurities, so that they are partially conducting even without the external influence of heat or radiation. This is due to the fact that additional energy levels of the impurity atoms occur in the region between the energy bands of the basic material. The atoms of the impurity may be either present in the natural material or added in controlled amounts so as to fit into the crystal lattice where they replace some of the atoms of the basic material. To be useful, they must further have a critical valency and possess energy levels lying either just below the band of empty energy states or immediately above the band of filled levels of the basic material. Semiconductors corresponding to these conditions are known as *"donor"* and *"acceptor"* types respectively.

When, for instance, phosphorus is present in silicon as impurity or is added in proper minute amounts, it replaces silicon atoms in the lattice. As phosphorus belongs to the fifth group of the periodic table, it has five valence electrons or one more than is needed to saturate the four valence bonds with neighboring silicon atoms. Hence the fifth valence electron is only weakly bound to the phosphorus atom, so that even at room temperature it is set free by the thermal vibrations of the crystal lattice. The electrons thus liberated contribute to the electrical conductivity. Such an impurity is called a *"donor"* since it donates electrons to the basic semiconductor. If the energy levels of the donor type of impurity lie just below the unfilled band of the basic material, the fifth unbound electrons can easily gain thermal energy from the lattice and pass over to that band, where they are then capable of gaining additional energy and momentum from the applied electric field and so carry current. In such semiconductors the current flow is entirely due to electron movement and they are called the *n*-type, the *n* signifying the negative polarity of the electrons.

If, on the other hand, **boron** is present in silicon as impurity, the semiconductor will be of the acceptor type. For boron is trivalent and so has one electron less than needed to saturate the valence bonds with four neighboring silicon atoms. The valence bonds lacking an electron may shift from one bond to another by motion of electrons in the opposite direction. Such an impurity is called an *"acceptor"*, since its energy levels lying just above the filled band of the basic material are furnished with electrons that make upward transitions from the filled band on gaining thermal energy from lattice vibrations. The holes thus created in an otherwise filled energy band make possible the flow of current, for, electrons in this band may gain energy from the applied electric field and enter the holes. This transition creates other holes and the process may repeat indefinitely. In such semiconductors the current flow may be considered as due to the motion of the holes in the nearly filled band of energy states, even though the carriers of current are actually electrons as confirmed by gyromagnetic experiments. These semiconductors are called *p*-type, the *p* indicating the positive polarity of the hole, which is the ultimate cause for conduction.

214

The amount of impurity present in or added to the semiconductor to produce the above results is amazingly small—about 1 impurity atom in every 10 million pure semiconductor atoms.

The electron-hole activity in a semiconductor is accelerated at high temperatures, so that the conductivity of the semiconductor increases with temperature, which means that the resistance of the semiconductor has a negative temperature coefficient unlike a true conductor.

But the density of impurity atoms is so low that at high temperatures (above 600°K), where electrons pass readily across the energy gap E_G by thermal excitation, impurity atoms contribute a negligible amount to the conductivity. However, at. room temperature and below, the electrical conductivity depends almost entirely on impurity content, the impurities contributing electrons or holes that permit additional current flow, the resistivity decreasing with increase in the amount of impurity. The ionized impurity atoms in the lattice also contribute to the electrical resistance, since on account of their charge they can scatter electrons and even capture them. But the coulomb interaction between an electron and a singly charged impurity atom is greatly reduced by the dielectric constant of the basic material. As the temperature falls in this region, the resistance increases, since electrons or holes are captured by positive or negative impurity ions and the thermal vibration energy is too small to re-ionize them. In the intermediate temperature range between pure and impure types of semiconductor behavior, the resistivity increases with rising temperature, due to the increased thermal vibration of the lattice causing greater scattering of the Broglie waves associated with the electrons or holes. When the temperature reaches a value such that intrinsic conductivity becomes important the relative effect of lattice vibrations becomes negligible.

It may be noted that the conduction process in a semiconductor is slower than in a true conductor. This is because the electrons in a semiconductor move slowly as they encounter obstructions due to crystal imperfections. The holes move even slower because of their progress by jumps through gaps. The mobility of the electrons is greater than that of the holes and hence greater current will be conducted by the electron-drift than by the shift of the holes.

It may also be noted that the rapid variation of resistivity of semiconductors With temperature makes them important for temperature measurement and control. Devices made of mixtures of NiO with Mn_2O_2 and Co_2O_3, known as *thermistors,* are available for this purpose. The electrical behavior of this mixture of oxides depends only slightly on impurity content, so that thermistors may be manufactured as a reproducible product having several technical applications.

In practice, the two types of semiconductors are manufactured as follows.

Semiconductor materials are prepared in very pure forms by various melting techniques under controlled atmospheres. The semiconductor is grown as a single crystal by dipping a small seed crystal into the melt and slowly withdrawing it under rotation. While the crystal is being pulled from the melt, controlled amounts of impurities may be added at proper times to give the *n*- or *p*-type. What is of immediate interest to us here is to form suitable junctions of the two types, which make the assembly acquire properties of a rectifier or diode and of a triode or transistor.

Semiconductor diodes are obtained by forming junctions of the *n*- and *p*-type semiconductors, which may be done in several ways. For instance,

(i) In *"grown junction"*, the *n* and *p* regions arc grown in the same semiconductor by adding controlled amounts of donor and acceptor impurities to a single crystal during its formation.

(ii) A *"diffused junction"* is formed by welding a pellet of donor impurity on one face of a wafer of an acceptor type of semiconductor, that the pellet melts and diffuses upto a short distance into the wafer, thereby creating a region of *n*-type in intimate association with the *p*-type in the same semiconductor.

(iii) In the *"point contact"* type, the junction is produced near the contact by passing a heavy current through the point, causing a change in the *n*-type base material ,near the point which apparently converts it to *p*-type material.

In a junction thus formed, electrons and holes are available in the *n* and *p* regions respectively of the semiconductor as carriers of current. They are spoken of as *majority carriers,* since in each case they far outnumber the carriers of opposite sign. Each region is electrically neutral when the total charges of all the atoms are considered.

When a potential is applied to make the p-region positive, the situation is altered. The holes in the p-region are repelled by the positive field there, while the electrons in the n-region are also repelled by the negative field there. Hence both the holes and the electrons drift towards the junction. The electrons move across the junction and fill the adjacent holes. At the same time the holes that migrate to the junction replenish the hole supply at that point. Hence a current flows in the circuit as long as the *e. m. f.* is applied.

Under these conditions the junction is said to be biased in the *forward direction* offering low *resistance* to the passage of current.

If, on the other hand, the p region is made negative and the n region positive, by reversing the direction of the applied *e. m. f.,* both holes and electrons are attracted towards the respective terminals and away from the junction. This leaves the junction region devoid of charges and consequently the current flowing in the circuit is very low. The junction now appears to offer *high resistance* to the passage of current and is said to be biased in the *reverse direction* or to have developed a *potential barrier.*

It is to be noted that a junction exhibits rectifying property, since it is able to pass more current in one direction than in the other, like a *diode* rectifier, and this constitutes one of the many uses of the semiconductor device. Such rectifiers are now produced commercially; one type will pass 2.5 amperes with a potential drop of less than 1 volt and will withstand 600 volts in the reverse direction.

Semiconductor triodes or transistors
It is possible to join two semiconductor diodes to form a semiconductor triode, called a *transistor.* This has been achieved, in practice, in the following different ways, similar to the junction formation processes

(a) *Junction types.*
Two types of junction transistors are available. A *grown-in-type* which is made by forming a crystal with a very thin (about 0.001 inch or 25.4 micron thick) p-layer between two n-layers of greater thickness. Thus by "doping" the same tetravalent germanium (as the single. crystal is grown from the melt) with the addition of small amounts first of trivalent gallium and then of five valency antimony a n-p-n transistor is produced. A crystal may also be grown with p-n-p layers to give similar performance. A *diffused type* is obtained when, for instance, a pellet of indium (boron or gallium) is melted on each face of a thin wafer of a n-type single crystal germanium, so that some of the indium, diffuses into the faces of the wafer, thereby forming a p-n-p transistor.

(b) *Point contact types*
These may be considered as a special version of the junction type. Two fine pointed wires make pressure contact with a face of a n-type germanium wafer, separated by a very small distance of the order of 0.002 inch or 51 microns. When a large current is momentarily passed across the points of contact, the heat produced drives a few electrons away from the regions of the points of contact, leaving holes and thus converting into p-type a small volume of germanium immediately under and around the points of contact, so that a p-n-p transistor is effectively produced. In a similar number; point contact n-p-n transistor can be made.

(c) *Surface barrier types*
There are produced by electrolytic etching and plating processes. Two fine streams of indium sulphate solution are played upon axially opposite points on the surfaces of a n-type germanium wafer. At the same time a direct current is passed through the wafer and solution in such a direction as to remove germanium, electrolytically from the faces of the wafer. The tiny sprayed areas are gradually etched away. When the desired thickness has been reached, the etching process is stopped abruptly by a sudden reversal of current. This reversal causes an indium metal dot to be plated on each of the opposite faces of the wafer in the etched out portion, thus forming a p-n-p type transistor.

Transistors can therefore be classified into two categories, the n-p-n and p-n-p types. Their *principle of action* is illustrated in Fig. 220.

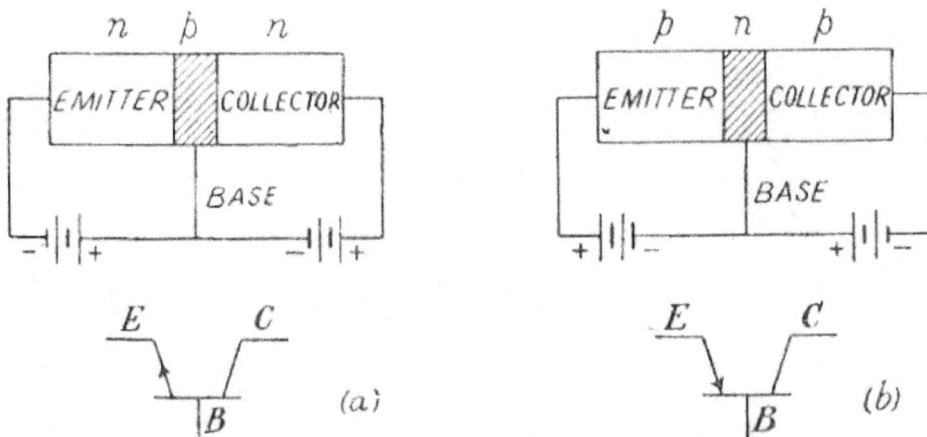

Fig. 220. Principle of action of a transistor.

The circuit connections are made as shown in the figure.

With the *n-p-n* type [Fig. 220 (a)], when the diode formed by the left hand *n-p* junction is biased in the "forward direction" by applying *e. m. f.* as shown in the figure, it constitutes a "low resistance" and electrons from the *n*-layer readily enter the hole-rich *base* region of the *p*-layer. It is then said that the *n*-layer emits electrons and it is named the *emitter*. Upon entering the *p* region, some of the electrons combine with holes, but most of them come under the influence of the positively biased *n*-layer of the right hand *n-p* junction and are attracted or collected by it, whence the *n*-layer of that junction is called *collector*. Because the hole-rich central *p*-layer is so thin, only a few of the electrons are able to combine with holes in the base and 95 to 99% of the emitted electrons reach the collector and cause a collector current to flow from base to collector. The ratio of the collector to the emitter currents is, therefore, of the order of 0.95 to 0-99.

It is to be noted that (1) the *emitter* is *always biased in the forward direction, while the collector in the reverse direction* and (2) the *transistor is formed by joining two semiconductor diodes back-to-back.*

With the *p-n-p* type [Fig. 220 (b)], the circuit connections are reversed, so that the left hand *p-n* junction is *forward biased,* so that its *p*-layer constitutes the emitter according to the general rule stated above, while the right-hand *p-n* junction is *back-biased* and its *p*-layer is the collector. The positive emitter injects holes into the *n*-type base region (or attracts electrons from the base). Those holes attract electrons from the negatively biased collector and cause a collector current to flow from collector to base. A few of the injected holes combine with electrons in the base region so that once again the ratio of collector to emitter currents will be of the order of 0.95 to 0-99.

The diagrammatic symbols for *n-p-n* and *p-n-p* transistors are shown below their respective circuit diagrams in the figure. The direction of the arrow on the emitter terminal in each case shows direction of easy current flow. Apart from the different polarities and directions of current flow, the *p-n-p* transistor resembles the *n-p-n* transistor. From what we have said above, it is readily seen that a transistor must have the following three basic components: (i) *emitter* (ii) *base* and (iii) *collector.*

The emitter injects carriers into the base (electrons in the *n-p-n* type and holes in the *p-n-p* type) and is the counterpart of the cathode in a thermionic valve. The collector attracts the carriers through the base and causes a current to flow through the transistor; hence it corresponds to the plate of a thermionic valve. The base is biased to control the flow of changes through the transistor and is the counterpart of the grid in a valve. Thus we have a semiconductor triode capable of performing the functions of a triode valve.

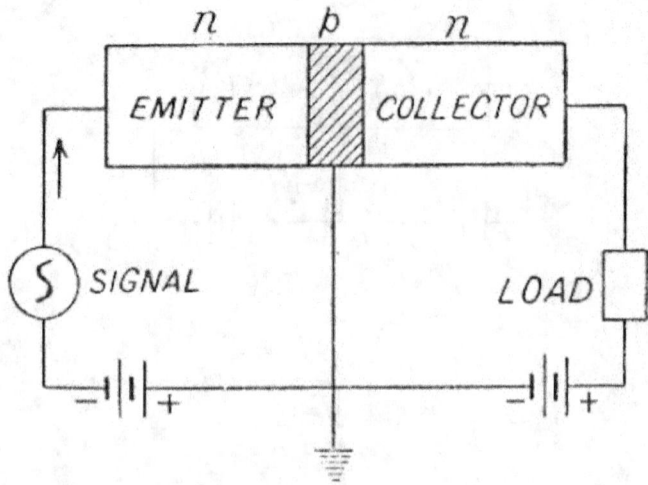

Fig. 221. *n-p-n* transistor as an amplifier

Amplification produced by a transistor

If the potential of the base with respect to the emitter is varied by the signal introduced in the emitter circuit as in Fig. 221 the flow of electrons will be modulated in accordance with the varying base-to-emitter potential. Although the ratio of the collector output current to the emitter input current is less than unity (of the order of 0.95 to 0.99), as explained above, the transistor is nevertheless capable of both *voltage and power amplification,* because the emitter or input current flows in a low-resistance circuit, while the nearly equal collector or output current flows in a high resistance circuit. Referring to the emitter-base low resistance circuit a very small change in base-to-emitter voltage will produce a large change in emitter current with a correspondingly large change in collector current. To obtain this same change in collector current by means of a voltage applied between base and collector would require for this high resistance circuit a much larger voltage change, of the order of 1000 times greater than the base-to-emitter voltage change. This implied voltage amplification may be realized by placing a high resistance of the order of 10,000 ohms in series with the collector-base circuit as load. The changing collector current produced by the signal introduced in the emitter circuit will develop a much larger output voltage across this load. Since the output voltage is greater than the input voltage (whereas the input and output currents are nearly equal) it is evident that the output power will also be greater than the input power. The resistances or impedances of the two circuits are such that power gains as great as 50 decibels are possible with emitter currents as low as 10 microamperes at 0.1 volt.

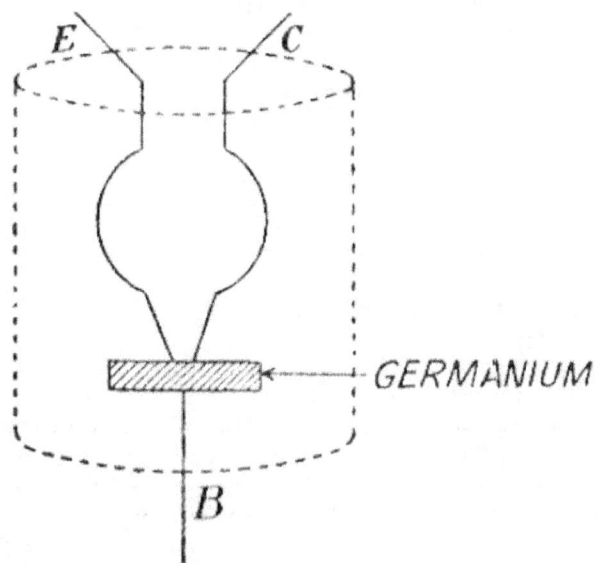

Fig. 222. Point contact. transistor.

Current amplification α is defined as the ratio of the change in the collector current (dI_c) to a given change in emitter current (dI_e) which produces it, and for junction type transistors this is of the order of 0.95 to 0.99, as already stated. In the point-contact (*p-n-p*) transistors, shown in Fig. 222, much higher values for α are possible. For, on account of the close spacing of the contacts (0.002 inch or 51 micron) some of the holes injected by the emitter (*E*) flow directly across to the collector (*C*) and there create a positive space charge which is attractive to electrons. This increases the flow of electrons into the base from the negative collector. In consequence the collector current may be several times that flowing in the emitter circuit and values for a of 2 or 3 are to be expected.

A comparison of triode and transistor amplifiers is illustrated in Fig. 223.

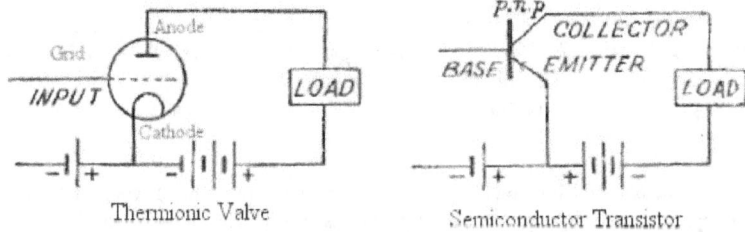

Fig. 223. Comparison of triode and transistor amplifier.

In the *p-n-p* transistor shown, the base requires a positive power supply voltage with respect to both the emitter and collector. The negative voltage bias on the base is low compared with that of the collector.

For the *n-p-n* transistor the polarity of these voltages is reversed.

The input impedance of a transistor is low, about 600 ohms, compared with about 1 mega-ohm for a valve. The transistor output collector impedance is high.

It is to be noted that in transistors signal energy is required to modulate the injection of carrier into the semiconductor in order to modulate the carrier current.

Hence a signal source employed with a transistor must deliver current and the transistor is a *"low-impedance current-actuated"* device. Herein the transistor basically differs from the thermionic valve, which is ordinarily a *"high- impedance voltage-actuated" device,* where the electron current is modulated electrostatically by a signal voltage, so that no signal current is required.

It is to be remarked also that three basic circuits can be used with transistors:

(1) the *common-base* (or grounded base) *circuit* corresponding to a grounded grid circuit of its valve counterpart,

(2) the *common-emitter circuit* corresponding to a grounded cathode, and

(3) the *common-collector circuit* corresponding to a grounded.

As in the case of the analogous valve circuits, the characteristics of three basic circuits differ. The common-emitter circuit is the one most frequently used, as in the case of valves.

Multi-stage amplification can be obtained with transistors as with valves using either resistance-capacity or transformer coupling. Resistance-capacity coupling has the advantages of compactness, light weight and relatively low cost. Transformer coupling has the advantages of higher gain, less loss, lower d.c. voltage supply and better operating point stability; but its disadvantages are the greater size, weight and cost. Both types of coupling are used in practice. As with valve amplifiers, push-pull circuits are often used for the output stages of transistor audio amplifiers.

Any transistor amplifier can be made self-oscillatory by feeding a portion of its output to the input circuit in proper phase, as in the case of vacuum tubes. Transistor oscillators are used when low output, power is needed.

Thus most operations employing valves can also be performed with transistors. Particularly at audio frequencies, transistors possess many practical advantages which will often dictate their use instead of valves. Their small size, light weight, long life, ruggedness are important advantages, and their low power requirements make battery operation feasible and economical. Tests indicate that the life of a transistor is several tens of thousands of hours, which means that a transistor, under favorable conditions, can operate continuously for several years. Unlike valves, transistors are practically free from damage due to mechanical shock and vibration. A transistor has no filament to be heated. The voltage and current at which it operates are considerably lower than in valves. This means that heat generation during operation is negligible. Further there is no heating up period ; a transistor begins to function as soon as the voltage is applied.

Transistors have certain shortcomings as well. For instance, transistor *noise* level is higher than that of a valve and increases as the number of amplifying stages is increased. *Frequency response* is lower than with valves. Transistors are quite *temperature sensitive;* at temperatures above 50°C, their efficiency is considerably reduced, with decrease of amplification and increase of noise. *Careful circuit design* is required to prevent extreme variations in gain and to allow interchangeability of transistors.

At audio frequencies junction transistors are used almost exclusively. Compared with the point-contact type, junction transistors have higher gain, better efficiency, less distortion, greater stability and less noise. Their poorer high-frequency performance which limits their use at radio frequencies is not a serious handicap at audio frequencies. New types of transistors have become possible and they are now being produced with such names as "drift", "mesa" and "tunnel diodes". In this last case, the transistors are no longer satisfied with absorbing very little current, they absorb none at all, because their conductivity has become negative.

Applications of the transistor
On account of the several decided advantages over the vacuum tubes, transistors will be ideal for a number of applications where thermionic valves have been found unsuitable. We shall now briefly state some of the applications already realized:

(a) *Pocket radio sets.*
Portable, all transistor superlight receiving sets, very small in overall size , but giving full loudspeaker volume, have replaced in thermionic radio sets.

(b) *Electronic computers* are devices in which the counting is performed with pairs of thermionic valves in a "flip-flop" circuit which has two stable states and can be tipped by a pulse from one to the other. Any single electronic computer usually employs some thousands of such circuits, so that the heat generated becomes quite a problem and sets a practical limit to the number of valves that can be used in the apparatus. By the use of transistors in the place of valves, this difficulty is entirely removed, besides the considerable saving in space and the extension of the upper limit secured.

In the last fifty years, most of the limitations of transistors have been overcome. Their maximum power has risen from 2 to 50 kw, frequency from 50 million to 3000 million cycles per second. This means that not only the field of television but also the realm of ultra high frequency is now open to transistors.

FINE STRUCTURE OF SPECTRAL LINES

Since a spectral line arises due to the transition of the atom from one energy state to another, in the interpretation of the fine structure of a spectral line, one must necessarily start with the determination of the different possible energy states. Then by the use of adequate selection rules the states between which transitions actually take place are picked out. Such transitions should account for the observed fine structure.

The different energy states of the atom between which transitions can take place have come to be known by a special name in spectroscopic study, *viz., spectral terms.* The problem of the fine structure is thus essentially a problem of spectral terms. When the latter is solved, the former is readily understood. Hence in the application of the vector atom model to the fine structure phenomenon, it must, first of all, be seen how the model can be used to fix up the spectral terms involved.

For the sake of clearness, the fine structure in optical and X-ray spectra will be considered separately, as the notations used in the two cases are different.

OPTICAL SPECTRA

 Spectral terms. In the classification of the spectral terms by the use of the vector atom model, it is convenient to divide atoms into two main categories, viz., *one-electron system* and *many-electron system.*

The atoms belonging to the first class have *only one valence* or *optical electron, e.g.,* hydrogen and hydrogen-like atoms, such as the alkalis. Even if the atom of this system actually contains several electrons, all of these, except one, are interlocked in closed shells, which, in consequence, do not contribute to the total angular momentum of the atom, according to Pauli's principle. Hence the single free electron in the outermost shell is alone effective in determining the spectral properties of the atom.

The atoms of the second class have *more than one valence* or *optical electron, i.e.,* several free electrons not interlocked in closed shells. They will therefore contribute their share to the total angular momentum of the atom and hence become effective in fixing the spectral properties, *e.g.,* the alkaline earths, belonging to the two-electron system, aluminum, scandium, etc., belonging to the three electron system, titanium to the four-electron system and so on.

Considering the **one-electron system,** since the state of the atom as a whole is determined by that of a single free electron, the values of l, s and j referring to the electron are equal to those of L, S, J defining the whole atom. Hence S $= s = \pm 1/2$ and the multiplicity of state given by $r = (2S +1)$ is two. This system **will therefore give rise only to doublet terms with the exception of the ground term** which corresponds to the normal state of the atom, *i.e.,* to L =0. For, in the present case, the possible values of J are only two, viz., J = (L +1/2) and (L -1/2). Now if L = 0, J = ± 1/2; but since the net angular momentum of the atom is always a positive quantity, the only value of J in this case is J = +1/2, which means a *singlet.* If L = 1, J = 1 ± 1/2, i.e , 3/2 or 1/2 (*doublet*). For L = 2, J = 2 ± 1/2, i.e., 5/2, or 3/2, (*doublet again*) and so on.

In the **many-electron system,** on the other hand, the state of the atom will be determined by the states of several free electrons. In consequence, the vectors L, S and J defining the state of the atom will be the resultant of the vectors l, s and j of those electrons. Hence the value of S is not always 1/2, but may be 0, 1/2, 1, 3/2, 2, etc., according to the number of free electrons involved and their orientations, parallel or anti-parallel and the multiplicity r is no longer two in all cases, but is

1 when S = 0,
2 when S = 1/2,
3 when S = 1,
4 when S = 3/2,
5 when S = 2,

and so on. Thus the possible values of J for a given value of L may be one; two, three, four, etc., which means **each state may be a singlet or multiplet.**

In the *two-electron system,* since S = 0 or 1, the state can be only a *singlet or a triplet.*
In the *three-electron system,* as S = 1/2 or 3/2, each state can be a *doublet or a quartet.*
In the *four-electron system* since S = 0, 1 or 2, *singlet, triplet and quintet* are possible. In general, *odd electron -systems have even number of terms, while even electron systems have odd number of terms.* Accordingly, the number of spectral terms of successive atoms in the periodic table of electrons will alternate between even and odd.

But the ground term is always a singlet in this system also. For, it corresponds to L = 0 and J = S. Now, since S can assume values 0, 1/2, 1, 3/2, etc., S is never less than L. For S = L = 0, the possible value of J is one; when S > L, the multiplicity of the state is given by (2L +1) which leads again to the possible value of J as one.

Limitation of term multiplicity due to Pauli's principle

Pauli's exclusion principle causes, however, the multiplicity of terms not to attain always the maximum (2S + 1) values which are to be expected from the number of free electrons involved. Taking, for instance, the case of an atom with 6 electrons in the L shell, at first sight, one expects the multiplicity $r = (2S + 1)$ to attain the maximum value of 7 on the assumption that the spins of all the six electrons are similarly directed thereby leading to a value of 3 for S. But such an arrangement of spins cannot take place without violation of Pauli's principle. For, in this *case, n-= 2, l = 0* and 1 ; m_l can have only four values, viz., 0 corresponding to *l*= 0 and 1, 0, - 1, corresponding to *l*= 1. Now if we associate with these 4 values of m_l six *like* values of m_s, *i.e.,* all either + 1/2 or -1/2 , the result would be that more than one electron will have identical quantum number as shown below:

L_I $(n = 2, l = 0)$	L_{II} $(n = 2, l = 1)$
$m_l = 0$	$+ 1, 0, - 1.$
$m_s = + 1/2, +1/2$	$+ 1/2, + 1/2, + 1/2, + 1/2$

The *correct arrangement* is

$m_l = 0$	$+ 1, 0, - 1$
$m_s = + 1/2, -1/2$	$+ 1/2, + 1/2, + 1/2, - 1/2$

Hence only four of the six electrons can have similarly directed spins, (*i.e.,* same sign for m_s) while the other two must be directed oppositely. In consequence, the value of S reduces to 1, from which it follows that the actual multiplicity is only 3. Another inference is that, as a shell is gradually filled by the addition of successive electrons, the term multiplicity first rises and then falls, the variation being symmetrical with respect to the middle of the shell.

Notations of spectral terms

The states of the atom, in which the values of its L vector, are 0, 1, 2, 3, 4, etc., are symbolically represented by the capital letters S, P, D, F, G, etc. respectively. The value of J is appended to this symbol as a subscript and the multiplicity of the state r written as a left superscript.

Thus the spectral terms corresponding to L =1 and S =1/2 are written as $^2P_{1/2}$ and $^2P_{3/2}$. For, since L =1, the capital letter which represents the state is P, the multiplicity of the state $r = (2 \times \frac{1}{2} +1) = 2$ which is put as a superscript; since J = L $\pm$S = 1 ± 1/2, *i.e.,* 1/2 or 3/2, these are put as subscripts.

To take another example, the notation $^4D_{5/2}$ specifies a spectral term, belonging to a state whose L =2, J = 5/2, multiplicity $r = 4$ and hence S = 3/2 (from $r = 2S +1$). Since the possible values of J are four, viz. 7/2, 5/2, 3/2, 1/2, the above notation refers only to one of the four terms of the given state.

In the atoms belonging to the one-electron system, since every state is a doublet, except the ground state, the notations for the spectral terms in these atoms are

$$^2S_{1/2}, \ ^2P_{1/2}, \ ^2P_{3/2}, \ ^2D_{3/2}, \ ^2D_{5/2}, \ ^2F_{5/2}, \ ^2F_{7/2}, \ etc.$$

In the many-electron system atoms, the notations will be more complex, but they can be readily written, provided the necessary data are available.

It may be noted that *the multiplicity symbol of the system is used on all terms even if all of them are not present.* Thus the ground term of the alkali atoms is written as $^2S_{1/2}$, although the multiplicity is one and we should strictly speaking write $^1S_{1/2}$. But this method has the advantage of indicating to which system the ground term belongs; in the present case, for instance, to the one-electron system, giving rise to doublet terms in general.

It is likewise important to note that the vector model, where only the vectors L, S and J are involved, cannot determine the magnitude of the energies corresponding to the different spectral terms, since the total quantum number n, necessary for such a determination, does not directly enter in the fixing up of the spectral terms. Yet the three vectors, L, S and J characterize the different energy states in such a definite manner that they may be used, without the intervention of n, to predict which transitions between the various energy states are permitted and hence which spectral lines may be expected.

Fine structure

As spectral lines arise due to transitions of the atom from one spectral term to another, it is easy to understand, in a general way, how complex lines may appear when there are several spectral terms for a given atomic state defined by the vector L.

In the atoms belonging to the one-electron system all the states except the ground one are doublets. Transitions between these doublet states give rise to the fine structure. To account fully for the actually observed fine structure, as regards their number and intensity, the selection and intensity rules must be taken into consideration.

In the case of atoms appertaining to the many-electron system the fine structure may be expected to be much more complex on account of the greater term multiplicities involved. But here also, knowing the spectral terms and applying the selection and intensity rules, the observed fine structure can be accounted for. It is to be remarked that on account of the special property of closed shells the free electrons responsible for the spectra are limited to a small number. Hence, the complexity of the fine structure does not go on increasing from one end to the other of the periodic table. The selection rules also, in their turn, put a limit on the complexity of fine structure. Likewise the limitations of term multiplicity arising from Pauli's exclusion principle may reduce the complexity still further.

These general principles may now be applied to the interpretation of the fine structure in the two simple cases, already mentioned, viz, the sodium D line and the hydrogen H_α line, to serve as illustrations.

FINE STRUCTURE OF THE SODIUM D-LINE

As we have already seen, ten out of the eleven electrons of the normal sodium atom are interlocked in closed shells, so that the eleventh free electron alone is responsible for the spectrum. The spectral lines arise from the various orbits which this single electron may occupy beyond the two closed shells and the state of the atom as a whole is determined by the state of that electron. For the same reason, all the states of the sodium atom except the ground one are doublets.

With the simple theory of Bohr it has been shown that the D line belongs to the principal Series which are due to transitions from the P state to the S state. Now for the *upper P state*, L = 1, J = 3/2 or 1/2 so that two terms are possible: $^2P_{3/2}$ and $^2P_{1/2}$. For the *lower S state*, L = 0, J = 1/2, so that only one term is possible: $^2S_{1/2}$.

The transitions that can take place between the two terms of the P state and the single term of the S state are two:

$$^2P_{1/2} \ \rightarrow \ ^2S_{1/2} \ and \ ^2P_{3/2} \ \rightarrow \ ^2S_{1/2}$$

Now applying the selection rules $\Delta L = \pm 1$ and $\Delta J = 0$ or ± 1 (excluding 0 →0), both the transitions are allowed, which explains the doublet fine structure of the sodium D line. The D_1 component (5896 A°) is due to the transition $P_{1/2} \rightarrow S_{1/2}$ and the D_2 component (5890 A°) to $P_{3/2} \rightarrow S_{1/2}$ (Fig. 224).

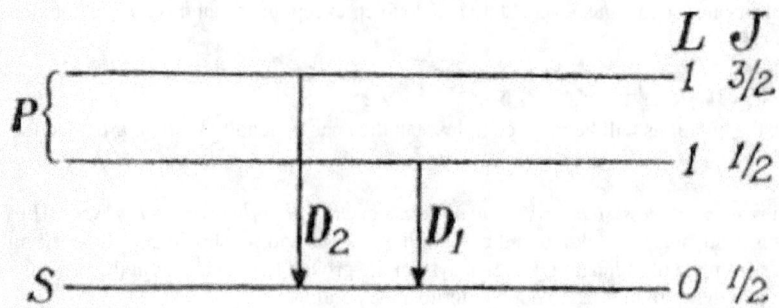

According to the general intensity rules, D_2 will be more intense than D_1, since for D_2, $\Delta L = -1$ and $\Delta J = -1$, while for D_1, $\Delta L = -1$, but $\Delta J = 0$.

Applying the total intensity $(2J + 1)$ rule it can be shown that the ratio of the intensities of D_2 to D_1 is $2 : 1$, which has been fully confirmed by observation.

Note. *Fine structure of lines in the other series of sodium.* As the *sharp* series arises due to transitions between the S and P states, the lines in it will be *doublets,* as in the principal series.

The *diffuse* series arises due to transitions between the D and P states and hence the lines in it should be *triplets,* since both the states have two sub-terms each: $D_{5/2}$ and $D_{3/2}$, $P_{3/2}$ and $P_{1/2}$. Four transitions are therefore possible:
$D_{5/2} \rightarrow P_{3/2}$,
$D_{5/2} \rightarrow P_{1/2}$,
$D_{3/2} \rightarrow P_{3/2}$,
$D_{3/2} \rightarrow P_{1/2}$,

Of these, $D_{5/2} \rightarrow P_{1/2}$ is forbidden, since $\Delta J = 2$.

The lines
$D_{5/2} \rightarrow P_{3/2}$,
$D_{3/2} \rightarrow P_{1/2}$,
will be intense, since $\Delta L = -1$ and $\Delta J = -1$,

while
$D_{3/2} \rightarrow P_{3/2}$,
will be weak, since $\Delta L = -1$ and $\Delta J = 0$.

Experimentally, with ordinary resolving power instruments the weak line is not seen. When it was first discovered, it was called a "satellite" of the apparent doublet formed by the two brighter lines.

Thus the chief spectrum of the alkalis should consist of doublets and triplets and theory accounts for them very well.

FINE STRUCTURE OF THE H_α LINE

As the hydrogen atom, in whose spectrum the H_α line appears, belongs to the one-electron system, $L = l$, $S = s$ and $J = j$ and all the quantum states are doublets, except the singlet ground state. According to the simple theory of Bohr, the H_α line, the first of the Balmer series, arises due to transition from the third quantum state $(n = 3)$ to the second $(n = 2)$.

For the *upper state* $(n = 3)$, L can take any of the values 0, 1, 2. Using the relation $J = L \pm S$, since S in this case has the value $1/2$, the possible number of terms are five, *viz.,*

$^2D_{5/2}$

$^2D_{3/2}$
$^2P_{3/2}$,
$^2P_{1/2}$,
$^2S_{1/2}$,

For the *lower state* (n = 2), L can have two values 0 and 1, which gives the possible number of terms as three, viz.,
$^2P_{3/2}$,
$^2P_{1/2}$,
$^2S_{1/2}$,

Fifteen transitions are theoretically possible between the five terms of the upper state and the three terms of the lower state. But the selection rules $\Delta L = \pm 1$, and $\Delta J = 0$ or ± 1 reduce the permitted transitions to the following seven:
$D_{5/2} \rightarrow P_{3/2}$,
$D_{3/2} \rightarrow P_{3/2}$,
$D_{3/2} \rightarrow P_{1/2}$,
$P_{3/2} \rightarrow S_{1/2}$,
$P_{1/2} \rightarrow S_{1/2}$,
$S_{1/2} \rightarrow P_{3/2}$,
$S_{1/2} \rightarrow P_{1/2}$,

Two pairs in separate cases of these seven allowed transitions are identical, viz.
$D_{3/2} \rightarrow P_{1/2}$,
$P_{3/2} \rightarrow S_{1/2}$,
$P_{1/2} \rightarrow S_{1/2}$,
$S_{1/2} \rightarrow P_{1/2}$,

since they represent transitions between *coincident levels, i.e.,* levels whose L value differ by unity, but J values are the same.

Taking into account these identical lines, *the fine structure of the H_α line should have five components.*

Applying the intensity rules, the two strongest components are:
$D_{5/2} \rightarrow P_{3/2}$,
and
$D_{3/2} \rightarrow P_{1/2}$,
$P_{3/2} \rightarrow S_{1/2}$,

since they are of the type $\Delta L = -1$, $\Delta J = -1$.

The two components
$D_{3/2} \rightarrow P_{3/2}$,
$P_{1/2} \rightarrow S_{1/2}$,

are less intense, being of the type $\Delta L = -1$, $\Delta J = 0$, The component
$S_{1/2} \rightarrow P_{3/2}$,
is still weaker, since it is of the type $\Delta L = +1$, $\Delta J = +1$.

The weakest line would be
$S_{1/2} \rightarrow P_{1/2}$,
($\Delta L = +1$, $\Delta J = 0$), but it is fused with
$P_{1/2} \rightarrow S_{1/2}$,
for reasons given above.

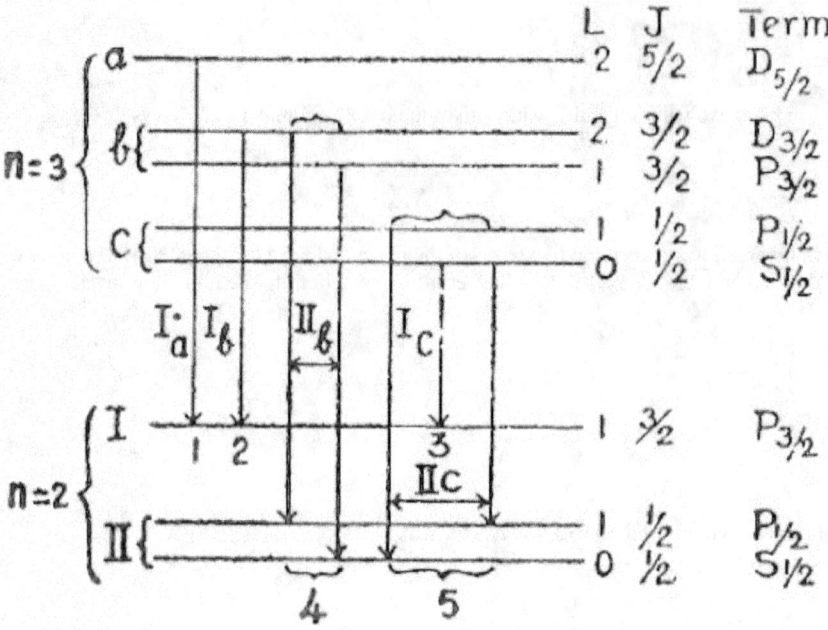

Energy level diagram: (Fig. 225).

The coincidental levels are represented by horizontal lines drawn close There are two such levels in the upper state and one in the lower.

The other lines represent the *neighboring* levels. If we call those vertical lines which terminate at the upper level of the lower state as group I and those which terminate at the lower level as group II and if we number the initial levels of the upper state as a, *b* and *c,* in so far as they do not coincide, then the two identical pairs are II_b and II_c. The five components are marked 1, 2, 3, 4 and 5 corresponding to I_a, I_b, I_c, II_b and II_c respectively.

According to the intensity rules, the strongest components are I_a and II_b (1 and 4). The components I_b and II_c (2 and 5) are less strong. I_c (3) is still weaker.

In practice, not all the five components but only doublets can be observed. This is due to the fact that the five components are merged together by the Doppler broadening caused by the thermal motion of the molecules. Very careful experiments conducted by R. C. Williams, in 1938, where the Doppler effect was reduced to a great extent by the use of a liquid air-cooled discharge tube containing deuterium, revealed three components.

X-RAY SPECTRA

Spectral terms. We have seen that

each X-ray energy level represents a state of an atom which has one electron missing from a closed shell.

Pauli pointed out that in a configuration, in which an electron is missing from a completed shell, the spectral term is the same as if that one electron alone occupied the shell. This means that the **term scheme for X-ray spectra corresponds to the one-electron system,** such as hydrogen, alkali metals, etc. Moseley's researches on the characteristic X-rays also suggest the same conclusion, viz., that the terms involved in X-ray spectra are of the hydrogen type.

The notations used are, however, different. The X-ray energy states are designated by the capital letters K, L, M, etc. specifying the shells to which they refer. According to the vector model, not only each of these states is characterized by a total quantum number n, (being 1 for the K state, 2 for the L state, 3 for the M state, etc.) but also all the states, except K, are further spilt into subgroups, L into two, M into three, N into four, etc.

For, using the fact that an atom emitting X-rays is assimilated to the one-electron system, it follows that in the mechanism of X-ray emission, $L = l$, $S = s$, $J = j$, so that $J = j = l \pm 1/2$, and that every *distinct state of an atom except the lowest is a doublet.*

Hence the K *state* ($n = 1$, $l = 0$) corresponding to the lowest ground state is a *singlet.*

The L *state* ($n = 2$, $l = 0$) has two sub-groups. The first sub-group ($n = 2$, $l = 0$) gives one term, while the second sub-group ($n = 2$, $l = 1$) *two terms* corresponding to the two different values of s, *viz.*, 3/2 and 1/2. The L state thus having three possible terms is a *triplet.*

The M *state* ($n = 3$, $l = 0, 1, 2$) with three sub-groups gives rise to *quintet,* one corresponding to $l = 0$, and $j = 1/2$, first subgroup), two others to $l = 1$ and $j = 3/2$, 1/2 (second sub-group) and the last two to $l = 1$, and $j = 5/2$, 3/2 (third sub-group).

The N *state* ($n = 4$, $l = 0, 1, 2, 3$) with four sub-groups is a *septet* composed of a singlet for $l = 0$ and three doublets for $l = 1, 2, 3$. The term multiplicity does not, however, go on increasing. Pauli's principle putting a limit, the next *O state* is a *quintet* and *P state a triplet.*

The term multiplicities of the different X-ray energy states thus derived from theoretical considerations are confirmed by experimental observations such as the spectra of the secondary, corpuscular rays and the X-ray absorption limits. **There are in all 24 different terms which are quite sufficient to represent completely the X-ray spectra of all the elements.** They are denoted by the Roman numerals attached to the capital letters representing the states, *e.g:*

$$\text{K, L}_\text{I} \text{ L}_\text{II} \text{ L}_\text{III}, \text{M}_\text{I} \text{ M}_\text{II} \text{ M}_\text{III} \text{ M}_\text{IV} \text{ M}_\text{V}, \text{ etc.}$$

They indicate all the possible combinations of l and j for given values of n.

Fine structure. The X-ray lines, in general, will arise due to transitions among the 24 different X-ray terms. But by the selection rules transitions can only occur subject to the conditions $\Delta L = \pm 1$ and $\Delta J = 0$ or ± 1, so that the number of possible transitions between two states is thereby considerably reduced and in consequence the number of fine structure components of the X-ray lines will be limited.

If the transition ends in the K state the corresponding radiation is called a K line; if it ends in the L state, a L line and so on. Fine structure of these general classes are distinguished by small Greek letters in accordance with the following recognized convention:

If the transition takes place between the different terms of the L state and the single term of the K state, the lines are called $K_{\alpha 1}$, $K_{\alpha 2}$, etc.;

If from the different terms of the M state to the K state, $K_{\beta 1}$, $K_{\beta 1}$, etc. Thus a general classification is made as K, L, M lines etc., according to the final state of transition, while the particular classification of the individual fine structure components of each line is made according to the terms of the initial state of transition.

These points are illustrated in Fig. 226. The term scheme of the four states K, L, M, N alone is drawn. The values of n, l and j, for each term are indicated.

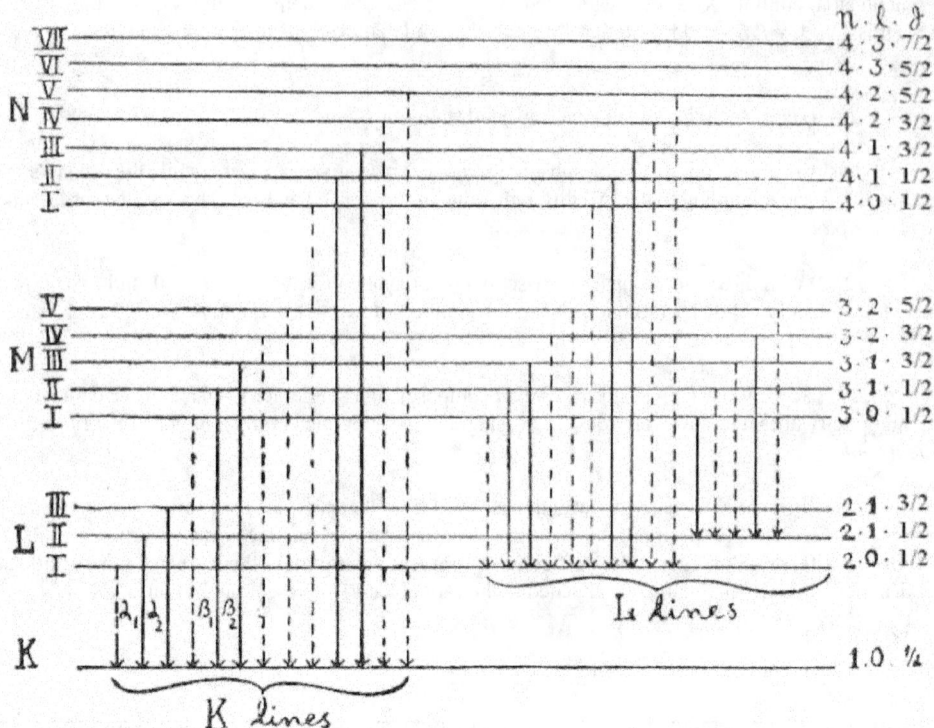

Fig. 226. Diagram of X-ray fine structure lines.

Some of the fine structure of the K and L lines are marked. The continuous lines represent the permitted transitions giving rise to the fine structure components while the dotted lines indicate transitions forbidden by the selection rules. As one proceeds to elements of higher and higher atomic number, the number of terms involved and consequently the number of fine structure lines will increase. This clear and simple interpretation of the fine structure of X-ray lines in a single scheme was obtained in 1927, thanks chiefly to the work of four physicists, Smekal, Wentzel, Coster and Bohr.

Thus the fine structure of spectral lines, both in the optical and X-ray regions, finds an adequate explanation on the basis of spectral terms determined by the vector model and of certain selection rules. It should be emphasized that the fine structure of spectral lines can be satisfactorily accounted for only by the introduction of electron spin.

Quantitative treatment of the effect of electron spin on the fine structure of spectral lines

We shall deal here with the case of one *electron system* alone, *i.e.,* where there is only one free electron responsible for the optical spectrum, such as the hydrogen, ionized helium and alkali atoms. The vector model adequately explains the fine structure of such a system, but only *qualitatively,* by pointing out the *doublet* nature of the spectral terms involved on the assumption that the electron spin *s* may be either parallel or antiparallel to the orbital vector *l. We* are now going to estimate *quantitatively* the effect of spin on the fine structure by considering the energy change caused by the spill action.

According to Sommerfeld's theory of fine structure, the magnitude of the relativity correction term is given by the expression:

$$\triangle E_R = -\frac{8\pi^4 m Z^4 e^8}{c^2 h^4 n^4}\left(\frac{n}{k}-\frac{3}{4}\right) \qquad \dots \quad (1)$$

228

where ΔE_R represents the energy change due to relativity correction, n is the total quantum number, $k = n_\varphi$ is the azimuthal quantum number, related to the orbital quantum number l by $l = k - 1$ and the other symbols have the usual significance.

The introduction of the electron spin necessitates the addition of another correction term which may be called the 'spin correction'. For, the magnetic field produced by the orbital motion of the electron and having the direction of l (or k) will act on the small magnet arising from electron spin and tend to pull it into its direction. But this turning action of the orbital field is opposed by the rotational inertia of the spinning electron and the resulting effect is a precessional motion of the spin axis about the axis of the orbital field.

The relative motion of the l and s vectors to each other, is, however, governed by another condition, viz., the l and s vectors together must form the resultant j vector which, for its part, is fixed in space and is of invariable value. It follows therefore that the l and s vectors rotate together as a rigid system and perform a precessional motion about j. The effect of the precession of the spin axis in the orbital field is to cause a change in the energy of the atomic system, which will introduce an additional convection term.

The change in energy due to spin action may be computed as follows:

Consider an electron of mass m and charge e describing an elliptical orbit around a nucleus charge Ze, situated at one of the foci, which is taken as the origin of polar coordinates (Fig. 227).

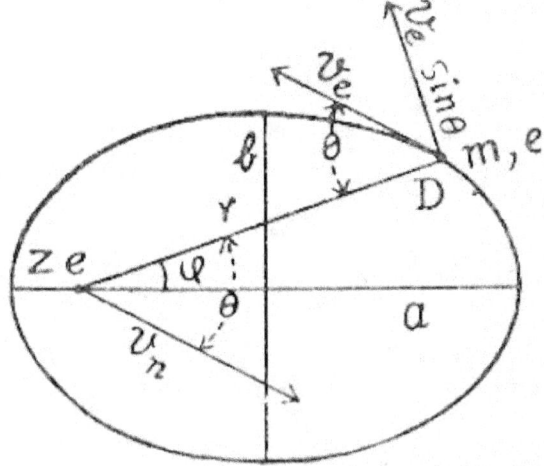

When the electron is at D (of coordinates r and φ) let its instantaneous velocity be v_e, marking an angle θ with the radius vector r. The angular velocity ω of the electron at the instant considered is given by $\omega = (v_e \sin \theta)/r$ and its angular momentum p_φ, assumed constant, by

$$p_\varphi = I\omega = mr^2 . (v_e \sin \theta / r) = m\, r\, v_e \sin \theta.$$

Although we have postulated a stationary nucleus and a moving electron, the result would be the same if the electrons were treated stationary at the instant and the nucleus moved with a velocity v_n equal in magnitude but opposite in direction to v_e. The motion of the nucleus will produce at D a magnetic field H_D, whose *magnitude* is readily obtained by the Biot and Savart law as

$$H_D = \frac{Z\,e\,v_n\, \sin\, \theta}{c\, r^2},$$

and whose *direction* will be parallel to that of the azimuthal quantum vector k. Further, since $v_n = v_e$, the instantaneous value of the orbital field, which is also H_D, is given by

$$H_D = \frac{Z\, e\, v_e\, \sin\theta}{c\, r^2} = \frac{Z\, e}{m\, c\, r^3} \cdot m\, r\, v_e\, \sin\theta$$

$$= \frac{Z\, e}{m\, c\, r^3} \cdot p_\varphi = \frac{Z\, e}{m\, c\, r^3} \cdot k\, \frac{h}{2\pi} \qquad \ldots(2)$$

On account of the magnetic moment arising from electron spin, the spin axis will experience a torque in this magnetic field H_D. The effect of this torque is to cause the spin axis to precess round the field direction, which results in a change in the energy of the system.

This energy change may be obtained, by analogy, as follows:

Taking the simplest case of an electron describing a circular orbit of radius r in a plane at right angles to a magnetic field H, since the orbital motion of the electron gives rise to a mechanical momentum p_1 as well as a magnetic moment μ_1, the electron will execute a precessional motion round the field direction. By Larmor's theorem, the effect of such a precession may be considered as merely a change in the angular velocity of the electron by an amount
$O = eH/2mc$,
known as the *Larmor precessional velocity*.

If v_o and ω_o be the linear and angular velocities of the electron in the absence of the field, the kinetic energy E_o of the electron in zero field is given by

$$E_o = (1/2)\, mv_o^2 = (1/2)\, mr^2\, \omega_o^2$$

If v_H and ω_H are the changed linear and angular velocities under the action of the field H, the kinetic energy E_H of the electron in the field is

$$E_H = (1/2)\, mv_H^2 = (1/2)\, mr^2\, \omega_H^2$$

By Larmor's theorem,

$\omega_H = \omega_o + O$,
where $O = eH/2mc$, therefore,

$$E_H = (1/2)\, mr^2\, (\omega_o + O)^2$$

Hence the change in energy ΔE_l due to the Larmor precession of the electron is

$$\Delta E_l = E_H - E_o = mr^2\, \omega_o . O,$$

neglecting second order quantities.

Since the orbital momentum p_1 is equal to $mr^2\, \omega_o$

$$\Delta E_1 = O \cdot p_1$$

We have assumed the axis of p_1 to be parallel to the direction of H. If, however, it is inclined at an angle to the field direction, the above relation becomes

$$\Delta E_1 = O \cdot p_1 \cdot \cos(p_1, H)$$

where $(p_1, H))$ is the angle between the orbital axis and the field direction.

By analogy, we may now write the change in energy due to the precession of the spin axis in the orbital magnetic field:

$$\triangle E_s = \overline{O_s \cdot p_s \cdot \cos(p_s, H_D)} \qquad \ldots \qquad \ldots (3)$$

the bar over the right-hand expression indicating that a time average over a complete cycle is to be taken, since O_s which depends on H_D, varies from point to point in the electronic orbit.

To evaluate O_s: Since in the case of orbital motion, $\mu_1 / p_1 = e/2mc$, we have

$$O = - (\mu_1 / p_1) \cdot H.$$

Again, by analogy, we may write

$$O_s = - (\mu_s / p_s) \cdot H_D.$$

the negative sign being introduced since H_D refers to the moving nucleus, while here we want the effect due to the moving electron, whose velocity $v_e = v_n$.

On account of the Landé's "g" factor for the electron spin being 2,
$(\mu_s / p_s) = e / mc.$
and hence
$O_S = - (e / m\ c) \cdot H_D$ .... (4)

But, as was first pointed out by Thomas (Nature 1926, 107, p 514), a rigorous relativity treatment of the problem leads to a value for the precessional velocity of the spin axis which is only half as great as (4) and has the reverse sign, *i.e.,* $+ (e/2\ mc)\ H_D$.

Taking this into consideration, as well as introducing the values of H_D and of $p_s = s.h/2\pi$ and replacing $\cos(p_s, H_D)$ *by* $\cos(k, p_s)$ since H_D is parallel to k, we finally obtain

$$\triangle E_s = \overline{\frac{e}{2mc} \cdot \frac{Ze}{mcr^3} \cdot k \cdot \frac{h}{2\pi} \ \frac{s.h}{2\pi} \cdot \cos(k, p_s)}$$

$$= \frac{Ze^2h^2}{8\pi^2m^2c^2} \cdot \frac{1}{r^3} \cdot ks \cdot \cos(k, p_s)$$

Since all quantities in the expression except r remain constant as the electron traverses the ellipse, we may write

$$\triangle E_s = \overline{(1 / r^3)} \cdot \frac{Ze^2h^2}{8\pi^2m^2c^2} \cdot ks \cdot \cos(k, p_s),$$

the bar being put only over $(1/r^3)$, thus indicating that in computing $\triangle E_s$, we have to take the time average of $1/r^3$ over a complete cycle.

To evaluate $(\overline{1/r^3})$:
It can be shown that $\overline{1/r^3} = 1/b^3$

where b is the semi-minor axis of the ellipse, which, according to Sommerfeld's theory of elliptical orbits, is given by

$$b = n_\varphi(n_\varphi + n_r) \cdot (h^2/4\pi^2 m Ee)$$

Putting $\quad n_\varphi = k,\ (n_\varphi + n_r) = n$ and $E = Ze$

$$b = n\,k\,(h^2\,/\,4\pi^2 m Ze^2)$$

$$\therefore \quad \overline{(1/r^3)} = \frac{64\pi^6 m^3 Z^3 e^6}{n^3 k^3 h^6} \qquad \dots (5)$$

To *evaluate cos* (k, p_s):

Representing the orbital angular momentum (p_φ) and the spin angular momentum (p_s) by the vector k and s respectively, the resultant vector j is obtained vectorially as

$$j^2 = s^2 + k^2 + 2ks \cdot \cos(s, k)$$

$$\therefore \quad \cos(k, s) = (j^2 - s^2 - k^2)/2ks$$

We can now write the final expression for ΔE_s:

$$\Delta E_s = \frac{64\pi^6 m^3 Z^3 e^6}{n^3 k^3 h^6} \cdot \frac{Ze^2 h^2}{8\pi^2 m^2 c^2} \cdot ks \cdot \frac{j^2 - s^2 - k^2}{2ks}$$

$$= \frac{8\pi^4 m Z^4 e^8}{n^4 c^2 h^4} \cdot \frac{n(j^2 - s^2 - k^2)}{2k^3} \qquad \dots (6)$$

The total correction ΔE arising from both the relativistic variation of the electronic mass and the electron spin is

$$\Delta E = \Delta E_R + \Delta E_s.$$

Hence using eqns. (1) and (6),

$$\Delta E = \frac{8\pi^4 m Z^4 e^8}{n^4 c^2 h^4} \left\{ -\frac{3}{4} + \frac{n}{k} - \frac{n(j^2 - s^2 - k^2)}{2k^3} \right\} \dots (7)$$

This expression is found to hold good approximately for large values of n, j and k, but fails when these numbers are small. But wave mechanics leads to an expression for ΔE in much better agreement with experimental data for small values of n, j and k, which is very similar to relation (7), but differs from it in the following way:

k is replaced by $(l + 1/2)$,
k^2 by $l(l + 1)$,
k^3 by $l(l+1/2)(l+1)$,
j^2 by $j(j + 1)$ and
s^2 by $s(s + 1)$.

The wave-mechanical expression is

$$\Delta E = -\frac{8\pi^4 m Z^4 e^8}{n^4 c^2 h^4} \left\{ -\frac{3}{4} + \frac{n}{(l + 1/2)} \right.$$

$$\left. -n \cdot \frac{j(j + 1) - l(l + 1) - s(s + 1)}{2l(l + 1/2)(l + 1)} \right\} \qquad \dots (8)$$

232

For the one-electron system, $s = 1/2$ and j can have either of the two values ($l + 1/2$) or ($l - 1/2$). Substituting these values

$$\frac{\triangle E}{(j = l + 1/2)} = -\frac{8\pi^4 m Z^4 e^8}{n^4 c^2 h^4} \left\{ -\frac{3}{4} + \frac{n}{l + 1} \right\}$$

$$\frac{\triangle E}{(j = l - 1/2)} = -\frac{8\pi^4 m Z^4 e^8}{n^4 c^2 h^4} \left\{ -\frac{3}{4} + \frac{n}{l} \right\} \qquad \dots (9)$$

Comparing these equations with that of Sommerfeld's relativity correction alone, viz.

$$\triangle E_R = -\frac{8\pi^4 m Z^4 e^8}{n^4 c^2 h^4} \left\{ -\frac{3}{4} + \frac{n}{k} \right\} \qquad \dots (10)$$

we see that the relativity correction term alone splits each energy level of a given total quantum number n into n component levels, each component being displaced by an amount which depends on the ratio n/k, or $n/(1 + 1/2)$. When the spin correction is included, as in eqns. (9), it is readily seen that a given level of total quantum number n splits into twice as many component levels as those predicted by the older theory, since for each value of l, $\triangle E$ has two values corresponding as j is equal to ($l+1/2$) or ($l - 1/2$). i.e., according as the spin vector is parallel or anti-parallel to the orbital vector.

A closer scrutiny of the equations, however, shows that a component for a given value of l and $j = (l - 1/2)$ coincides exactly with the component for ($l - 1$) and $j = (l +1/2)$, and that these coincident levels are identical with that given by the relativity correction alone for $k = l$. Thus the introduction of the spin correction does not appear to improve matters, since the estimation of $\triangle E$, even by the wave-mechanical methods, does not change either the number or the magnitude of the displaced component levels for a given value of n.

But, luckily the selection rules come to the rescue, since they are quite different in the two cases. In the older theory, only those transitions are permitted for which $\triangle k = \pm 1$ while, according to the vector model, the permitted transitions are governed by $\triangle l = \pm 1$ and $\triangle j = 0, \pm 1$. Hence one can expect the number of fine structure components predicted by the two theories to be different. Observations show certain components predicted only by the wave mechanical theory. Also other experimental data, such as "doublet separation" and "screening effect" in alkali-like elements, are in favor of the new theory. We shall illustrate these points by a few concrete examples:

(a) *The fine structure of the ionized helium line = 4686 A°*

This line arises from the transition $n = 4$ to $n = 3$. According to the older theory the upper level ($n = 4$) is split into four component levels, while the lower ($n = 3$) into three. On the other hand, the wave mechanical theory will lead to twice as many component levels, i.e., eight for the upper and six for the lower. Yet the resulting effective levels will be the same due to the coincident levels, as stated above. In both the theories, the quantity outside the bracket in the final relation, being the same, may be written as (Z^4/n^4) M, where M = $8\pi^4$ m $e^8/c^2 h^4$ is a constant which can be readily evaluated from the known values of π, m, e, c and h and is found to be 5.82 cm^{-1}. The value of Z is 2, as we are dealing with helium.

Using relation (10), *the relativity correction alone* ($\triangle E_R$) can be computed for the component states of the two levels involved remembering that for the upper level ($n = 4$), $k = 4, 3, 2, 1$, while for the lower ($n = 3$), $k = 3, 2, 1$. Thus knowing n and k in each case, the corresponding value of the term within the bracket ($- 3/4 + n/k$) is found. Multiplying this by the already evaluated (Z^4 / n^4) M, $\triangle E_R$, is calculated.

Similarly, with the two relations (9) the *relativity plus spin correction* ($\triangle E$) can be estimated for the different components, remembering that for the upper level ($n = 4$), $l = 3, 2, 1, 0$, while for the lower ($n = 3$), $l = 2, 1, 0$. Evidently there will be two sets of values for $\triangle E$ corresponding to $j = l + 1/2$ and $j = l - 1/2$.

In the two cases, the several component levels can be obtained by subtracting from the uncorrected energy of the two levels ($n = 4$ and $n = 3$) the respective corrections, ΔE_R in one case and ΔE in the other. With these data the energy level diagrams according to both theories can be drawn. It is then found that, although due to coincidence of levels the effective component levels are the same in the two cases, yet due to the different selection rules of the two theories the older one gives *five* fine structure lines, while the new wave mechanical theory *eight* fine structure lines.

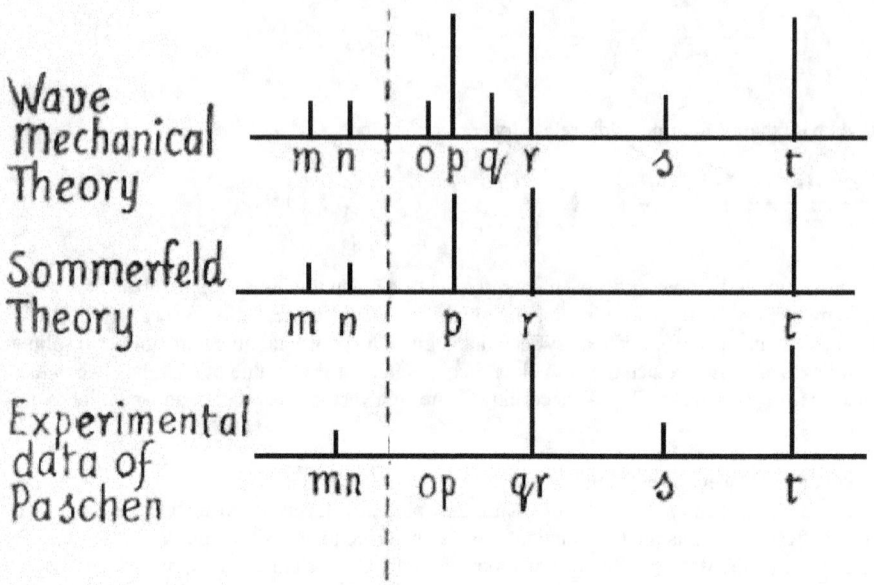

Fig. 228.

Comparison of these theoretical results with, the experimental data obtained by Paschen is shown in Fig. 228. The dotted line indicates the position of the single line of ionized helium A = 4686 A°. The heights of the component lines give their relative intensities qualitatively. Although experiment gives only five components and thus may seem, at first sight, to favor the older theory, yet on closer examination it is found that Paschen's observations support the new wave mechanical theory.

Components *m* and *n* predicted by both theories are coincident with the observed unresolved line *mn*.

Component *Q* of the new theory is so close to *p* that it probably accounts for the experimental unresolved line *op*.

Component *q* similarly is not resolved from *r*. Paschen's observations therefore do not distinguish between the two theories so far as components *o, p, q* and *r* are concerned. But Paschen got a fairly strong component *s* which is predicted only by the new theory and not by the old. This finding seems to decide in favor of the new theory.

(b) The alkali-like doublets

The doublet characteristic of spectral terms of alkali-like elements is due to the fact, that for a given value of *l*, *s* may be either parallel or anti-parallel to *l*. This gives rise to doublet terms, such as ($P_{3/2}$, $P_{1/2}$), ($D_{5/2}$, $D_{3/2}$), etc.

The numerical values of the separations of such doublets can be estimated using the relations derived above. But, for this it is necessary to determine the net effective nuclear charge Ze around which the single optical electron of the alkali-like system describes its orbit.

Taking Li, for example, since $Z= 3$, $Ze = + 3e$. Two of these combine with $- 2e$ of the two electrons of the first completed K shell, so that we have approximately a net effective nuclear charge $+ e$, round which the third free electron may revolve. The two K electrons may be considered to "screen" the nucleus and their "screening effect", is assumed to be 2. In general, the screening effect of n electrons around a nucleus would be less than n except for distances from the atom very large compared to atomic dimensions.

If S_n is the screening effect of n electrons, the effective nuclear charge will be $(Z - S_n)e$ or the effective atomic number is $(Z - S_n)$.

By substituting $(Z - S_n)$ for Z in equations (9), these equations may be applied to those n, l orbits which do not penetrate the "core" of the atom. The difference in energy ΔE_d between two doublets is given by

$$\Delta E_d = \Delta E_{(j = l + 1/2)} - \Delta E_{(j = l - 1/2)}$$

$$= \frac{8\pi^4 m e^8}{c^2 h^4} \cdot \frac{(Z - S_n)^4}{n^4} \cdot \frac{n}{l(l+1)}$$

Expressing this energy difference in wave-number units

$$\Delta \bar{v}_d = \frac{\Delta E_d}{ch} = \frac{8\pi^4 m e^8}{c^3 h^5} \cdot \frac{(Z - S_n)^4}{n^4} \cdot \frac{n}{l(l+1)} \text{ cm.}^{-1}$$

Since $8 \pi^4 . m\, e^8 / c^2\, h^5$ is equal to 5.82 cm^{-1}

$$\Delta \bar{v}_d = 5 \cdot 82 \frac{(Z - S_n)^4}{n^4} \cdot \frac{n}{l(l+1)} \text{ cm.}^{-1}$$

Considering the doublet separation between the two terms $P_{3/2}$ and $P_{1/2}$ of lithium,

$S_n = 2$, $n = 2$, $l = 1$, and we get

$$\Delta \bar{v}_d = 0 \cdot 364 \text{ cm.}^{-1}$$

Bowén and Millikan obtained experimentally the value of 0.338 cm^{-1} for this separation in good agreement with theory.

They investigated also the doublet separation of $(P_{3/2} \rightarrow P_{1/2})$ for a number of lithium-like atoms, such as Be$^+$, B^{++}, C^{+++}, N^{++++} and O^{+++++}. The values of $\Delta \bar{v}_d$, calculated theoretically, as above, on the assumption that the screening effect constant for the two K electrons was 2, were found consistently smaller than the observed value, unlike in the case of lithium. This shows that the effective nuclear charge for these ions is greater than $+ (Z - 2)e$, which, in turn, means that the screening effect of the 2 electrons decreases as the atomic number increases, as might be expected from the fact that the P orbits come closer to the nucleus with increasing atomic number.

Considering the sodium atom $(Z = 11)$, it has two completed shells, the K shell with 2 electrons and the L shell with 8, around which the eleventh free optical electron revolves. If the doublet separation between the terms $P_{3/2}$ and $P_{1/2}$ is calculated on the assumption that the screening effect of the ten interlocked electrons in 10, the value of $\Delta \bar{v}_d$ comes out to be 0.108 cm^{-1} whereas the actually observed value is 17.18 cm^{-1}.

In order to explain this great discrepancy, the P orbit in sodium is to be considered as an ellipse which "penetrates" the core of the atom, so that the *"screening effect"* must be much less than 10. Calculating S_n from the observed value of

$\overline{\Delta v}_d$, a value of 7.45 is obtained instead of 10. For "penetrating" orbits of this type, Laudé modified the theoretical formula as

$$\Delta\, \overline{v}_d = 5\cdot 82\ \frac{Z_i{}^2 z^2}{(n^*)^3\, l(l\,+\,1)}\ cm^{-1}.$$

where Z_i is the effective atomic number inside the outermost completed shell (*e.g.,* inside the L shell for sodium), z is the valency of the ion (*e.g.* for sodium $z = 1$, for magnesium++, $z = 2$) and n* the effective quantum number of the orbit in question.

One should expect then that the orbits of large value for n and of low eccentricity would be completely outside the core of the atom, i.e., they would be non-penetrating. In such cases, the original expression would apply even in the case of sodium-like ions. This conclusion was proved to be true by the observations of Bowen and Millikan who found, for example, that the doublet separation $D_{5/2}$ - $D_{3/2}$ for the sodium-like Al^{++} ions was 1.28 cm^{-1} , which gave $S_n + 9.97$, very close to 10. The ten electrons of the first two completed shells screen the nucleus with its + 13**e** charge, so that the net effective nuclear charge is + 3e.

These facts taken together decide in favor of the wave mechanical theory.

ZEEMAN EFFECT

Among the several magneto-optic phenomena, the Zeeman effect is one of very great importance, as it provides a link between the spectroscopic and magnetic properties of elements and throws much light on atomic magnetism.

The experimental arrangement for observing the effect has already been described. It has been pointed out there that while with very strong magnetic fields a spectral line splits up into a comparatively simple *triplet* called the *normal effect,* with ordinary weak fields a more complex resolution of a greater number of components, known as the *anomalous effect,* is obtained. Thus, for instance, in the case of the sodium D line, whose normal triplet resolution we have already studied, the Zeeman pattern becomes complex when the magnetic field is weak

The D₁ line, viewed *transversely,* splits into *four components,* symmetrically situated, all of them plane-polarized, but the inner pair parallel to the magnetic field (π components) while the outer pair perpendicular to the field (σ components). When viewed longitudinally the two outer components alone are seen and are circularly polarized in opposite directions (Fig. 229).

The D₂ line, viewed *transversely,* splits into *six components* symmetrically situated, all plane-polarized, the innermost two parallel to the field and the remaining four perpendicular to the field. In the *longitudinal* view, the innermost pair is absent, while the four outer components appear circularly polarized (Fig. 229).

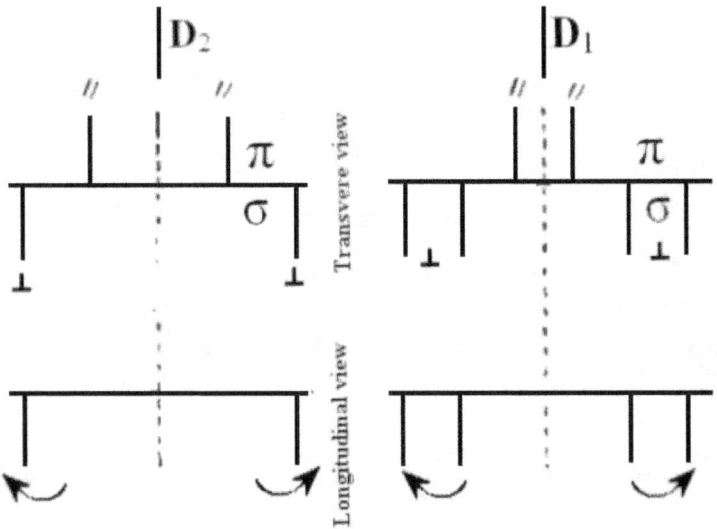

It may be noted that the classification of the effect into *normal* and *anomalous,* chiefly the nomenclature, is somewhat inappropriate, since it might lead to the wrong idea that what is termed as normal is the most common occurrence, while the anomalous is a rare exceptional phenomenon. In reality, the more complex anomalous pattern is the general rule in atomic spectra, while the simple triplet pattern is the exception which occurs under very special conditions, such as abnormally high magnetic fields, the original line being a singlet, etc. The distinction arose from the fact that Lorentz's classical explanation, although applicable to the splitting of lines into triplets and hence to the normal effect, failed to account for the more complex (anomalous) pattern. From another point of view, however, the classification is justifiable, as it indicates the fact that the anomalous effect is a weak-field effect, while the normal effect is a strong-field effect. *The actual transition from the anomalous-weak-field pattern to the normal-strong-field pattern* can be observed in many cases, as was done for the first time by Paschen and Back in 1913 and hence called the *Paschen-Back effect.*

Application of the vector model to the Zeeman effect

Lorentz was able to explain quite satisfactorily the normal Zeeman effect on the basis of the classical electron theory, as we have already seen. Later, Debye was able to interpret the same using the vector atom model where, however, the concept of electron spin was not taken into account. But neither the theory of Lorentz nor that of Debye could explain the anomalous effect. To achieve this, the idea of the spin of the electron had first to be introduced, the magnetic moment bound up with the mechanical spin of the electron had next to be considered and compared with the orbital magnetic moment and the difference between them established. Then only an adequate explanation of the anomalous effect was possible.

In order to appreciate fully the importance of these remarks, we shall first of all deal with. Debye's interpretation of the normal effect and then proceed to the consideration of the anomalous effect.

Peter Joseph William Debye (March 24, 1884 – November 2, 1966)

DEBYE'S EXPLANATION OF THE NORMAL ZEEMAN EFFECT

Leaving out the spin of the electron, the only angular momentum possessed by the electron is that due to its orbital motion.

The magnitude of this orbital angular momentum which is quantized, is given by

$$p_1 = l. \ h/2\pi \qquad \ldots\ldots\ldots \quad (1)$$

The corresponding magnetic moment is given by

$$\mu_1 = l. \ e \ h / 4\pi \ mc = e \ p_1 / 2 \ mc \qquad \ldots (2)$$

The vectors l and μ_1 will act along the same line, viz., perpendicular to the orbital plane.

In the presence of a magnetic field of uniform intensity H which acts not necessarily in the direction of the orbital vector l, this vector will execute a Larmor precession round the field direction as axis.

The effect of this Larmor precession is to alter the energy of the system by an amount dE, which may be computed as follows:

We have already seen that dE = O. p_1. cos $(p_1$.H) where O is the Larmor processional velocity. Bu t

$$\mu_1 /p_1 = e / 2 \ mc = \text{O/H, since O} = e\text{H} / 2 \ \pi \ mc$$

Therefore

O. $p_1 = \mu_1$ .H

As the directions of p_1 and μ_1 are parallel, dE = μ_1. H cos θ, where θ is the angle between the axis of the orbital magnet and the field direction.

If m_l is the magnetic orbital quantum number, then $\cos\theta = m_l / l$ since m_l is the projection of l in the field direction. Hence

$d\text{E} = \mu_l.\,\text{H}\,(m_l / l)$

Substituting the value of μ_l from equation (2)

$d\text{E} = (e\,h\,/\,4\pi\,m c).\,\text{H}.m_l$ .... (3)

The quantity $v_L = (e\text{H}\,/\,4\pi\,m c)$ is known as the *Larmor frequency* ($v_L = \text{O}/2\pi$). In terms of this quantity $d\text{E} = h\,.\,v_L\,.\,m_l$.

Now change in energy means that each energy level of the undisturbed state of the atom is split up under the action of the field into sub-levels. Further, since m_l restricted to $(2l + 1)$ integral values, at unit intervals, the number of sub-levels are $(2l + 1)$ separated from one another by an amount hv_L.

For a given value of H, v_L is constant, so that the separation of the sub-levels is equidistant.

If E_o represents the energy of a state of the atom of orbital quantum number l before the application of the field and E its energy after applying the field, then

$\text{E} = \text{E}_o + h.\,v_L\,.\,m_l$ .... (4)

Considering another state, for which the orbital quantum number is l', if E_o' and E' are the energies of that state before and after applying the field respectively, then

$\text{E}' = \text{E}'_o + h.\,v_L\,.\,m_l'$ .... (5)

Under the influence of the field the two states will be split up into $(2l + 1)$ and $(2l' + 1)$ components respectively.

If transitions take place between these two multiplet states, a multiplet group of lines will result. The frequency v_H of any one of these lines is obtained from (4) and (5) using Bohr's frequency condition:

$v_H = (\text{E}' - \text{E})\,/\,h$
$\quad = (\text{E}'_o - \text{E}_o)\,/\,h + (m_l' - m_l)\,v_L$
$\quad = v_o + \Delta m_l\,.\,v_L.$

where v_o represents the frequency of the line in absence of the magnetic field and
$\Delta m_l = (m_l' - m_l)$.
Applying the appropriate selection rule, *viz.*, $\Delta m_l = 0$ or ± 1, it is seen that the permitted transitions are limited to three, so that

$v_H = v_o$ for $\Delta m_l = 0$, and
$v_H = v_o \pm v_L$ for $\Delta m_l = \pm 1$.

This means that a given line of frequency v_o will be split up into three components whose respective frequencies are

$(v_o + v_L)$, v_o, and $(v_o - v_L)$

Substituting the value of $v_L = (e\text{H}\,/\,4\pi\,m c)$, the three components are

$(v_o + e\text{H}\,/\,4\pi\,m c)$, v_o, and $(v_o - e\text{H}\,/\,4\pi\,m c)$

The frequency shift dv produced by the field is therefore $\pm e\text{H}./4\pi m c$, the same as that derived on the classical theory for the normal Zeeman effect.

Note.

(*i*) Although eqn. (3) applies to the case of an atom belonging to the single-electron system, it can be generalized to hold good for the many-electron system also, provided p_l is replaced by the resultant orbital angular momentum p_L of all the effective electrons and M_L, the total resultant magnetic moment of the system is substituted for u_l.

(*ii*) In the foregoing discussion we have not taken into consideration the spin of the electron. Even if we had done so, nothing in the above derived relations would be changed if we associated with the spinning electron a magnetic moment μ_s, bearing to the mechanical spin momentum p_s the same ratio as the orbital magnetic moment μ_l does to the corresponding mechanical momentum p_l, *i.e.*, if we assumed that $\mu_s/p_s = \mu_l/p_l = e/2mc$. For the total angular momentum would then be p_j and the total magnetic moment

$$\mu_j = j.\, e\, h/\, 4\,\pi\, m\, c.$$

Thus j and μ_j would have the same direction and would set themselves in the magnetic field in accordance with spatial quantization or process round the field direction in common. The single difference would be that now, not $(2l + 1)$ but $(2j + 1)$ possible orientations exist and that therefore every undisturbed energy level is split by the magnetic field into $(2j + 1)$ terms, but in such a way that the amount of splitting would be exactly the same as before. In the spectrum there would be no difference at all, as the selection rule in the present case is the same as the previous case: $\Delta\, m_j = 0$ or ± 1. This shows that the vector model even with the use of the mechanical spin momentum can explain only the normal Zeeman effect, if the magnetic moment due to the electron spin is disregarded. Hence the difference that exists between the orbital and spin magnetic moments, conveniently expressed by the Landé g-factor, is essential to the understanding of the anomalous Zeeman effect.

EXPLANATION OF THE ANOMALOUS ZEEMAN EFFECT

With the introduction of electron spin, the total angular momentum of the atom J becomes the vector sum of the orbital and spin angular momenta, L and S, *i.e.*,

$$J = L + S \quad \text{----------------} \quad (1)$$

The physical interpretation of this relation is that the whole vector figure, representing the atom, precesses about the direction of J which, for its part, is fixed in space and of invariable value.

The orbital magnetic moment M_L is given by

$$M_L = L\, .eh/\, 4\pi\, mc = (e\, /\, 2mc)\, P_L\ldots\ldots\ldots\ldots\ldots (2)$$
and the spin magnetic moment M_S by

$$M_S = 2S\, .\, eh/\, 4\pi\, mc = (e\, /\, mc)\, P_S\ldots\ldots\ldots\ldots\ldots (3)$$

Since the vectors L and S precess about J, M_L and M_S must also precess about J.

The total magnetic moment M of the atom is the vector sum of M_L and M_S *i.e.*,

$$M = M_L + M_S \ldots\ldots\ldots\ldots\ldots\ldots (4)$$

From equations (2) and (3) it is readily seen that the ratios M_L/p_L and M_S/p_S are not equal; the latter is twice as great as the former.

On account of this inequality of the two ratios it immediately follows that *the direction of M does not coincide with that of J. For,* considering the case diagrammatically (Fig. 230), let L and S represent the orbital and spin angular momenta respectively. The total angular momentum is then represented by J. Let the orbital magnetic moment M_L be supposed to be twice as large as L. Then the spin magnetic moment M_S must be four times as large as S, since the ratio M_S/p_S is twice as great as M_L/p_L. The directions of M_L and M_S though along L and S respectively, will be oppositely directed to L and S as shown in the figure on account of the negative charge of the electron. Compounding M_L and M_S it is seen that the resultant M does not lie along J. If the ratios were made equal, then M would coincide with J in direction, as can be readily made out from the figure.

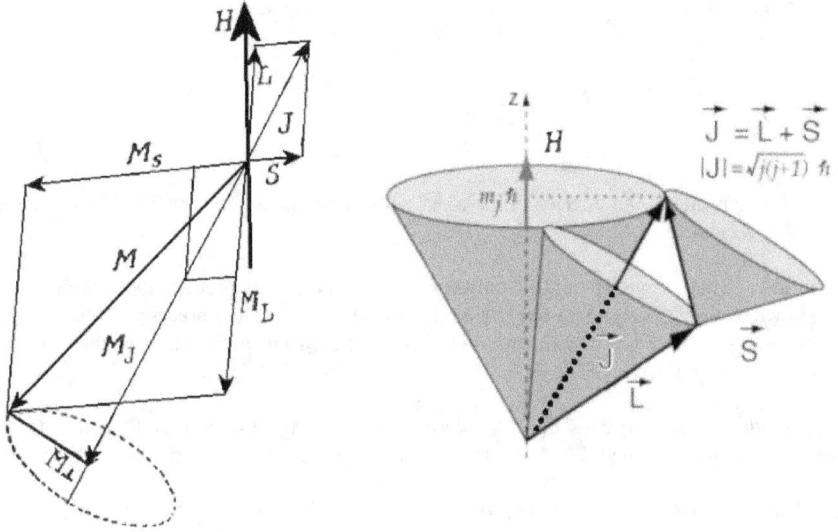

Fig. 230

Since any vector not in the direction of J will precess round it, as the vector figure of the atom precesses about J the magnetic axis of the atom as a whole, *i.e.*, the M vector, will also precess about the same direction.

Resolving the vector M into two components, one parallel to J (M_J) and the other perpendicular to it ($M\perp$), if the precession of M about J occurs sufficiently rapidly, the value of the perpendicular component ($M\perp$) will be constantly changing in direction, so that its time average over a period will be zero. Under these conditions the effective magnetic moment of the atom will be M_J whose magnitude is given by the vector sum of the components of M_L and M_S along J.

Hence

$M_J = M_L \cos (L, J) + M_S \cos (S, J)$,

where (L, J) and (S, J) represent the angles between the directions of L and J and S and J respectively.

Substituting the values of M_L and M_S from (2) and (3),

$M_J = (eh/4 \pi mc) [L \cos (L, J) + 2S \cos (S, J)]$

From the figure it is seen that

$S^2 = L^2 + J^2 - 2 L J \cos (L, J)$
and
$L^2 = S^2 + J^2 - 2 S J \cos (S, J)$
from which we get
$\cos (L, J) = (L^2 + J^2 - S^2) / 2 L J$
and
$\cos (S, J) = (S^2 + J^2 - L^2) / 2 S J$

Substituting these values

$$M_J = \frac{eh}{4\pi mc}\left\{\frac{L^2 + J^2 - S^2}{2J} + \frac{S^2 + J^2 - L^2}{J}\right\}$$

$$= \frac{eh}{4\pi mc}\left\{\frac{3J^2 + S^2 - L^2}{2J}\right\}$$

$$= \frac{eh}{4\pi mc} \cdot J \cdot \left\{1 + \frac{J^2 + S^2 - L^2}{2J^2}\right\} \qquad \ldots (5)$$

This expression defines what happens in the interior of the atom undisturbed by any external field and hence refers to what is known as the *"inner precession"*.

If the atom is placed in an external magnetic field of uniform intensity H, it will precess about the direction of the applied field, which is known as the *"outer precession"*. Although the atom as a whole precesses about the direction of H, still the nature of the precessions of the constituent vectors will depend on the strength of the external field as compared with that of the field in the interior of the atom.

(i) *If the external field strength is small compared with the internal,* the coupling between L and S is intact, so that there is physical significance in compounding L and S into J and then considering the precession of J about H.

(ii) *If, on the other hand, the external field strength is large compared with the internal,* then the coupling between L and S is loosened or even broken. In this case the vector J loses its significance and it becomes necessary to consider the precession of L and S about H independently.

The distinction between "weak" and "strong" external fields as compared with the internal field is of great importance in the explanation of the anomalous Zeeman effect which is produced only when the external field is weak. For, in such a case alone, one can legitimately assume that the precessions of L and S (whose coupling is unaffected) about J occur much more rapidly than that of J about H, which means that the perpendicular component $M_\perp$ may be neglected when averaged over a period and M_J may treated as the effective magnetic moment of the atom. The opposite case of "strong" field precess much more rapidly round H than round J. This is the Paschen-Back effect, of which we shall speak later.

Considering now the case of relatively "weak" field, proper to the anomalous Zeeman effect, the change in energy due to the action of the field on the atomic magnet is given by

$dE = M_J . H.\cos (J, H)$ ....(6)

The vector J according to the principle of spatial quantization, sets itself with respect to the field direction in such a way that its projection on the field direction m_J can have only (2 J + 1) values. The energy difference of these (2J +1) possible settings is given by the above equation (6).

Since
cos (J, H0 = m_J /J,
$dE = M_J . H . m_J / J$

Substituting the value of M_J from equation (5),

$$dE = \frac{eh}{4\pi mc} J \cdot H \cdot (m_J / J)\left\{1 + \frac{J^2 + S^2 - L^2}{2J^2}\right\}$$

$$= \frac{eh}{4\pi mc} \cdot H \cdot m_J \left\{1 + \frac{J^2 + S^2 - L^2}{2J^2}\right\}$$

Replacing $\left\{1 + \dfrac{J^2 + S^2 - L^2}{2J^2}\right\}$ by 'g' and $\dfrac{eH}{4\pi mc}$ by ν_L.

we get

$$dE = h\,\nu_L.\,m_J.\,g \text{ ------------------------ (7)}$$

This relation represents the fact that an undisturbed energy level of the atom is split up by the magnetic field into ($2J$ +1) levels, with equidistant separation given by $h\nu_L.g$.

Comparing this quantity with the corresponding one obtained in the normal Zeeman effect, viz., $h\nu_L$, it is seen that the difference between the normal and anomalous effects lies in the factor 'g'. In other words, in the anomalous Zeeman effect, although an undisturbed energy level is split up in a magnetic field into ($2J + 1$) equidistant levels as in the normal effect, yet the magnitude of the separation is equal not to $h\nu_L$ as in the normal, but to $h\nu_L.g$. This difference is precisely due to the assumption made in the present case shat the ratios M_L/p_L and M_S/p_S are not equal. Now, since the factor that affects these ratios is the Landé.'y' factor, it being equal to 1 for M_L/p_L and 2 for M_S/p_S, the 'g' that appears in equation (7) is evidently the same as the Landé 'g' factor. Hence it follows also that without the intervention of the Landé 'g' factor the anomalous effect cannot be understood.

The expression
$g = \{1+ (J^2 + S^2 - L^2)/2\,J^2\}$
which refers to the atom as a whole, indicates that the value of 'g' depends on those of L, S and J. It will, in consequence, vary from term to term and is thus the determining cause of the anomalous behavior of the atom in Zeeman effect.
The pattern will not be the normal triplet; many more lines will occur in accordance with the fact that the energy differences corresponding to the selection rules $\Delta m_J = 0$ or ± 1 are not now, as they were in the normal effect, the same for all values of m_J, If the electron spin is disregarded, $S = 0$ and the total angular momentum J is equal to the orbital angular momentum L. Then the quantity $(J^2 + S^2 - L^2)/2\,J^2$ reduces to zero, so that $g = 1$. In such a case, the separation of energy level is reduced to $h\nu_L$ which is the normal effect. This means that if spin and Landè factor are not introduced in the problem, only the normal effect can be explained.

The value of g has been deduced above according to the older quantum theory. In the new wave mechanics this representation is still permissible, provided that L^2, S^2, J^2 are replaced by L (L + 1), S(S + 1), J (J +1) respectively. With this improvement,

$$g = 1+ \{ J (J+1) + S (S+1) - L (L+1) \}/2\,J (J+I)$$

which will be used in considerations that follow.

Testing the validity of the theory proposed

The theoretical explanation of the anomalous Zeeman effect developed above may be tested in the following three ways:

1. Analysis of the anomalous pattern of some concrete lines such as the **D** lines of sodium, the green (5461 A°) and violet (4358 A°) lines of mercury.

2. Experimental verification of the "g" formula.

3. Measurement of the atomic magnetic moment.

1. ANOMALOUS ZEEMAN PATTERNS OF SOME CONCRETE SPECTRAL LINES

A. **Anomalous Zeeman patterns of the D$_1$ (5896 A°) and D$_2$ (5890 A°) lines of sodium.**

As sodium belongs to the one-electron system, the D line in its principal series is split into two components even in the absence of the external field, hence caused by the "inner" precession. Further, as we have already seen, the D_1 component is due to the transition $P_{1/2} \rightarrow S_{1/2}$, while the D_2 component to $P_{3/2} \rightarrow S_{1/2}$. The total number of terms involved is therefore three: $P_{3/2}$, $P_{1/2}$, and $S_{1/2}$.

According to the theory, under the influence of the external magnetic field, each of these terms will be split into component levels whose energy difference from no-field value is given by

$$dE = (e\,h\,/\,4\pi\,mc).\ H.m_J\ .g$$

The quantity $(e\,h\,/\,4\pi\,mc)$. $H = M_B$. $\mathbf{H}$, which represents the separation in the normal effect, is known as a *Lorentz unit* and it is customary to express energy differences in the present case in terms of this unit. Hence $dE = m_J$. g in *Lorentz units*. The splitting factor m_J. g is readily obtained by calculating the values of g and m_J for the three terms involved. Thus, for instance, taking the term $P_{3/2}$, $L = 1$, $S = 1/2$ and $J = 3/2$.

$$g = 1 + \{\ J\,(J+1) + S\,(S+1) - L\,(L+1)\ \}\ /\ 2\,J\,(J+1)$$
$$= 1 + \{\ (3/2)\,[(3/2)+1] + (1/2)\,[(1/2)+1] - 1\,(1+1)\ \}\ /\ 2\,(3/2)\,[(3/2)+1]$$
$$= 4/3$$

m_J can have $(2J+1)$ values at unit intervals. Since $J = 3/2$, m_I has four valves, viz., $3/2$, $1/2$, $-1/2$ and $-3/2$.

Hence $m_J.g$ has values: 2, $2/3$, $-2/3$ and -2.

In a similar manner the values of m_J for the other two terms can be computed and the results obtained are listed in the table below:

Term	L	S	J	g	m_J	$m_J \cdot g$
$P_{3/2}$	1	1/2	3/2	4/3	$3/2,\ 1/2\ -\ 1/2,\ -\ 3/2$	$2,\ 2/3, -2/3,\ -\ 2$
$P_{1/2}$	1	1/2	1/2	2/3	$1/2,\ -\ 1/2$	$1/3,\ -\ 1/3$
$S_{1/2}$	0	1/2	1/2	2	$1/2,\ -\ 1/2$	$1,\ -\ 1$

As **the D_1 line** arises from $P_{1/2} \rightarrow S_{1/2}$, the possible number of *transitions* between the two split levels of the upper term $P_{1/2}$ characterized by $1/3$ and $-1/3$ for the $m_J.g$ values and the two split levels of the lower term $S_{1/2}$ with 1 and -1 for the $m_J.g$, values *are four*.

Applying the appropriate selection rules, viz. $\Delta m_J = 0$ or ± 1, it is seen that *all the four transitions are allowed*.

Hence the D_1 line will be split up into four components, symmetrically situated on either side of the original line at distances $+4/3$, $+2/3$, $-2/3$, $-4/3$ in Lorentz units. The two inner components $+2/3$ and $-2/3$ governed by $\Delta m_J = 0$ are the π components, polarized parallel to the field and visible only in the transverse view. The outer components $+4/3$ and $-4/3$ governed by $\Delta m_J = \pm 1$ are the σ components, plane polarized but perpendicular to the field in the transverse view and circularly polarized in the longitudinal view.

The D_2 line arises from $P_{3/2} \rightarrow S_{1/2}$, Hence in the Zeeman effect, *transitions* can take place between the four split levels 2, $2/3$, $-2/3$, -2 of the upper term $P_{3/2}$ and the two split levels 1 and -1 of the lower term $S_{1/2}$, which means that there are *eight possible* transitions. Of these, two are forbidden by the selection rules $\Delta m_J = 0$ or ± 1, since in those two cases $\Delta m_J = \pm 2$, so that *only six transitions* are *allowed*.

The D_2 line is therefore split up into six components which can be easily shown to be symmetrically situated about the normal position at distances +5/3, + 3/3, +1/3, -1/3, - 3/3 and -5/3. The innermost pair +1/3 and -1/3 corresponding to $\Delta m_j = 0$ are the π components polarized parallel to the field and visible only in the transverse view. The other four components corresponding to $\Delta m_j = \pm 1$ are the σ components, plane polarized perpendicular to the field in the transverse view and circularly polarized in the longitudinal view.

It maybe noted that the shift caused by the field is not always equal to that in the normal effect. Thus the components of the D_1 line are 2/3 and 4/3 of the normal shift, while in the case of the D_2 lines, the shifts are 1/3, 3/3, and 5/3, so that the middle pair alone coincides with the normal shift.

The Zeeman components of the D_1 and D_2 lines that appear in the transverse view are diagrammatically represented in Fig. 231.

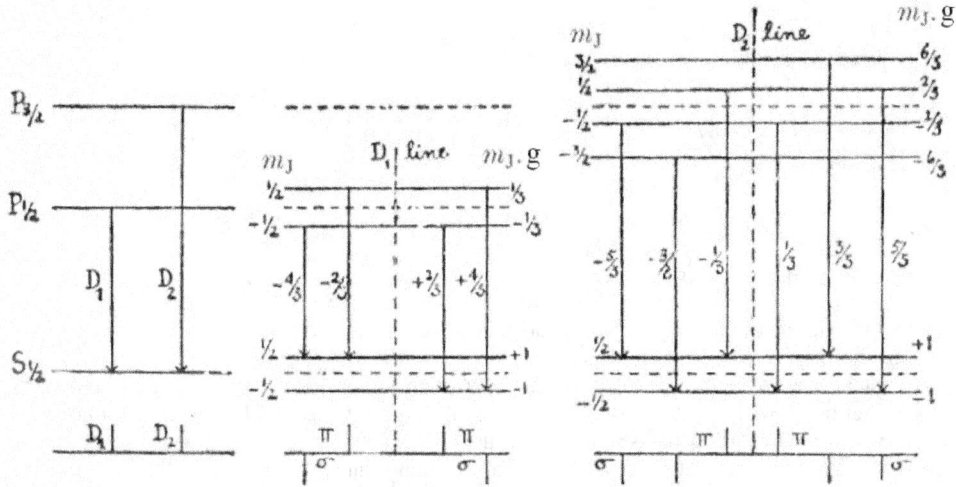

Fig. 231. Diagram representing the anomalous Zeeman patterns of the sodium D lines.

B. Anomalous Zeeman patterns of the green (5461 A°) and violet (4358 A°) lines of mercury.

The mercury atom belongs to the "two electron" system and from spectroscopic study it is known that the green line arises from the transition $^3P_2 \rightarrow {}^3S_1$ and the violet line from $^3P_1 \rightarrow {}^3S_1$. The term multiplicity r in both the cases under investigation is 3, which gives the value of spin S = 1, since 3 = 2S + 1.

Taking first the **green line** ($^3P_2 \rightarrow {}^3S_1$) let us find the g, m_J and $m_J.g$ values for the two terms involved:

For $^3P_2$: L =1, J = 2, S = 1 and m_J has (2J +1) or 5 values: 2, 1, 0, -1, -2

$g = 1+ \{ J (J+ 1) + S (S+ 1) — L (L + 1) \} / 2 J (J + 1)$
   $= 1+ (2 (2+ 1) + 1 (1+ 1) — 1 (1 + 1) \} / 4 (2 + 1)$
   $= 3/2$
Therefore, $m_J.g = 3, 3/2, 0, - 3/2, - 3$

Hence the upper $^3P_2$ is split into 5 components, separated from the no-field position by the above values in Lorentz units.

For $^3S_1$: L =0, J = 1, S = 1 and m_J has (2J +1) or 3 values: 1, 0, -1

$g = 1+ \{ J (J+ 1) + S (S+ 1) — L (L + 1) \} / 2 J (J + 1)$

$= 1+ \{((1+1)+1\,(1+1) — 0\,(0+1)\,\}\,/\,2\,(1+1)$
$= 2$

Therefore, $m_J.g = 2,\ 0,\ -2$

The lower state $^3S_1$ is therefore split into 3 components with separations as given above in Lorentz units.

The possible number of transitions between the five components of the upper state and the three components of the lower state is evidently fifteen. Of these, the selection rule $\Delta m_J = 0$ or ± 1 forbids the following six

$(2 \rightarrow 0)$,
$(2 \rightarrow -1)$,
$(1 \rightarrow -1)$,
$(-1 \rightarrow 1)$,
$(-2 \rightarrow 1)$,
$(-2 \rightarrow 0)$,

Hence the green line will be split into nine components with the following characteristics:

$$\left.\begin{array}{c}\text{Separation}\\ \text{given by } \triangle m_J \cdot g\end{array}\right\} -4/2, -3/2, -2/2, -1/2, 0, +1/2, +2/2, +3/2, +4/2$$

$$\left.\begin{array}{c}\text{Nature}\\ \text{given by } \triangle m_J\end{array}\right\} \begin{array}{ccccccccc} -1, & -1, & -1, & 0, & 0, & 0, & +1, & +1, & +1 \\ \sigma, & \sigma, & \sigma, & \pi, & \pi, & \pi, & \sigma, & \sigma, & \sigma \end{array}$$

The innermost three components are of the π type, plane polarized parallel to the field and visible only in the transverse view. The others are of the σ type, plane polarized perpendicular to the field in the transverse view and circularly polarized in the longitudinal view. One of the π components appears in the no-field position, due to the fact that, although Δm_J for it is given by $(0 \rightarrow 0)$, yet its ΔJ value is $(2 \rightarrow 1)$. Such a component will be absent *for lines arising from terms of odd multiplicity, if* $\Delta J = 0$ $(0 \rightarrow 0)$, as in the case of the 4358 violet line, as we shall see presently and also *for lines arising from even multiplicity,* as in the case of the D lines of sodium, since in this case J will have half-integral values, so that m_J cannot be zero and Δm_J $(0 \rightarrow 0)$ is not possible. Hence the two conditions simultaneously required for the appearance of a component in the no-field position are

$\Delta J \neq (0 \rightarrow 0)$ and $\Delta m_J = 0 \rightarrow 0$

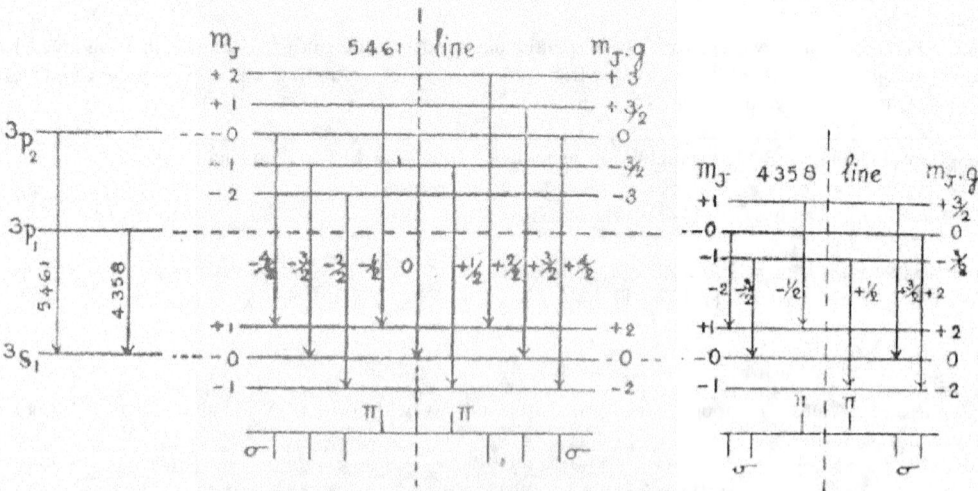

Fig. 232. Diagram of the anomalous Zeeman patterns of the mercury 5461 and 4358 lines.

Considering the violet line ($^3P_1 \rightarrow {}^3S_1$), the values of g, m_J and $m_J\,g$ of the two terms involved are:

246

For $^3P_1$: $L = 1$, $J = 1$, $S = 1$ and m_J can have 3 values : 1, 0, -1
Therefore,

$g = 1 + \{ J(J+1) + S(S+1) - L(L+1) \} / 2 J(J+1)$
$= 1 + ((1+1) + 1(1+1) - 1(1+1)) / 2(1+1)$
$= 3/2$
$m_J \cdot g = 3/2, 0, -3/2$.

For $^3S_1$: As calculated above,

$g = 2$,
$m_J = 1, 0, -1$
and
$m_J \cdot g = 2, 0, -2$.

Hence both the upper and lower states are split into three components each; the separations of the components of the upper state from the no-field position are 3/2, 0, and -3/2, while those of the lower 2, 0, and - 2, The possible number of transitions is nine. But the selection rule forbids three of them, *viz.*
(1 → -1),
(0 → 0),
(-1 → 1),

The 4358 line is therefore resolved into six components with the following characteristics:

Separation
($\triangle m_J \cdot g$) } $- 4/2, \quad - 3/2, \quad - 1/2, \quad + 1/2, \quad + 3/2, \quad + 4/2$

Nature
($\triangle m_J$) } $\begin{array}{cccccc} -1, & -1, & 0, & 0, & +1, & +1 \\ \sigma, & \sigma, & \pi, & \pi, & \sigma, & \sigma \end{array}$

The inner two are of the π type, while the outer four of the σ type. There is no component in the position of the parent line, since the transition is between terms of odd multiplicity and $\Delta J = (0 \to 0)$.

These conclusions are represented diagrammatically in Fig. 232.

The theoretical results are in exact agreement with the experimental patterns obtained, both as regards the complexity of the components and the magnitude of their separations. The Zeeman pattern of any spectral line can be analyzed by the same procedure and complete agreement between theory and experimental results can be established.

2. EXPERIMENTAL VERIFICATION OF THE "g" FORMULA

Theory gives the separation of energy levels as $dE = h.v_L.g$.
The frequency difference between the split up energy levels is therefore given by

$dv = dE / h = v_L \cdot g = (e H/ 4 \pi mc) \cdot g$

Therefore

$g = dv \cdot 4 \pi c / (H e/m)$

The frequency difference dv between, successive levels can be obtained by measurements made on the Zeeman pattern produced with a known magnetic field H. Assuming the values of c and e/m, g can be calculated as an experimental

247

result with the help of the above relation. On the other hand, substituting the values of L, S and J, (appropriate to the state, from which the split up levels are formed), in the "g" formula, a theoretical value of g is obtained. If the two agree, then the validity of the proposed theory is established.

So great is the variety of atomic states, so great the number of different values of L, S and J represented among them, that the study of even a single element produces many different checks on the validity of the "g" formula. Considering the fact that many different elements have been studied and in all cases experiment has fully confirmed the correctness of the formula with spectroscopic accuracy, one is compelled to conclude that the theoretical explanation of the anomalous Zeeman effect based on Landé 'g' factor represents the actual truth.

Special values of 'g'
Additional confirmatory evidence is obtained by a consideration of some special values of g. Thus, for instance.

(a) *when S = 0 and J = L,* we are dealing with the single terms of a two-electron system, such as, for example, Zn. Substituting these values of L, S and J in the "g" formula, g is found to be equal to 1, which means that only *a normal triplet will be obtained even with "weak" fields.* This is actually found to be the case, as seen from the Zeeman pattern of zinc singlet;

(b) *when J = 0 and S = L,* we are dealing with terms like ${}^1S_0$, ${}^3P_0$, etc. Since J = 0, $m_J = 0$, and $dE = 0$, which means that these terms do not split at all. The g value in such a case is 0/0, hence indeterminate;

(c) *in a very few cases when J $\pm$ 0, g = 0,* which occurs when $[\{S (S + 1) - L (L + 1)\} / 2J (J + 1)] = -3/2$. For example, the term ${}^4D_{1/2}$ corresponding to L = 2, J =1/2 and S = 3/2, and found in a three-electron system, does not split.

Therefore lines resulting from transitions between terms for which g is indeterminate (case *b)* and zero (case *c)* should show no Zeeman effect, which is confirmed by experiment.

3. EXPERIMENTAL MEASUREMENT OF THE ATOMIC MAGNETIC MOMENT

The resolved component of the magnetic moment of the atom in the field direction is given by

$\mu = M_J \cos (J, H)$,
so that
$dE = \mu . H$

But from theory, we have

$dE = (e h / 4 \pi m c) . H. m_J . g$
Therefore

$\mu = (e h / 4 \pi m c) . m_J . g$

Since $(eh / 4\pi mc)$ is a Bohr magneton,

$\mu = m_J . g$ in *Bohr units,*

i.e., the values of $m_J . g$ give the possible resolved components of the magnetic moment of the atom in the field direction in terms of Bohr units.

Now, considering the normal state of the atom which corresponds to the spectroscopic ground (or S) term, and choosing an atom of the one electron system, we have L = 0, J = S =1/2 so that g = 2 and m_J can have (2J + 1) or 2 values, viz., $\pm$ 1 /2.

Hence μ for the normal state of an atom belonging to the one electron system is given by $\mu = \pm 1/2 \times 2 = \pm 1$ Bohr magneton.

μ can be experimentally determined by the Stern and Gerlach experiment for such atoms in the normal state, and in all cases so far analyzed, such as the alkalis and noble metals, μ is found to be equal to ± 1 Bohr magneton.

This result, in its turn, establishes the correctness of the theory proposed.

Note.
(i) Since the total quantum number n does not occur in the "g" formula, the *Zeeman pattern will be the same for all the lines of* a given series. This fact is of great assistance to the spectroscopist in classifying lines that belong to the same series.

(ii) *The greater the fine structure of a spectral line the more complicated will be its anomalous Zeeman pattern.* Thus, for instance, the Balmer lines, H_α, H_β, H_γ, etc., will show quite a complicated Zeeman pattern.

(iii) *The magnetic field that could produce anomalous Zeeman effect should be "weak", not absolutely, but relative to that of the internal atomic field which causes the fine structure.* Thus, for instance, the Zeeman *separation* (about 2 cm^{-1}) of the D lines in a field of 30,000 gauss being several times smaller than the fine structure separation (17.8 cm^{-1}) of the D_1 and D_2 lines, the field used is to be considered "weak" and will produce anomalous effect. On the other hand, in the case of hydrogen or lithium, whose spectra consist of very close doublets, much smaller fields would be considered "strong" that will not produce the anomalous effect.

PASCHEN-BACK EFFECT

Paschen and Back found that, whatever the Zeeman patterns of a given group of lines might be in a weak field, they always approximated to the triplet characteristics of the normal effect in a strong field. This transition phenomenon is known as the Paschen- Back effect.

If the *magnetic field is gradually made more intense, the outer precession will gradually attain the order of magnitude of the inner one.* This means that we are no longer able to calculate as above, as if J remains fixed and the inner precession of L and S took place about J. Rather the external field will appreciably disturb the internal field and will loosen the magnetic coupling between L and S, thus converting the previously uniform precession about J into an irregular precession. In consequence, we can no longer replace the magnetic moment M of the atom by the parallel component M_J , since $M \perp$ will not now disappear when averaged over a period. We are now in the region of the Paschen-Back effect.

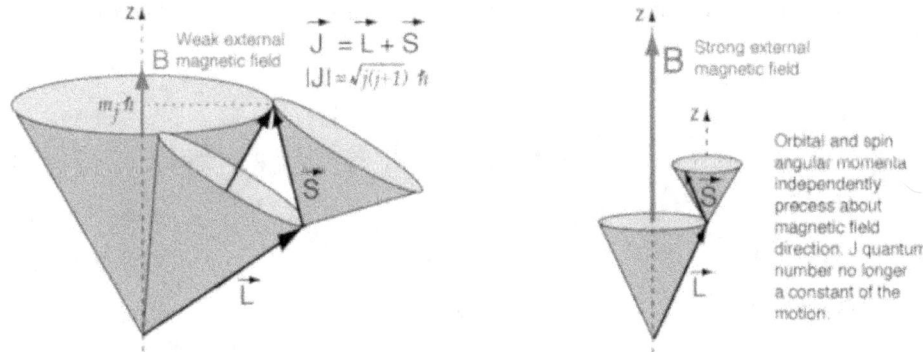

If the *magnetic field strength is raised still further, so that it predominates considerably over the internal field,* the term resolutions produced by the external field become much greater than the natural "field-free" resolutions. In such a case, the coupling between L and S becomes practically annulled and the orbital and spin vectors L and S precess separately about the field direction, independent of each other, as shown in Fig. 233.

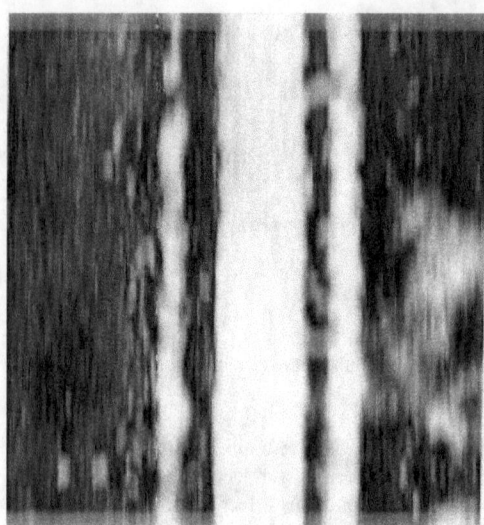

Paschen-Back effect of H_α line

The energy change due to the presence of the field will then be made up of two parts, one arising from the precession of L about H and the other from the precession of S about H.

Hence

$d\text{E} = (d\text{E})_\text{L} + (d\text{E})_\text{S}$
$\quad = \text{M}_\text{L} . \text{H} . \cos (\text{L, H}) + \text{M}_\text{S}. \text{H} . \cos (\text{S, H})$
$\quad = (eh / 4\pi mc).\text{H} . \{ \text{L} \cos (\text{L, H}) + 2\text{S} \cos (\text{S, H})\}$
$\quad = (eh / 4\pi mc).\text{H} . \{ m_\text{L} + 2 m_\text{S} \}$

The quantity ($m_\text{L} + 2 m_\text{S}$) is known as the *"strong field quantum number"*.
m_L can take only integral values, while m_S may be either integer or half integer.
The term ($m_\text{L} + 2 m_\text{S}$) can therefore have only integral values.
Further, since it can be shown that S and therefore m_S cannot change during a transition, the change in ($m_\text{L} + 2 m_\text{S}$) is due to the change in m_L, which according to the selection rules is 0 or ± 1. This means that in a strong magnetic field, a given line will be split into three components only, with equidistant separation of ($eh / 4\pi m$.c)H or $h\nu_\text{L}$, the same as in the normal effect.

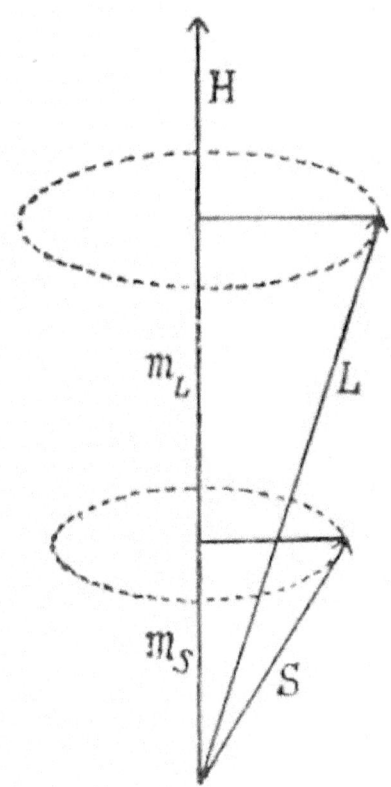

Fig. 233.

To illustrate, let us consider the *Paschen-Back effect in the principal series of lithium.* The fine structure of this series is a *doublet, as* in the case of the principal series of sodium, to which the D lines belong. Hence, the doublet .considered arises from transitions $P_{3/2} \rightarrow S_{1/2}$ and $P_{1/2} \rightarrow S_{1/2}$. In the field-free atom, the value of $J = L \pm S$ of the P state may be either 3/2 or 1/2, but in a sufficiently strong magnetic field, the L-S coupling is broken and the two sub-states $P_{3/2}$ and $P_{1/2}$ no longer have a definite meaning. Rather, we must think of the strong field magnetic quantum number ($m_L + 2 m_S$), according to which there will be 5 components for the upper P state and 2 components for the lower S state:

	State	L	S	m_L	m_S	$(m_L + 2m_S)$
	$P_{3/2}$	1	$+ 1/2$	1	$+ 1/2$	$+ 2$
P State		1	$+ 1/2$	0	$+ 1/2$	$+ 1$
		1	$+ 1/2$	$- 1$	$+ 1/2$	0
	$P_{1/2}$	1	$- 1/2$	1	$- 1/2$	0
		1	$- 1/2$	0	$- 1/2$	$- 1$
		1	$- 1/2$	$- 1$	$- 1/2$	$- 2$
S State	$S_{1/2}$	0	$+ 1/2$	0	$+ 1/2$	$+ 1$
		0	$- 1/2$	0	$- 1/2$	$- 1$

Note.

251

The two states for which $(m_L + 2 m_S) = 0$ constitute a single component of the P state.

Between the *five* components of the P state and the *two* components of the S state, *ten* transitions are possible. But the appropriate selection rule $\Delta(m_L + 2 m_S) = 0, \pm 1$ does not permit four out of the ten.

The allowed six transitions constitute *three* pairs of coinciding lines, since their $\Delta(m_L + 2 m_S)$ coincide, so that only three lines will be observed, as in the normal effect, with nearly the same amount of separation. These facts are represented in the diagram (Fig. 234).

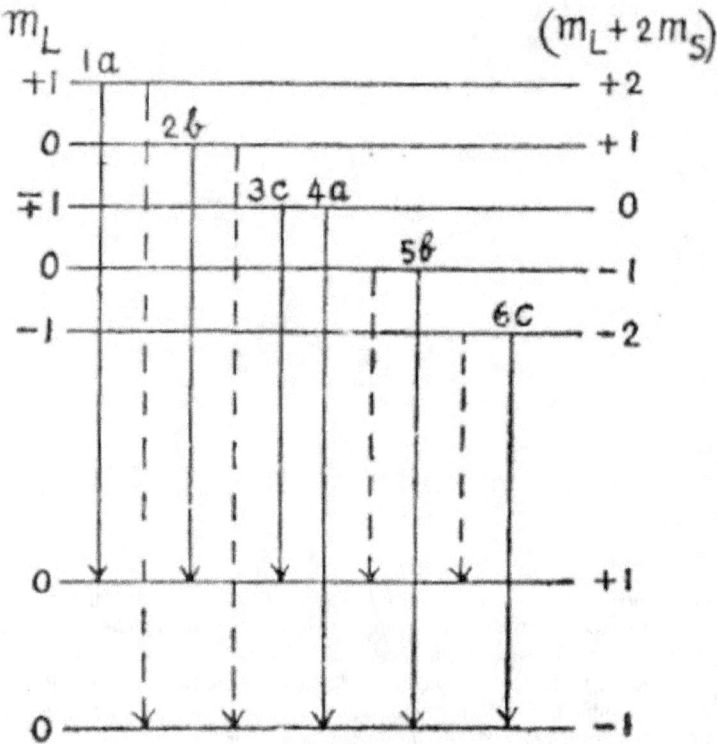

Fig. 234. Diagram representing the Paschen-Back effect in the principal series of lithium.

The dotted lines indicate the forbidden transitions, for which $\Delta(m_L + 2 m_S)$ is not 0 or ± 1. Of the six permitted transitions indicated by continuous lines, (1a and 4a), (2b and 5b) and (3c and 6c) coincide, since the values of $\Delta(m_L + 2 m_S)$ for them are $+ 1$, 0, and $— 1$ respectively.

The case considered here refers to the final state where the magnetic resolutions have become considerably great compared with the field-free fine structure resolutions. But it is the transitional zone from the anomalous weak-field pattern to the normal strong-field pattern, which is usually referred to as the Paschen-Back effect, and which, in consequence, will be somewhat more complicated than that suggested by the above simple discussion.

Conclusion.

The *Zeeman, effect, therefore, really provides a link between spectroscopic and magnetic properties of the atom.*

For, the splitting of the energy levels or terms of an atom in a magnetic field is explained adequately only by taking into account the directing force of the external magnetic field on the atomic magnet constituted by the orbital and-spin motions of the electrons in the atom. *Likewise, the atomic and quantum nature of magnetism is also clearly brought out in this effect.* The atom as a whole is a magnet—the resultant of two magnets, one due to the orbital motion and the other due to the spin of the electron. Further, the atomic magnet is quantized both and direction.

STARK EFFECT

Introduction.

The Stark effect is the electrical analogue of the Zeeman effect.

Soon after the discovery of the Zeeman effect in 1896, a similar resolution of spectral lines in an applied *electric* field was looked for. But it was only in 1913 that Stark succeeded for the first time in demonstrating such a phenomenon in the spectrum of the hydrogen atom. The chief practical difficulty, which was also the cause of failure of earlier investigators, was the absence of a proper technique. The aim was to subject hydrogen atoms emitting spectral lines to a powerful electric field. This was not, however, possible with the ordinary Geissler tube containing hydrogen, since

the gas in such a tube is comparatively a good conductor and hence incapable of maintaining **a** strong electric field.

Experimental study. Stark in Germany and Lo Surdo in Italy were able to overcome the difficulty stated above, by the use of special devices and obtained interesting results.

Stark's method

It consists in applying a strong electric field to the luminescent layer just behind the perforated cathode in **a** canal ray tube and analyzing the resulting effect with a high resolution spectroscope.

The arrangement used by Stark is shown in Fig. 235. The canal rays are produced in an ordinary glass discharge tube provided with a perforated cathode C. When the pressure in the tube is not very low, discharge takes place between the anode A and cathode C maintained at a suitable P.D., and the canal rays stream through the perforations in the cathode and form behind the cathode narrow cylindrical bundles of luminous rays. A third electrode F is placed parallel and close to C at a distance of a few millimeters and a very strong electric field of several thousand volts per cm. is maintained between F and C. The shortness of the space between F and C is a very important factor, since it not only favors the production of the resulting great potential drop, but also prevents the spontaneous discharge between F and C. The P.D. between F and C is great enough to influence effectively the ions in that region and distort perceptibly their electronic orbits.

The effect produced can be studied both *transversely* and *longitudinally, i.e.,* at right angles and parallel to the direction of the field. For the transverse view, the light from the canal rays enters the spectroscope as in (a); for the longitudinal view the arrangement is as shown in (b).

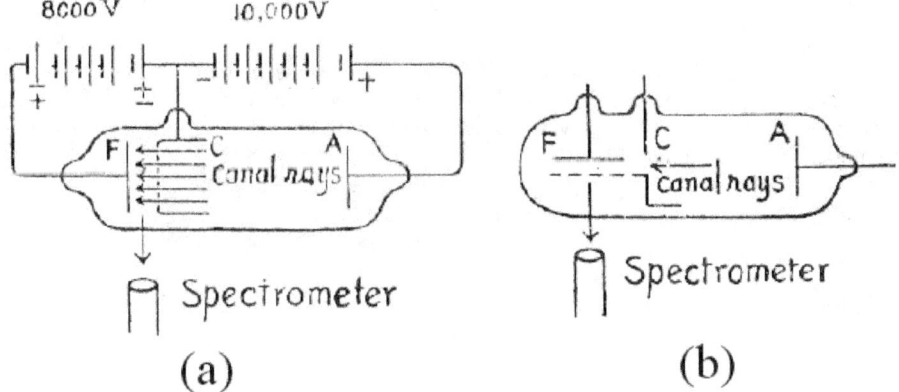

Fig. 235. Stark's apparatus.   (a) transverse view,   (5) longitudinal view.

Stark observed that the lines in the spectrum emitted by the canal rays of hydrogen were split up, under the action of the applied field, into numerous sharp components, somewhat after the manner of the Zeeman effect, the components being polarized, some parallel to the field (π components), others perpendicular to it (σ components).

Lo Surdo's method

A little after the announcement of his discovery by Stark, Lo Surdo (1914) was able to observe the effect, without using an additional field, except that of the discharge tube itself, in the faint light in front of the cathode. In the discharge tube we know the drop of potential is very great in the region of the cathode glow. Lo Surdo enhanced this effect by making the glass walls of the discharge tube near the cathode constricted in the form of a capillary tube and by adjusting the pressure in such a way that the length of the cathode dark space was between 1 to 10 mm. The light emitted just in this part of the tube showed the Stark splitting.

This method has the disadvantage that the exact magnitude and distribution of the field are not known, while in the Stark device, a uniform and measurable electric field is obtained. Further, the separation of the lines is not the same at all points due to the variation of the field and only transverse observations could be made. Thus this method sacrifices *quantitative* definiteness, but it offers special advantages for the purpose of *qualitative* observations. For this reason, many workers, particularly in Japan, have used Lo Surdo's method.

The successful and accurate investigations of Stark, Rausch von Traubenberg, Mark and Wierl and others on hydrogen lines were done chiefly with the Stark's method, while the effect on helium lines was studied by Foster with a modified Lo Surdo's arrangement.

Results

The results obtained by the two methods are chiefly in connection with the lines of the Balmer series of the hydrogen spectrum, although the effect in the spectra of other elements, such as He, Li, etc., has been investigated to a certain extent. The results obtained with the Balmer lines may be summarized as follows:

(i) Every line is split up into *a number* of *sharp components, their number increasing with the series number of the parent line.* Thus the number of components of H_β line is greater than that of the H_α line, similarly the components of H_γ are greater than those of H_β.

(ii) The components are *polarized* in the following manner: In the transverse view they are linearly polarized, some parallel to the direction of the field (π components) and others perpendicular (σ components). In the longitudinal view the π components are absent, while the σ components are unpolarized. The *intense π components, in general, lie outside, while the intense σ components lie inside.*

(iii) The components *lie symmetrically on both sides of the original line.* The distances of the components from the central line are *integral multiples* of the distance of the least displaced component from the centre. If this smallest line interval or *'resolution'* is measured in the scale of wave numbers, it is found to be the same for all the lines of the Balmer series.

(iv) Up to fields of about 100,000 volts per cm. the *resolution increases proportionate to the field strength.* In this region, therefore, we have what is known as the *'linear'* Stark effect. In the case of more intense fields, more complicated effects, the so-called *'quadratic'* Stark effect and even of *higher order* are observed in addition to the linear effect. Rausch von Traubenberg has succeeded in observing and analyzing these higher order effects by using electric fields of strength over a million volts per cm.

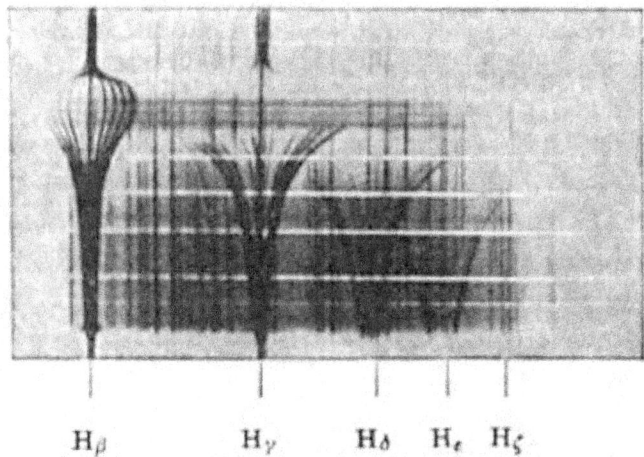

H$_\beta$ H$_\gamma$ H$_\delta$ H$_\epsilon$ H$_\zeta$

Lo Surdo photograph of Balmer lines

(v) The Lo Surdo photograph of the Balmer lines taken by Bausch von Traubenberg and his collaborators, reproduced here, is very instructive. The electric field increases from bottom upwards, from about 100,000 to 1.14 million volts per cm., and at the very top suddenly becomes zero. The Magnitude of the splitting of the lines confirms the above-mentioned variation with the intensity of the field. In general the increase in resolution with the field shows the effects of higher order together with the linear effect. The photo has a particular significance in the fact that it exhibits the fading out of the different lines for different field strengths, the *higher members of the series disappearing at lower intensities.* The H$_\epsilon$ line ceases to exist at a lower field that the H$_\delta$ line, and the latter sooner than H$_\gamma$. It is seen also that *the long-wave components of the Stark effect of any given line disappear at a lower field strength than the corresponding short-wave components.*

Theoretical explanation

Unlike the Zeeman effect, the Stark effect cannot be explained by the classical theory, even in an .elementary manner. But it was a happy coincidence that in the same year, 1913, when Stark discovered the effect, Bohr's quantum theory of the spectral lines of the hydrogen atom was proposed and was elaborated by Sommerfeld far enough, with the additional conception of elliptical orbits, quantized not only as regards their shape and size, but also as regards their spatial position, that it became possible to grapple with the problem of the electrical splitting of the hydrogen lines. The explanation was obtained in 1916 by Schwarzchild and Epstein almost simultaneously and along essentially similar lines. Later, Bohr proposed an alternative explanation based on the *method of perturbations,* which brought out the physical nature of the process much more vividly. We shall limit ourselves here to a brief statement of Epstein's explanation which accounts practically for all the observed facts, a complete treatment being extremely laborious.

According to the Bohr-Sommerfeld atom model, with the idea of spatial quantization included, the circular and elliptical orbits of the single electron revolving round the nucleus in the hydrogen atom will be defined by *three* quantum numbers referring to their *shape, size* and *orientation* respectively, so that a given total quantum number will be made up of three component quantum numbers. The application of an external electric field will distort these electron orbits in a rather complicated way; they will be no longer circular or elliptical. The problem consists, in the first place, in determining the nature of these distorted orbits of the electron, which constitutes the *mechanical part* of the solution. Then, applying the quantum theory to such deformed orbits, we have to select those that are distinguished by having quantum values for the triple aspect mentioned above and express the orbital energy as a function of the three quantum numbers involved. This forms the *quantum theory part* of the solution. Corresponding to each such quantum triplet in the initial and the final orbits, we have, in general, a different component in the Stark fine structure.

The *mechanical problem* consists in defining the trajectories of an electron subjected to the action of a uniform electric field of intensity F, in addition to the field of the nuclear charge Ze (Z = 1 for hydrogen). Such a case is contained in the more general one, much used in astronomy, viz., how does a point-mass move under the influence of two arbitrarily fixed centers of Coulomb attraction? Jacobi has shown that the appropriate coordinates for the treatment of the general problem are the parameters of the families of confocal ellipse and hyperbolae that are described about the centers as foci

255

together with the angle reckoned from the line connecting the centers. The present case can be considered as a particular application of the general one, for, if one of the attracting centers is removed to infinity and its attractive power correspondingly increased, it can effectively replace the uniform external field. Then, the systems of confocal ellipse and hyperbolae resolve into two families of confocal parabolas whose common focus is the second fixed attracting centre, the nucleus, and whose common axis gives the direction of the uniform external field F. Now the most convenient coordinates to work with are the parabolic coordinates, viz., the parameters of the two parabolic systems ξ and η and the angle ψ counted from the direction of the external field F be along the X-axis (Fig. 236). Let the electron find itself at P at any instant. The position P of the electron in the *Cartesian* system is represented by (x, y) considering the (x, y) plane. In the *parabolic* system, the same point P is represented by the intersection of two confocal parabolas whose common focus is at O and common axis along the X-axis. It can be shown that these two parabolas are given by the equations:

$y^2 / \xi + 2x = \xi$ (const.) .... (1)
$y^2 / \eta - 2x = \eta$ (const.) .... (2)

Adding (1) and (2),

$y^2 / \xi + y^2 / \eta = \xi + \eta$ or $y^2 (1/\xi + 1/ \eta) = \xi + \eta$

Therefore,
$y^2 = \xi \eta$ or $y = (\xi \eta)^{1/2}$ .... (3-a)

From (1),
$2x = \xi - y^2 /\xi = \xi - \eta$ .... (3-b)

These relations for x and y express the fact that the point P can very well be represented by the parameters ξ and η of the above specified parabolas. The parabola of parameter, ξ = constant, has its axis in the negative direction of the X-axis, while that of parameter, η = constant, in the positive direction of the same axis.

Passing from the (x, y) plane to three-dimensional space, which has to be considered if the plane of the figure be rotated about the X-axis, we shall have to introduce a third parameter to fix the point P in space. This third parameter is ψ, the angle of rotation of the whole figure with respect to a fixed plane. Hence the position P of the electron in space at any instant can be defined in terms of the *parabolic coordinates* ξ , η and ψ, instead of x, y and z of the Cartesian system.

Using these parabolic coordinates, let us now find an expression that represents the motion of the electron. For this, let us suppose that the electron moves to a point P' in a short time dt and let PP' $=ds$. This *line-element 'ds' in space* can be expressed as a function of the differentials $d\xi$, $d\eta$, and $d\psi$, and we get the relation:

$$(ds)^3 = {}^1/_4 \, (\xi + \eta) \left\{ \frac{(d\xi)^2}{\xi} + \frac{(d\eta)^2}{\eta} \right\} + \xi\eta \, (d\psi)^2 \qquad \dots (4)$$

The *kinetic energy of the electron* is given by

$$E_{Kin} = {}^1/_2 \, m \left(\frac{ds}{dt} \right)^2$$

where m is the mass of the electron and (ds/dt) its velocity at the instant considered. Using eqn. (4), $(ds/dt)^2$ can be expressed in terms of the parabolic coordinates, and this gives

$$E_{Kin} = \frac{m}{2} \left[{}^1/_4 (\xi + \eta) \left\{ \frac{d\xi/dt)^2}{\xi} + \frac{(d\eta/dt)^2}{\eta} \right\} + \xi\eta \left(\frac{d\psi}{dt} \right)^2 \right] \dots (5)$$

Where $(d\xi/dt)$, $(d\eta/dt)$, and $(d\psi/dt)$ are the *parabolic velocity* coordinates.

Taking into account the momenta in terms of the parabolic coordinates, *i.e.*, p_ξ, p_η and p_ψ (which are obtained by differentiating relation (5) with respect to the parabolic coordinates), it can be shown that

$$E_{kin} = \frac{1}{2m(\xi + \eta)} \left\{ 4\xi\, p_\xi^2 + 4\eta\, p_\eta^2 + \left(\frac{1}{\xi} + \frac{1}{\eta} \right) p_\psi^2 \right\} \quad \dots (6)$$

The *potential energy of the electron* is given by

$$E_{pot} = -\frac{Ze^2}{r} + e\, Fx,$$

where $(-Z\, e^2 / r)$ is the P. E. of the electron at a distance $r = OP$ from the nucleus and $(+\, e\, F\, x)$ is the P. E. due to the force of the electric field of uniform intensity F at a distance x from the X-axis.

Using relations (3), r and x can be expressed in parabolic coordinates;

$$r^2 = x^2 + y^2 = \tfrac{1}{4}(\xi - \eta)^2 + \xi\eta = \tfrac{1}{4}(\xi + \eta)^2$$
$$\therefore \quad r = \tfrac{1}{2}(\xi + \eta) \text{ and } x = \tfrac{1}{2}(\xi - \eta)$$

Substituting these values of r and x,

$$E_{pot} = -\frac{Ze^2}{\tfrac{1}{2}(\xi + \eta)} + \tfrac{1}{2}\, eF\, (\xi - \eta) \qquad \dots (7)$$

Hence the *total energy* W of the electron, given by $E_{kin} + E_{pot}$, a *Hamiltonian function*, which remains invariant during the whole motion, is obtained:

$$W = \frac{1}{2m\,(\xi + \eta)} \left\{ 4\xi\, p_\xi^2 + 4\eta p_\eta^2 + \left(\frac{1}{\xi} + \frac{1}{\eta} \right) p_\psi^2 \right\}$$
$$- \frac{2Ze^2}{(\xi + \eta)} + \tfrac{1}{2}\, eF\, (\xi - \eta)$$

This relation can be put in the form

$$2m\,(\xi + \eta)\, W = 4\xi\, p_\xi^2 + 4\eta p_\eta^2 + \left(\frac{1}{\xi} + \frac{1}{\eta} \right) p_\psi^2$$
$$- 4mZe^2 + meF\, (\xi^2 - \eta^2)$$

In Hamiltonian mechanics it is shown that

$$p_\xi = \frac{\partial S}{\partial \xi}, \; p_\eta = \frac{\partial S}{\partial \eta} \text{ and } p_\psi = \frac{\partial S}{\partial \psi},$$

where S is known as the *action function*, given by the relation:

$$S = 2 \int E_{kin}\, dt$$

Since ψ is a *cyclic* coordinate, giving the position of the figure with respect to the fixed plane and hence varying from 0 to 2π,
$p_\psi = (\partial S / \partial \psi) = $ a constant.
Remembering this and substituting for p_ξ and p_η, $\partial S/\partial \xi$ and $\partial S/\partial \eta$ respectively, we get the *Jacobi's partial differential equation* for the motion of the electron as:

$$4\xi \left(\frac{\partial S}{\partial \xi}\right)^2 + 4\xi \left(\frac{\partial S}{\partial \eta}\right)^2 = 2m\,(\xi + \eta)\,W + 4mZe^2$$

$$- me\,F\,(\xi^2 - \eta^2) - \left(\frac{1}{\xi} + \frac{1}{\eta}\right) p_\psi^2 \qquad \ldots (8)$$

Separating the variables in this equation, since the separated parts, according to a known rule in mathematics, must be equal to the same constant which may be conveniently represented by — $2m\beta$, we get

$$4\xi \left(\frac{\partial S}{\partial \xi}\right)^2 - 2m\xi W - 2mZe^2 + meF\xi^2 + \frac{p_\psi^2}{\xi}$$

$$= -4\eta \left(\frac{\partial S}{\partial \eta}\right)^2 + 2m\eta W + 2mZe^2 + me\,F\eta^2 - \frac{p_\psi^2}{\eta}$$

$$= -2m\beta \text{ (a constant)} \qquad \ldots (9)$$

With the variables thus separated, a complete solution becomes possible. Without entering into the complicated details of the solution, we may pass on to the final result concerning the nature of the orbits described by the electron, which the solution offers.

It is found that both the coordinates ξ and η vary continuously between a maximum and a minimum value. Hence taking the (x, y) plane, i.e.; a plane where ψ is constant and considering the four parabolas
$\xi = \xi_{min}$, $\xi = \xi_{max}$, $\eta = \eta_{min}$, $\eta = \eta_{min}$

given by the solution of equation (9), we get a *curved quadrangle* ABCD enclosed by the four parabolas (Fig. 237).

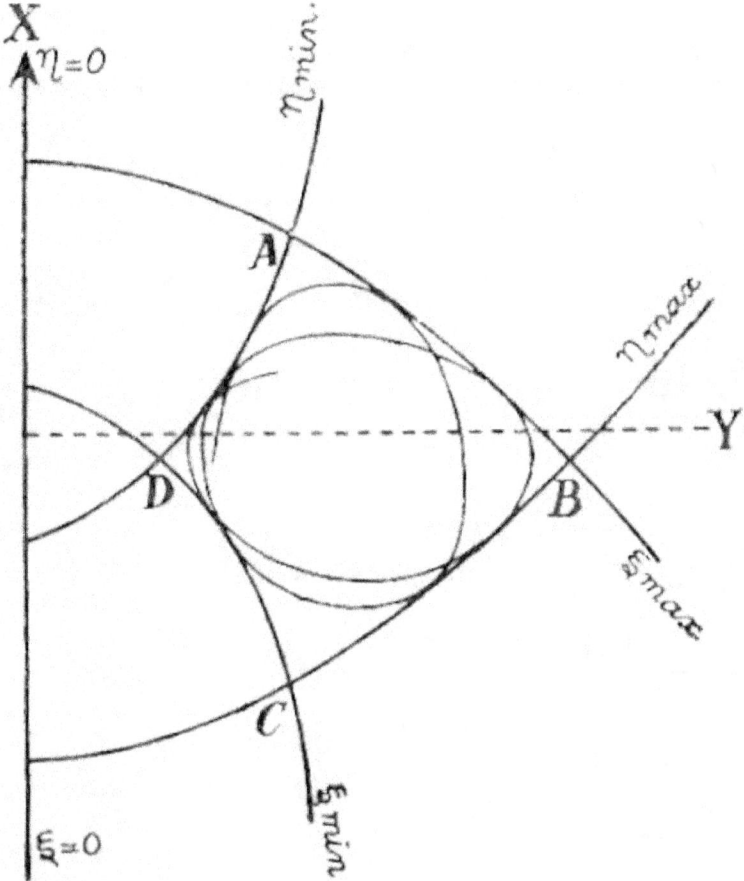

Fig. 237.

The orbital curve of the electron is contained within that curved quadrilateral; the curve alternately touches the parabolas of the ξ and η groups, and in the course of time closely covers the whole of the curved quadrilateral. Moreover, the plane containing the complicated orbital curve rotates in space about the direction of the electric field with a rotational velocity $d\psi/dt$, which is variable within certain limits, p_ψ however remaining constant. Thus the electronic orbit is a spatial curve which continuously coils round the direction of the field.

The quantum part of the problem. The first step in the treatment of the quantum aspect of the problem is to apply Sommerfeld's quantum conditions to the three parabolic coordinates:

$$\oint p_\xi \, . d\xi = n_\xi h \, ;$$

$$\oint p_\eta \, . d\eta = n_\eta h \, ;$$

$$\oint p_\psi \, . d\psi = n_\psi h \qquad \dots (10)$$

where n_ξ, n_η, and n_ψ are integral numbers, the first two being called *parabolic quantum numbers*, while the third *equatorial quantum number*.

The three quantum numbers are related to the total quantum number n as:

259

$$n = n_\xi + n_\eta + n_\psi \qquad \qquad \dots (11)$$

The next step is to express the total energy W in terms of the above-stated quantum numbers. To do this, taking eqn. (9), the *first* term maybe written as:

$$p_\xi^2 = \frac{1}{4\xi}\left\{ 2m\xi W + 2m(Ze^2 - \beta) - meF\xi^2 - \frac{p_\psi^2}{\xi} \right\}$$

$$= \frac{1}{4}\left\{ A_1 + \frac{2B_1}{\xi} + \frac{C_1}{\xi^2} + D_1\xi \right\} \qquad \dots (12)$$

where
$A_1 = 2\,mW,$
$B_1 = m\,(Ze^2 - \beta),$
$C_1 = - (p_\psi)^2$ and
$D_1 = -\,m\,e\,F.$

The *second* term, likewise, may be written as

$$p_\eta^2 = \frac{1}{4\eta}\left\{ 2m\eta W + 2m(Ze^2 + \beta) + meF\eta^2 - \frac{p_\psi^2}{\eta} \right\}$$

$$= \frac{1}{4}\left\{ A_2 + \frac{2B_2}{\eta} + \frac{C_2}{\eta^2} + D_2\eta \right\} \qquad \dots (13)$$

Where
$A_2 = 2\,mW,$
$B_2 = m\,(Ze^2 + \beta),$
$C_2 = - (p_\psi)^2$ and
$D_2 = +\,m\,e\,F.$

It is seen that $A_1 = A_2$, $C_1 = C_2$ and $D_1 = -D_2$. Also, both p_ξ and p_η are given by expressions that contain W as one of the coefficients.

In order to express W in terms of the quantum numbers involved we must work upon the relations (10) giving the quantum conditions.

Taking

$$\oint p_\psi \,.d\psi = n_\psi\, h,$$

since
$p_\psi = n_\psi\, h/2\pi,$ is a constant, and
ψ varies from 0 to 2π,
so that
$C_1 = C_2 = -(n_\psi\, h/2\pi)^2.$

Considering the other two phase integrals, the integration extends over a whole range of values of ξ and η, *i.e.*, from to ξ_{min} to ξ_{max} and back again, for ξ; similarly for η . Hence the integration is rather complicated. From relations (12) and (13) we write

$$p_\xi = \tfrac{1}{2}\left(A_1 + \frac{2B_1}{\xi} + \frac{C_1}{\xi^2} + D_1\xi \right)^{1/2}$$

$$\therefore \oint p_\xi \cdot d\xi = \oint \tfrac{1}{2}\left(A_1 + \frac{2B_1}{\xi} + \frac{C_1}{\xi^2} + D_1\xi \right)^{1/2} \cdot d\xi = n_\xi h.$$

Similarly,

$$\oint p_\eta \cdot d\eta = \oint \tfrac{1}{2}\left(A_2 + \frac{2B_2}{\eta} + \frac{C_2}{\eta^2} + D_2\,\eta \right)^{1/2} \cdot d\eta = n_\eta h.$$

Combining these two to assume the general form of an elliptical integral, we may write :

$$\oint \left(A + \frac{2B}{r} + \frac{C}{r^2} + Dr \right)^{1/2} \cdot dr = 2nh \qquad \ldots (14)$$

where r stands for ξ and n for n_ξ in one case,
while r for η and n for n_η in the other.
Likewise, the coefficients A, B, C, D denote
A_1, B_1, C_1, D_1 in the first case,
and
A_2, B_2, C_2, D_2, in the second.

This elliptical integral can be worked out and it leads to the evaluation of A_1, B_1, C_1, D_1 and A_2, B_2, C_2, D_2, in the two cases. But what interests us here is the value of the coefficient $A = A_1 = A_2$, which contains W. It turns out to be equal to

$$\frac{4\pi^2 (mZe^2)^2}{(n_\xi + n_\eta + n_\psi)^2 h^2} - \frac{3h^2F}{4\pi^2 Ze}(n_\eta - n_\xi)(n_\xi + n_\eta + n_\psi)$$

neglecting terms involving higher powers of F.
Since $A = A_1 = A_2 = 2mW$, then $W = A/2m$

$$\therefore \quad W = \frac{2\pi^2 mZ^2 e^4}{n^2 h^2} - \frac{3h^2 F}{8\pi^2 mZe} \cdot n \cdot (n_\eta - n_\xi) \qquad \ldots (15)$$

replacing $(n_\xi + n_\eta + n_\psi)$ by n.

The *first* term on the right-hand side gives the energy in the *undisturbed* state, which shows that the calculation in parabolic coordinates leads to the same result as in the earlier calculation with polar coordinates, *(cf.* p. 618) and thereby attests to the essential soundness of the method used here.

The *second* term gives the *perturbation of the first order* arising from the electric field, since we have neglected terms with higher powers of F. Remembering that, when F = 0, W reduces to the first term only, the second term evidently gives the energy change dW produced by the electric field. Hence

$$d\mathrm{W} = -\frac{3h^2 F}{8\pi^2 mZe}(n_\eta - n_\xi) \cdot n \qquad \ldots (16)$$

This is known as *Epstein's formula* which is capable of interpreting adequately all the observed facts in the Stark effect of the Balmer lines.

Limitation of the theory proposed

The above method of deriving the energy value is only an approximate one. Actually the energy of the atom in the electric field is to be expressed in the form of a series in ascending powers of F, as

$$W = W_0 + W_1 F + W_2 F^2 + W_3 F^3 + \ldots$$

261

where
W$_o$ represents the energy in the *undisturbed state*,
W$_1$F may be termed the energy of the *first order* effect,
W$_1$F^2 of the *second order* effect and se on.

The theory given above takes only the linear term W$_1$F into account, and can explain only the *linear Stark effect* which occurs when the *applied electric field is sufficiently strong*, say within the limit of 5,000 to 10,000 volts/cm. in the case of hydrogen. The simplest and most direct way of deciding whether a field is strong or weak is, as in the case of the Zeeman effect, to compare the resolution produced by the electric field with the field-free fine structure separation.

H.A. Kramers, in 1920, worked out the case of *very weak* fields, but it is only of theoretical interest, as the Stark splitting, under these circumstances, can hardly be observed. In his final formula for energy change, the term of the first order effect is absent, while that of the second order appears. This can be explained as follows:

The natural doublet separation of a Balmer line arises, as we have seen, from the precessional rosette motion caused by the relativistic variability of the mass of the electron. Now, if the field is relatively weak, the above-mentioned precessional motion is not suppressed and this causes a displacement of the electric centre of gravity. When the field is so strong that the precession due to it is much larger than the relativity-precession, the electric centre always lies on the same side of the plane through the nucleus normal to the field. When, however, the field is so weak that the field-precession is very small compared to the relativity-precession, the electric centre will lie on one side of this plane during one half of the rosette motion and on the other side during the next half, so that the first order term is alternately positive and negative and its time average is zero, and hence only the very small quadratic term appears.

It may also happen that the applied field becomes of the same order of magnitude as the nuclear field, *i.e.*, when the resolution due to the external field is about the same as the field-free fine structure separation. In such a case, on that portion of the orbit where the external field is opposed to the nuclear field, the electron will be subject to practically no force and the orbit becomes indeterminate. Under these conditions, there can be no energy transition from or to the orbit. After the fine structure limit is passed, the resolution is, up to a certain range, proportional to F and we have the linear effect. If the field is still further increased the resolution is no longer represented by the approximate theoretical relation derived above. Terms of the second and higher orders become appreciable, and hence not to be neglected. The *second order* effect can be considered as due to the *polarization of the atom* induced by the electric field, *i.e.*, the field induces in the atom a proportional dipole moment $\mu = \alpha F$.

Testing the validity of the theory proposed

(i) Considering relation (16), *viz,*

$$dW = - \frac{3h^2F}{8\pi^2 mZe} (n_\eta - n_\xi)\, n,$$

we see that any energy level of total quantum number n is resolved into sub-levels of different energies given by all possible values of $(n_\eta - n_\xi)$, which may be called the *Stark effect quantum number*. Further, since n_η, n_ξ and n are all whole numbers, the energy separation of the sub-levels will be integral multiples of a minimum value equal to

$3h^2F / 8\pi^2 mZe.$

As the field strength F increases, this quantity will also increase, so that the *energy change increases proportionally* to F. As the value of the total quantum number *n* increases, the number of quantum triplets into which *n* may be resolved also increases and with it the number of sub-levels, which accounts for the *increase in complexity of the components with the total quantum number,* in agreement with observed facts.

(ii) If dW_1 and dW_2 denote the changes of energy produced by the electric field in the initial and final states of total quantum numbers n_1 and n_2 respectively, *the resolution of the component lines* in terms of frequency is calculated by applying Bohr's frequency condition:

$$d\nu = \frac{dW_1 - dW_2}{h}$$

Now,

$$dW_1 = -\frac{3h^2 F}{8\pi^2 mZe}(n_\eta - n_\xi)_1\, n_1$$

And

$$dW_2 = -\frac{3h^2 F}{8\pi^2 mZe}(n_\eta - n_\xi)_2\, n_2$$

The indices 1 and 2 are used for the Stark effect quantum numbers, since the values of $(n_\eta - n_\xi)$ will be different in the two cases.

$$d\nu = \frac{3hF}{8\pi^2 mZe}\left\{ (n_\eta - n_\xi)_2\, n_2 - (n_\eta - n_\xi)_1\, n_1 \right\} \quad \ldots \quad (17)$$

From this relation we see directly that, since the minimum value of the term within the brackets is unity, *all the line resolutions in a series are whole multiples of a minimum line-interval, a = 3hF / $8\pi^2 mZe$.*, which is also directly proportional to F. This accounts for the observed *increase in resolution proportionally to* F (linear Stark effect).

(iii) Taking relation (11), $n = n_\xi + n_\eta + n_\psi$ the equatorial quantum number n_ψ is conveniently replaced by another, say λ, the relation between the two being given by

$\lambda = n_\psi - 1$ where $\lambda = 0, 1, 2, 3 \ldots..$

since, according to wave mechanics, the quantity λ, like l, the orbital quantum number, has to be a positive integer, zero being included.

With this change, the three parabolic quantum numbers n_ξ, n_η and λ have the common characteristic of being able to assume all positive integral values including zero, viz., 0, 1, 2, ... $(n-1)$.

Expressing n_ψ. in terms of λ,

$n = n_\xi + n_\eta + \lambda + 1$

From this relation, it follows that the total quantum number n can assume all positive integral values except zero, i.e., $n = 1, 2, 3\ldots..$

According to the vector model, any level of total quantum number n consists of a number of sub-levels having characteristic values of l and j. In the present treatment, no account has been taken of l and j, but new quantum numbers have been introduced, whose interpretation is entirely different. Still, it is easy to realize that λ represents the projection of l on the field direction and thus it corresponds to m_l used in the Zeeman effect. Hence by correspondence principle, *the selection rule for λ is the same as for m_l, viz.,* $\Delta\lambda = 0$ or ± 1 which is further confirmed by wave mechanics.

Representing the total quantum number n by $(n_\xi\ n_\eta\ \lambda)$, if the transition $(n_\xi\ n_\eta\ \lambda)_1 \rightarrow (n_\xi\ n_\eta\ \lambda)_2$ is possible, then the converse transition $(n_\xi\ n_\eta\ \lambda)_2 \rightarrow (n_\xi\ n_\eta\ \lambda)_1$ must also be possible. If the former leads to a component at a distance $+d\nu$ from the original line, then the latter leads to a component at a distance $-d\nu$. Hence the *displacement of the Stark components is symmetrical on both sides of the parent line*, again in agreement with experimental facts.

(iv) Also *the polarization of the lines will be symmetrical*, since polarization depends only on λ, as shown by correspondence principle and since λ is left unchanged in the two transitions symmetrically disposed about the original line. As in Zeeman effect, here also, transitions corresponding to $\Delta\lambda = \pm 1$ give rise to lines that are circularly polarized in the: plane *normal* to the electric vector (σ components). In the *transverse* view, they appear in all circumstances *plane polarized perpendicular to the direction of the field.* When view-ed *longitudinally,* they would appear as circularly polarized if only one process of emission. were observed. In reality, however, every observation represents an average of many elementary processes. Of their total number, the transitions $\Delta\lambda = +1$ occurs as frequently as the transitions $\Delta\lambda = -1$. If the former leads to left-handed circular polarization, the latter leads, just as often, to right-handed circular polarization. The superposition of these two oppositely polarized waves of the same frequency brings about that *no polarization of the components is observed in the longitudinal view.* This is in contradistinction to Zeeman effect, where the right-handed and left-handed components have different frequencies and hence can be distinguished from each other.

On the other hand, the transitions that correspond to $\Delta\lambda = 0$ give rise to lines linearly polarized in the direction of the field (π components). When they are viewed *transversely,* they appear as *plane polarized parallel to the field ;* the same are *invisible* when viewed *longitudinally.*

In actual observation, since both types of transition, viz., $\Delta\lambda = \pm1$ and 0 are present, we have both σ and π components which appear in the transverse effect, while in the longitudinal effect the σ components are unpolarized, while the π components are invisible.

(v) We have seen that when the two fields, external and nuclear, are of comparable strength, the electron orbit becomes indeterminate, which excludes energy transitions from or to it. Now, since the size r of the orbit varies as n^2, the force of the nuclear field is inversely proportional to n^4. Hence, as n increases, the nuclear field will decrease rapidly and the external field strength, required to oppose the nuclear field and make the orbit indeterminate, will be correspondingly less. This explains why *certain transitions are suppressed and lines disappear, as the field strength increases, the disappearance occurring at lower values of F for the lines with higher values of* n.

Calculation shows also that

the *short-wave components* of the Stark Pattern of each line correspond to transitions starting from orbits for which the applied electric field opposes the nuclear attraction as the electron passes through its position of closest approach to the nucleus (perihelion).

The *long-wave components,* on the other hand, are due to transitions from orbits for which the greatest weakening of the nuclear attraction occurs when the electron is at its maximum distance from the nucleus (aphelion). This provides an explanation of the experimental fact that the *long-wave components of any given line disappear at lower field strengths than the corresponding short-wave components.*

THE STARK PATTERN OF THE H_α LINE

We shall now consider a concrete example, viz., the Stark pattern of the H_α line and see how the theory proposed can be applied to study it.

Taking relation (17) and putting $(3hF / 8\pi^2 mZe) = a$, we may write :

$$d\nu = a \{(n_\eta - n_\xi)_2, n_2 - (n_\eta - n_\xi)_1, n_1\}$$

If we measure the displacements of the components from the parent line in terms of the unit a, the displacement Δ of any component in question is given by

$$\Delta = d\nu / a = (n_\eta - n_\xi)_2, n_2 - (n_\eta - n_\xi)_1, n_1$$

For the H_α line, $n_1 = 3$ and $n_2 = 2$, and hence,

$$\Delta = 2 (n_\eta - n_\xi)_2 - 3 (n_\eta - n_\xi)_1 \ldots\ldots\ldots\ldots (18)$$

Considering the *upper level (n$_1$ = 3)*; the number of sub-levels is given by $(n_\eta - n_\xi)_1$. Since $n_1 = 3$, the values of n_η are 2, 1, 0 and those of n_ξ 2, 1, 0. With each one of the three values of n_η the three different values of n_ξ can be combined. Thus:

$n_\eta = 2$, $(n_\eta - n_\xi)_1 = 0, +1, +2$
$n_\eta = 1$, $(n_\eta - n_\xi)_1 = -1, \ 0, +1$

$n_\eta = 0, \quad (n_\eta - n_\zeta)_1 = -2, -1, \; 0$

We see that $(n_\eta - n_\zeta)_1$ can take five different values: + 2, +1, 0, -1 and - 2. There are thus *five distinct sub-levels in the upper state.* [It may be noted that any given sub-level given by a particular value of $(n_\eta - n_\zeta)$ may arise from different combinations of n_η and n_ζ. In general, for any given value of n, $(n_\eta - n_\zeta)$ can assume values ranging from $+ (n-1)$ to $- (n-1)$].

The separations of the five sub-levels in terms of the unit 'a' are given by

$3 (n_\eta - n_\zeta)_1 = + 6, + 3, 0, -3, -- 6$

The values of the equatorial quantum number λ_1 of these sub-levels can be found from the relation: $n = n_\eta + n_\zeta + \lambda + 1$. In the present case, $\lambda_1 = 3 - (n_\eta + n_\zeta + 1)$, from which the different values of λ_1 are found:

n_η	n_ζ	$n_1 (n_\eta - n_\zeta)_1$	λ_1
2	0	+ 6	0
2	1	+ 3	− 1
2	2	0	− 2
1	0	+ 3	+ 1
1	1	0	0
1	2	− 3	− 1
0	0	0	+ 2
0	1	− 3	+ 1
0	2	− 6	0

It is seen that $\lambda_1 \leq 2$, *i.e.,* $\geq (n_1 - 1)$ in absolute value; also that if a particular sub-level arises in a number of ways, there are an equal number of values for λ_1 and corresponding to any positive value of λ_1 there is always a negative value. The distinct values of λ_1 are +2, +1, 0, -1 and - 2.

Taking the *lower level* $(n_2 = 2)$, the number of sub-levels is given by $(n_\eta - n_\zeta)_2$. Since $n_2 = 2$, the values of $n_\eta = 1, 0$ and those of n_ζ are 1, 0. Combining with each one of the two values of n_η the two different values of n_ζ, *three distinct sub-levels* $(n_\eta - n_\zeta)_2$ are obtained as + 1, 0, -1, whose separations in terms of a are given by $(n_\eta - n_\zeta)_2 = + 2, 0, -2$. The corresponding λ_2 values are obtained from the relation $\lambda_2 = 2 - (n_\eta + n_\zeta + 1)$:

n_η	n_ζ	$n_2 (n_\eta - n_\zeta)_2$	λ_2
1	0	+ 2	0
1	1	0	− 1
0	0	0	+ 1
0	1	+ 2	0

The values of λ_2 are +1, 0, -1, i.e., $\lambda_2 \leq 1$, or $\leq (n_2 - 1)$.

Having thus determined the sub-levels of the two states between which transitions take place, giving rise to the Stark components, the displacements Δ of these components are readily obtained from equation (18). The appropriate selection rule, viz., $\Delta\lambda = 0, \pm 1$ is used to eliminate the forbidden transitions. The nature of the polarization of the components is determined by the rule : $\Delta\lambda = 0$ gives rise to the π components, while $\Delta\lambda = \pm 1$ to σ components:

265

Lower state		Upper state		Δ	$\Delta\lambda = (\lambda_1 - \lambda_2)$	Nature of Components
$2(n_\eta - n_\xi)_2$	λ_2	$3(n_\eta - n_\xi)$	λ_1			
$+2$	0	$+6$	0	-4	0	π
	0	$+3$	± 1	-1	± 1	σ
	0	0	$0, \pm 2$	$+2$	$0, [\pm 2]$	π
	0	-3	± 1	$+5$	± 1	σ
	0	-6	0	$+8$	0	π
0	± 1	$+6$	0	-6	± 1	σ
	± 1	$+3$	± 1	-3	$0, [\pm 2]$	π
	± 1	0	$0, \pm 2$	0	$\pm 1 [\pm 3]$	σ
	± 1	-3	± 1	$+3$	$0, [\pm 2]$	π
	± 1	-6	0	$+6$	± 1	σ
-2	0	$+6$	0	-8	0	π
	0	$+3$	± 1	-5	± 1	σ
	0	0	$0, \pm 2$	-2	$0, [\pm 2]$	π
	0	-3	± 1	$+1$	± 1	σ
	0	-6	0	$+4$	0	π

Separating the π and σ components and rearranging according to the increasing order of Δ, we get

eight π components : $+8, +4, +3, +2, -2, +3, -4, -8$ and
seven σ components : $+6, +5, +1, 0, -1, -5, -6$.

They are symmetrically arranged on either side of the parent line, as indicated by the same positive and negative values of Δ.

The *energy level diagram* of the Stark pattern of the H_α line, as predicted by the theory, is given in Fig. 238.

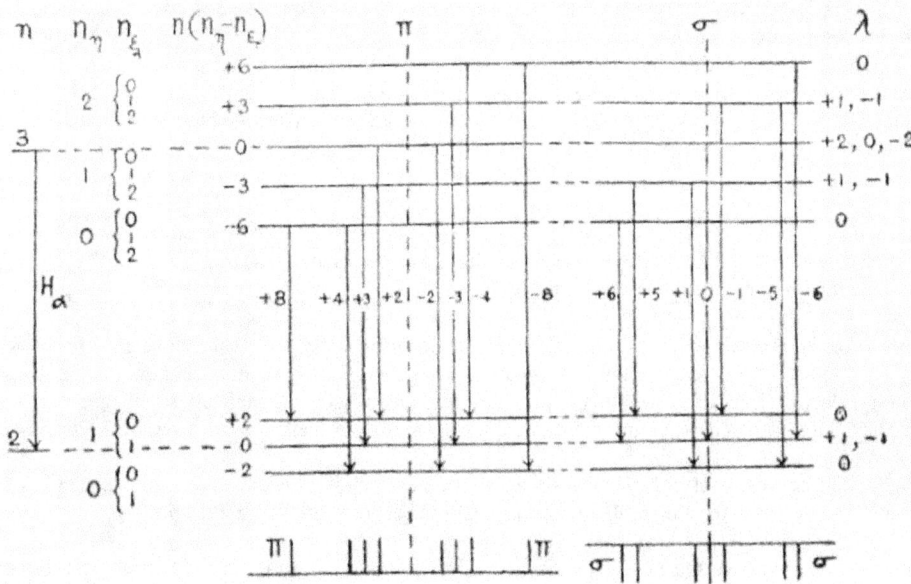

Fig. 238. Stark pattern of the H_a line.

Comparison with experiment

If experimental data obtained by Stark and others on the H_a line are arranged on the scale $\Delta = dv / a$ (in wave numbers), they appear as shown in Fig. 239. We see, up to $\Delta = 4$, the predictions of the theory agree fully with experimental results. For instance, the positions 0 and 1 are free from π components and are occupied by σ components, whereas the reverse is the case at the positions 2. 3, and 4.

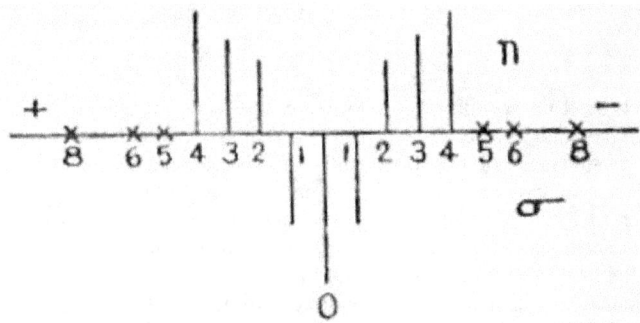

Fig. 239

A more refined wave mechanical theory indicates a few other components of greater resolution, at position 8, π components, at 5 and 6, σ components, which are not observed. But this is due to the small intensity of these lines. Schrödinger has shown by a wave mechanical calculation that the π components in question must be a thousand times weaker and the σ components a hundred times weaker than the average intensity of the observed components.

The absolute value of the interval between the components can be calculated from the constant $a = (3hF/8\pi^2 \, mZe)$ and compared with the experimental value. But this quantity is not very accurately determined by experiment. Hence, as Epstein and others had done, the field intensity F can be calculated from a measured resolution and compared with the field strength actually used. This shows good agreement between theory and experiment.

Conclusion

As far as the Balmer lines are concerned, the experimental data, are found to be in complete accord with the theory proposed. But in the development of the theory, the relativity effect as well as the spin-interaction energy, whose inclusion, as is done in more refined calculations, gives only a small additional splitting of certain energy levels, have been neglected. This is, however, justified, because the field ordinarily used is so great that the Stark separations are many times larger than the field-free fine structure separations, as also the Zeeman splitting obtained with the usual fields.

For this very reason,

Stark effect is of minor importance in the study of *atomic spectra*. But the reverse is the case in *molecular spectra*, since most molecules have ground electronic state, *i.e.,* they have no unpaired electrons, no electronic angular momentum or spin, and no electronic magnetic moment.

With a few significant exceptions (such as O_2, NO) the magnetic moments are of very low value. The electric dipole moments, however, are of appreciable strength. Hence, while magnetic fields of many thousands of gauss often cause only very small splitting, electric fields of a few hundred to a few thousands volts my completely resolve the individual line components. We have already noted the great importance attached to the Stark effect in microwave spectroscopy.

The theory of the Stark effect in rotational spectra was worked out already in the early days of quantum mechanics, but lay idle for many years, because the resolution in the optical region was inadequate for its application.

In a crystal, the Stark effect arising from the electric fields exerted by neighboring ions is very important, as it forms the basis of the theory of magnetic properties of solids. The Stark effect has been utilized also in the study of dielectric properties in the exploration of conditions in electric discharges and in astrophysics.

FORBIDDEN LINES IN ATOMIC SPECTRA

Forbidden lines are those which result from transitions that are forbidden by the selection rules for L, S and J of the vector model. As we have already noted, such transitions do sometimes occur, giving rise to lines, normally with very small intensity.

But sufficiently intense forbidden lines are found in nebular and auroral spectra, which were first wrongly attributed to elements unknown in terrestrial sources.

Note.
The excitation of **atoms** and ions to metastable levels by electron collision is followed by cascade to lower levels which, in the process, emit the so-called forbidden quanta. The transition probabilities of **spectral lines** are quite few by comparison to the allowed transition. The allowed transitions are electric **dipole** radiations, whereas forbidden transitions correspond to magnetic-dipole and/or electric-quadruple radiations. There are three types of transitions which are the result of collisional excitation: nebular, auroral, and transauroral. All the upward transitions are due to collisional excitation only; however, the downward transitions can be one of two types, i.e., superelastic collisions, or radiation of forbidden lines. The level density and atomic constants determine which of the latter transitions is likely to take place in depopulating the level. Also, the forbidden spectra are observed only for ions whose metastable levels lie a few electron volts above the ground state. Collisionally excited lines are observed in low lying levels of the spectra of CIII, CIV, NIII, NIV, NV, SIIII, etc., in the far ultraviolet.
The study of forbidden lines is one of the major areas of investigation in gaseous nebulae since they dominate the spectra of most gaseous nebulae.

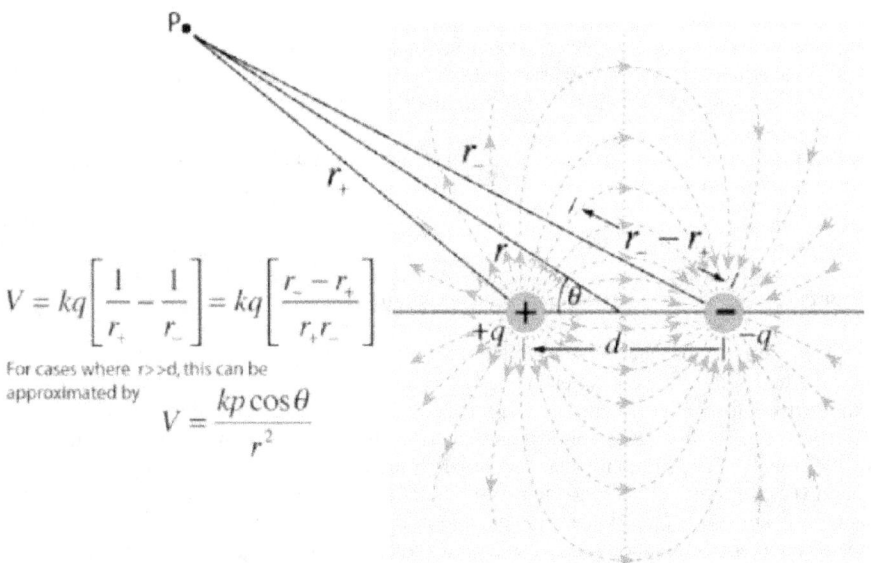

$$V = kq\left[\frac{1}{r_+} - \frac{1}{r_-}\right] = kq\left[\frac{r_- - r_+}{r_+ r_-}\right]$$

For cases where r>>d, this can be approximated by

$$V = \frac{kp\cos\theta}{r^2}$$

Dipole, literally, means "two poles," two electrical charges, one **negative** and one positive. Dipoles are common in **atoms** whenever electrons (-) are unevenly distributed around nuclei (+), and in molecules whenever electrons are unevenly shared between two atoms in a covalent bond.

When a dipole is present, the atom or covalent bond is said to be polarized, or divided into negative and positive regions. This is indicated by the use of partial negative (δ-) and partial positive (δ+) signs. The magnitude and direction of the electrical charge separation is indicated by using an arrow, drawn from the positive pole in a **molecule** to the negative pole.

In covalent bonds, permanent dipoles are caused when two different atoms share their electrons unevenly. The atom that is more electronegative—the one that holds electrons more tightly—pulls the electrons closer to itself, creating a partial negative charge there. The less electronegative atom becomes partially positive as a result because it has lost partial possession of the electrons. The electric strength of a dipole generally increases as the electronegativity difference between the atoms in the bond increases. This strength, called a dipole moment, can be measured experimentally. The size of a dipole moment is expressed in Debye units in honor of the Dutch chemist, Peter Debye (1884-1966).

The dipole moments of a series of molecules are listed below:

Molecule Dipole Moment (in Debye units, D)

HF	1.91 D
HCl	1.03 D
HBr	0.78 D
HI	0.38 D

The measurement of dipole moments can help determine the shape of a molecule. The net dipole moment of a **water** molecule (H_2O) represents the overall electrical charge distribution in that molecule. (See Figure 1.)

The H_2O molecule is bent. Its dipole vectors do not cancel. The water molecule has a net resultant dipole moment of 1.87 D. If the molecule were linear, the measured dipole moment would be **zero**. Its individual dipoles in the two oxygen-hydrogen covalent bonds would have cancelled each other out.

Individual atoms (and ions) will be naturally polarized if their electrons happen to move irregularly about their nuclei creating, at least temporarily, lopsided looking atoms with δ+ and δ-portions. Natural collisions occurring between atoms can induce this temporary deformity from an atom's normal spherical, symmetric shape. Larger atoms are considered to be "softer" than smaller, "harder" atoms. Larger atoms are then more likely to be polarized or to have stronger dipoles than smaller atoms.

The presence of dipoles helps to explain how atoms and molecules attract each other. Figure 2 shows how the electrically positive side of one xenon atom (Xe) lines up and pulls towards the negative side of another xenon atom. Likewise, the positive side of one H-Cl molecule is attracted to the negative side of another H-Cl molecule. When many atoms and molecules are present in **matter**, these effects continue on indefinitely from atom to atom and molecule to molecule.

Dipole forces tend to organize matter and pull it together. Atoms and molecules most strongly attracted to each other will tend to exist as solids. Weaker interactions tend to produce liquids. The gaseous state of matter will tend to exist when the atoms and molecules are **nonpolar**, or when virtually no dipoles are present.

We shall first of all state the different causes for the appearance of forbidden lines and then account for the presence of such lines in nebular and auroral spectra.

Forbidden lines may arise due to the following causes:

(i) *The selection rule under consideration may be true only to a first approximation.*

We have an example of this in the selection rule $\Delta S = 0$ which holds unconditionally only under the assumption of vanishing coupling between L and S, as in the case of elements of low atomic numbers, (such as helium and some alkaline earths), where the terms of the triplet system (S = 3) practically do not combine with the terms of the singlet system (S e= 0). This selection rule, however, becomes less and less rigorous, as the coupling between L and S increases, as in the case of elements of higher atomic numbers or larger multiplet terms. In such cases, forbidden transitions do occur, though very weak as compared to the allowed transitions. Lines that arise out of these forbidden transitions are known as *inter-combination lines,* whose number and intensity increase with increasing atomic number. The best known case of such an inter-combination line is the mercury resonance line $\lambda = 2537$, arising from the transition $^3P_1 \rightarrow {}^1S_o$, rather intense, but considerably weaker, than the corresponding non-inter-combination line at $\lambda = 1849$, arising from the transition $^1P_1 \rightarrow {}^1S_o$. It should be noted that the selection rule for J, *viz.,* $\Delta J = 0$ or ± 1 holds also for the inter-combination lines.

(ii) *A transition which is forbidden as 'electric dipole' radiation, may be permitted as 'electric quadrupole' radiation or 'magnetic dipole' radiation.*

An *electric dipole* is constituted by two equal and opposite electric charges + *e* and - *e,* separated by a distance *d* and the moment of such a dipole is given by *e* x *d*.

According to the classical electromagnetic theory of radiation, the simplest model capable of electromagnetic radiation is an oscillating electric dipole, such as the Hertzian oscillator.

Electromagnetic waves are radiated with the same period and hence the same frequency with which the electric charge flows back and forth in such a dipole, while the intensity of radiation depends upon the magnitude of the alteration of the dipole moment.

Now, according to wave mechanics, the selection rules already stated hold good only for the electric dipole radiation, and the above general rules have been further reduced to the special form, known as the *Laporte rule* which states:

"Even terms can combine only with odd and odd only with even."

A direct consequence of this rule is the *prohibition of the combination of two terms of the same electron configuration,* since in such a case both terms will be either even or odd.

An *electric quadrupole* is formed by an assemblage of charges such as that shown in Fig. 240.

$$E_r = \frac{2q}{4\pi\varepsilon_o}\left[\frac{1}{r^2} - \frac{r}{(r^2+d^2)^{3/2}}\right]$$

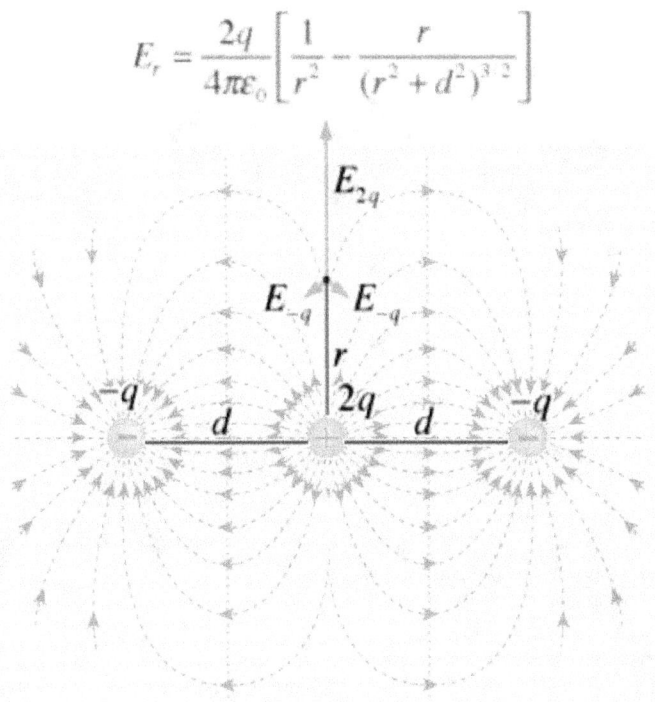

Fig. 240.

In general a quadrupole is much more complicated than the one represented here, but the simple case is sufficient for our .purpose. Although such a system has no dipole moment, yet it gives an external field which, however, falls off more rapidly with increasing distance than that of the dipole which itself decreases faster than that of the monopole. The potentials in three cases are proportional to $1/r^3$ for quadrupole, $1/r^2$ for dipole, and $1/r$ for monopole, respectively.

The action of the quadrupole must therefore be characterized by a quadrupole moment which is not zero. Now, just as a variable dipole moment leads to radiation, so also a variable quadrupole moment leads to radiation which is aptly termed *quadrupole radiation*. According to wave mechanics, the transitions which are strictly forbidden for dipole radiation, may occur, though quite weakly, due to quadrupole radiation. The ratio of the transition probability of ordinary dipole radiation to quadrupole radiation is found to be $1: 10^{-8}$.

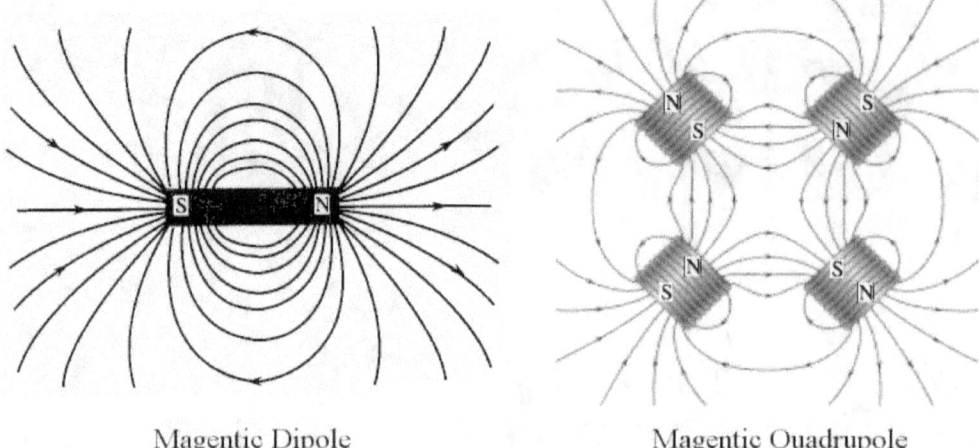

Magentic Dipole   Magentic Quadrupole

A *magnetic dipole* is constituted by two equal and opposite poles $+ m$ and $— m$ separated by a distance d and the moment of such a dipole is given by m x d. According to the classical theory, a variable magnetic dipole moment, such as that produced by an alternating current in a coil, gives rise to electromagnetic radiation which is called the *magnetic dipole radiation.*

Correspondingly, in wave mechanics, it gives rise to a transition probability which may be different from zero, even if the ordinary electric dipole transition probability is zero. Thus the transitions which are strictly forbidden for electric dipole radiation may occur, though weakly, due to magnetic dipole radiation. The transition probability of the magnetic dipole radiation is also small compared with that of the electric radiation (10^{-5} : 1).

It is shown in wave mechanics that similar relations hold good both for the electric quadrupole and magnetic dipole radiations. Hence we have *for these two types of radiations selection rules exactly the opposite to the Laporte rule* governing electric dipole radiations, *viz.,*

"Even terms combine only with even and odd only with odd."

A direct consequence of this is that *the combination of two terms of the same electron configuration is permitted* in the present case. It follows also that electric dipole radiation on the one hand and quadrupole or magnetic dipole radiation on the other cannot take part simultaneously in one and the same transition.

The following selection rules have been obtained:

For the *electric quadrupole radiation,*

for L : $\Delta L = 0, \pm 1, + 2$ ($0 \rightarrow 0$ excluded);
for J : $\Delta J = 0, \pm \mathbf{1}, \pm 2$, with the addition that $J_1 + J_2 \geq 2$, where J_1 and J_2 are the values of J in the upper and lower states, which means that $0 \rightarrow 0$, $1/2 \rightarrow 1/2$, $1 \rightarrow 0$ are excluded ; *for* S : $\Delta S = 0$, the same as in the electric dipole radiation to the same degree of approximation.

For the *magnetic dipole relation,*

for L : $\Delta L = 0, \pm 1$,
for J : $\Delta J = 0, \pm 1$ ($0 \rightarrow 0$ excluded)
for S : $\Delta S = 0$, the same as for the other typos of radiation.

Actually, cases have been observed in which transitions that are strictly forbidden by the electric dipole selection rules take place due to electric quadrupole or magnetic dipole radiation. Thus, for instance, the forbidden lines of the alkalis $^2S \rightarrow {}^2D$) (note that $\Delta L = \pm 2$. and even terms combine with even) have been observed with small intensity. It is important here to note that the selection rules for the Zeeman effect for the quadrupole and magnetic dipole radiations differ from those for ordinary electric dipole radiation and also from one another. Consequently an investigation of the Zeeman effect gives an unambiguous criterion for the kind of transition under consideration, Segré and Bakker have shown, from the study of the Zeeman effect of the above-mentioned forbidden lines that they are undoubtedly due to quadrupole radiation.

(iii) *A third cause for the appearance of forbidden lines may be that the selection rules for dipole radiation hold good in the absence of electric or magnetic fields but may be, however, transgressed when such fields are applied externally or are produced by neighboring atoms or ions.*

This is known as *enforced dipole radiation,* in which case, the intensity of the forbidden lines may even become comparable to the intensity of the allowed lines. The selection rules for the Zeeman effect of these enforced dipole radiations differ, as in the previous case, from those for ordinary dipole radiation, which, in consequence, serve as a sure means of detecting these lines and distinguishing them from others. Kuhn has observed the higher members of the transition $^2S \rightarrow {}^2D$ of the alkalis in the presence of an external magnetic field, which are not seen in the absence of the field. Thus we are dealing with enforced dipole radiations here and not with the quadrupole radiations mentioned above, which occur in the absence of the field and are restricted to the lower members of the series.

In emission spectra, transitions due to enforced dipole radiation are sometimes observed in electric discharges, where electric fields are always present (external or ion fields). Here also it is chiefly the higher members of the series that appear, since they are influenced more strongly by the field than the lower. (*Cf.* Stark effect).

(iv) *Nuclear spin.*

There occur also lines which contradict the selection rules of all the different kinds of radiation, electric dipole, quadrupole and magnetic dipole. Thus, in the absorption spectrum of mercury vapor we get a forbidden line $\lambda = 2269.8$ corresponding to the transition $^1S_o \rightarrow {}^3P_2$, and in the emission spectrum the line $\lambda = 2655.6$ corresponding to $^3P_o \rightarrow {}^1S_o$. According to Bowen, the transition is apparently due to the influence of nuclear spin.

Forbidden lines in nebular and auroral spectra

Spectra of nebulae

Telescopic observation shows that in addition to the stars in the sky there are vast regions of space filled with attenuated gases, some of which get hot enough to emit light; these have been classified as *nebulae.* It is believed that stars and suns are formed by condensation out of huge nebulae. There are great numbers of nebulae visible in the telescope and these are at an average distance of two million light years, while the farthest nebula is at a distance of 140 million light years. Our own sun is one of many million stars within a nebula, which we know as the *Milky way,* while the Milky way is one of the several million nebulae. There are also what are known as *spiral nebulae* or globular star clusters, one of the nearest of which is the *nebula in Andromeda,* just visible to then naked eye. Mention may be made also of a grouping of about hundred nebulae in the neighborhood of the Polo star.

In the spectroscopic study of the nebulae, one of the very important discoveries is the presence of forbidden lines. They were first believed to be produced by an element, unknown on earth and in the other celestial bodies, to which the name *'nebulium'* was given and the lines themselves were called 'nebulium lines'.

Auroral spectra

The aurora is a luminous effect visible in the polar regions; in the northern regions it is called the *'aurora borealis'* and in the southern regions *'aurora australis'.* Their appearance is very varied, but the frequent and common form consists of an arch of pale light with characteristic quivering rays or streamers of a pale rose color radiating from the polar centre.

Aurorae are usually accompanied by 'magnetic storms' and outbursts of energy in the photosphere of the sun. Aurorae appear in greater frequency at periods of about 11.5 years, which agrees pretty well with the cycles of maximum of magnetic storms and of sunspots. The phenomenon is believed to be due to the passage of electric discharge through the upper rarefied regions of the atmosphere. The earth is exposed to a continuous bombardment by the electrons emitted by the sun. The earth's magnetic field tends to deflect these electrons and move them towards the polar regions. This taking place in the rarefied upper regions of the atmosphere, the phenomena of the discharge tube are reproduced. A torrent of electrons constitutes an electric current of strength sufficient to produce considerable magnetic effects and thus the connection of aurorae with magnetic storms is understood. The spectroscopic analysis of auroral light shows it to be due to gaseous matter, its spectrum consisting of a few bright green and red lines, apparently characteristic of the gas contents of the upper atmosphere.

In order to explain adequately the appearance of forbidden lines in nebular and auroral spectra, certain terms connected with the emission of spectral lines, such as *mean life, transition probability* and *metastable state* have first to be considered. If an electron in an atom is displaced from its normal to higher quantum state, the atom which is excited thereby, will return to the normal condition with the emission of radiant energy or spectral lines. The time during which the electron will remain in the higher quantum state before dropping into lower states is known as the *mean life* of the excited state of the atom. This has been the subject of extensive researches by Wien, Dempster and others and their experiments have shown that the mean life of the excited state is of the order of 10^{-8} sec., which is very great compared with the time of transition from one state to another, this time being computed to be 10^{-15} sec., i.e. of the same order of magnitude as the period of radiation emitted. The *transition probability* is the probability for an atom to drop from a higher quantum state to a lower one, giving rise to a spectral line. Those states from which the atom cannot go to a lower state with the emission of radiation and correspondingly those states which cannot be reached by the atom from a lower state by absorption of radiation are called *metastable.*

According to quantum mechanics, it can be shown that *the mean life is inversely proportional to the transition probability.* Hence, when the probability of a given transition is extremely small, the corresponding upper state has a very long life and can be treated as metastable. Further, the intensity of emission or of absorption of light by a. large number of atoms depends on the magnitude of the transition probability. The greater the probability the more intense will be the radiation emitted or absorbed. On the other hand, when the probability is extremely small, the corresponding upper state will be metastable and an atom in such a state, before it could radiate spontaneously, has the opportunity to collide many times with other atoms and thus lose its excitation energy without radiating.

We are now in a position to explain the presence of forbidden lines, of even appreciable intensity, in nebular and auroral spectra. In the case of ordinary terrestrial light sources, the forbidden transitions are very difficult to observe in emission, since, the probability of such a transition being extremely small, the corresponding upper state has a very long life and an atom in such a metastable state will more readily lose its excitation energy by collisions without radiating. The influence of collisions may, of course, be kept sufficiently small under special conditions, for example, at very low pressures or by the addition of gas whose atoms or molecules either are not able to remove the excitation energy of the metastable state, or can remove it only with difficulty. Since, however, the life of a state which is actually metastable to dipole radiation is of the order of seconds as compared with the 10^{-8} sec. mean life of an ordinary excited state, it is almost impossible in terrestrial light sources to reach a pressure so low as to avoid the effect of collisions, especially since, at low pressures, collisions with the walls of the vessel lead to the loss of excitation energy. More favorable conditions are, however, present in cosmic light sources, as in the nebulae and aurorae, which make the metastable atoms lose their excitation energy not through collisions, but by emitting radiations, when they return to normal.

Bowen, in 1927, was the first to show that the so-called nebulium lines arise from forbidden transitions between the low or deep terms of ionized oxygen and nitrogen. Transitions between such low terms involving dipole radiation are strictly forbidden by the Laporte rule, as these terms belong to the same electron configuration. Bowen established that the wavelengths of the forbidden lines, calculated from the combination of these terms, agreed vary well with the wavelengths of the so far unexplained nebulium lines and that, therefore, it was no longer necessary to assume the presence of a new element in the nebulae. Actually, the conditions in the nebulae are extremely favorable for occurrence of these forbidden lines. For, it is estimated that the densities in the nebulae are of the order of 10^{-17} to 10^{-20} gm./c.c. Assuming a reasonable value for the temperature (10,000° K approximately) the time between two collisions suffered by an atom becomes very great, 10 to 10^4 secs. Thus, when the ionized oxygen and nitrogen atoms, which are certainly present in the nebulae, go into the low metastable states by allowed transitions from higher states, they remain there uninfluenced, until they radiate spontaneously.

A large fraction of the more highly excited ions must come eventually into these states and practically every ion goes from them to the ground state by radiation. This explains why the forbidden lines are very intense in nebulae, whereas they are not observed in terrestrial light sources, in which the other allowed

In a similar manner, Mc Lennan and Paschen have explained the green and red auroral lines as corresponding to forbidden transitions of the neutral oxygen atoms. The auroral lines have also been obtained in the laboratory in suitable light sources by Mc Lennan and Shrum, and Paschen, for example, in discharges through argon with a small addition of oxygen. The destruction of the metastable state of the oxygen atoms is considerably prevented by the argon atoms.

EXPERIMENTAL CONFIRMATION OF THE VECTOR MODEL

The three fundamental concepts on which the vector atom model is built are:

(i) spatial quantization,
(ii) spinning electron and
(iii) *a* quantized atomic magnetic moment.

Although the phenomena of the fine structure of spectral lines and Zeeman effect may be considered as experimental evidence of the validity of the vector atom model, they are still only *indirect proofs*, since the various characteristics of the vector model have been introduced empirically, based on them. This leaves the desirability of having an independent and direct experimental confirmation of the above-mentioned fundamental postulates of the proposed model.

STERN AND GERLACH EXPERIMENT

The experiments on the deflection of a beam of "atomic rays" in a non-homogeneous magnetic field, first devised by Stern and Gerlach and since then elaborated in many directions by different workers, have achieved this purpose of offering *a very direct and most convincing confirmation of the essential features of the vector atom model.* Not only the theoretical significance of the results obtained, but also the delicacy and ingenuity of the measurements involved make the Stern and Gerlach experiment one of outstanding interest and importance.

Otto Stern (17 February 1888 – 17 August 1969), Germany

Walter Gerlach (1 August 1889 - 10 August 1979), Germany

Principle

The atom, with its magnetic moment arising from the orbital and spin motions of the electrons in it, may be regarded as an elementary magnet, whose dimensions, though small, are yet finite. If this atomic magnet be placed in a magnetic field, it will be acted upon by the field. But the nature of this action will depend on the nature of the field.

If the field is homogeneous, the atomic magnet will experience merely a couple which will rotate its axis into the direction of the field, since the external field acts upon its poles with equal and opposite forces. *If the atomic magnet moves in such a field* in a direction normal to the field, although its magnetic axis will be rotated into the direction of the field, it *will trace a straight line path* without any deviation, *i e.,* without any translational displacement.

On the other hand,

if the field is non-homogeneous, the forces on the two poles will not be equal, which would result in a translatory displacement of the atom as a whole, over and above the rotation of its axis into the direction of the field.

If then, the atomic magnet flies across such a non-homogeneous field normal to the field direction, *it will lie deviated away from its rectilinear path.* An expression for the amount of the deviation thus produced may be obtained as follows:

Let the magnetic field be non-homogeneous along the X-direction, so that the field gradient is dH$/d$X and is positive (Fig. 241).
Let the atomic magnet, whose pole strength is p, length l and magnetic moment M *be placed* in the non-homogeneous field with its axis inclined at an angle θ to the field direction.

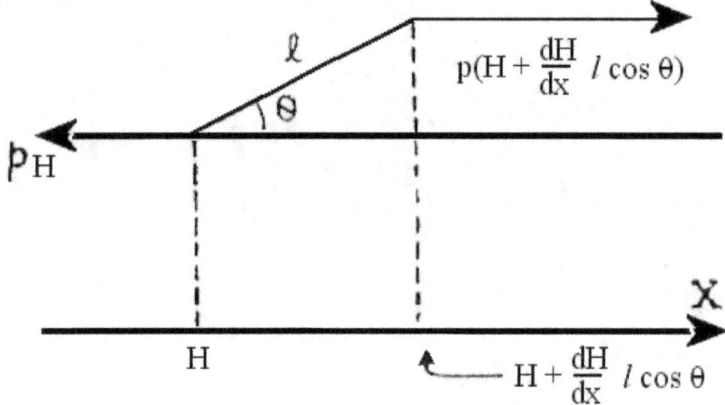

Fig. 241.

If the field strength at one of the poles be H, then at the other it will be

H + (dH/dX) l cosθ.

276

Hence the forces on the two poles of the atomic magnet due to the field are

pH and $p\{$ $H + (dH/dX)$ $l\cos\theta$) respectively. This means that, over and above the equal and opposite forces pH constituting the rotating couple, there is an extra force equal to and p (dH/dX) $l\cos\theta$ acting on one of the poles alone; it is this force which displaces the atom as a whole; calling this translatory force we have

$$F_x = p\, l \left(\frac{d H}{d X} \right) \cos\theta = M \cos\theta \left(\frac{d H}{d X} \right) \qquad \dots (1)$$

Supposing that the attune magnet *flies* across the non-homogeneous field at right angles to the lines of force, it will be displaced from its straight path in the field direction. To find the amount of this displacement, let v be the velocity of the atomic magnet of mass m as it enters the field, L the length of its path in the field and t the time of flight through the field.

The acceleration α_x imparted to the atom along the field direction by the translatory force F_x is given by F_x/m.

The displacement D_x of the atom along the field direction at the end of time t is given by

$D_x = (1/2)\,\alpha_x t^2 = (1/2)\,(F_x/m)\,(L/v)^2$,

since $t = (L/v)$, the velocity of the atom perpendicular to the field being unaffected by the force in the field direction.

Substituting the value of F_x from (1),

$D_x = (1/2)\,(M \cos\theta\,/\,m)\,(dH\,/dX)\,(L\,/\,v)^2$

If μ be the resolved component of the magnetic moment in the field direction, then

$D_x = (1/2)\,(\mu\,/\,m)\,(dH\,/dX)\,(L\,/\,v)^2$ .... (2)

From this relation we see that if D_x, m, $dH\,/dX$, L and v are known, μ can be calculated. If $dH/dX = 0$, *i.e.*, if the field is made homogeneous, $D_x = 0$, *i.e.*, there is no deviation.

Since $D_x \propto dH/dX$, the amount of deviation is determined by the degree of inhomogeneity of the field. In fact, to produce even a small measurable displacement, the inhomogeneity must be so marked that the field changes even within the small length (of the order of 10^{-8} cm.) of the atomic magnet. Stern and Gerlach succeeded in producing a sufficient degree of non-homogeneity by a suitable construction of the pole-pieces of a magnet, one piece being shaped as a knife-edge, while the other had a flat face provided with a groove, as shown in the figure. The magnetic lines of force, in consequence, crowd together at the knife-edge, so that the field strength is considerably greater there than at the other pole-piece.

Apparatus. The experimental arrangement devised by Stern and Gerlach is shown diagrammatically in Fig. 242.

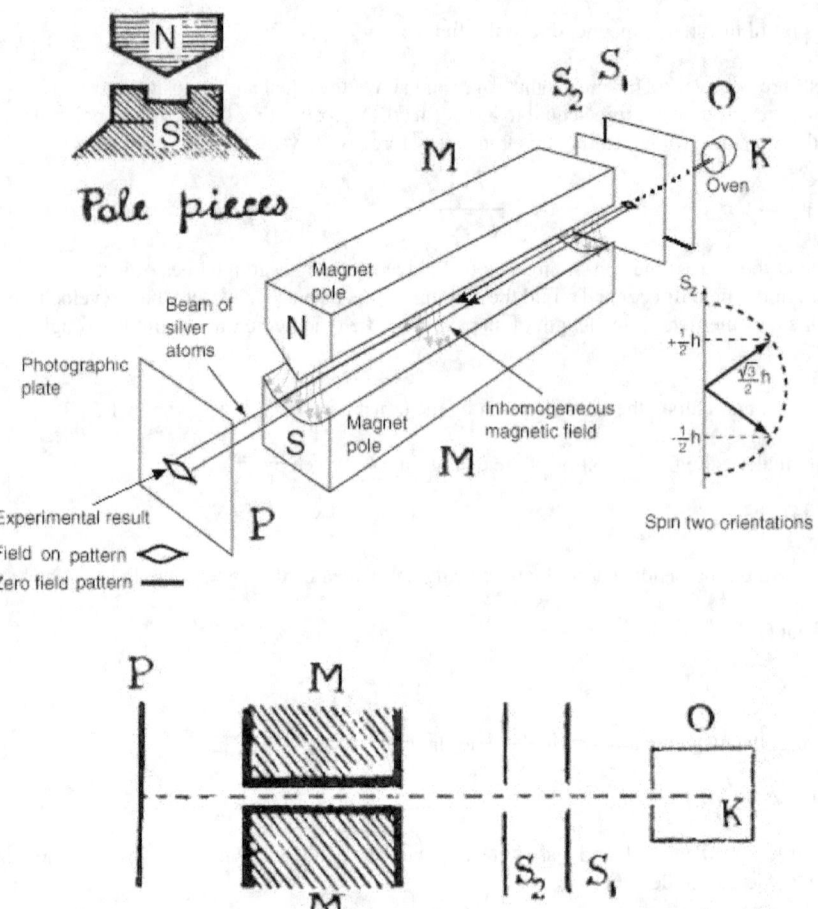

Fig. 242. Apparatus used in the Stern and Gerlach experiment.

A sample K of the substance to be examined is heated in an electric oven O and, in consequence, sends out atomic rays in all directions with a velocity corresponding to the temperature of vaporization. By the use of slits S_1 and S_2 a sharp linear beam of atoms is obtained, which then passes between the specially shaped pole-pieces of the electromagnet MM. The magnetic field is made as intense and non-homogeneous as possible, its lines of force being perpendicular to the direction of the beam. Finally the beam is made to strike a suitably prepared plate P, placed normal to the initial direction of the beam. The entire apparatus is in an evacuated chamber.

The arrangement, though simple in design, involves a very delicate technique in which a large number of details have to be carefully attended to. The different parts of the apparatus are quite small. In some cases, the pole-piece is about 5 cms long, the oven 2 cms long and the recording plate 3 mm square. The distance between the knife-edge and the plane of the slotted pole-piece is about 1 mm.

Experimental procedure
In performing the experiment, the following precautions have to be taken:

(a) As the deflections obtained are small, the different parts of the apparatus must be carefully and accurately aligned.

(b) The oven must be capable of withstanding a high temperature and this must be under delicate control as regards constancy over long periods.

(c) The velocity of the atoms depends on the oven temperature which must be measured thermoelectrically or optically. For atoms in thermal equilibrium, the kinetic theory of gases gives the relation

$$(1/2)\, mv^2 = (3/2)\ kT \text{ or } v = (3\, kT\, /m\,)^{1/2}$$

where m and v are the mass and velocity of the atom, k is Boltzmann's constant and T the absolute temperature. Some preliminary experiments, however, showed that the mean velocity of the atoms issuing through slits was somewhat greater than this; and hence the average value is to be taken as

$$v = (3.5\, kT\, /m\,)^{1/2}$$

 (d) In order to prevent collisions of the atoms in the beam with the atoms and molecules of any residual gas, the chamber must be exhausted for a long time, liquid air and charcoal being used.

(e) Accurate measurement of dH/dX is not easy. It is obtained by measuring the repulsion, at various points between the polo-pieces, of a thin bismuth wire mounted parallel to the edge of the knife-edge pole-piece. A correction is always found necessary for the variation in the value of dH/dX along the path of the deflected beam, as the beam is nearer the pole-piece at entrance into the field than at exit.

(f) Another problem is the development of the traces to render them visible. Receiving plates of different kinds are used for the study of different atoms. On account of these several technical difficulties to be overcome, it is not surprising that of the many experiments begun few were carried to a successful conclusion.

Silver was the first to be investigated very completely, many series of experiments being made. With other elements, it was then possible to test whether the apparatus was functioning properly by interpolating experiments with silver.

Meissner and Scheffers improved the technique in the following manner which enabled the deviation D_x to be measured very accurately. Working with alkali metals, they caused the deflected beam to impinge on a heated tungsten filament instead of the receiving plate. The atoms of the alkali metals, on striking the hot filament, lost their electrons and were liberated as positive ions; the ionization current thereby produced was measured in different positions and from the positions of the maximum current the magnetic deviation was accurately found.

Phipps and Taylor successfully carried out the difficult but important experimental determination of the magnetic moment of the hydrogen atom. The hydrogen atoms were produced in a long discharge tube under appropriate conditions or by dissociating the molecules by means of a hot filament. A stream containing a fairly high proportion of atomic hydrogen was passed through three special slits in a glass tube, between which two high vacuum pumps were operating so that the emergent beam contained effectively free atoms. This atomic beam after passing through a powerful non-uniform magnetic field was received on a plate coated with molybdenum trioxide. The atoms reduce the oxide and so produce a visible trace. A sharply defined blue line against a white background was the result in the absence of the field, while when the field was on, the beam was separated into two, and two traces were obtained. For determining the exact positions of the deviated lines microphotographs were taken. From the deviation thus measured the magnetic moment of the hydrogen atom was calculated to be one Bohr magneton within the limits of experimental error.

Results
In the case of silver, with the field off, a fairly sharp line was traced by the atomic beam on the receiving plate. On establishing the non-homogeneous field a double trace was obtained, one on either side of the original trace. The traces obtained were somewhat as shown in Fig. 243.

(α)   (b)

Fig. 243. Traces obtained with silver atoms. (a) without field (b) with field

The irregularity in the right-hand trace is due to the irregularity of the magnetic field near the knife-edge pole-piece. The convergence at the top and bottom arises from the fact that the field gradient decreases transversely. The traces actually obtained are much smaller than what is shown here, but they can be magnified for easy measurement. A certain diffusiveness is unavoidable owing to the velocity distribution (Maxwellian) of the atoms, but the magnetic splitting of the beam is relatively great enough, that this diffusiveness becomes of secondary importance. The mean deviation D_x of the beam caused by the field can be measured from the traces with fair accuracy. Thus D_x being known, and the other quantities such as dH/dX, L and v being readily measured, μ can be calculated. In the case of silver, it was found to be equal to one Bohr magneton. The results, obtained with other elements experimented upon may be summarized as follows:

 Cu, Au, H, Li, Na, K:
All gave double traces, the separation leading to the value of $\mu = \pm 1$ Bohr magneton, like silver.

Zn, Cd, Hg, Sn, Pb:
No effect was produced by the field. Hence $\mu = 0$.

Tl (Thallium):
The trace was divided into two, there being a definite absence of undeviated atoms. The deviation gave the value of $\mu = \pm 1/3$ Bohr magneton.

Sb, Bi:
Sb showed only an undeviated trace.
The Bi trace obtained by Stern and Gerlach was remarkable in showing a one-sided continuous broadening indicating not only undeviated but also strongly attracted carriers. But later, a symmetrical splitting was obtained by Lou who showed that the distribution of density was consistent with the presence of atoms having values for μ in the ratio of 1 :3.

Ni, Co, Fe:
Nickel showed curious results in that three clearly defined traces appeared, one of them in the undeviated position itself, which meant that some of the atoms have no magnetic moment ($\mu = 0$). For the deviated traces μ was somewhat greater than one magneton. In addition to the three pronounced traces, there were indications of the presence of atoms of moments considerably greater. Iron, after repeated trials, gave indications that its atoms might have a value of μ greater than six. In the case of cobalt, a value of $\mu = 6$ could just be established.

Interpretation of results
It is important to consider, in the first place, whether the beam consists of single atoms or molecules, since the experiments directly give the resolved moments of the carriers, whether atoms or molecules. At the temperatures employed, there is little doubt that the vapors of Cu, Ag, Au, Sn, Pb and Tl are monatomic, so that the values of μ deduced for them do correspond to atomic moments, Ni, Co, Fe, are also probably monatomic. But with Bi and Sb there is large proportion of molecules in the beam.

Considering only the sure cases, it is possible to show how the results obtained offer a beautiful confirmation of the fundamental postulates of the vector model:

(a) Spatial quantization. The *classical theory* lays no restriction on the orientation of the atomic magnet with respect to the field direction, all values from $0°$ to $180°$ for the angle θ in the relation $\mu = M \cos \theta$ being permissible. Hence the displacement of the atoms in the non-homogeneous field should cover a continuous range, so that a *continuous diffuse band* should be obtained on the receiving plate. Further, on account of the Maxwellian distribution of velocities of the atoms in the beam, the densest part of the trace should be in the centre, since the number of atoms whose axes make angles between θ and $(\theta + d\theta)$ with the field is nearly proportional to $\sin \theta$ and hence a large fraction of the atoms would be oriented nearly transversely to the field and so would suffer very small deflections.

On the other hand, *according to the spatial quantization theory*, not all settings are possible for the atomic magnet with respect to the field direction, but only a certain discrete number. If the total angular momentum quantum number of the atom is J the permitted orientations are $(2J + 1)$. Hence on the plate instead of obtaining a continuous band, $(2J + 1)$ distinct traces should be obtained. Further, if the atom under test belongs to the *one-electron system* and is in the *ground state*, L = 0, S = 1/2 = J, so that $(2J + 1) = 2$. In such a case, therefore, only two orientations are permitted and hence two traces should be obtained on the plate. In the case of *many electron system*, the atoms in the *ground state* can have higher values of J, since although L = 0, S can assume values 0, 1, 1/2, 3/2. 2, etc., and J = S. The number of permitted orientations may be greater than two, so that more complex setting of the beam could be expected: *e.g.*, for atoms of $^6S_{5/2}$ term, J being equal to 5/2, six traces should be obtained on the plate.

The separation between the traces should be proportional to the "g" values, since $\mu = m_J.g$ in Bohr units, and neighboring traces will differ by unity in the values of m_J in every case.

All these conclusions of the spatial quantization theory are exactly verified by the results of the Stern and Gerlach experiment. Thus

(i) *H, Li, Na, K, Cu, Ag, Au*, all belonging to the *first column of the periodic table*, have in the ground state the spectral term $^2S_{1/2}$, which means that $2J + 1 = 2$. They all give only two traces as expected from theory. As soon as the atoms of these elements enter the magnetic field their magnetic axes orient themselves in the two permitted directions parallel and antiparallel to the field direction, no intermediate position being allowed. Further, in this case $g = 2$ and $m_J = \pm 1/2$, so that $\mu = \pm 1$ Bohr magneton, actually obtained from experimental data.

(ii) *Zn, Cd, Hg*, belonging to the *second column*, have two s electrons outside closed shells, and their normal state, as indicated by their spectra, is $^1S_0$. This means that J = 0 and hence the atom as a whole has no resultant angular momentum and no magnetic moment. This is why no effect is produced on the atoms of these elements by the field.

(iii) *Tl* belonging to the *third column* has one p electron in addition to completed groups. This gives a $^2P_{1/2}$ ground term, *i.e.*, L = 1, S = 1/2, J = 1/2, $2J + 1 = 2$. Hence it gives rise to two traces. The value of $m_J.g$ calculated from the *"g"* formula is $\pm 1/3$, which agrees with the experimental result.

(iv) *Sn, Pb*, belonging to the *fourth* column, have two p electrons not forming a completed group, but the total magnetic moment vanishes in their normal $^3P_0$ state. This is why the atoms of these elements are undeviated by the field in the experiment.

(b) Electron spin. Considering the well established case of silver, experiment clearly shows that the atomic beam of silver is split up by the magnetic field into two parts of approximately equal intensity, Further, the value of the magnetic moment obtained is ± 1 Bohr magneton. From theory it is known that for the normal state of the silver atom L = 0. Hence, if the spin of the electron does not exist, S would be zero, so that J also would be equal to zero and there would be no splitting of the atomic beam into two components. If, on the contrary, the existence of electron spin is admitted and the value of 1/2 be given to S in this case, there is perfect agreement between theory and experiment.

The case of hydrogen is much more interesting. The experimental result obtained by Phipps and Taylor, *viz.*, the magnetic moment of the hydrogen atom is one Bohr magneton, could be accounted for in a straightforward manner on the orbital theory, without taking into account the spin of the electron at all. But according to quantum mechanics, the electron field of the hydrogen atom in the normal state possesses spherical symmetry and must, therefore, under all circumstances, show perfect isotropy in collisions with other atoms. This was tested by Fraser in a most ingenious experiment, in which he determined whether there was a change in the relative number of neutral atoms and protons which passed through a 'resting gas', when a magnetic field was applied. To within one or two per cent there was no change. This indicates that

> as far as the electric charge is concerned the hydrogen atom is spherically symmetrical. This leads to zero magnetic moment for the normal state, whereas experiment gives ±1 Bohr magneton. Hence the spin *of the electron must exist,* since it is the spin which orients in this case, so that the corresponding magnetic moment equals one Bohr magneton, while the orbital magnetic moment instead of being oriented at random, would set itself at right angles to the field.

(c) Atomic and quantum nature of magnetism. The idea of atomic nature of magnetism goes back to the days of Weber. The view that

> *paramagnetic* substances have a permanent molecular magnetic moment, while *diamagnetic* substances possess no such moment

is a long established fact in the physical theory of magnetism. Weber was the first to develop this idea on the molecular current hypothesis of Ampere. It was next rendered certain by Langevin's treatment of paramagnetism based on the kinetic theory of gases and of diamagnetism on the basis of the classical electron theory. Weiss, in 1911, from experimental data then available, concluded that there was a fundamental unit of magnetic moment, of which all atomic and molecular moments were multiples. This unit was called after him the **Weiss magneton, whose** value was estimated to be 1123.5 gauss-cm. per gram-atom or per gram-molecule, which gives 1.85×10^{-21} gauss-cm. per atom or per molecule. With the advent of the quantum theory, another fundamental unit, known as the **Bohr magneton,** came to be recognized. Its value was found to be 5,590 gauss-cm. per gram-atom or 9.21×10^{-21} gauss-cm. per atom, which is, therefore, almost five times as large as the older unit.

The Stern and Gerlach experiment not only confirms the general conclusions of the classical theory about para- and diamagnetism, but goes much further and establishes the quantum nature and atomic origin of magnetism. For, in the first place, according to the results of the experiments, substances which are found to be diamagnetic have no atomic magnetic moment ($\mu = 0$), *e.g.,* Zn, Cd, Hg, Pb, Sn, etc. Substances which are paramagnetic are made up of atoms with one valency electron in the ground state, thereby giving rise to a magnetic moment of 1 Bohr magneton due to the spin of the single free electron: e.g., H, Li, the alkalis, Cu, Ag, Au. Ferromagnetic substances (Fe, Ni, Co) are made up of atoms with intermediate incomplete electronic shells and have a large value for the atomic moment.

Secondly, the results establish also that the true fundamental unit is the Bohr magneton and not the Weiss magneton. The latter based on the classical theory, according to which all possible orientations of the atomic axis with respect to the field are permissible, has only an apparent existence. The former is the one given by the experiment of Stern and Gerlach and conforms to the quantum theory; it alone, therefore, has a real existence.

If the Bohr magneton is the true fundamental unit, one would expect the magnetic moments of different atoms to be integral multiples of the Bohr magneton. But the values obtained experimentally are seldom such multiples. Pauli, in 1920, was the first to attempt an explanation of this difficulty as follows:

According to the classical theory of Langevin, any orientation of the elementary magnet is possible, so that in the relation

$C_M = (1/3) (\sigma_0^2 / R)$ the factor 1/3 is the result of taking the average value of $\cos^2 \theta$, *i.e.,* $\overline{\cos^2 \theta} = 1/3$. But in the quantum theory, the orientations of the elementary magnet are quantized and can assume only a few discrete positions with respect to the field direction. In consequence, the average value of $\cos^2 \theta$ should be taken only for a discrete number of values of θ and $\cos^2 \theta$ will have a different value for every different quantum number. Pauli showed that

$$\overline{\cos^2 \theta} = (1/3) (n + 1) (2n + 1) / 2 n^2 \; ..... \quad (1)$$

where n is the quantum number which determines the number of discrete orientations and hence the number of Bohr magnetons in the elementary magnet.

[It may be noted that relation (1) is derived from the fact that the sum of the squares of the integers from 1 to n has the value

$$\sum_{n=1}^{n} x^2 = (n/3)(n+1)(n+\tfrac{1}{2})$$

From relation (1) we see that, as n approaches infinity, $\overline{\cos^2 \theta}$ tends to 1/3 , the value according to the classical theory. For n =1, $\theta = 0$ and one direction alone is possible for the elementary magnets, all of which will be oriented parallel to the field.

For $n = 2$, $\theta = 0$ or 60°, both directions being equally probable and $\overline{\cos^2 \theta}$ is equal to $(1/2)(1 + 1/4) = 5/8$. For $n = 3$, there are three permitted orientations, viz., $\cos \theta = 1$, 2/3 or 1/3. Thus the number of Bohr magnetons in an elementary magnet is not an integral number, as it would be if the elementary magnet could assume all possible orientations, but a fractional number, the value of which depends upon the quantum number defining the permitted orientations.

The general expression for the magnetic moment per gram-molecule given by

$$\sigma_o = (RC_M / \overline{\cos^2 \theta})^{1/2} ,$$

$\overline{\cos^2 \theta}$ being defined by (1). Considering the simplest case, $n = 1$, $\overline{\cos^2 \theta} = 1$ and σ_o has the value $(RC_M)^{1/2}$ according to the quantum theory; but in the classical theory

$$\sigma_o = (3 RC_M)^{1/2} ,$$

Hence the value of the magneton should be $3^{1/2}$ times that of the Weiss magneton. As the value of Bohr magneton is approximately five times that of the Weiss magneton, the value of the former should be $5(3^{1/2}) = 8.7$ times the Weiss magneton. If $n = 2$, $\cos^2 \theta = 5/8$ and $\sigma_o = (8 RC_M / 5)^{1/2}$, so that the value of the magneton is $(15/8)^{1/2}$ times that of the Weiss magneton and the value of the Bohr magneton is $5(15/8)^{1/2}$ times the Weiss magneton. Further, since, in this case, two orientations are permitted, the final value of the Bohr magneton should be $2 \times 5(15/8)^{1/2} = 13.7$ times the Weiss magneton.

Precise measurements by Weiss and Piccard on the two paramagnetic diatomic gases, oxygen (O_2) and nitric oxide (NO) have given the values for their magnetic moments as 14.2 and 9.2 Weiss magnetons respectively, which agree closely with the theoretical value for $n = 2$ and $n = 1$. In the case of oxygen, the two atoms must be joined in such a way that their individual moments add up, while with nitric oxide the two atoms are aligned parallel to the field direction so that the magnetic moment of the molecule is the same as that of each of the atoms. Unfortunately, there exist no monatomic paramagnetic gases for a further checking up of this point.

Pauli's theory was modified by later workers, particularly Sommerfeld (1925), in the light of the vector atom model with its additional conceptions of spatial quantization, electron spin, Landé's g factor, etc.

In the classical theory, the resolved magnetic moment is equal to $\mu \cos \theta$ and can assume all possible values from $+\mu$ to $-\mu$. On this basis, Langevin obtained the relation

$$I/I_s = a \overline{\cos^2 \theta}$$

If $\overline{\mu}$ is the mean resolved magnetic moment in the field direction, $\overline{\mu}/\mu = I/I_s$ and the above relation can be written as

$$\overline{\mu}/\mu = a \overline{\cos^2 \theta} \quad \dots\dots\dots\dots (2)$$

According to the vector atom model, the resolved magnetic moments in the field direction are confined to $m_J.g$ values (in Bohr units), where m_J can take only $(2J + 1)$ values, from J to $-J$ at units intervals, viz., J. (J — 1), (J — 2)... -J and g is the Landé's splitting factor, a function of J, L and S. Under these conditions relation (2) can be replaced by

$$\overline{\mu}/\mu = a\left(\overline{\frac{m_J}{J}}\right)^2 \quad \ldots\ldots\ldots\ldots (3)$$

since $\cos\theta = \dfrac{m_J}{J}$.

As there are $(2J+1)$ valves for m_J, from $+J$ .... To $-J$

$$\frac{\overline{\mu}}{\mu} = a \cdot \frac{1}{2J+1} \cdot \frac{1}{J^2}\left[J^2 + (J-1)^2 \ldots (-J^2) \right]$$

$$= a\,(J+1)\,/\,3J. \qquad\qquad \ldots (4)$$

Substituting $\mu H/kT$ for a, and gJ for μ, the value of μ, in terms of the Bohr unit, is given by

$$\overline{\mu} = \frac{\mu_B{}^2\,H}{3kT} \cdot \frac{g^2 J^2\,(J+1)}{J}$$

If M_B be the Bohr unit per gram-atom and N_M the Avogadro's number, the atomic susceptibility is given by

$$\chi_M = \frac{\overline{\mu}\cdot N_M}{H} = \frac{g^2 J(J+1)\,M_B{}^2}{3\,R\,T} \qquad \ldots (5)$$

If M_w is the value of the Weiss magneton and p the number of Weiss magnetons per gram-atom,

$$p = \sqrt{3\,R\,T\,\chi_M}/M_W$$

Substituting for from (5),

$$p = (M_B\,/\,M_W)\,g\,\sqrt{J(J+1)}$$

Since $M_B\,/\,M_W = -\,4.97$,

$$p = 4\cdot97\,g\,\sqrt{J\,(J+1)} \qquad \ldots (6)$$

Thus, the number of Weiss magnetons per gram-atom can be calculated for any atom or ion. The simplest case is that of atoms in an S state, when the magnetic moment is entirely due to the electron spins, i.e., $g = 2$ and $J = S$, so that

$$p = 4\cdot97\,\sqrt{4S\,(S+1)} \qquad \ldots (7)$$

The permitted values of S are 0, 1/2, 1, 3/2 .... corresponding to 0, 1, 2, 3 .... unbalanced electrons, the magnetic moment being 0, 1, 2, 3 .... in Bohr units.

From relation (7), therefore, the p values corresponding to integral multiples of the Bohr units may be calculated. Thus:

Spectroscopic state	$^1S_0$	$^2S_{1/2}$	^3S	$^4S_{3/2}$	$^5S_2$	$^6S_{5/2}$
Bohr magneton	0	1	2	3	4	5
p	0	8·6	14·1	19·3	24·4	29·4

These p values were found to agree closely with the experimental values for the rare earth ions. But for the ions of the first transition group, the theoretical values were completely different from those observed. Since atoms and ions are not necessarily in S states, the more complete expression (6) has in general to be used. Van Vleck (1928) in his treatment of paramagnetism on the basis of quantum mechanics, has obtained similar relation for p, which accounts for these discrepancies to a certain extent. We shall come back to this point in the next section dealing with the wave mechanical atom model.

GYROMAGNETIC EFFECT

The existence of electron spin is experimentally confirmed also by the phenomenon known as the gyromagnetic effect. According to the vector model, the spin and the corresponding magnetic moment being parallel vectors, magnetization of a substance means ultimately the alignment of the electron spin in the direction of the magnetizing field.

Hence, magnetizing a piece of iron, for instance, should cause it to rotate, and conversely a simple rotation of the piece of iron without any magnetizing agent should magnetize it. These two complementary phenomena together are known as the gyromagnetic effect, which though inappreciably small under ordinary conditions, has been actually observed and measured under both aspects by delicate experimental technique.

Thus, in 1915, Einstein and de Haas subjected a small iron or nickel bar, suspended by a wire and capable of rotation about its longitudinal axis, to a powerful alternating magnetic field, the period of which was adjusted to equal the period of free oscillation of the suspended bar. Under these conditions, the rod executed forced oscillations of frequency equal to that of the .alternating current in the magnetizing solenoid. From the amplitude of oscillation the mechanical angular momentum J was deduced. The magnetic moment M was known from the intensity of the current in the solenoid. The ratio J/M known as the *gyromagnetic ratio* was then calculated. It was found to be in close agreement with the theoretical value of $mc/e = p_s / M_s$, exactly half that for an electron describing an orbital motion, which is $2 mc/e = p_l / M_l$. This shows that

all magnetic phenomena depends on the *existence of elementary magnets, originating not from the orbital motion but from the spin of electrons.*

This conclusion has been further confirmed by the study of ionic paramagnetism which gives considerable support to the view that the orbital moment, though present, may not be fully effective, as we shall see in the next section when we deal with the wave mechanical theory of paramagnetism. The result obtained establishes also that the spin of the electron is 1/2.

The converse effect to the one described above was discovered by Barnett in 1915 and hence is sometimes called the *Barnett effect*. Rotating very rapidly a cylindrical iron rod about its axis, the magnetic moment thereby acquired by the rod was measured, and the gyromagnetic ratio was determined which had the same value mc/e as in the previous case.

Thus the different essential characteristics of the vector atom model are confirmed by the experiments of Stern and Gerlach and the gyromagnetic effect.

CONCLUSION

There is no doubt that the vector model is a superb and many-sided conception, which has enabled us to solve most of the important and intricate problems concerning atomic structure and spectral phenomena. It has offered a rational interpretation of the electronic structure in the atom, leading up to a logical classification of elements into a periodic system; it has succeeded in giving a complete interpretation of atomic spectra and in particular, a cogent explanation of the complex "fine structure" of spectral lines as well as of the Zeeman effect for such lines with anomalous characteristics.

But the *great drawback of this model* is that it does not present itself as a single unified system, containing the necessary theoretical justification of its principles and postulates. These are introduced somewhat piecemeal, partly on empirical data and partly by analogy.

No adequate reasons are given for the use of spatial quantization and of electron spin with half integer value;

The assumptions of discrete energy states and of emission of radiation when the atom passes from one of those states to another are arbitrary as in the older models;

The selection and intensity rules adopted are derived primarily by trial/ and error from experimental results.

The model is, therefore, at best an empirical one, which requires a sound theoretical foundation. most recent theory known as the wave mechanical atom model/supplies this want and presents a picture of the peripheral structure of the atom, the best that can be had in the present state of our knowledge.

THE WAVE MECHANICAL ATOM MODEL

A detailed study of this model is beyond the scope of this book. A masterly treatment by Dirac of the relativistic theory of the electron based on wave mechanics offers, almost without any special effort, the theoretical justification of the postulates and conclusions of the vector model. We shall limit ourselves to a brief survey of the essential features of this new conception, indicating the great advance made over the older models.

According to wave mechanics, in order to have a complete picture of the electron, we should consider it under the double aspect of *particle* and *wave*. The chance of finding a moving electron at a given point is therefore governed by a wave equation which necessarily involves a certain degree of uncertainty about its exact position. The older atom models, which placed the electron at a precise point in a well-defined orbit, did not take into account the correct nature of the electron and to this fact all their shortcomings are ultimately to be traced.

The wave mechanical theory of the atom can be explained in a simple manner, *i.e.*, without entering into complicated mathematical treatment, by analogy with other forms of waves.

Free electrons, emitted from some source and spreading outwards through a vacuum are like the ripples on the surface of a large pond, which advance from the source of disturbance in ever-widening circles, becoming feebler as the distance from the starting point increases, until finally they are hardly perceptible.

On the other hand, electrons in an atom, experiencing the electrostatic attractive force of the nucleus and hence prevented from leaving its neighborhood, are like the stationary waves set up by the reflection of the ripples at the walls of a narrow vessel in which the ripples are produced, or better still, like stationary waves set up in a fixed string with a number of loops and nodes giving the various modes of vibration. On this analogy, the electron waves may be said to be reflected back when they reach the boundary of the atom, so that a system of stationary waves may be imagined within the atom. With such a picture of the atom a number of important conclusions may be drawn:

(i) **Removal of individualistic particle picture of the electrons in the atom.** Giving to the electron a fixed position in a definite orbit is no longer correct. The electron probability is distributed over the whole range of the atom, though corresponding to the loops and nodes in the stationary wave system the distribution is not uniform. For the same reason, when several electrons are contained in an atom, they lose their individualities to a certain extent, as it is not possible to say that any particular part of any of the "probability waves" representing the electrons belongs to any particular electron.

(ii) Nebulous picture of the atom. The atom loses mach of its former vividness of the older models with their well-defined electronic orbits, for the electron probability does not vanish sharply at any distance from the nucleus; it becomes zero only at infinity, so that theoretically there is a chance of finding the electron anywhere in space. But there is a region where the probability falls steeply to a small value and this corresponds roughly to what is ordinarily called the boundary of the atom.

(iii) Quantum stationary states of the atom. The stationary electron waves can vibrate in different modes which imply different wavelengths and frequencies and therefore different amounts of energy in the system. In the mathematical treatment of the problem, this means that

the equation representing the stationary waves in the atom can be solved only when certain coefficients are given appropriate values, which must be whole numbers or occasionally simple fractions, like the different harmonics of a fundamental vibration.

These coefficients correspond to the *quantum numbers* of the older models, but, they appear here quite naturally as characteristic solutions of the wave equation, whereas in the older theories they were introduced somewhat artificially, with the only reason that they were found necessary for a satisfactory interpretation of the experimentally observed atomic and spectral phenomena. The veracity of these remarks can be easily established by the application of Schrödinger's wave equation to the simple case of the one-electron atom.

SOLUTION OF SCHRÖDINGER'S EQUATION FOR THE ONE-ELECTRON ATOM

In this problem an electron of charge e is at a distance r from the nucleus of charge Ze, so that the potential energy of the electron in the field of the nucleus is

$$V = -Ze^2/r \qquad \dots (1)$$

Taking into account the motion of the nucleus of mass M, the electronic mass m is to be replaced by the reduced mass m' given by

$$m' = m\,M/(m+M) \quad \dots\dots (2)$$

Substituting these values in Schrödinger's wave the wave-equation of the one-electron atom:

$$\nabla^2 \psi + \frac{8\pi^2 m'}{h^2}\left(W + \frac{Ze^2}{r}\right)\psi = 0 \qquad \dots (3)$$

In the solution of this equation, wave functions are sought, subject to the restrictive *boundary conditions* that ψ and its first derivatives are everywhere finite, single-valued and continuous.

Further, for solving second-order partial differential equations of this type, it is found more convenient to use three-dimensional polar coordinates (r, θ, φ) instead of Cartesian coordinates (x, y, z). For, then, it is found possible to replace the single equation by three total differential equations, each containing one of the three variables, $r, \theta,$ or φ. The solutions of the three equations, with the use of appropriate boundary conditions, lead to the following results:

(a) The appearance of three coefficients, n, l and λ, all integers governed by the condition

$$|\lambda| < l < n \quad \dots\dots\dots\dots(4)$$

with the possible values of n as 1, 2, 3, 4, 5 ...... of l as 0, 1, 2, 3, 4, up to $(n-1)$ and of λ all integral values from $-l$ through 0 to $+l$ and hence $(2l+1)$ values. These numbers, n, l and λ serve as convenient labels for the wave functions and are commonly called "quantum numbers", *i.e.,* the wave function ψ is dependent upon the values assigned to n, l and λ and is designated as $\psi_{nl\lambda}$.

(b) While there exist solutions for all positive values of the total energy W, solutions are possible for negative values of W only when W has the discrete proper values, given by

$$W = -\frac{2\pi^2 m' e^4 Z^2}{n^2 h^2} \qquad \dots (5)$$

It is readily seen that equation (5) is identical with the expression for the energies of the stationary states characterized by the *total quantum number n* in Bohr's theory.

The case W = 0 corresponds to the energy of the atom when the electron has been removed to an infinite distance from the nucleus;

Hence W < 0 corresponds to the case of elliptical orbits of Bohr's theory, where energy must be supplied to remove the electron from the bondage of the atom.

The case \V > 0 corresponds to Bohr's hyperbolic and parabolic orbits giving rise to the continuous spectrum beyond the series limit.

(c) Stationary states
Considering the expression (4) it is realized that *while n corresponds to the total quantum number as stated above, l is associated with the orbital quantum number and λ with the magnetic orbital quantum number m_l of the vector atom model.*

In this way wave mechanics accounts for all the stationary states of the older models, without any arbitrary hypothesis or assumption being introduced in the solution of its fundamental equation.

Although discrete energy levels defined by the proper wave functions are no longer associated with electron orbits as such, yet *there is a definite correspondence between the radii of these orbits and the distances at which the electron probability function has its maxima.* For, the quantity $|\psi|^2$. *dv* being defined as the probability of finding the electron in the volume element *dv*, let us consider the probability of finding the electron at a given distance *r* from the nucleus. If $|\psi|^2$ is multiplied by the surface area of a sphere, $4\pi r^2$, then the term $|\psi|^2$. $4\pi r^2$ *dr* represents the probability of finding the electron within a spherical shell of radius *r* and thickness *dr*.

Plotting $|\psi|^2$. $4\pi r^2$ against *r* for quantum numbers $n = 1, l = 0$, n 2, $l = 1$, and $n = 3, l = 2$, which evidently correspond to the circular orbits in Bohr's theory, curves as shown in Fig. 244 are obtained.

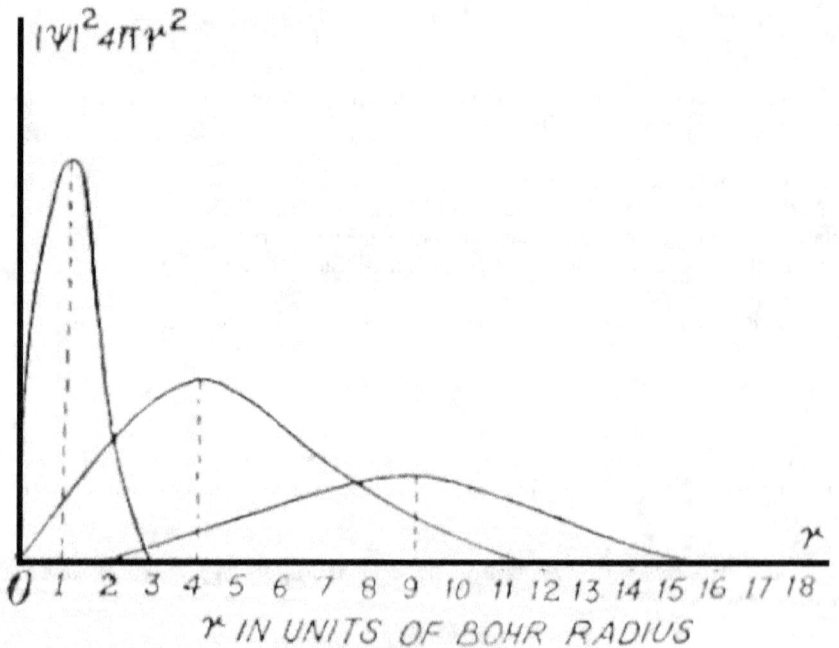

r IN UNITS OF BOHR RADIUS

From these curves it is seen that the probability of finding the electron at any distance from the nucleus has maximum value at a distance equal to the radius of the corresponding Bohr orbit, which is proportional to n^2.

The quantum number l determines the number of nodal surfaces, depending on the angular coordinates. Hence, for each permitted value of the energy W only certain classes of vibrational modes are possible, the actual modes becoming more complicated the greater the value of the quantum number n. These can be analyzed graphically by plotting $|\psi|^2$. $4\pi r^2$ against r for different values of n and l. The curves obtained demonstrate that the elliptical orbits correspond to electron probability distributions with various maxima at different distances, and that the property of eccentricity in the ellipse finds its analogue in the complexity of the radial distribution of the probability function in the wave mechanical model. It is to be noted that in spite of their complicated probability distribution the wave mechanical vibration corresponding to Bohr ellipses of high eccentricity possess complete *spherical symmetry.*

(d) Spatial quantization

It can also be shown that a given wave function $\psi_{n1\lambda}$ represents a particular state of the electron, in which its *angular momentum* about the axis of the polar coordinates has the definite value $\lambda. h./2\pi$. This means that in the $(2l +1)$ quantum states that exist for a given n and l, the angular momentum λ about the polar axis ranges by integral steps from $+l.h./2\pi$ through 0 to $—l.h./2\pi$. Such a variation of the angular momentum from one state to another is represented, if we imagine that the angular momentum of the electron has a fixed value of l units, each of magnitude $h/2\pi$, but is inclined at various discrete angles to the axis. This corresponds to the *spatial quantization* of time vector model.

Wave mechanics proves further that the total angular momentum of the atom in a state of quantum number l is not equal to $l.h/2\pi$ but to

$$[l(l +1)]^{1/2}.h/2\pi,$$

and has no fixed vector direction as in the classical theory.

(e) Electron spin

Dirac, in 1928, in order to bring the wave mechanical theory into harmony with the theory of relativity, adopted a wave equation, different from that of Schrödinger, but conformable to the relativistic postulate of symmetry. This new equation which has four wave functions instead of one, when solved, leads automatically, without any special assumption, to results equivalent to those deduced from *electron spin.*

According to Dirac's theory, the electron behaves as if it had an internal angular momentum or spin and an associated magnetic moment. This might be understood by a consideration of the fact that,

if a wave packet is formed representing the electron, there is in it, in general, something like a closed current or eddy of probability that can be regarded as analogous to the spin motion.

The complete Dirac theory with the four wave functions is seldom used in spectroscopic study. A more simple form, first suggested by Pauli in 1925, in which only two wave-functions are used, one for each direction of spin moment, is ordinarily employed. The spatial probability density is then

$$|\psi_+|^2 + |\psi_-|^2$$

where ψ_+ and ψ_- stand for the two functions. Both wave functions have the same form as regards space coordinates for the one-electron atom. In such a case, it is convenient to take account of spin merely by adding two more quantum numbers s and μ. The first of these which corresponds to the spin quantum number s has always the fixed value 1/2 ; the value of μ which corresponds to the magnetic spin quantum number m_s is +1/2 for one of the two wave functions and —1/2 for the other. Since s never changes, it need not be indicated; a. wave function is therefore completely defined by n, l, λ, μ and is designated as $\psi_{n1\lambda\mu}$. The spin angular momentum is analogous to the orbital angular momentum and is given by

$$[s(s +1)]^{1/2}.h/2\pi.$$

According to wave mechanics, therefore, definite values for a set of these four quantum numbers n, l, λ and μ, are required to define completely the mode of a wave system which correspond to a particular energy level.

(iv) Non-radiative character of the stationary states. In a stationary wave system, the amplitude at any given point is constant and does not change with time. Now, considering the wave mechanical atom from the point of view of what is known as the *electronic cloud* round the nucleus or the density distribution of the electric charge in the atom, which also represents the electron probability, and interpreting the square of the amplitude as measuring the electric charge density, it follows that the charge density at any point must be independent of time. This means that

> for any stationary state of the atom, no matter how great the proper energy value, the charge distribution must be static, *i.e.,* must not vary with time.

There being no movement of electric charge in the permitted energy states, the atom in such states cannot radiate electromagnetic energy in consonance with classical electrodynamics. Thus we arrive quite naturally at a result, which in older theories had to be assumed as an arbitrary postulate contrary to the classical theory.

Considering the case analytically, for a given state, the probability of finding the electric charge e in a volume element dv is $e\,|\psi|^2\,dv$. In the classical theory the radiation emitted by the atom may be calculated from the time-rate of variation of the electric dipole moment p of the atom.

By the correspondence principle, the connection should be true in wave mechanics also. The dipole moment p according to wave mechanics is given by

$$p = e \int r \, |\psi|^2 \, dv,$$

where r is the radius vector from the nucleus to the point of integration.

It can be shown that the above integral vanishes for all states of the atom, so that the dipole moment also vanishes. In consequence, the radiation emitted is equal to zero. This means that

> the stationary states do not radiate, due to the fact that the radiations from the individual moving elements of the electronic cloud cancel each other by interference.

(v) Emission of radiation governed by Bohr's frequency condition. When two vibrations of somewhat different frequencies are superimposed they give rise to the so-called "beat phenomena", in which the amplitude waxes and wanes with a frequency, equal to the difference between the two superposed frequencies. In a similar manner, when two different modes of the wave system corresponding to two permitted energy states of the atom can be excited simultaneously so that they are superimposed, radiation, whose frequency is given by the number of beats per second between the frequencies of the two electron waves superposed, will be emitted. If we once again interpret the square of the amplitude of vibration in the wave mechanical model as a measure of the electric charge density, it is readily seen that, when different energy states are simultaneously excited, the charge density at any given point is no longer constant but varies with time. Hence, there is an oscillation of the electric charge which must give rise to the emission of electromagnetic waves. The frequency of these emitted waves will be the beat frequency. If v_L be the frequency of light emitted and v_1 and v_2 be the frequencies of the two superimposed vibrations of electron waves, then

$v_L = v_1 - v_2$

According to the wave mechanical theory the energy W of a stationary state is given by W = $h\nu$, where ν is the frequency of the electron waves corresponding to that state. Hence

$v_1 = W_1/h$
and
$v_2 = W_2/h$

where W_1 and W_2 are the energies of the two states involved. Therefore,

$v_L = (W_1 - W_2)/h$

Thus we arrive logically at Bohr's frequency condition which was introduced arbitrarily as a fundamental postulate of his theory, at variance with the classical standpoint.

Reviewing the case analytically, in analogy with the density function $|\psi|^2$ of a definite state, we can find the "transition density" corresponding to the transition from a state n to another state m. The dipole, moment corresponding to this transition is given by

$$p_{nm} = e \int r \, |\psi_{nm}|^2 \, dv.$$

Further, as the transition corresponds to the "beat phenomenon", its rhythm is given by the time factor

$$e^{-(2\pi i/h)(W_n - W_m)t}$$

of the transition density, and its frequency by

$$\nu_{nm} = \frac{W_n - W_m}{h}$$

$$\therefore \qquad p_{nm} = e \, r_{nm} \, e^{-2\pi i \nu_{nm} t}$$

According to the classical theory, such a dipole moment will radiate per unit time the energy

$$J = \frac{2e^2}{3c^2} \left(2\pi \nu_{nm} \right)^4 |r_{nm}|^2$$

where e is the electronic charge, c the velocity of light, ν_{nm} the frequency of the radiation emitted and r_{nm} the matrix clement of the vector coordinate r.

Thus, according to wave mechanics, the emitted radiation is obtained by calculating, in purely correspondence fashion, the radiation emitted by an oscillating dipole, on classical principles. From this it follows automatically that the origin of spectral lines due to electronic jumps from one quantum state to another is not contrary to the classical theory, which the older theories failed to recognize.

It is not to be understood, however, that both states n and m are excited simultaneously when emission takes place; it is rather a matter of their virtual presence.

Radiation takes place when the wave system changes from vibrating in one mode to vibrating in another.

(vi) Intensity and selection rules.

The intensity of a spectral line is the product of two factors, the number of excited atoms and the radiating strength J of an individual atom.

Exact calculation of the intensity J of the individual elementary act, depending upon the evaluation of the matrix elements, leads to the selection rule of

$\Delta l = \pm 1$ and $\Delta \lambda = 0, \pm 1$

in the case of the one-electron atom, since all matrix elements which do not correspond to one of the transitions mentioned, vanish, and with them vanishes also the radiation of the corresponding frequencies. There are no selection rules for the total quantum number n, but the probability of transition from a state n_2 to a state n_1 becomes smaller with increase of

$\Delta n = n_2 - n_1$

This means that the intensity of the lines of a series falls off progressively as one passes towards higher members of the series. It is to be noted that the selection and intensity rules, assumed as *ad hoc* hypotheses in the older theories, are obtained by a. rigorously deductive method in the wave mechanical theory.

Thus wave mechanics is able to account adequately for all the postulates assumed by the older theories. Further, it has the supreme merit, in a scientific theory, of reducing the number of fundamental postulates.

It must be emphasized, however, that the analogies used in explaining the wave mechanical atom model are very imperfect. For the simplest one-electron system, the wave equation may be considered literally to mean the electron probability in three dimensions round the nucleus as centre of force. But for more complex systems the equations involve more than three dimensions so that the situation is no longer one that can be visualized. Wave equations can, however, be constructed for as many dimensions as we wish, though the problem will then be solved with greater difficulty and with fewer concrete pictures to help and guide. Thus, for instance, in the applications of wave mechanics to many-electron atoms, very little progress has been made except by means of a method of approximation known as the *perturbation theory.* Hence, although the wave mechanical theory has displaced the older theories for purposes of rigorous treatment, required for a detailed interpretation of observed facts and for predicting new ones, it is still found exceedingly convenient and fairly reliable to work with the vector atom model.

SOME APPLICATIONS OF THE WAVE MECHANICAL THEORY

The important applications of the wave mechanical theory may be classified as follows:

1. The hydrogen and helium atoms.
2. Magnetic properties of materials.
3. Optical and X-ray spectra of atoms and molecules.
4. Zeeman and Stark effects.
5. Phenomena involving interaction between matter and radiation, such as dispersion, thermionic and photoelectric effects, Compton effect, Raman effect, Auger effect, etc.
6. Alternating intensities in band spectra—para- and ortho-hydrogen.
7. Scattering of electrons, protons, neutrons and alpha particles in their passage through matter.
8. Alpha and beta disintegrations of radioactive substances.
9. Artificial transmutations.
10. Nuclear properties, such as nuclear spin and magnetic moment, nuclear statistics, nuclear stability, etc.

The list is not exhaustive. We have already indicated some of these applications. We shall here deal only with the first two, viz, applications to the hydrogen and helium atoms and to the magnetic properties of materials, since they are directly related to the peripheral electronic structure of the atom, the subject matter of this chapter. Most of the other important applications are considered in their appropriate places.

1. HYDROGEN AND HELIUM ATOMS

A. The hydrogen atom
A complete, but rather complicated, wave mechanical theory of the hydrogen atom is worked out on lines indicated above, viz., solution of Schrödinger's equation for the one-electron atom. We shall now propose a simplified treatment, which holds good for the *hydrogen atom in the ground state.*

The Schrödinger wave equation for the hydrogen atom is

$$\nabla^2 \psi + \frac{8\pi^2 m'}{h^2} \left(W + \frac{e^2}{r} \right) \psi = 0 \qquad \ldots (1)$$

since the potential energy V of the electron in the field of the nucleus is $- e^2/r$.

For the ground state under consideration e^2/r is spherically symmetrical. Hence there will be a class of solutions for ψ which are also spherically symmetrical. Further r depends only on the coordinates x, y, z and is given by

$$r = (x^2 + y^2 + z^2)^{1/2}$$

So that

$$\frac{dr}{dx} = \frac{x}{(x^2 + y^2 + z^2)^{1/2}} = \frac{x}{r}$$

If it be assumed that ψ depends only on r and not on the polar angles

$$\frac{d\psi}{dx} = \frac{d\psi}{dr} \cdot \frac{dr}{dx} = \frac{d\psi}{dr} \cdot \frac{x}{r}$$

$$\therefore \quad \frac{d^2\psi}{dx^2} = \frac{1}{r} \cdot \frac{d\psi}{dr} + \frac{x^2}{r^2} \cdot \frac{d^2\psi}{dr^2} - \frac{d\psi}{dr} \cdot \frac{x}{r^2} \cdot \frac{x}{r}$$

$$= \frac{1}{r} \cdot \frac{d\psi}{dr} + \frac{x^2}{r^2} \cdot \frac{d^2\psi}{dr^2} - \frac{x^2}{r^3} \cdot \frac{d\psi}{dr}$$

Similarly

$$\frac{d^2\psi}{dy^2} = \frac{1}{r} \cdot \frac{d\psi}{dr} + \frac{y^2}{r^2} \cdot \frac{d^2\psi}{dr^2} - \frac{y^2}{r^3} \cdot \frac{d\psi}{dr}$$

And

$$\frac{d^2\psi}{dz^2} = \frac{1}{r} \cdot \frac{d\psi}{dr} + \frac{z^2}{r^2} \cdot \frac{d^2\psi}{dr^2} - \frac{z^2}{r^3} \cdot \frac{d\psi}{dr}$$

Adding, we get

$$\nabla^2\psi = \frac{3}{r} \cdot \frac{d\psi}{dr} + \frac{x^2 + y^2 + z^2}{r^2} \cdot \frac{d^2\psi}{dr^2} - \frac{x^2 + y^2 + z^2}{r^3} \cdot \frac{d\psi}{dr}$$

$$= \frac{3}{r} \cdot \frac{d\psi}{dr} + \frac{d^2\psi}{dr^2} - \frac{1}{r} \cdot \frac{d\psi}{dr}$$

$$= \frac{2}{r} \cdot \frac{d\psi}{dr} + \frac{d^2\psi}{dr^2}$$

We can therefore rewrite equation (1) as

$$\frac{d^2\psi}{dr^2} + \frac{2}{r} \cdot \frac{d\psi}{dr} + \frac{8\pi^2 m'}{h^2}\left(W + \frac{e^2}{r}\right)\psi = 0 \quad \cdots (2)$$

Let the solution of this equation be
$$\psi = e^{-ra} \quad \ldots\ldots\ldots\ldots (3)$$
where a is constant which satisfies the boundary condition, viz., $\psi \approx 0$ when r is large.

Differentiating (3) twice,

$$\frac{d\psi}{dr} = -a\, e^{-ra}$$

$$\frac{d^2\psi}{dr^2} = a^2\, e^{-ra}$$

Substituting these in (2),

293

$$a^2 e^{-ra} - \frac{2a}{r} \cdot e^{-ra} + \frac{8\pi^2 m'}{h^2}\left(W + \frac{e^2}{r}\right).e^{-ra} = 0$$

or
$$a^2 - \frac{2a}{r} + \frac{8\pi^2 m'W}{h^2} + \frac{8\pi^2 m'e^2}{h^2 r} = 0$$

Regrouping,

$$a^2 + \frac{8\pi^2 m'W}{h^2} + \frac{1}{r}\left(\frac{8\pi^2 m'e^2}{h^2} - 2a\right) = 0$$

Equating separately to zero the terms independent of r and the terms involving r,

$$a^2 + \frac{8\pi^2 m'W}{h^2} = 0 \quad \text{or} \quad W = -\frac{a^2 h^2}{8\pi^2 m'}$$

$$\frac{8\pi^2 m'e^2}{h^2} - 2a = 0 \quad \text{or} \quad a = \frac{4\pi^2 m'e^2}{h^2}$$

$$\therefore W = -\frac{16\pi^4 m'^2 e^4}{h^4} \times \frac{h^2}{8\pi^2 m'} = -\frac{2\pi^2 m'e^4}{h^2} \quad \cdots (4)$$

This represents the energy of the hydrogen atom in the ground state ($n = 1$).

The probability of finding the electron in a volume dv is given by

$\psi^2. dv = e^{-2ar} dv.$

If dv is the volume contained between two spheres of radii r and $r + dr$, the probability of finding the electron is

$e^{-2ar} 4\pi r^2 . dr.$

This shows that the probability ($r^2 e^{-2ar}$) is small both when r is small due to the r^2 term and when r is large due to the e^{-2ar} term.

The most probable value of r is obtained differentiating the expression $r^2 e^{-2ar}$ and equating to zero, with the result that r is found to be equal to $1/a$.
Hence
$r = h^2/4\pi^2 m'e^2.$
This gives the radius of the first Bohr orbit.

The energy and radii of the other orbits of the hydrogen atom can be derived by a treatment, somewhat more complicated, and it can be shown that the energy of the nth orbit is given by

$$W_n = -\frac{2\pi^2 m'e^4}{n^2 h^2}$$

and its radius by

$$r_n = \frac{n^2 h^2}{4\pi^2 m'e^2}$$

It is to be noted that the result obtained with the simplified treatment of the hydrogen atom in the ground state is of great significance; for, it emphasizes the fact that the hydrogen atom possesses perfect spherical symmetry in its lowest energy state, which the older theories could not account for.

B. The helium atom

Observations on the helium spectrum show that there are two distinct states in the helium atom, each with its own series of energy levels and associated spectral lines. They are known as *ortho-helium* and *para-helium.* The triplet system of the helium spectrum belongs to orthohelium, while the singlet system to paralielium. It has been found that no inter-combination of the two systems is possible and that the ortho terms are three times as numerous as the para terms. As a result of the existence of these two different states, the calculated energy of the ground state comes out very different from the observed value. The older theories are incapable of explaining these facts, while an adequate solution is obtained in the wave mechanical theory.

The Schrödinger wave equation for the helium atom can be readily formed by substituting for the potential energy

$$V = - \frac{2e^2}{r_1} - \frac{2e^2}{r_2} + \frac{e^2}{r_{12}},$$

where r_1 and r_2 are the distances of the two electrons from the nucleus and r_{12} is the distance between the two electrons. The quantities $-2e^2/r_1$ and $-2e^2/r_2$ represent the potential energies of the electrons in the field of the nucleus ($+2e$), while $+e^2/r_{12}$ the interaction between the two electrons. Neglecting, for the present, the interaction term, the wave equation for the helium atom is

$$\nabla^2 \psi + \frac{8\pi^2 m'}{h^2} \left(W + \frac{2e^2}{r_1} + \frac{2e^2}{r_2} \right) \psi = 0. \qquad \dots \ (1)$$

This equation can be solved only approximately and the proper energy value is found to be

$$W = - 4 R \left(\frac{1}{n_1^2} + \frac{1}{n_2^2} \right) \qquad \dots \ (2)$$

where n_1 and n_2 are positive integers associated with the respective proper values for the two electrons, which in this case can be obtained separately, one electron being in the n_1^{th} Bohr orbit and the other the n_2^{th}.

The energy of the ground state is obtained by putting $n_1 = n_2 = 1$ and from relation (2) is equal to -8 R, *i.e.,* $-8 \times 13.6 = -108.8$ volts. The experimentally observed value, however, is only -78.5 volts.

The difference indicates that the interaction term has a large effect and hence we are not justified in neglecting it.

The interaction ($+e^2/r_{12}$) between the two electrons has been adequately explained on the basis of a *resonance effect* which is quite peculiar to wave mechanics and is due to the identity of all electrons. The mathematical treatment of the problem of anatomic system containing two electrons given by Heisenberg leads to two proper functions and two different sets of energy states, corresponding to the spins of the two electrons being parallel (orthohelium) or antiparallel (parahelium). In the ground state, however, where $n = n_1 = n_2 = 1$, there is only one finite solution. This is in agreement with observation, for, the lowest term of the orthohelium system has $n = 2$. There is an energy difference of 19.8 volts between the lowest term of helium (^1S) and the next higher (2^3S).

It may be remarked that,

> according to the wave mechanical theory, *whenever two identical particles interact, there are two possible proper functions and two different energy values for the pair considered as a whole system.*

This is known as the *wave-mechanical resonance effect* and has been successfully applied in the interpretation of other phenomena also, such as para and ortho hydrogen, alternating intensity in band spectra, covalent binding *(exchange force)* in the hydrogen molecule, ferromagnetism, etc.

2. MAGNETIC PROPERTIES OF MATERIALS

A. Diamagnetism

As already remarked atoms and ions will be diamagnetic, only if the resultant spin and orbital momentum of the electrons in them is zero, or in general, if they are in a $^1S_0$ state, as are the inert gases (He, Ne, A, etc.) and ions of other elements which have their electronic configurations similar to those of the inert gases (*e.g.* the alkali ions Na^+, K^+, Rb^+, Cs^+, the alkaline earth ions Mg^{++}, Ca^{++}, Sr^{++}, Ba^{++}, the halogen ions F^-, Cl^-, Br^-, I^-, the metal ions Cu^+ Ag^+, Au^+, etc.). The diamagnetic susceptibilities have been measured carefully for several of the above-cited elements.

Pauli was the first, in 1920, to attempt a theoretical interpretation of the diamagnetism of the inert gases, based on the symmetry of their electronic structure. Calculating the value of R^2, the mean square of the radius of the atom, he proved definitely that

the experimentally measured atomic susceptibility, though agreeing with the theoretical value as regards the order of magnitude, was far greater than any reasonable theory would allow.

With the introduction of the quantum orbital theory, the problem was reconsidered and the following expression for R^2 was obtained:

$$R^2 = a_0{}^2 \frac{n^2}{Z^2} \left(\frac{5}{2} n^2 - \frac{3}{2} k^2 \right) \qquad \cdots (1)$$

where R refers to an *n*, *k* orbit, a_0 is the radius of the innermost (1, 1) orbit in the hydrogen atom and Z the effective nuclear charge.

This relation gives the contribution, which the (*n*, *k*) orbit makes to the atomic diamagnetic susceptibility χ_A, as

$- 2.85 \times 10^{10} a_0{}^2$

or, substituting the value of $a_0 = 0.532 \times 10^{-8}$,

$$- 0.81 \times 10^{-6} \frac{n^2}{Z^2} \left(\frac{5}{2} n^2 - \frac{3}{2} k^2 \right) \qquad \cdots (2)$$

If the effective nuclear charge Z is calculated from this relation for helium with two electrons, using the experimentally observed susceptibility (-1.88×10^{-6}) it comes out as 0.93, which is much too small, since it should be about 1.7, as estimated from the ionization potential; this means, in other words, that the theoretical value of susceptibility is much smaller than the experimental value.

Van Vleck, in 1926, using the new quantum mechanics, obtained the relation

$$R^2 = a_0{}^2 \frac{n^2}{Z^2} \left[\frac{5}{2} n^2 - \frac{3l\,(l+1)-1}{2} \right] \qquad \cdots (3)$$

where *l* is the orbital quantum. number, equal to $(k-1)$.

This gives the gram-atomic susceptibility

$$\chi_A = - 0.81 \times 10^{-6} \, \Sigma \, \frac{n^2}{Z^2} \left[\frac{5}{2} n^2 - \frac{3l\,(l+1)-1}{2} \right] \qquad \cdots (4)$$

where the summation is to be taken over all the electrons in the atom the appropriate effective Z being used for each.

For the $n = 1$, $l = 0$ hydrogen orbit, relation (4) gives a result three times as great as relation (2). The effective nuclear charge of He from the observed value of susceptibility comes out as 1.61, which is quite satisfactory.

Linus Carl Pauling (February 28, 1901 – August 19, 1994), USA.

Pauling, in 1927, using the wave mechanical theory with the charge distribution conception of the atom, deduced the relation

$$\chi_A = -0{\cdot}804 \times 10^{-6}\, \Sigma\, \frac{n^2}{(Z-S)^2}\left[\frac{5}{2}n^2 - \frac{3l\,(l+1)-1}{2}\right] \quad \dots (5)$$

which is practically the same as relation (4) except for the explicit introduction of. a "screening constant" S that determines the action of the electrons which are nearer the nucleus than the electron under consideration and which partially neutralize the nuclear field .He evaluated S and hence the effective nuclear charge (Z - S) for the different groups of electrons in atoms and using the above formula calculated the diamagnetic susceptibilities for the inert gases as well as for inert gas-like ions. The calculated values agree fairly well with those observed, except for ions containing a large number of electrons for which the calculated values are too high. Pauling's calculations necessarily involved a number of rough approximations.

Stoner, in 1929, using Hartree's 'self-consistent' field method which is applicable to spherically symmetrical atoms and which enables the charge distribution satisfying the Schrödinger equation to be evaluated much more precisely, has calculated the diamagnetic susceptibilities for He, K^+, Na^+, Rh^+ and Cl^-. In the first four cases the theoretical values are in good agreement with those observed. For Cl^-, however, the calculated value is too high, which has been explained as due to either small errors in the charge distribution for large values of orbital radius or different 'boundary conditions' for an ion in solution or in a crystal, the theory proposed being applicable only to a free ion.

Thus the wave mechanical theory, under its various aspects has well nigh solved, in a constant manner, the problem of the inner mechanism of the diamagnetic effect, viz., that it is a universal property, arising from the action of an applied magnetic field on the individual electronic orbits in the atom.

B. Paramagnetism
Wave mechanics has been able to introduce further refinements in the theory of paramagnetism concerning two points chiefly, viz..

(i) the p values of the ions of the first transition series of elements and
(ii) the time required for orientation of the atomic magnets in the field direction.

(i) *The p values in the first transition group of elements*
Experimental study has shown that paramagnetism is a characteristic property of ions which have an incomplete group of electrons, hence of ions of the various transition series of elements.

The ions of the first transition series, *viz.,* the iron series with 18 to 28 electrons, have been very completely investigated. The rare earth ions also have been studied in detail. But very little is known about the magnetic properties of the second and third transition series, of which the most characteristic elements are Ru, Rh, Pd, Os, Ir and Pt.

Considering, therefore, only the two well analyzed cases, in the first transition series, groups of electrons are incomplete for which $l = 2$ (*d* electrons— the complete group has 10), while with the rare earths, groups for which $l = 3$ (*f* electrons—the complete group has 14) are incomplete.

Experimentally, it is found that, in the above two cases, although there is variation in the *p* values for some ions, these values for ions with the same number of electrons, group round fairly well-defined averages. As already remarked these experimental *p* values agree well with those calculated (using the theoretical relations based on the vector model) in the case of the rare earth ions, but not so for the ions of the first transition group.

The wave mechanical theory has been able to throw some light on the above-mentioned discrepancy. Van Vleck in his quantum mechanical theory of paramagnetism, distinguishes the following two cases:

(1) *When the multiplet intervals are large, i.e ,* for $\Delta v >> kT$, the spin and orbital moments are firmly coupled together to give a resultant J and the expression for *p* turns out to be the same as that obtained in the vector model, viz.

$p = 4.97 \, g \, [\, J(J + 1)]^{1/2}$
which reduces to
$p = 4.97 \, g \, [\, 4 \, S(S + 1)]^{1/2}$
for atoms in an S state.

(2) *When the multiplet intervals are small, i.e.,* for $\Delta v << kT$, the spin and orbital moments are not strongly coupled, so that L and S are to be quantized separately with respect to the field and the expression for *p* is then

$p = 4.97 \, g \, [\, 4 \, S(S+1) + L(L+1)]^{1/2}$

The two cases are limiting cases, the first holding good for $\Delta v \to \infty$ or $T \to 0$, while the second for $\Delta v \to 0$ or $T \to \infty$.

Considering the case of the ions of the first transition series, if the *p* values calculated from the three expressions given above are represented graphically, we get curves as shown in Fig. 245.

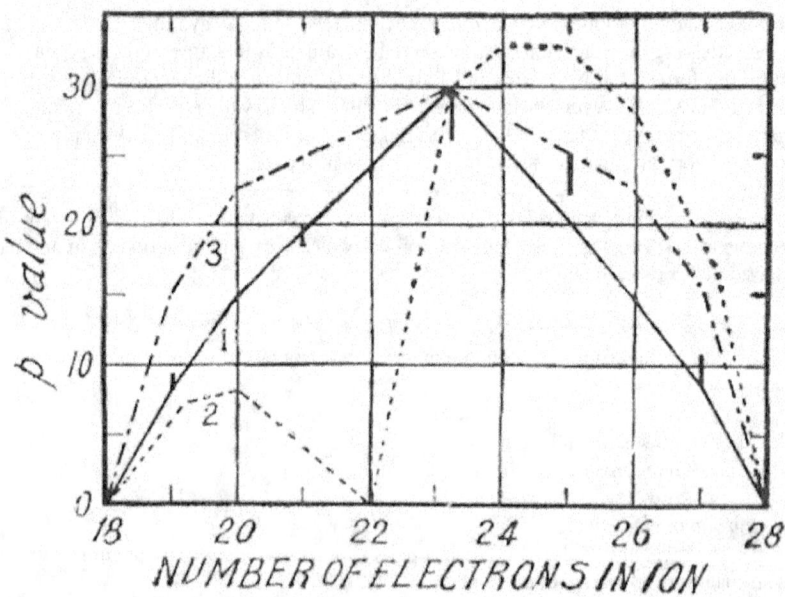

Curves 1, 2, 3 represent the p values calculated from the expressions

$p = 4.97 \, g \, [\, 4 \, S(S + 1)]^{1/2}$
$p = 4.97 \, g \, [\, J(J + 1)]^{1/2}$
$p = 4.97 \, g \, [\, 4 \, S(S+1) + L(L+1)]^{1/2}$

respectively. The short thick vertical lines represent the range of observed values.

It is readily seen that the observed values for the ions first transition group do not agree at all with those given by curve 2, $\{p = 4.97 \, g \, [\, J(J + 1)]^{1/2} \}$, which is, however, satisfactory for the rare earths, and thus argues to the existence of large multiplet intervals in their case. Sommerfeld and Laporte suggested that this might be due to the relative smallness of the multiplet intervals in the ions of the first transition group. But, for vanishingly small multiplet intervals, the observed values should lie on curve 3, $\{ p = 4.97 \, g \, [\, 4 \, S(S+1) + L(L+1)]^{1/2} \}$, which is not the case either.

Bose has developed the idea that the paramagnetism of the ions is entirely due to the spin moment, the orbital moment playing no part. This leads to curve 1, $\{ p = 4.97 \, g \, [\, 4 \, S(S + 1)]^{1/2} \}$. Again, the agreement is not satisfactory, chiefly for ions in the second half of the group: also no reason is given as to why the orbital moment is ineffective in the first transition series, while it definitely plays a part in the rare earth ions.

Stoner, observing that the experimental values lie between the curves 1 and 3 as limiting cases, has suggested the following explanation in terms of a definite interaction between the paramagnetic ion and surrounding ions and molecules:

The interaction will effect primarily the electrons in the group of the highest total quantum number. In the ions of the first transition series, a certain number of the electrons responsible for the paramagnetism are of very high total quantum number, unlike in the rare earth ions. Now, according to Van Vleck's treatment, the spin and orbital moment should be considered separately, unless they are firmly coupled. The symmetry of the electric charge distribution of an ion depends on the orbital moment L and the definite way in which ions arrange themselves in a crystal suggests that there may be a strong L-interaction. At the same time the S moment depending on the intrinsic spins of the electrons may be relatively free. In the case of a strong interaction, the S moment only may be affected by an applied magnetic field, the p value then being given by
$p = 4.97 \, g \, [\, 4 \, S(S + 1)]^{1/2}$,
When the interaction is weak, however, the L moment will also be affected, and the p value will then be given by
$p = 4.97 \, g \, [\, 4 \, S(S+1) + L(L+1)]^{1/2}$
This would account for the intermediate position of the observed value between curves 1 and 3.

(ii) *Time of orientation of the atomic magnets in the field direction.*
This problem, which could not be satisfactorily explained by the older theories, receives a new light from the principle of indeterminacy of the wave mechanical theory. The determination of the resolved component of the atomic magnetic moment μ in the field direction is equivalent to a determination of energy, since $E = \mu H$. Now the accuracy with which energy determinations can be made in the magnetic deviation experiments depends upon the wavelength associated with the atom ($\lambda = h \, /mv$). There is thus a probable error in the energy determination. There is also a probable error, depending upon the length of the magnetic field and the velocity of the atoms, in fixing the time t, to which this determination applies. By the principle of indeterminacy, the product of the ranges of indeterminateness of E and t, like other conjugated pairs (momentum and position for instance) is of the order of h, Planck's constant. The latter thus becomes the limiting barrier in the consideration of the time involved in the orientation of the atomic magnets in the field direction. It is, therefore, not necessary to imagine that a sudden change of orientation takes place as an atom enters a magnetic field; theory and experiment determine a *probability distribution* of the atoms among different energy states, which varies as the atoms pass through the field.

C. Ferromagnetism
The explanation of ferromagnetism has long remained a puzzle. Weiss, postulating an intrinsic molecular field, was able to account for and coordinate several experimental facts connected with ferro- and para- magnetics. But the problem of the *origin of such a molecular field* was not *easy* to solve. It was soon realized that

Weiss's molecular field could not originate in the *purely magnetic interaction* of the carriers of magnetic moments, since the magnetic forces between the carriers could only provide fields which were insignificant compared with the molecular field.

This conclusion was confirmed by a number of effects observed with ferromagnetics, such as magnetostriction, large changes of volume and of specific heat on passing through the Curie point, etc.

Attempts to explain the molecular field as due to an associated *electrostatic interaction* also failed. In fact on a classical basis, no adequate explanation could be found of those phenomena which are formally correlated by the molecular field hypothesis. Even *the quantum theory, in its initial stages, could not grapple with the problem,* for its inadequacy was particularly apparent where interaction was concerned, as found in the case of the helium atom.

Heisenberg, in 1928, suggested the molecular field phenomena might arise from a new type of *quantum mechanical resonance,* corresponding to an interchange interaction, similar to the case of the two electrons in the helium atom, which he had just previously solved satisfactorily.

Later, Heitler and London employed the analogous wave mechanical method to the hydrogen molecule and showed that its stable formation with two hydrogen atoms was due to a similar interchange interaction of the electrons in the two atoms. Heisenberg's quantum mechanical interpretation of Weiss molecular field, which is ordinarily known as *Heisenberg's theory of ferromagnetism,* is too complicated and too long to be given in full here; hence only the main lines of development of the theory will be indicated.

The problem is best approached by a consideration of the interaction of two normal hydrogen atoms, as developed by Heitler and London. A system consisting of two hydrogen atoms at a given distance apart, has two different possible energy states, one corresponding to an attraction (at distances not too small) and the other to a repulsion. The difference in energy between the two states is large compared with the energy of purely magnetic interaction due to the electron spins. The state of greater energy (the repulsion) corresponds to the anti-symmetrical orbital wave function (ignoring their spins) and the state of lower energy {the attraction) to the symmetrical orbital wave function. The complete wave functions (including the spins of the electrons) must, of course, be anti-symmetrical in the electrons (Pauli's principle) and therefore the greater energy corresponds to a solution symmetrical in the spins of the electrons and the lesser to an anti-symmetrical one which corresponds to the formation of the stable hydrogen molecule. The difference in energy between the state (of lower energy) and one in which only the ordinary electrostatic interaction of the two atoms is considered is due to a perturbation which arises from the fact that the electron in the first atom may also be regarded as forming part of the system of the second atom, and that of the second as forming part of the system of the first. There may be an interchange of identity between the electrons of the atoms without any actual transfer of an electron from one atom to the other. This interchange interaction gives rise to a term in the expression for the energy, which is of the form

$$ J_0 = \frac{1}{2} \int\int \psi_k^\kappa \ \psi_k^\lambda \ \psi_l^\kappa \ \psi_l^\lambda \left\{ \frac{2e^2}{r_{kl}} + \frac{2e^2}{r_{\kappa\lambda}} - \frac{e^2}{r_{\kappa k}} - \frac{e^2}{r_{\kappa l}} - \frac{e^2}{r_{\lambda k}} - \frac{e^2}{r_{\lambda l}} \right\} d\tau_k \ d\tau_l $$

$$ \dots (1) $$

where k and l refer to the two electrons, κ, λ to the two nuclei, ψ is the unperturbed wave-function for the specified electron near the specified nucleus and $d\tau$ is an element of configuration space of the specified electron.

This new type of interaction is peculiarly characteristic of quantum mechanics and is of very great importance in the study of the formation of molecules and crystals.

Heisenberg, assuming a similar type of interchange interaction for a satisfactory explanation of the intrinsic molecular field phenomena, has worked out the general effect of the interchange interaction on the magnetic properties of a crystal representing the ferro-magnetic substance and consisting of a lattice of similar atoms, each with one interacting electron. These electrons alone are supposed to contribute to the magnetization. Each state of the *system as a whole* is characterized by a definite spin moment $S.h/2\pi$ depending on the number of unpaired electrons. Although, it is not possible to specify the energy of all the states corresponding to a given value of S, the mean energy of these states can be calculated. When an external field H is applied there will be an additional energy depending on the projection m_s of S along the field direction. A complicated statistical treatment is then required to determine the most probable value of m_s, and hence of the magnetic moment. Without proceeding further in the theoretical calculations, we shall give the result, in a form suitable for comparison with the Weiss equations for ferromagnetism.

300

The two simultaneous equations determining the state of magnetization in Heisenberg's theory are

$$y = \tanh (x) \quad \ldots\ldots\ldots\ldots\ldots\ldots\ldots\ldots\ldots\ldots\ldots \quad (2)$$
$$x = c + y (b - b^2 / z) + b^2 y^2 / 4z^2 \quad \ldots\ldots\ldots\ldots\ldots \quad (3)$$

where
$y = n'/n$,
n being the total number of electrons and
n' the number of unpaired electrons,
$c = \mu H/kT$;
μ being the magnetic moment of the electron, and
$b = zJ_0 /kT$,
z being the number of neighbors surrounding each atom and
J_0 an interaction integral of the type given in (1).

To correlate (2) and (3) with the Weiss equations, it is to be noted that when the magnetization is due to electron spin, σ/σ_0 becomes equal to n'/n, and as there are only two orientations for the spin, the Langevin expression;
$\coth (a) - 1/a$,
has to be replaced by
$\tanh (a)$.

When an external field is applied, the Weiss equations, with the above introduced change, become

$$y = \tanh (a) \quad\quad\quad \ldots (4)$$

$$a = \frac{\mu (H + \beta I)}{kT} = c + \frac{\sigma_0{}^2 \beta \rho}{RTM} \, y = c + \frac{\beta I_0 \mu}{kT} \, y \quad\quad \ldots (5)$$

where I_0 is the saturation intensity of magnetization, so that βI_0 is the maximum molecular field, say, H_m. Comparing the two sets of equations, viz., (2) and (3) with (4) and (5), as a rough approximation,

$$\frac{\beta I_0 \mu}{kT} \, y = \frac{b}{2} \, y, = \frac{zJ_0}{2kT} \, y$$

$$\therefore \quad\quad \beta I_0 = H_m = \frac{zJ_0}{2\mu} \quad\quad\quad\quad \ldots (6)$$

This relation (6) indicates that the molecular field is not a field in the ordinary sense, but one that arises from the interchange interaction of the electrons and is proportional to its magnitude. When there is no external field, the Heisenberg equations (neglecting terms in y^3) become

$$y = \tanh (x) \quad\quad\quad\quad \ldots (7)$$

$$x = \frac{1}{2} \left(b - \frac{b^2}{z} \right) y \quad\quad\quad\quad \ldots (8)$$

The graphs of these equations are somewhat similar to those corresponding to the Weiss equations. For spontaneous *magnetization* to occur the slope of (8) must be less than the slope of the tangent at the origin of (7). This gives

$$b - b^2/z \geq 2 \ \ldots (9)$$

The expression on the left is a maximum when $b = z/2$. As an approximation, for ferromagnetism to occur the following condition must therefore hold:

$$z/2 - z^2/4z \geq 2 \quad\quad \text{or} \quad\quad z \geq 8 \ \ldots\ldots\ldots (10)$$

This means that each atom in the lattice must have at least eight neighbors. This condition is satisfied by the ferromagnetic elements which have either face-centred or body-centred cubic lattices in the ferromagnetic state.

The theory thus shows that the molecular field phenomena both in ferro- and para- magnetics can be attributed to interchange interaction, the value of θ in the Curie-Weiss law; $\chi = C /(T - \theta)$, for paramagnetics, or the Curie point for ferromagnetics being approximately proportional to the interaction factor $(z\, J_o)$. For ferro-magnetism to occur, it is necessary that J_o should be positive, which is always the case when the quantum numbers of the interacting electrons are high, n being equal to at least 3, according to Heisenberg's estimation. This condition is again fulfilled by the ferromagnetic elements.

The part of the theory regarding the numerical magnitude of J_o is little developed as yet, owing to the mathematical difficulties. Fowler and Kapitza, in 1929, have given a satisfactory qualitative interpretation of magneto-striction and change of specific heat at the Curie point on the basis of Heisenberg's theory. According to them, however, the model with one interacting electron per atom, assumed by Heisenberg, is too simple to correspond to any actual ferromagnetic. In iron, for example, the saturation intensity and the change of specific heat at the Curie point show that there must be two or three interacting electrons per atom. Hence the interaction which affects the magnetic properties may be only a small part of the total interaction.

There is also no explanation in Heisenberg's theory as to why ferromagnetism is peculiar to only three of the elements.

As the valence electrons of metals contribute only a feeble paramagnetism, the electrons responsible for ferromagnetism should occupy inner shells of the atoms. But electrons in completed shells or subshells cannot be effective for ferromagnetism, since in these the spins are oppositely oriented, two by two, according to Pauli's principle. Hence only incomplete subshells need be considered. The electronic configurations of the elements near iron in the periodic table are:

$$
\begin{array}{lllll}
\text{Mn} \rightarrow 1s^2 & 2s^2 2p^6 & 3s^2 3p^6 3d^5 & 4s^2 & (25) \\
\text{Fe} \rightarrow 1s^2 & 2s^2 2p^6 & 3s^2 3p^6 3d^6 & 4s^2 & (26) \\
\text{Co} \rightarrow 1s^2 & 2s^2 2p^6 & 3s^2 3p^6 3d^7 & 4s^2 & (27) \\
\text{Ni} \rightarrow 1s^2 & 2s^2 2p^6 & 3s^2 3p^6 3d^8 & 4s^2 & (28) \\
\text{Cu} \rightarrow 1s^2 & 2s^2 2p^6 & 3s^2 3p^6 3d^{10} & 4s^1 & (29)
\end{array}
$$

Ten electrons are required to complete the $3d$ subshell. This number is present in Cu and so this metal is not ferromagnetic. The rest of the elements have vacancies in the $3d$ shells, but of these only Fe, Co, Ni are ferromagnetic, so that vacancies in the subshell are a necessary but not a sufficient condition for ferromagnetism. It is to be noted also that whether the number of $3d$ electrons is even or odd is not a determining factor.

Slater, in 1930, has made an important progress in discovering a reason for the unique position of Fe, Co and Ni as ferromagnetics by importing into the theory the idea that the probability of interaction increases as the ratio of the inter-atomic distance to the diameter of the incomplete inner electron shell increases.

Bethe, in 1933, has shown that this ratio must exceed a critical value of about 1.5 for ferromagnetism to occur.

Fig. 246 shows Bethe's calculated values of the exchange interaction energy as a function of this ratio. It is seen that the ferromagnetic materials Fe, Co, Ni, all have the positive exchange energy required to produce ferromagnetism, whereas Mn has a negative exchange energy. The rare earth elements, gadolinium and dysprosium, in which the electron vacancies occur in the $4f$ subshell, also have positive exchange energies, and below their Curie points of 289°K and 88°K are weakly ferromagnetic. Ferromagnetism exhibited by Heussler alloys, which contain Mn along with Cu, Al and other metal can be accounted for on the same principle, viz., alloying introduces changes in crystal structure to make the ratio for Mn, which is very near the critical condition for ferromagnetism, exceed the value of 1.5 and thereby cause a change in sign of the Heisenberg exchange force from negative to positive.

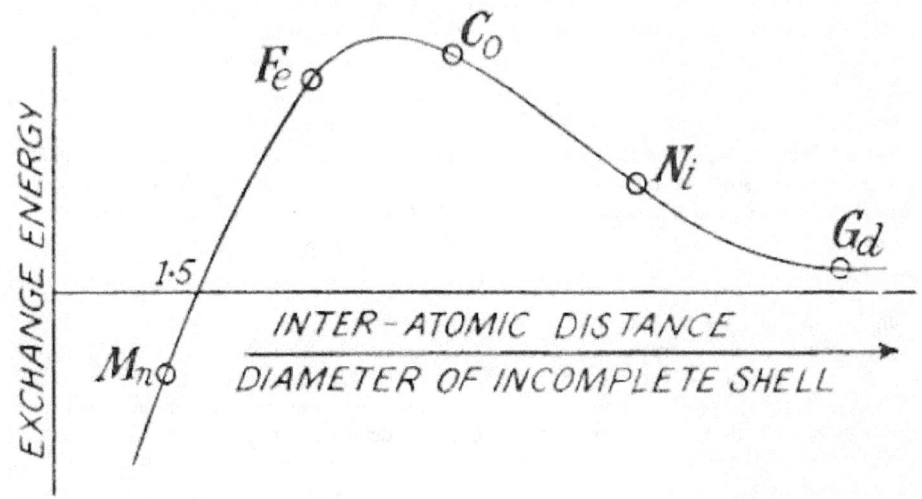

These considerations show that Heisenberg's theory must be refined further before it can be considered completely satisfactory. But it is essentially correct in its finding that ferromagnetism is possible, only if the interaction integrals are positive and even then only under very special conditions of crystalline arrangement.

CHAPTER 12

MOLECULAR SPECTRA

Introduction

A study of molecular spectra finds its proper place at this stage. For, in the first place, the postulates and principles involved in the elucidation of atomic spectra can be used also, *mutatis mutandis,* in the interpretation of the different characteristics of molecular spectra. Secondly, analysis of this type of spectra leads to the realization of certain important structural properties of the nucleus, such as the isotopic constitution, spin, etc., with which we have to deal in the following chapters.

For a clear understanding of molecular spectra, a knowledge of the structure and properties of molecules is very helpful, although many points concerning molecular structure have been actually gathered from a scrutiny of molecular spectra. The first section of this chapter is therefore devoted to a brief statement of molecular structure. In the next section, we take up the study of molecular spectra. Then, two important allied phenomena, viz., isotopic effect in molecular spectra and Raman effect are considered. Finally, in a supplementary section, some details are given concerning solid state spectroscopy.

1. MOLECULAR STRUCTURE

A molecule may be defined as the smallest particle of a substance (element or compound) which is capable of existing permanently in a free state. In 1808, Gay-Lussac discovered the *law of gaseous volumes,* which states that when gases combine together, the volume of the product, if gaseous, bears a simple ratio to the volumes of the reacting gases. In 1811, Avogadro, in order to explain this law which was untenable on a purely atomic basis, put forward the idea of the molecule as an aggregate of a definite number of atoms, capable of existing in a free state.

General classification of molecules. According to the number of the constituent atoms, molecules are *monatomic* (one atom), *diatomic* (two atoms), *triatomic* or *polyatomic* (three or more atoms).

Monatomic molecules can have only translatory motion, while diatomic and polyatomic molecules can have also rotational and vibrational motions,

since the constituent atoms can rotate about an axis passing through the common centre of gravity, as well as vibrate along the line joining their nuclei.

Molecules have been classified from considerations of their *rotational motion.* Rotational motion gives rise to moment of inertia.

Let A, B and C be the moments of inertia about three principal axes of the molecules and let $A < B < C$.

If the moments of inertia about the internuclear axis is zero and the other two about axes perpendicular to the internuclear axis are equal, we have the *diatomic,* or a *linear polyatomic molecule, e.g.,* H_2, D_2, N_2, O_2, HCl, HBr, CO_2, N_2O, HCN, etc.

If $A = B = C$, we have *spherical molecules,* which have no preferred axis, *e.g.,* CH_4 (methane), SiH_4 (silane), where the four hydrogen atoms are at the corners of a regular tetrahedron.

If two of the moments of inertia are equal, the molecule is called *symmetric rotator* or *top.* The motion of this type can be represented as a pure rotation about the axis of either the greatest or least inertia, with the axis making a precession round that of total angular momentum.

If the two smaller moments of inertia are equal, *i.e.,* $A = B$, we have the *oblate symmetric top* (*a* disc in the extreme case).

If the two larger are equal, *i.e.,* $B = C$, we have the *prolate symmetric top* (spindle-shape). A large and important group of molecules, such as NH_3, methyl halides, ethane, etc., are *symmetric rotators.*

Finally if $A \neq B \neq C$, we have *asymmetric rotator* or *top* which can be described as executing pure rotation about the axis of either the greatest or least inertia, while this axis makes a precession (the motion of the axis of a spinning body, such as the wobble of a spinning top, when there is an external force acting on the axis) and a nutation (a wobble in a spinning gyroscope or other rotating body) round the axis of total angular momentum. It is a spinning top whose angle to the vertical is oscillating. The majority of molecules found in nature belong to this class. Simple triatomic molecules, such as H_2O and SO_2 are *asymmetric tops.*

The details of such a classification of molecules have been studied with the help of the observed molecular spectra and appropriate theoretical considerations. The available experimental data in this study have greatly increased in volume and importance with the advent of the microwave. technique of analysis of pure rotational spectra of molecules.

Molecules have been classified into different categories with respect to the *vibrational motions* of the constituent atoms. Thus we have

(i) the *triatomic linear molecules* with two sub-groups viz. those with central symmetry, like CO_2 and CS_2 and these without central symmetry, like N_2O, COS, HCN, ClCN, etc.,

(ii) the *triatomic triangular molecules* (YX_2) a special. and interesting case of which is the type of molecule having an axis of symmetry that bisects the angle XYX, *e.g.*, H_2O, H_2S, O_3, NO_2, SO_2, etc.,

(iii) *the four-atom linear molecules,* (Y_2X_2) a well-known example of which is acetylene (HCCH) that has also a centre of symmetry,

(iv) *the four-atom pyramidal molecules,* (YX_3) such as ammonia (NH_3) in which the N atom is at the apex of the pyramid and the H atoms at the corners of an equilateral triangle form the base,

(v) the *five-atom molecules* (ZYX_3) *e.g.*, methyl halides (CH_3F, CH_3Cl, CH_3Br, CH_3I), and methane (CH_4) as a special case, where the H atoms are at the corners of a tetrahedron and the C atom at its centre, with a consequent high degree of 'spherical' symmetry, and so on. More complex molecules with a greater number of atoms may be regarded as made up of simpler molecules coupled together.

This method of differentiation of molecules is based on the analysis of the vibrational frequencies of molecular spectra. The general principle used may be stated as follows:

If a molecule has n atoms and three degrees of freedom are assigned to each atom, the total number for the whole system is $3n$. Of these, three are accounted for by the translational motion of the molecule as a whole and three more are referred to its free rotation. The remaining $3n - 6$ degrees of freedom must therefore represent modes of vibration of the system.

In the case of linear molecules, since only two degrees of freedom are involved in rotation, ($3n - 5$) will be the degrees of freedom for their vibrations. According to this postulate, linear triatomic molecules ($n = 3$) should have four fundamental frequencies, triangular triatomic molecules three fundamental frequencies, four-atom linear molecules ($n = 4$) seven, pyramidal molecules six, five-atom molecules ($n = 5$) nine and so on. These internal modes of vibrations are substantiated with data obtained from the observed molecular spectra, due attention being paid to degenerate cases arising from symmetry conditions.

Electronic structure in molecules. When atoms combine to form molecules, they bring with them their peripheral electrons and the governing factor in the stable formation of molecules is the valencies of the constituent atoms, determined by the number and nature of the electrons in their outermost incomplete shells. Hence a molecule also has a configuration of electrons which form one coherent system, as in the case of an atom, with closed shells, each containing a definite number, as permitted by Pauli's principle; the electrons outside the closed shells act as the 'optical' and 'valency' electrons, responsible for the observed molecular spectra and the formation of more complex molecules, elements and compounds. But it appears that commonly the constituent atoms retain the electrons in their K-shells and that the electronic structure external to these is best regarded as a large number of shells, each containing two electrons only and each having a distinct ionizations potential or term value.

Linkages between atoms in molecules. A very important consideration in molecular structure is to see how the individual separate atoms are bound together into a molecule. This is ordinarily known as *chemical binding*. Quite early in the study of the proper ties of chemical compounds, it was found necessary to distinguish two types of molecules.

The first type, well represented by common salt (NaCl), forms electrically conducting aqueous solutions, in which the molecule splits up into ions, *(e.g.* Na^+ and Cl^-). Even in the solid state, the charges are distributed unequally between the constituents, as seen from X-ray analysis of crystals like NaCl. Substances of this type usually have relatively high boiling points and typically salt-like properties.

The second type, which is represented by HCl, BeO, CO, CN and organic compounds like methane (CH_4) and benzene (C_6H_6,), is characterized by low electrical conductivity, relative volatility and complete lack of salt-like properties. Abegg, the chemist called the first type *heteropolar* and the second *homopolar*. It is to be noted that this classification fails to cover 'elementary' molecules as H_2, N_2, C_2, O_2, Cl_2, I_2 etc., which are *non-polar*.

Nowadays, following Franck, according to whom the decisive criterion is whether a molecule in dissociating tends more readily to split up into ions or into atoms, the first type is called *ionic molecule* and the second *atomic molecule*. The above-mentioned elementary molecules can now be put under the category of atomic molecules.

Consideration of the interchange of electrons by the constituent atoms in the formation of the molecule has enabled us further to designate the ionic type as *electrovalent* and the atomic type as *covalent*. There are compounds which have intermediate types of binding; others cannot be assigned to either of the above two classes. Of these, the following two are now sufficiently well recognized:

(i) *cohesion binding, i.e.,* loose binding without saturation of valency, due to the Van der Waals forces and

(ii) *metallic binding* which is effective in the lattice formation of metals.

We shall first consider, somewhat in detail, the two prominent types, viz. the ionic and atomic molecules and then state briefly the chief characteristics of the other two, as they are also of interest and importance from the physical standpoint.

Electrovalent ionic molecules

The existence of free ions in the solid, liquid or solution state of ionic compounds, such as the alkali, metal halides, shows that *the linkage involved is electrostatic in nature*. Kossel was the first to suggest that the formation of ionic molecules is due to the tendency for atoms to build up complete electron shells or sub-shells and in particular to assume the stable electron configuration of an inert gas. For example, when Na combines with Cl to form NaCl, the Na atom, being electropositive and monovalent, gives it valency electron to the Cl atom which is electronegative and monovalent. After the transfer, the Na atom has 2K, 8L, electrons and is like a Ne (neon) atom; the Cl atom has 2K, 8L, 8M electrons and is like an Ar (argon) atom. In the NaCl molecule thus formed the nucleus of Na carries a charge $+ 11e$, so that the net charge of Na is now $[+11\ e -10\ e] = +\ e$; the nucleus of Cl carries a charge $+17e$ and hence the net charge of Cl is $[+17e — 18e] = —\ e$; thus the Na and Cl atoms in the NaCl molecule are ionized as Na^+ and Cl^- and, in consequence, attract each other in accordance with Coulomb's law.

By itself, this would lead to the absolute coalition of the two ions; however, at small distances, repulsive forces come into play, which can be understood as follows:

If the distance between a positive ion and a negative ion is progressively diminished, the positive ion will penetrate into the electron atmosphere of the negative ion. Under these circumstances, that part of the electron atmosphere which lies at a greater distance than the positive ion from the nucleus of the negative ion will, to a first approximation, exert no force on a charged particle inside it. Hence the force of attraction on the positive ion will be diminished by its penetration into the electron atmosphere of the negative ion. As the distance between the ions is further reduced, a position of equilibrium will finally be reached, in which the repulsion between the similarly charged nuclei is just balanced by the attraction between the positive ion and the inner region of the electron atmosphere Of the negative ion. To represent the repulsive forces, a law of the form b/r^n has been tried, with good success. The position of equilibrium is given by the value of r for which the sum of the energy of attraction. $(—e^2/r)$ and the energy of repulsion $(+ b/r^n)$, has a minimum value. Quantum mechanical treatment, gives approximately an exponential law

$be^{-r/\rho}$

for the repulsive forces which leads to better results.

This explanation of the binding in ionic molecules can be conveniently tested in the case of highly symmetrical crystals, such as rock salt, where the deformation of the electron atmosphere of the two ions is reduced to a minimum on account of the symmetry. The *lattice energy* U, *i.e.*, the energy required to break up the crystal lattice completely, is given by

$$U = \sum (\pm e^2/r + b/r^n),$$

the summation being taken over all the lattice points;
r refers to the lattice constant, obtained readily from X-ray analysis;
b's are the constants that can be determined using two relations, one referring to the equilibrium condition, *viz.*
$dU/dr = 0$, (*i.e.*, the lattice energy is minimum) and the other relating to the compressibility of the crystal viz. d^2U/dr^2, which is the force required to compress the crystal by a certain amount and hence can be measured experimentally.
U can, therefore, be calculated using the theoretical formula.

Now, the lattice energy can be experimentally determined by means of thermal dissociation, as was done by J. Mayer. The experimental value agrees fairly well with the theoretical value in the case of several salts. The theory can also be tested indirectly by calculating the *electronic affinity* E, (*i.e.*, the energy set free when the electron settles in its place, in the formation of the molecule) of a halogen derived from different salts, say of Cl from the compounds NaCl, KCl, RbCl, etc., with the help of a *cyclic process* suggested by Born and Haber. The value of E of the same halogen deduced from different salts is found to be nearly the same.

It may be noted that, in practice, it is not easy to decide whether a molecule is of the ionic or atomic type. Such molecules as BeO, HC1, etc., might from the numbers of electrons satisfying the valency bonds, be thought to be ionic. Spectroscopic evidence, however, classifies them as atomic. For, spectroscopically, ionic molecules are always capable of dissociation by absorption of light into two *unexcited atoms,* while atomic molecules usually dissociate under similar conditions into *one exited atom* and *one neutral atom.* It is upon this spectroscopic evidence that the hydrogen halides are classified as homopolar and differentiated from the alkali metal halides which are hetoropolar.

Covalent atomic molecules
Lewis, in 1923, pointed out, on an empirical basis, that the formation of stable covalent molecules is frequently associated with the existence of an *electron pair,* which is such that each of the pair is shared by, or belongs to both of, the two constituent atoms. In the N_2 molecule, for instance, the linkage appears to be due to the completion of an L shell by the electrons from both constituent atoms; for, the similarity of the spectrum of N_2 molecule to that of an alkaline earth indicates that two of the fourteen electrons are relatively loosely bound and obviously belong to the M shell. The electrons of the nitrogen molecule can therefore be arranged as 2K, 2K, 8L, 2M. In many cases, however, complete shells appear to be formed around the constituent atoms by the sharing of electron pairs; e.g., Cl_2. This is especially true of shells of eight electrons. Such a pairing is readily understood with the help of Pauli's principle, since a pair of electrons with opposite spins represents complete occupation of any level. On account of their opposite spins, the two electrons neutralize each other. Those electrons, which possess no partner in the above sense, may obviously become neutralized by pairing off with a new electron.

It is these unpaired electrons which are responsible for chemical combination. Covalent binding usually occurs when the constituent atoms can exist in a state of high valency.

An important example is a structure with a valence of four, as in carbon, silicon, etc., which occur in a crystalline form where each atom is surrounded by four nearest neighbors located at the corners of a regular tetrahedron. In this structure each of the four valence electrons of an atom enters into an electron-pair or covalent bond with electrons from nearest neighboring atoms and all valences are saturated, As a result, such substances have high electrical resistance and are very hard and strong.

Assuming therefore that the covalent linkage arises from the sharing of an electron pair by the constituent atoms, we have next to inquire into the inner mechanism of such a sharing in order to understand the nature of the linkage involved.

It cannot be electrostatic in nature, as in the case of ionic molecules, since, according to the classical theory, the pair of electrons should always repel each other and hence cannot form a link between the two atoms.

Nor can the linkage be affected by the magnetic forces due to the spin of the electrons, which, calculation shows to be too weak.

It may be recalled here that we met with a similar difficulty in connection with the large energy difference between the term-system of the helium atom and that Heisenberg solved it by invoking an *exchange energy of wave-mechanical resonance* that gives rise to a strong interaction between the two electrons. Heitler and London, in 1927, employing the same wave-mechanical resonance principle, were able to account satisfactorily for the covalent linkage in the H_2 molecule.

Wave mechanical theory of covalent linkage

Heitler and London considered the H_2 molecule, which is built up of two equal nuclei and two electrons, as the simplest case of covalent linkage that can be analyzed with relative ease. The structure of H_2 molecule is assumed to involve the two nuclei, still separate, but surrounded by a common electron atmosphere corresponding to the two electrons. The application of wave mechanical resonance to such a system is easily understood by the following analogy:

If two electrical oscillating circuits, having the same frequency v_o, are brought close to each other, the coupling throws them to some extent out of tune, v_o being split up into two different frequencies, one a little greater than v_o and the other a little lower, and beats are produced. The conditions in the hydrogen molecule are similar: the electrons revolving round the nuclei of the two separate hydrogen atoms correspond to the two identical oscillating circuits. When the two atoms approach close to each other and are coupled, the coupling puts them out of the tune a little, giving rise to two frequencies, one slightly higher than v_o and the other slightly lower. Since, to every frequency v_o corresponds an energy hv_o, the total undisturbed energy $2E_o$ of the two separate hydrogen atoms gives rise on combination to a somewhat lower and a somewhat higher energy of the coupled system,

$$E_1 = 2E_o - W_1 (d),$$
$$E_2 = 2E_o + W_2 (d) \dots (1)$$

where $W_1 (d)$ and $W_2 (d)$ represent the coupling energies which depend on the distance d between the two atoms.

On the principle that less energy means greater stability, the state represented by E_1 possesses less energy than the state of the two separate atoms and therefore corresponds to a possible binding; E_2 is greater than $2E_o$ and hence denotes repulsion between the two atoms. Those conclusions may be graphically represented by plotting the variation of the potential energy of the system as a function of the distance between the nuclei, when curves of the type shown in Fig. 247 are obtained.

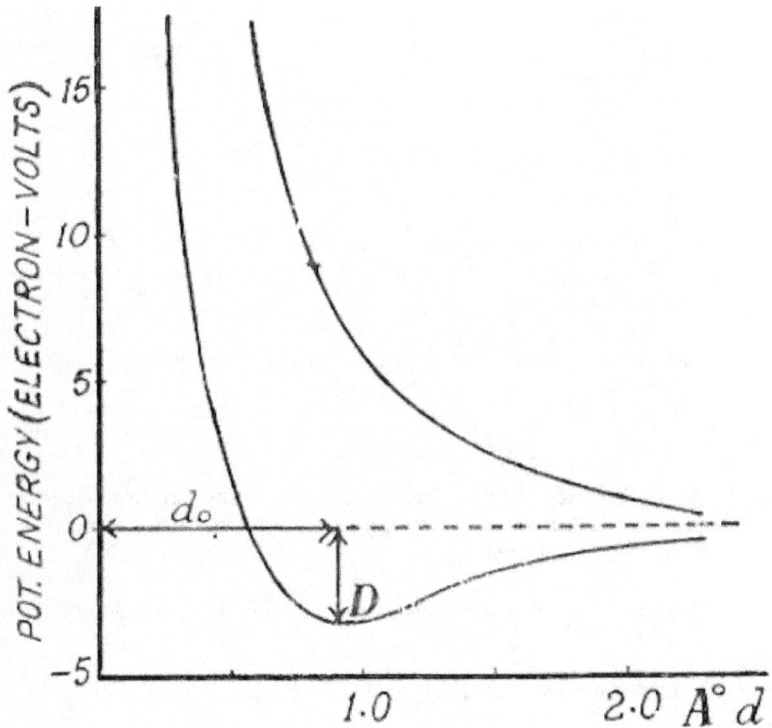

Fig. 247. Potential energy of two hydrogen atoms as a function of the distance between their nuclei.

The upper curve corresponds to the case of permanent repulsion between the two atoms, in which the potential energy increases monotonically as the distance between the nuclei is diminished. The minimum in this curve occurs only when the internuclear distance is infinity. This means also that work must always be done to the nuclei together. The lower curve, on the other hand, represents the case of possible binding of the two atoms. For, here, the potential energy decreases to a minimum when the nuclei are at a distance d_0 apart. This minimum represents a state of stable equilibrium (actually the ground state of the H_2 molecule). The energy represented by D is that set free as the internuclear distance is diminished from infinity to d_0. Hence this must also be the energy of dissociation, *i.e.*, the energy to be supplied in order to dissociate the hydrogen molecule into two separate hydrogen atoms. From the minimum at d_0 the potential energy rises very steeply as d becomes smaller, the Coulomb repulsion between the two nuclei preponderating here. For internuclear distances greater than d_0, the curve flattens and asymptotically approaches the zero level; this corresponds to the case of nuclei which are far apart, so that the molecule is completely split into its constituent atoms.

Considering the case that corresponds to a binding in relation (1), the energy $W_1(d)$ is due to the frequency with which the deformation of the wave functions, caused by the interaction, oscillates from one atom to the other. Since the square of the amplitude of a wave function represents the probability of finding a particle at a given place, this can be interpreted as an "exchange" of the two electrons. Hence $W_1(d)$ is called the *exchange energy*.

The specific nature of the state that leads to the binding of the two hydrogen atoms into a stable hydrogen molecule may be determined as follows:

If the wave functions of the two electrons 1 and 2, associated with the two hydrogen nuclei a and b respectively, are $\psi_a(1)$ and $\psi_b(2)$; the two possible states of the hydrogen molecule are represented by the products

$\psi_a(1)\,\psi_b(2) = C$ and
$\psi_a(2)\,\psi_b(1) = D$ ....(2)
since the two electrons are indistinguishable.

309

The *orbital part of the wave function* of the hydrogen molecule, for small distances between the nuclei, can be shown to be approximately one of the following linear combinations:

$$\psi_S = (C + D) / 2^{1/2}$$
$$\psi_A = (C - D) / 2^{1/2} \quad \dots\dots\dots\dots\dots\dots\dots\dots\dots\dots\dots\dots\dots\dots\dots\dots(3)$$

where
ψ_S and ψ_A represent the symmetric and anti-symmetric orbital functions respectively and
$1/2^{1/2}$ the approximate normalizing factor ensuring that the total probability of the existence of an electron in the whole of space is unity.

Since the probability of the occurrence of an electron between the nuclei is proportional to ψ^2, rewriting relations (3) as

$$\psi_S^2 = (C^2 + D^2 + 2CD) / 2$$
$$\psi_A^2 = (C^2 + D^2 - 2CD) / 2 \quad \dots\dots\dots\dots\dots\dots\dots\dots\dots\dots\dots\dots\dots(4)$$

we see that $\psi_S^2 > \psi_A^2$, since C and D are positive. This means that the electron density corresponding to a possible binding is greater for the symmetric function than for the anti symmetric one as shown in Fig, 248, in which the contour lines of the electron density are plotted round the nuclei.

ψ_A^2 is represented by (*a*) where the two atoms repel each other at all distances apart, so that a collision between them will be elastic.

ψ_S^2 is represented by (*b*) where the atoms attract each other at all distances apart up to a certain definite minimum, below which mutual repulsion sets in.

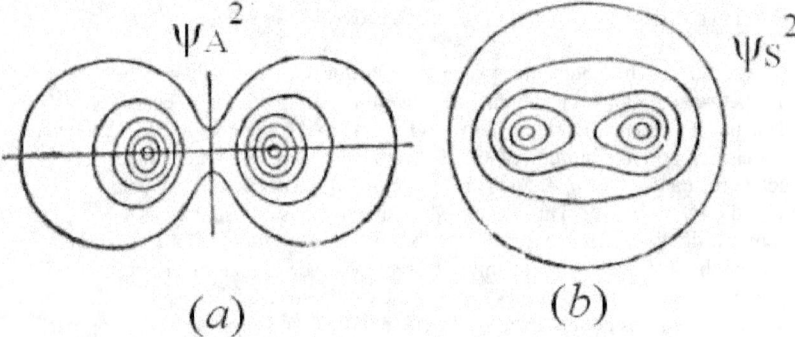

Fig. 248. Contour lines of electron density of:
(*a*) two mutually repelling hydrogen atoms;
(*b*) two hydrogen atoms forming a stable hydrogen molecule.

Thus *the symmetric orbital wave function corresponds to binding properties and the anti-*symmetric *orbital wave function to the anti-binding properties of the electrons.*

Taking into account the spins of the two electrons, let S_S and S_A be the *spin functions* when the electron spins are parallel and antiparallel respectively. S_S is symmetrical, since its sign is unchanged when the electrons are interchanged and S_A is anti-symmetrical, since its sign is changed by the interchange of the electrons.

The *complete wave function* is the product of the orbital and spin functions. In the present case, there are *four possible complete wave functions, viz.*

$\psi_S\ S_S,\ \psi_S\ S_A,\ \psi_A\ S_S,\ \psi_A\ S_A$

since a hydrogen molecule can be formed only when the orbital function (ψ_S) is symmetrical, the third case (ψ_S S_S) is excluded. Pauli's principle, which in its wave mechanical aspect implies that the complete wave function of a system (atomic or molecular) cannot be symmetrical in respect of similar particles (electrons or nuclei), excludes the first and the fourth, *i.e.*, $\psi_S S_S$ and $\psi_A S_A$. Hence the second case ($\psi_S S_A$), the only one left over, should be the complete wave function representing the H_2 molecule.

This means that two hydrogen atoms can combine to form a stable molecule only when the spin function is anti-symmetrical (S_A), i.e., when the two electrons have their spins anti-parallel, so that the electron spin of the stable hydrogen molecule is zero. Thus, the state that leads to binding is characterized by the balancing of the electronic spins and the *covalent linkage is interpreted as due to a pair of electrons with opposite spins.*

It appears also that in general the *spin may be regarded as a physical substitute for the. chemical valency. The* number of spins net compensated is equal to the number of the free valencies of the atom. In other words, the *true (homopolar) valency of an atom is equal to the number of unpaired electrons in it, i.e., electrons with spins not saturated.* Since *the* number of such electrons is 2S, where S is the total spin of the atom and the multiplicity r of the energy levels in the atom is given by $r = 2S + 1$, it follows that *the valency of an atom must be one less than the multiplicity of its energy levels.* This conclusion offers a means of testing the validity of the theory.

It is found that there is good general agreement with observation concerning this point. For example, the alkalis, which have a double state, are monovalent; the alkaline earths, which possess a triplet system, are divalent. But the rule does not seem to apply for the *grand states* of the atoms. Since the ground state in all atoms is a singlet, the atoms should have zero valency, according to the rule. This difficulty is overcome by the fact that the *atoms usually react in excited states, to which chiefly, the rule must be applied.* Since in the excited states the electrons are raised to levels in which the full possibilities of occupation are not realized, *increase of excitation is in general accompanied by an increase of valency.* Thus, for example, the tetravalence of carbon corresponds to an excited state. The carbon atom possesses, close above the ground state, an excitation level in which it *is* tetravalent and the ordinary chemical bonds of carbon are due therefore to that first excited state.

Formation of covalent linkages by valency saturation will manifest itself also in the spectrum of the molecule. This is actually found to be the case. For instance,

> when N combines with O to form NO, only two electron pairs are formed, one electron remaining unpaired. As is to be expected from the theory, the ground state of NO is found to be a doublet state.

> The CO Molecule, on the other hand, contains no such unpaired electrons and its ground state is accordingly a singlet state.

It may be noted, however, that the theory fails to account for the strong paramagnetism of O_2 molecule which has all its electrons paired apparently.

In conclusion, we may say that, although we are still far from having a complete and quantitative theory of covalent molecules, the wave mechanical resonance idea has indicated a general line of approach in the right direction, while its *special success* lies in the fact that it makes it possible to deal quantitatively with nuclear distances, heats of dissociation and vibrational frequencies of simple molecules like hydrogen.

Van der Waals type of binding
There exists a class of molecules of which the atoms are loosely bound, characterized by a small dissociation energy: e.g.. Hg_2 , Cd_2, Zn_2 and mixed molecules which these elements are capable of forming with alkali metals. The binding in such molecules is satisfactorily accounted for by the Van der Waals cohesion forces without any valency saturation—hence called *cohesion, binding.*

The Van der Waals attractive forces depend on the mutual deformation of the atoms and that in two different ways. First, the action of the field due to the electric moment of the molecules, permanent or induced, leads on the average to an attraction. This result was established by Debye and Keesom (1921) on the basis of the classical theory. From this, however, it would follow that spherically symmetrical atoms (as of the inert gases, He, Ne, etc.) or molecules should show no cohesion, which is contrary to the experimental fact that all gases can be condensed, even liquefied at sufficiently low temperatures.

London, in 1930, was able to solve this difficulty by showing that the mutual deformation of the atoms has a second effect which is characteristic of quantum mechanics. According to this theory

there exists a finite "zero-point energy" which implies that even in the lowest energy state, an atom or molecule has to be put in correspondence with a system of moving charges (electrons), so that it carries a dipole oscillating with electronic frequency. If then two such systems come near each other, the zero-point motions of the dipoles always act in such a way that on the average the result is attraction.

Calculation gives an interaction energy which is inversely proportional to the sixth power of the nuclear distance ($1/d^6$).

In many atoms or molecules without free valencies (spins), even before condensation, a sort of pairing occurs as a result of the balancing of Heitler-London repulsion and Van der Waals polarization attracting forces, since the latter will sometimes fall off more slowly with increase of internuclear distance than the former.

Using this theory of cohesion it is possible to calculate the heats of sublimation of molecular lattices from atomic properties, as London has done.

Metallic binding
The existence of metals cannot be explained by the types of binding mentioned above; a few cases, carefully studied, such as lithium and sodium, indicate that the positive atomic residues are held together by the free electron in the metal. Difficulties arise however in the case of diamond-type lattices, in which the binding apparently depends on valency saturation. Attempts have been made to account for them in the light of quantum mechanics.

Electric moments of molecules
A very important feature of molecules is their electric moment which is caused by an *unequal distribution of charge* in them. The condition for a molecule to have electric moment is that the centre of gravity of the negative (electronic) charges should not coincide with that of the positive (nuclear) charges. Hence a molecule, in which the positive and negative centres of charge are separated, behaves like an electric dipole and hence is said to possess a *permanent electric dipole moment,* denoted by the vector

$$p_o = \sum \overline{er} ,$$

where r gives the position vectors of the nuclei and the electrons, the bar signifies averaging over the electron motion and the summation is taken over all the nuclei and electrons.

Taking as a simple example the HCl molecule, the valency electrons spend, on an average, more time near the chlorine nucleus with its larger positive charge than near the hydrogen nucleus. The molecule thus behaves as if the chlorine end were negative and the hydrogen end positive, *i.e.,* as an electric dipole and is therefore considered to have a permanent electric dipole moment. Molecules have been classified as *polar* or *non-polar* according to the presence or absence of electric moment in them.

The great majority of molecules probably have an electric moment, *i.e.,* are polar. On the other hand, almost all 'elementary' molecules C_2, N_2, O_2, Cl_2, I_2, etc., are non-polar.

A molecule, in which the centres of positive and negative charges coincide on account of symmetrical distribution of charges, has no electric dipole moment. The electric behavior of such a molecule is determined to a first approximation by a *quadrupole moment* which is analogous to the mechanical moment of inertia and hence called also the *electrical moment of inertia.*

If a molecule is subjected to an *electric field,* it is deformed, the positive nuclei being attracted in the direction of the lines of force, the negative electrons in the opposite direction. Consequently *even when there is no permanent dipole moment, a dipole is induced.* The magnitude of this induced moment is directly proportional to the field strength. Such an induced dipole moment is usually represented by the relation
$p = a \, E,$
where a is called the *polarizability,* and has the dimensions of volume.

For spherically symmetrical molecules, a is a constant, independent of direction. In general, however, a depends on the direction and is represented by an ellipsoid, the *ellipsoid of polarization.* When molecules can rotate freely, as in the case of gases, the ellipsoids of polarization corresponding to the individual molecules can assume all possible orientations in space, so that when an external field is applied, a mean polarization of the gas arises, which is given by

$$\overline{p} = \overline{a}\,E,$$

where $\overline{a}$ is called the *mean polarizability*.

Since the polarizability of a molecule is represented, in general, by an ellipsoid, the polarizability can be considered along the directions of the three principal axes of the ellipsoid. Molecules are distinguished as *isotropic,* if the polarizability is the same in all the three directions and *anisotropic,* if the polarizability is different in the three directions. This consideration has an important bearing on the polarization relations in the case of scattering *of light;* anisotropic molecules produce measurable *depolarization* effects, known as *optical anisotropy.*

In the scattering process, the electron atmosphere is caused to oscillate with the frequency of the incident light. If now the electron atmosphere is not equally polarizable in all directions, its oscillations will not always be in the same direction as those of the electric vector of the incident light and as a result the scattered light will be partially depolarized. By measuring the degree of depolarization, conclusions can be drawn as to the anisotropy of the polarizability of the molecules.

Size and shape of molecules

The kinetic theory of gases has indicated several methods, such as viscosity, mean free path, diffusion, Van der Waals equation, etc., for determining the sizes of molecules. These methods, which assume the molecules to be spherical and determine the distances up to which molecules can approach one another in ordinary thermal collisions, are bound to give only approximate results. In recent years, other more refined methods of great precision have been employed to obtain quantitative information about the distances between the centres of atoms in molecules or ions in crystal lattices, the modes of arrangement of these atoms or ions and the angles between the valency bonds linking atoms. The combination of the results obtained by the different methods has yielded detailed information about the size and shape of molecules. We shall briefly state the most important of these methods and the results obtained with them.

(i) *Diffraction, of X-rays and of electrons by solids, liquids and gases*

The X-ray analysis of crystal structure, giving the distances between the atoms and their modes of arrangement in the molecule has already been described. The X-ray method has also been used to investigate liquids, gases and vapors. The electron diffraction method has been applied in these cases with greater effectiveness and better results. The interatomic distances have been measured with accuracy; they are found to vary from about 1.5 to 3 A°.

Thus, for instance, in the case of CCl_4, the distance between the Cl atoms is 2.99 ± 0.03 A° and that between the C and Cl atoms 1.82 A°. The interatomic distance is important in deciding the bond character. The length of any particular chemical linkage is very nearly independent of the rest of the molecule in which it occurs. Indeed, it is possible to ascribe to each atom a definite effective radius in each type of combination, the so-called *normal covalency radius.*

Thus, in organic molecules we have the following values:

single bonds :	$H - 0{\cdot}3$ A°	$C - 0{\cdot}77$ A°	$N - 0.7$ A°
double bonds :		$C = 0{\cdot}65$ A°	$N = 0{\cdot}63$ A°
triple bonds :		$C \equiv 0{\cdot}58$ A°	$N \equiv 0{\cdot}55$ A°

From the experimental data conclusions, can be drawn also as to the *shape* of the molecules. Thus, for instance, the molecules

$HgCl_2$, $HgBr_2$, C_2N_2 , CS_2 are shown to be linear,
BCl_3 planar and triangular,
$SiCl_4$, $Ni(CO_4)$, P_4 tetrahedral,
PF_3 pyramidal,
SF_6 octahedral and so on;
F_2O, Cl_2O and SO_2 are bent at angles of 100°, 115° and 120° respectively.

(ii) *Optical methods: molecular spectra including Raman spectra*

The *infra-red absorption spectra* give information on the moments of inertia and hence the interatomic distances in the case of molecules having electric moments. More accurate results are obtained by the so-called *rotation-vibration* and *electronic band spectra,* the latter being applicable to molecules without permanent electric moments. The spectra of scattered light (Raman spectra) provide important complementary information, such as bond strengths which—can be deduced from the *restoring forces* involved in the vibrational frequencies of atoms in the molecule.

(iii) *Electric dipole moments*

Another important method of investigation is the determination of the electric dipole moments. A large class of molecules with unlike constituent atoms, *e.g.,* HCl, CO, NO, possess large electric moments.

The electric moments give useful information about the shape of molecules. The experimentally measured moment is a vector quantity and it may be calculated by adding the moments of the various atomic or ionic bonds by vector addition.

In this procedure, suitable bond angles must be assumed to give the correct resultant, agreeing with the experimental value, and from these angles some idea of the shape of the molecule can be obtained. Careful scrutiny of observed data shows that a definite dipole moment may be ascribed to each linkage between atoms even in complicated molecules. In the case of triatomic molecules of the type YX_2 it is possible to deduce whether the shape is: (a) linear and symmetrical, (b) linear and unsymmetrical or (c) bent. Thus, the absence of an electric moment for CO_2 CS_2, $HgCl_2$ indicates that these molecules have a linear symmetrical form. In order to discriminate between cases (b) and (c) other methods must be used such as analysis of the vibrational state of band spectra and the number and state of polarization of the Raman lines. The molecules H_2O, H_2S and SO_2 have thus been shown to have a bent shape, while NO_2 has a linear unsymmetrical structure.

Molecules with or without a permanent dipole moment acquire an *induced dipole* under the influence of an electric field. In this case, the electron atmosphere of an atom is displaced relative to the positive nucleus and the result is equivalent to the formation of a doublet.

This induced moment is stronger the more polarizable the atom or ion and the effect which can be measured from the refractive index (Kerr effect) also enable ionic radii to be estimated.

Dissociation of molecules: heat of dissociation

One of the very important quantities characterizing a molecule is the amount of work to be done in order to separate the constituent atoms. This is usually called *the heat of dissociation* and is expressed in calories. As was pointed out in the discussion on the formation of the H_2 molecule, the heat of dissociation is given by the height of the asymptote of the potential energy curve above the minimum. In the potential hollow of the curve, the nuclei vibrate as *quantum oscillators*. The shape of the curve close to the equilibrium position r_0 being approximately parabolic in form, for not too great amplitudes the nuclei vibrate like linear harmonic oscillators. We can therefore apply the wave mechanical formula

$E_n = h\nu_0 (n + 1/2)$

which leads to the energy levels of the nuclear motion. The natural frequency ν_0 is determined by the restoring force in the relation

$$\nu_0 = \frac{1}{2\pi} \sqrt{\frac{1}{m}\left(\frac{d^2 V}{dr^2}\right)_{r_0}}$$

Since the quantity $(d^2V/dr^2)\, r_0$ gives the curvature at the point $r = r_0$, we may draw the conclusion that greater the curvature at the equilibrium position, greater is the natural frequency of the nuclear motion and higher are the corresponding energy levels. As the above consideration holds good for small amplitudes only, low quantum numbers alone are involved.

For the more highly excited states with large quantum numbers, the deviation of the curve from the parabolic form does not allow the vibration to be treated as a harmonic oscillator. As the quantum number increases, the energy levels crowd more and more closely together up to the so-called *convergence limit*. This limit corresponds to the dissociation of the molecule; it requires a quantity of energy equal to the depth of the potential hollow below the zero level. When the molecule is excited by this or a greater amount of energy, it splits up into atoms or ions which then fly apart with a velocity given by the excess energy. In molecular spectra, this state of affairs is manifested by a band system occurring with a convergence limit which is immediately followed by a continuum. Such a spectrum indicates, as in the case of atomic spectra, that the molecule absorbs vibrational energy in a *quantized form* only up to a certain limit, equal to the energy of dissociation, beyond which the absorption of energy is continuous, the excess energy being converted into kinetic energy of the separated atoms, which is not subject to quantization.

Determination of the heat of dissociation
The existence of a convergence limit immediately followed by a continuum in molecular spectra offers a method of determining the heat of dissociation much more accurately than the thermal methods used in chemistry.

The absorption spectra of molecules are frequently observed to consist of a number of groups of lines which crowd together in the direction of increasing frequency and then pass over into a region of continuous spectrum.

The frequency v where the continuum starts gives the energy *hv* necessary to dissociate the molecule. Since the heat of dissociation is usually referred to *normal* atoms as dissociation products, corrections will have to be made according as one or both atoms are in excited states. The heat of dissociation is also ordinarily referred to one gram molecule. In this connection it is useful to remember that

1 electron volt = 8106 cm^{-1}.
From Avogadro's number
N = 6.064 x 10^{23}
we derive also
8106 cm^{-1}= 23,055 cal./ gm.mol.

In practice, the interpretation of data is not always simple, because the continuous absorption region is observed only in one type of molecular spectra, which involves an alteration of the electron configuration, thereby indicating that the frequency v of the limit of continuous spectrum corresponds to the dissociation of *excited* molecules. Hence, it is not possible, without further information, to say what the state of the products will be. In any particular case, it is necessary to determine the nature of the dissociated atoms, *i.e.,* normal or excited, by a principle, known as *Franck- Condon principle,* with which we shall deal presently. In the case of many *atomic* compounds, such as H_2, N_2, O_2, etc., the molecule breaks up into one normal atom and one excited atom.

Taking H_2 as an example, the region of continuous spectrum is observed to commence at 850 A° and the corresponding value of the heat of dissociation is 334 K cal. Since one of the atoms is excited, the excitation energy is to be calculated from the first member of the Lyman series, 1216 A°; it is found to be 234 K cal. The heat of dissociation into two normal hydrogen atoms is therefore 100, K cal., a value which is in fair agreement with that obtained by thermal methods (95-100 K cals).

In the case of *ionic* compounds, such as KCl, NaI, etc., it has been shown that the long-wave limit of the region of continuous absorption correspond to the energy required to dissociate the molecule into two *normal* atoms. For KCl, for instance, the observed value of λ, the long-wave limit, is 2800 A°, which gives for *hy* =101 K cal., agreeing with the value found by thermal methods (105 ± 2 K cal).

In cases where the convergence limit cannot be observed directly, as in emission spectra, Birge and Sponer have worked out a method of extrapolation, which determines the position of the limit fairly well. The progressive crowding together of the levels is expressed by an equation and the extrapolation is made for the energy at which the interval between successive levels becomes vanishingly small. This Method has been found to give satisfactory results.

Franck-Condon principle
Franck, in 1925, attempting to explain the mechanism of *photochemical dissociation, i.e.,* how a molecule could dissociate by absorption of light of suitable energy, deduced this important principle qualitatively by making use of potential energy curves. Condon was able to develop Franck's ideas quantitatively, first (in 1926) from the old quantum theory standpoint and later (in 1928) with wave mechanical principles.

The problem which Franck undertook to solve may be stated as follows:

In molecular spectra, three types can be distinguished:
(1) pure rotation spectra which lie in the far infra-red and are caused by quantum changes in the rotational energy,
(2) rotation-vibration spectra in the near infra-red, produced by quantum changes in both vibrational and rotational energies and

(3) electronic band spectra in the visible and ultra-violet, effected by quantum changes in the electronic, vibrational and rotational energies.

Now, when molecular spectra result from *light absorption,* the continuous absorption region, which is an indicator of the dissociation of the molecule, is observed only in the third type, *i.e.,* electronic bands, but not in the other two, viz., pure rotation or rotation-vibration bands.

Theoretically speaking, any adequate supply of energy should give rise to dissociation even by the increased rotation of the molecule which is responsible for the pure rotation spectrum or by increased vibration of the constituent atoms, which causes the rotation-vibration bands.

In practice, however, an electronic transition, which is responsible for the electronic bands, is required for the dissociation of the molecule *by* light absorption. Why cannot light-energy be absorbed as rotational or vibrational energy alone in sufficient quantity to dissociate the molecule ?

The solution proposed is as follows:

Molecules without an electric moment cannot be affected by the electromagnetic field of the incident light and hence their rotational energy cannot be increased by absorption at all.

Molecules with electric moment can be influenced by the incident light; but the selection rule governing the acquisition of rotational energy permits that only one rotational quantum can be absorbed, which is evidently not sufficient for the splitting up of the molecule.

Concerning vibrational energy, molecules with no electric moment cannot increase their vibrational energy by absorption of light for the reason given above; but a molecule with electric moment which may be regarded as an inharmonic oscillator (frequency varies with amplitude), can, in theory, absorb more than one vibration quantum, unlike in the case of rotation; but changes in vibrational energy of increasing magnitude diminish rapidly in probability, so that only the first few vibrational quanta are absorbed, which cannot apparently lead to dissociation.

The situation is altogether different in an electronic transition which is always accompanied by changes in vibrational and rotational energy, even in the case of molecules without electric moment. Considering first *vibrational energy,* by means of light absorption any number of vibrational quanta are transferred in one electronic transition, so that dissociation is realized. This can be proved by a consideration of the potential energy curves in the ground and excited states.

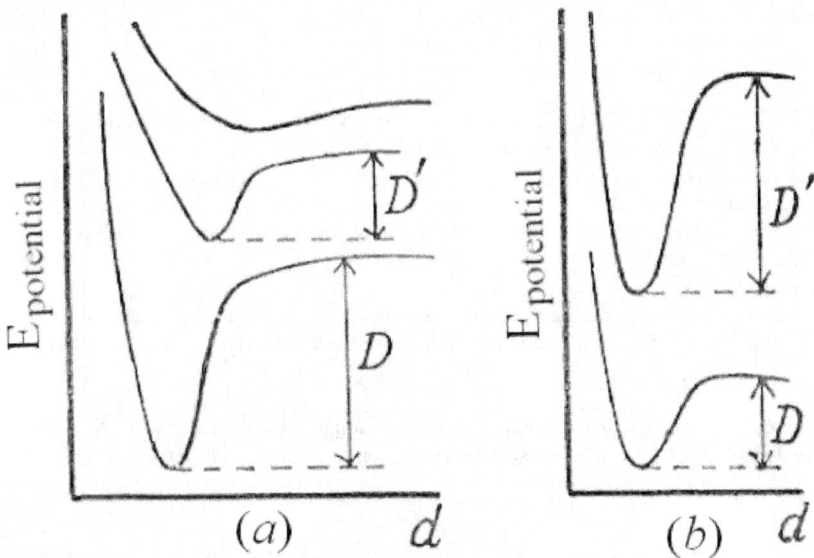

Fig. 249. Potential energy curves of diatomic molecules for ground and excited states.

In Fig. 249 (a), let the lowest curve represent the fundamental state. Absorption of light causes the molecule to pass into the excited electronic state represented by the next upper curve. Although the system of the *light* electrons is thereby altered, the relative positions of the *heavy* nuclei cannot alter during the transition. Hence in the act of absorption of light, the molecule passes from a state on the lower curve to a state *vertically above* on the upper curve. According to quantum mechanics, there is no state in which a molecule does not vibrate, but each molecule possesses a so-called *zero-point vibration.* Hence the internuclear distance cannot be determined by definite points but is represented by a probability curve. This means that the transition takes place, not merely from a given point on one potential curve to a given point on the other potential curve, but over a broad region with a maximum at the place where, according to ordinary mechanics, a transition alone would take place, *i.e.,* to the *nearest* quantized energy level: Further, since the vibrating nuclei spend a considerable part of their time at the turning points, where their velocity is least, it is more probable that the electronic transitions will take place from these points than from intermediate positions.

With this basic idea of Franck and Condon, the varying structure of molecular spectra can be readily understood. The potential curves given in Fig. 249 (a) are characteristic of covalent molecules, like H_2, O_2 and the halogens. The molecule is very stable in the fundamental state, but electronic transition is accompanied by a loosening of the linkage, *i.e.,* by a diminution of the heat of dissociation (D' < D). In the spectrum of these molecules, chiefly, those bands must occur, for which the change in vibrational quantum number is large. As the curve representing the higher state is rising steeply at a place where the lower curve is still flat, we may expect long series of bands leading from a lower state in which the vibrations are largely independent to a series of strongly vibrating upper states or even to a continuous region. If the transition takes place to a point on the upper curve which lies above the limit of dissociation, the molecule dissociates on absorption of light, which is shown spectroscopically by a purely continuous absorption region.

Fig. 249 (b) shows the opposite case, in which the linkage is strengthened by electronic excitation (D' > D). Such a case occurs with NO, HgH molecules which have a doublet band spectrum, indicating the presence of an unpaired electron that thus seems to weaken the linkage.

As regards *rotational energy* associated with electronic transition, a molecule can also increase its rotational energy to the extent of dissociation. The final decomposition may, however, arise from two causes:

(a) *Mechanical instability*
Oldenberg has given a strict criterion for the occurrence of mechanical instability, viz., when the rotations are large and the energy of dissociation small, and has applied it to Mercury hydride. The bands corresponding to transitions from

higher electronic states to the fundamental state all break off at a definite rotational quantum (different for the different vibrations) of the lower state. These rotational levels all lie much higher than the energy of dissociation of the state, as can be shown from the potential curves.

(b) *Pre-dissociation*

This term was first introduced by Henri, in 1924, to describe a phenomenon observed in the absorption spectrum of sulphur (S_2). The rotational fine structure became suddenly diffuse at shorter wavelengths. This means that for values of vibrational energy equal to or greater than a certain critical value, quantization of rotational energy ceases or becomes imperfect. Further researches by Henri and others showed that:

(i) the occurrence of pre-dissociation in absorption spectra is common with both simple and polyatomic molecules,

(ii) pre-dissociation may set in suddenly or gradually,

(iii) in many cases more than one region of pre-dissociation occurs and

(iv) the limit of pre-dissociation gives the upper limit for the energy of dissociation of the molecule. These points have been satisfactorily interpreted on the basis of the Franck-Condon principle, as due to *intersections of the potential curves.*

Franck-Condon principle is not limited to the optical dissociation of molecules. It has been applied also to chemical binding, intensity distribution in both absorption and emission spectra, continuous spectra, etc., into the details of which we cannot enter here.

2. MOLECULAR SPECTRA

Introduction

Molecular or *band spectra* are produced by molecules, *i.e.,* when the emitting substance is in the molecular state, as distinct from *atomic* or *line spectra* emitted by atoms, *i.e.,* when the emitter is in the atomic state.

The name "band spectra" was given at a time when spectroscopes of low resolving power alone wore available. With such instruments, the molecular spectra appeared as strips of graded intensity giving the impression of a continuous spectrum but divided into several bands.

The intensity in each band falls off from a definite limit called the *band-head;* the end of the weaker intensity is obviously hazy, so that the general appearance is often of channeled or "fluted" bands, on account of which molecular spectra are sometimes called also *fluted spectra.*

The fact that spectra of this typo are really due to molecules is readily established by the disappearance of the bands when the emitting substance is heated to temperatures at which the molecules dissociate into atoms.

With high resolving power instruments, molecular spectra disclose a three-fold structure:

(i) each band is composed of *a large number of lines* which are arranged With great regularity, often stretching in both directions from a gap in the band and crowding together at the band-head (which shows that the name 'line' spectra,. used to designate atomic spectra alone, is not quite correct);

(ii) several bands are found to follow one another in a regular sequence, constituting *a group of bands;*

(iii) a close and regular arrangement of different groups of bands, forming a *band system.* Several such band systems are comprised in a complete molecular spectrum; sometimes adjacent bands overlap and obscure the regularity of arrangement of lines, which makes the analysis very difficult.

Many substances are known to produce band systems in different spectral regions, *viz ,* the far infra-red (about 150μ to 30μ), the near infra-red (about 5μ to 1μ) and the visible and ultra-violet (7000 A° to 1000 A°). These three systems of bands, according to the accepted ideas of the mechanism of their production, are called:

1. *pure rotation bands* in the far infra-red, caused by the rotations of the molecules,

2. *rotation-vibration bands* in the near infra-red arising from the vibrations of the atoms inside the molecules, upon which are superposed the rotational motions mentioned above,

3. *electronic bands* in the visible and ultra-violet region, connected with electronic transitions, over which are superposed molecular vibrations and rotations.

Naturally, the pure rotation bands are the most simple, while the rotation-vibration bands are more complicated and the electronic bands the most complex.

Experimental study

The *pure rotation bands* in the far infrared were first noticed in the absorption spectra of the hydrogen halides. Czerny found in the case of HCl a series of seven absorption maxima in the region 120μ. to 44μ and was .able to show that the observed absorption lines could be represented by the relation:

$$\bar{v} = 20.8411\ m - 0.001814\ m^3 \quad, (m = 4, 6, 7, 8, 9, 10 \text{ and } 11)$$

For the lines of the other hydrogen halides also, he found similar relations:

HBr,
$$\bar{v} = 16.7092\ m - 0.001457\ m^3 \quad, (m = 5, 6, 7, 10, 12, 13 \text{ and } 14)$$

HI,
$$\bar{v} = 12.840\ m - 0.00082\ m^3 \quad, (m = 6, 7, 8, 9)$$

These relations show that the group of lines that constitute the rotation bands are separated by an approximately constant frequency difference, since *m* assumes successive integral values. The missing integral numbers for *m* correspond to lines not observed. The second term (in m^3) indicates that the frequency interval is not quite constant. The line for HI with $m = 1$ should lie at a wavelength of nearly 1 *mm., i.e.,* in the microwave region, as may be seen from the formula.

The molecules of water and ammonia also show complicated absorption spectra in this region. With the recent introduction of the microwave spectrometer, it has now become comparatively easy to observe and analyze the pure rotation bands in many cases and interesting results have been obtained.

The *rotation-vibration bands* in the near infra-red are also observed in the absorption spectra of the hydrogen halides. The general structure of a typical band of this sort (HCl) is shown in Fig. 250.

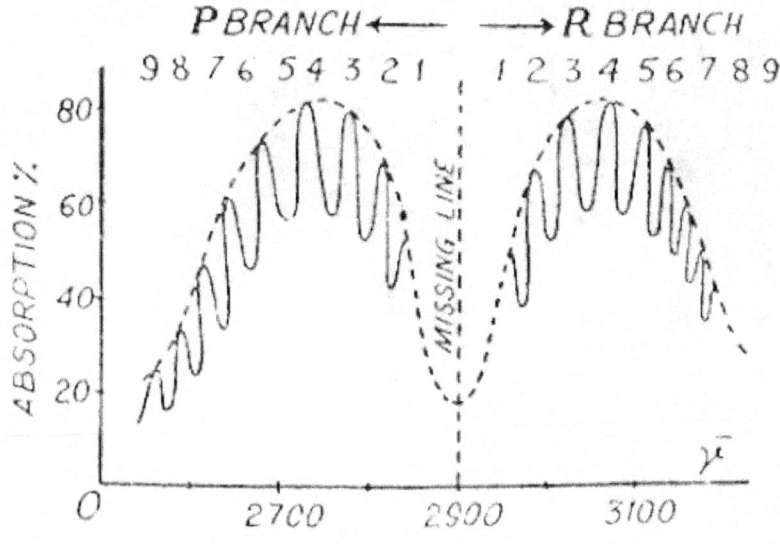

Fig. 250. Rotation-vibration band of HCl.

The form is seen to be double, the so-called *missing central line.* This feature of a double band with a central gap distinguishes the rotation-vibration spectra from the pure rotation spectra which have only a single band with no gap. The Maxwellian distribution of the lines on either side of the gap may also be noted; the corresponding lines in the two branches have approximately equal intensity.

The wave numbers for the maxima of the individual lines are well represented by the formula:

$$\bar{v} = 2886.2 + 20.5379\ m - 0.303181\ m^2 - 0\text{-}001814\ m^3$$

The first term indicates the region in which this type of spectra appears (2886 cm^{-1}), i.e., near infra-red; the second term gives the approximate distance between the lines, (20.5 cm^{-1}) which is nearly the same as the distance between the lines of the pure rotation spectra (20.8 cm^{-1}). This fact argues to the bands in the near infrared being due both to vibrations and rotations of the molecules— hence the name rotation-vibration bands. The other two terms are correction factors which show that the distance between the lines is not a constant, but decreases as we pass to higher wave numbers. The lines which are on the higher wave number side of the central missing line constitute the R branch, while those on the lower wave number side the P branch.

Rotation-vibration bands are found also with substances other than the hydrogen halides, but their structure is often more complex with no gap in the centre of the band. Thus the absorption spectrum of CO_2 shows many rotation-vibration bands from 1.46μ to 15μ and that of water vapor in the region 0.69μ to 6.26μ. These bands together with the rotational lines of water vapor are responsible for the marked absorption of the earth's atmosphere in the infra-red.

In order to exhibit rotation-vibration and pure rotation spectra, a molecule must possess an electric moment. Hence homonuclear diatomic molecules, such as H_2 , O_2, N_2, etc., which possess no moments, have no spectra of these two types. Gases composed of such molecules are entirely transparent in the infra-red region.

The *electronic bands* occur in the visible and ultra-violet regions both in emission and absorption. Probably the best known case of this type is the cyanogen (ON) bands in the red and violet regions, which appear in the spectrum of an electric are between carbon electrodes.

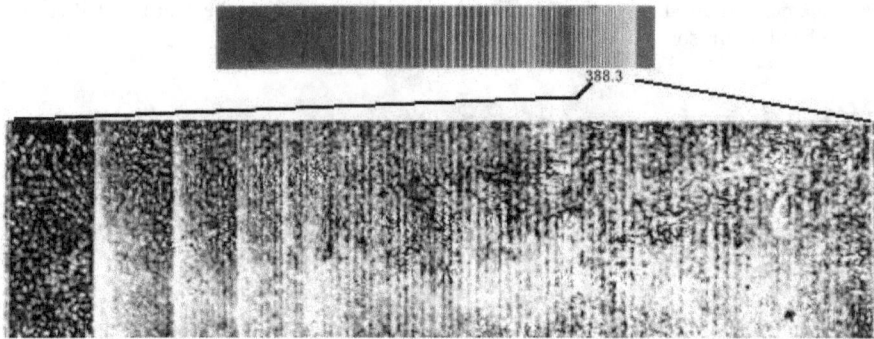

Cyanogen bands at about λ = 3883 A° (Graff)

The cyanogen bands in the violet region obtained by Graff.

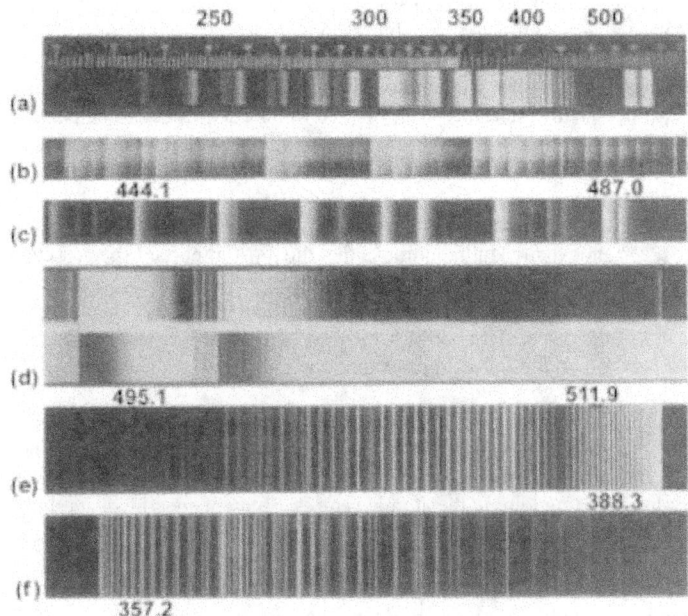

In the complete electronic band system there are several groups of bands, each group containing, in turn, several partial bands, sharply defined and brightest on the long wavelength side and fading away towards shorter wavelengths. Under greater resolution, the partial bands are found to be composed of a large number of lines of regular structure but crowded together at the band-head. Among these lines it has been possible to distinguish three branches, known as the P, Q and R branches, the central Q branch being absent in many bands. The distance between the partial bands is of the order of magnitude of the vibrational frequency of the molecule, while the spacing of the lines in the bands that of rotational frequency. It is, therefore, assumed that the bands are rotation-vibration bands associated with a simultaneous alteration of the electronic configuration of the molecule. The relatively large energy change involved in electronic transition determines the general position of the type, *viz.*, in the visible and ultra-violet and has suggested the name electronic bands.

The empirical relation discovered by Deslandres in connection with these bands is of the form

$$\bar{v} = A + B\,m + C m^2$$

where

$\bar{v}$ is the wave number of a line in a partial band,
A the wave number of the band-head,
B and C constants characteristic of the band and
$m = 1, 2, 3$, etc.

Theoretical explanation
The first step towards ordering band spectra and representing them by formulae was taken by Deslandres in 1885. The above formula, which he obtained empirically by considering a great mass of experimental data, became the model of all latter developments, just as Balmer's formula was the archetype of all series expressions in atomic spectra.

Bjerrum and Schwarzchild, in 1912, provided the foundation on which the empirical formula of Deslandres could be built and justified on the basis of the quantum theory. We are indebted to Imes and Heurlinger for testing the Bjerrum-Schwarzchild theory with actual experimental data.

Lenz improved upon Heurlinger's work of verification by taking a more comprehensive theoretical point of view. Finally, detailed papers on the band spectra of diatomic molecules by Kratzer closed the first phase of the theoretical investigation.

In the second phase of development, the problem of dependence of band structure on the electron configuration has been successfully attacked. Mecke and more particularly Mullikan were able to apply time ideas of the multiplet theory of atoms to molecules. The electron spin of the molecule determines the singlet, doublet, etc, character of time hand lines; there exists an alternation law for molecules exactly as for atoms, according as the number of electrons in the molecule is even or odd. Besides the electron spin, also the nuclear spin shows itself in the sequence of intensities of the band lines. Hund has set up general schemes for the coupling of the spin or orbital moments with the inner-nuclear axis in the case of diatomic molecules.

In the more recent and actual stage of theoretical interpretation, the more complicated polyatomic molecules have been analyzed by quantum mechanical methods to a certain extent. Mullikan thinks that the most important spectroscopic problems no longer lie in the realm of atomic spectra, but in that of molecular spectra.

Here, we shall limit ourselves to a simple theory of the molecular spectra, which explains the essential features of the three types stated above. The theory proposed is based on the following assumptions:

(i) Diatomic molecules alone, such as HCl, CN, etc., are considered. A diatomic molecule is supposed to be constituted with two atoms separated by a distance which is large in comparison with the atomic dimensions. The molecule as a whole can rotate about an axis passing through the centre of gravity of the system. There can also be relative vibratory motion of the two atoms inside the molecule along the line joining the atomic centres. Study of specific heats shows that at sufficiently high temperatures, in addition to the translatory motion of the molecules which gives the thermal energy, new degrees of freedom appear, such as rotation and vibration, mentioned above.

(ii) The total internal energy E of the molecule is assumed to be made up of three parts:
(a) rotational energy E_r due to the rotation of the molecule as a whole,
(b) vibrational energy E_v due to the relative vibrations of the atoms in the molecule and (c) electronic energy E_e due to the orbital motions of the electrons associated with the molecule. Hence
$E = E_e + E_v + E_r$.
From the spectral regions in which the three types of molecular spectra occur, it is readily seen that
$E_e \gg E_v \gg E_r$.

Here we have not included the mutual energy of the electrons and the intrinsic energies of the nuclei. Ordinarily these contributions have little influence on the molecular spectra. We shall have to consider the intrinsic energies of the nuclei when we come to study the isotopic effect in band spectra. Further, there are molecules, such as H_2,. where the effect of the magnetic moments of the nuclei becomes the dominating factor, as we shall see when we deal with nuclear spin.

(iii) The same quantum laws, as used in the interpretation of atomic spectra, are assumed to hold good also in the case of molecular spectra. Hence, the three components of the molecular energy can have only discrete, quantized values. Corresponding to each of the three types of energy, there is, therefore, a quantum number, m_e, m_v, and m_r, changes in which play the same role in band spectra as does the quantum number n in atomic spectra. Further, Bohr's frequency condition
$h \nu = W_1 — W_2$
is also adopted.

In framing the theory, we shall start with the simplest case of pure rotation bands, then proceed to the more complicated rotation-vibration bands and finally consider the most complicated system of electronic bands.

Theory of the pure rotation spectra
Considering a diatomic molecule to rotate as a whole about the axis passing through its centre of gravity.
Let ω be its angular velocity and I its moment of inertia about the axis of rotation.
Let m_1 and m_2 be the masses of the two atoms,
Let r_1, and r_1 their distances from the axis of rotation and d the distance between the centres of the atoms.

A simple theorem on centres of gravity gives
$m_1 r_1 = m_2 r_2$.
from which it can be shown that

$r_1 = d \left[m_2 / (m_1 + m_2) \right]$
and
$r_2 = d \left[m_1 / (m_1 + m_2) \right]$

The moment of inertia I is given by
$\mathbf{I} = m_1 r_1^2 + m_2 r_2^2$
Substituting the values of r_1 and r_2 and simplifying, we get

$I = d^2 \left[m_1 m_2 / (m_1 + m_1) \right] = \mu d^2,$

where μ, replacing the quantity within the double bracket, is known as the "resultant" or "reduced' ' mass.

The angular momentum p_φ of the rotating molecule = Iω.

Applying the quantum condition to this angular momentum,
$p_\varphi = m_r h/2\pi,$
where m_r is the *rotational quantum number.*

The angular velocity ω is given by

$\omega = p_\varphi / I = m_r h / 2\pi I.$

The kinetic energy of the rotating molecule is

$E_r = (1/2) I \omega^2 = (1/2) I (m_r h/2\pi I)^2 = (h^2/8\pi^2 I). m_r^2$

The rotational energy is therefore quantized, capable of assuming only the discrete values obtained by putting

$m_r = 0, 1, 2, 3$, etc.

When the energy changes from E_{r1} corresponding to a quantum state m_{r1} to corresponding to E_{r2} corresponding to a quantum state m_{r2}, a quantum of energy $h\nu_r$ is absorbed or emitted, which is given by

$h\nu_r = E_{r1} - E_{r2} = (h^2/8\pi^2 I) (m_{r2}^2 - m_{r1}^2)$

Hence, the frequency ν_r of the spectral line is given by
$\nu_r = (h/8\pi^2 I) (m_{r2}^2 - m_{r1}^2)$
$\quad = (h/8\pi^2 I) (m_{r2} + m_{r1}) (m_{r2} - m_{r1})$
$\quad = (h/8\pi^2 I) (m_{r2} + m_{r1}) \Delta m_r$

where $\Delta m_r = (m_{r2} - m_{r1}) = \pm 1,$
the + ve sign corresponding to absorption and the — ve sign to emission.

Considering only the case of absorption,

$\nu_r = (h/8\pi^2 I) (2 m_r + 1)$ ………………….. (1)

dropping the subscript.

This relation (1) predicts a spectrum of equally spaced lines with frequencies equal to integral multiples of a fixed quantity $(h/8\pi^2 I)$ and a constant frequency interval

$d\nu_r = h/4\pi^2 I.$

Wave mechanics leads to the same result, except that
m_r is to be replaced by $[m_r (m_r+1)]^{1/2}.$

On this basis, the energy of a given rotational state is

$(h/8\pi^2 I) [m_r (m_r +1)]$
$v_r = (h/8\pi^2 I) (2 m_r \pm 1)$
$dv_r = h/4\pi^2 I.$

The validity of the theory may now be tested.

According to Czerny's empirical relation for HCl, the lines in the pure rotation bands are given by $\bar{v}$ = 20.8 m. (cm^{-1}), omitting the small correction factor. In terms of frequency, since $\bar{v}$ = v/c, v_r = 20.8 $m \times c$.

As m stands for successive integers, it is readily seen that

$dv_r = h/4\pi^2 I = 20.8$ c
From this relation, I can be computed.

$I = h / (4\pi^2 \times 20.8c)$
$= 6.55 \times 10^{-27} /\{4\pi^2 \times 20.8 \times 3 \times 10^{10})$
$= 2.66 \times 10^{-40}$ gm. cm^2

[*Note.* The moment of inertia of the molecule can thus be determined from measurements on the molecular spectra.]

Now, since $I = d^2 [m_1 m_2 / (m_1 + m_1)] = \mu d^2$, where m_1 and m_2 are the masses of the two atoms, in the case of HCl, knowing the mass m_1 of the H atom and the mass m_2 of the Cl atom, d can be evaluated. As the mass of a single atom is its atomic weight divided by Avogadro's number (6.66 $\times 10^{23}$),

$m_1, = 1 / 6.06 \times 10^{23} = 1.65 \times 10^{-24}$ gm.
and
$m_2 = 35.5 / 6.06 \times 10^{23} = 5.86 \times 10^{-23}$ gm.

$d = [I (m_1 + m_2)/m_1 m_1]^{1/2}$
$= [2.66 \times 10^{-40} \times 60.25 \times 10^{24} / 1.65 \times 58.6 \times 10^{-48}]^{1/2} = 1.286 \times 10^{-8}$ cm.

Thus, the distance between the centres of two atoms H and Cl in the HCl molecule is of the order of 10^{-8} cm., which is in good agreement with the value of the molecular diameter obtained by other methods.

Hence the interpretation of the far infra-red spectra as due to the rotation of molecules is correct. The validity of the application of the general quantum conditions to the rotational transitions of molecules is also established.

Explanation of the correction term that appears in the empirical relation

As the velocity of rotation increases, the centrifugal force must increase the distance d and hence also the moment of inertia I, whereas in the theory proposed above, I is assumed to be constant and independent of the angular velocity ω. As a result of the variation of I, the energy levels corresponding to the higher m_r values must lie somewhat lower than that given in the theory. The lines corresponding to transitions involving higher m_r values must, therefore, have somewhat lower wave numbers than they would if I were constant. Actual calculation shows that the diminution in the wave number is proportional to m_r^3. This is why the correction term in the empirical relation contains m_r^3, m obviously standing for m_r.

Theory of the rotation-vibration spectra
In general, a diatomic molecule may rotate and vibrate at the same time, so that the total internal energy is partly rotational and partly vibrational:

$E_{v-r} = E_v + E_r$

Any given change of state may involve a simultaneous change of both types of energy, *i.e.*, the energy hv of a given radiation emitted by the molecule may come partly from the vibrational energy and partly from the rotational energy of the molecule.

Considering first the vibrational energy, it is evidently that of an oscillator and its value, according to wave mechanics, is given by

$$E_v = (m_v + 1/2) \, hv_0$$

in the case of a *harmonic oscillator*, (*i.e.*, frequency is independent of the amplitude) where m_v is *vibrational quantum number* which can take values 0, 1, 2, 3, etc., and v_0 the fundamental (natural) frequency of vibration.

Actually, the molecular vibration is *inharmonic*, (*i.e.*, frequency varies with the amplitude) and the expression for E_v is much more complicated as

$$E_v = (m_v + 1/2) \, hv_0 - (m_v + 1/2)^2 \, x \, hv_0 + \ldots\ldots$$

where x is a measure of the inharmonicity.

Assuming the simplest case of harmonic vibrations, transition of radiant energy given by from one vibrational state m_{v1} to another m_{v2}, will cause the emission of radiant energy given by

$$hv_v = hv_0 \, [(m_{v1} + 1/2) - (m_{v2} + 1/2)]$$
$$= hv_0 \, (m_{v1} - m_{v2})$$

Therefore,
$$v_v = v_0 \, (m_{v1} - m_{v2}) \ldots\ldots\ldots\ldots\ldots\ldots\ldots\ldots\ldots\ldots\ldots\ldots (2)$$

Upon the vibrational motion is superposed the rotation of the molecule; hence the frequency of the individual lines in the rotation-vibration band is given by

$$v = [\, (E_{v1} + E_{r1}) - (E_{v2} + E_{r2})] \, / \, h$$

where
E_{v1} and E_{r1} are the vibrational and rotational energies of the initial state and
E_{v2} and E_{r2} similar quantities of the final state.

Therefore,

$$v = (E_{v1} - E_{v2}) \, / \, h + (E_{r1} - E_{r2}) \, / \, h$$

Using equations (1) and (2),

$$v = v_0 \, (m_{v1} - m_{v2}) + (h/8\pi^2 \, I) \, (2 \, m_r + 1)$$
$$= v_v + v_r \ldots\ldots\ldots\ldots (3)$$
where
v_v is the contribution to the frequency from vibration and
v_r from rotation.

Thus we obtain an expression for the frequency of the lines in the rotation-vibration bands.

According to relation (2), the frequencies of the vibrational lines are integral multiples of a fundamental value. Actually this is found to be only approximately true. The reason for the discrepancy lies in the departure of the vibrations from pure harmonicity.

In the case of inharmonic oscillator, the Fourier analysis contains terms of all orders and hence transitions can occur in which m_{v1} and m_{v2} may have any integral value.

Taking the normal case of absorption in which the molecule in lower state ($m_{v2} = 0$) is raised to an upper state ($m_{v1} = 1$), it is seen from relation (2) that v_v is equal to v_o. This means that the position of the vibrational bands is determined by the mechanical vibrational frequency of the molecule, whose order of magnitude is such that the bands corresponding to $m_{v2} = 0 \rightarrow m_{v1} = 1$ will be found between 1000 and 4000 cm^{-1} for light molecules; this range is in the near infra-red where all the rotation-vibration bands are actually observed.

Since the vibrational energy is much larger than the rotational energy, every change in the former will be accompanied by a number of smaller changes in the latter. That is to say, for a given change in m_v, there will be many possible changes in m_r. In other words, for a given value of v_v, v_r has a series of values for the several integral values of m_r. Hence, in the rotation-vibration band, there will be a series of equidistant lines with a frequency interval
$dv = h/4\pi^2 I$.
This conclusion is in good agreement with experimental results, since, for instance, in the case of HCl, the frequency separation of the pure rotational lines in the far infra-red is practically the same as that of the rotation-vibrational lines in the near infra-red (about 20.5 cm^{-1} in both the cases).

The correction terms in *the empirical formula*

As in the case of pure rotational lines, here also the frequency interval between the lines is not constant, but decreases as we pass to higher members. The explanation for this is to be sought for along the same lines as in the case of pure rotational lines, viz, due to the alteration of the moment of inertia of the molecule.

The mean moment of inertia of a vibrating molecule must be expected to differ from that of the molecule not vibrating.

Moreover, the centrifugal force which comes into play during rotation must alter the strength of the linkage between the atoms and so must affect the vibrations. There must be therefore an interaction between rotation and vibration. This can, in fact, be calculated and the observed variations in the spacing of the lines can be well accounted for.

The central missing line and the P and R branches

In relation (3), even if we put $m_r = 0$ in the second term, this term does not vanish, but is equal to $h/8\pi^2 I$. Hence

it is impossible to have v equal to v_v alone, which means that the molecule cannot alter its vibrational energy without a simultaneous alteration of its rotational energy. This is the reason why the central line corresponding to $v = v_v$ is missing.

Changes in rotational energy for which m_r increases by unity give rise to an array of lines extending towards greater frequencies— the so-called positive R branch. Changes for which m_r decreases by unity are responsible for the other set of lines extending towards lower frequencies—the negative P branch.

Intensity distribution

We have seen that the observed intensities of the lines first increase with increasing value of m_r in both the branches and then fall off again. This is explained as follows:

Just as in the case of translational energy, where at each temperature there is a most probable velocity, so also there is a most probable rotational energy for each temperature; all the other rotational energies, larger or smaller, occur less frequently, The smaller the distance between the energy levels, the higher is the value of m_r corresponding to the most probable energy. Now, since the intensity of the lines is proportional to the frequency of the initial state in question, the higher the state, the farther will be the line of maximum intensity from the central gap; also lines corresponding to same values of m_r on either side will be equally intense.

This distribution of the rotational energy is also the reason why molecules do not rotate about the line joining the nuclei as axis. For, the moment of inertia I_x about this axis is extremely small in comparison with the moment of inertia I which refers to the perpendicular axis through the centre of gravity of the molecule. Hence, the rotational energy corresponding to I_x would be so high even in the lowest state ($m_r = 1$) that any appreciable number of molecules could be found in this state only at extremely high temperatures, *i.e.,* under conditions where the molecules would probably no longer be stable, but dissociate.

Thus the theory proposed gives a satisfactory explanation of all the observed facts about rotation-vibration spectra.

Theory of the electronic band spectra

Although this type is the most complex of molecular spectra, its explanation becomes relatively easy after what has been said about the other two types. Here the total energy of the molecule in a given quantum state is made up of all the three components of energy, viz., electronic, vibrational and rotational ($E_e + E_v + E_r$). Transitions between two energy states, for which

$$h\nu = (E_{e1} + E_{v1} + E_{r1}) - (E_{e2} + E_{v2} + E_{r2})$$

are those which give rise to the electronic band spectrum. The frequency of the individual lines in it is given by

$$\nu = (E_{e1} - E_{e2})/h + (E_{v1} - E_{v2})/h + (E_{r1} - E_{r2})/h$$
$$= \nu_e + \nu_v + \nu_r \quad \text{.......................(4)}$$

the first term representing the contribution to the frequency of a given line due to the change in the electronic energy, the second that due to the change in the vibrational energy, and the third due to rotational energy changes.

Since the electronic energy is of a greater magnitude than the vibrational or rotational energies ($E_e \gg E_v \gg E_r$), the position of the spectrum is determined by the change in E_e, which, as we know from atomic spectra, is in the visible and ultra-violet regions.

Further, every single change in E_e will be accompanied by a number of changes in E_v according to the many possible values of ($m_{v1} - m_{v2}$), which accounts for the existence of a number of partial bands in a group and of different groups in a band system, as we shall see presently. Each of the vibrational transitions will, in turn, be accompanied by a variety of rotational changes, according as m_r assumes different integral values, giving rise to the rotational lines in each band. Thus, just as the pure rotation spectrum can be. superposed upon a vibrational frequency, so also the whole rotation-vibration spectrum can be superposed on an optical frequency due to an electronic transition. This process produces a band system. The manifold of all such band systems connected with different electronic transitions in a molecule represents a complete electronic band spectrum.

In order to explain satisfactorily the complicated structure of a *single band system* comprising of several *groups of bands* each of which is composed of several *partial bands,* each partial band itself containing *an array of lines,* the following two general remarks are of importance:

(i) The effect on the rotational and vibrational states due to alterations in the electronic configurations is very much greater than the interaction between vibration and rotation in the rotation-vibration spectra. In consequence, the molecule has to be considered to have different strengths of linkage and different moments of inertia in the initial and final states. In other words, all constants that depend on the details of molecular structure, in particular, the moment of inertia; I, and the fundamental frequency of vibration; ν_o, become different in the initial and final states.

(ii) Further, the electronic motions, in general, possess a resultant angular momentum which combines vectorially with the rotation of the molecule as a whole and thereby permits transitions in the vibrational state without a simultaneous alteration of the rotational state, unlike in the case of rotation-vibration spectra. This means that the transition $\Delta m_r = 0$ is not forbidden, so that the central line will not be missing in the electronic band spectrum.

We shall, for the sake of clearness, first consider the rotational structure of each partial band, then the vibrational structure of a band-group as well as of the band-groups forming a band system and finally the electronic structure of a complete band spectrum comprising several systems.

(a) *Rotational structure*
Denoting the rotational quantum numbers of the initial and final states by m_r and m_r' respectively and the moments of inertia of the two states by I and I' respectively, the rotational energy of the initial state
$$E_r = (h^2/8\pi^2 I) \, m_r \, (m_r + 1)$$

and that of the final state
$$E_r' = (h^2 / 8\pi^2 I') \, m_r' \, (m_r' + 1)$$

Hence the contribution to the frequency due to this change in rotational energy is given by

$$(E_r - E_r') / h = (h / 8\pi^2) [\, m_r (m_r + 1) / I \; - m_r' \, (m_r' + 1) / I' \,]$$

The frequency of the individual lines in a partial band is, therefore, from relation (4)

$$\nu = \nu_e + \nu_v + (h / 8\pi^2) [\, m_r (m_r + 1) / I \; - m_r' \, (m_r' + 1) / I' \,]$$

Putting $(h / 8\pi^2 I) = B$ and $(h / 8\pi^2 I') = B'$ and replacing $(\nu_e + \nu_v)$ which represents the contribution due to the changes of electronic and vibrational energies, with which we are not dealing now, by a,

$$\nu = a + [\, B \, m_r (m_r + 1) - B' \, m_r' \, (m_r' + 1)]$$

Taking into account the three permitted transitions, viz.,
$\Delta \, m_r = +1, \, 0, \, — 1$, we can distinguish three cases:

Case I:
For $\Delta \, m_r = + 1$ or $m_r' = m_r + 1$,

$$\nu = a + [\, B \, m_r' \, (m_r' - 1) \; - B' \, m_r' \, (m_r' + 1)]$$
$$= a + (B - B') \, m_r'^2 - (B + B') \, m_r'$$

Putting $B + B' = 2b$ and $B - B' = c$,

$$\nu = a + 2b \, m_r' + c \, m_r'^2$$

Case II.
For $\Delta \, m_r = 0$ or $m_r' = m_r$

$$\nu = a + [\, B \, m_r' \, (m_r' + 1) \; - B' \, m_r' \, (m_r' + 1)]$$
$$= a + (B - B') \, m_r'^2 + (B - B') \, m_r'$$
$$= a + c \, m_r' + c \, m_r'^2$$

Case III.
For $\Delta \, m_r = -1$ or $m_r' = m_r - 1$

$$\nu = a + [\, B \, (m_r' + 1) \, (m_r' + 2) \; - B' \, m_r' \, (m_r' + 1)]$$
$$= a + (m_r' + 1) [\, B \, (m_r' + 1) \, (m_r' + 1) + B \; - B' \, (m_r' + 1) + B']$$
$$= a + (B + B') \, (m_r' + 1) \; + (B - B')(m_r' + 1)^2$$
$$= a + 2b \, (m_r' + 1) + c \, (m_r' + 1)^2$$

Thus we get three different expressions for ν according to
$\Delta \, m_r = +1, \, 0,$ or $— 1$.

They are known as the P, Q, R branches respectively of the rotational lines (nomenclature due to Hearlinger).

The frequencies of the individual lines in these branches are, therefore, given by
$$\nu_P = a - 2b \, m_r' + c \, m_r'^2 \; \dots\dots\dots\dots\dots\dots \; (5)$$
$$\nu_Q = a + c \, m_r' + c \, m_r'^2 \; \dots\dots \; \dots\dots\dots \; (6)$$
$$\nu_R = a + 2b \, (m_r' + 1) + c \, (m_r' + 1)^2 \; \dots\dots \; (7)$$

[*Note.* The quantum number of the final state may assume any positive integral value, including zero, except in the case of the P branch, where the smallest value of m_r' is clearly 1 (transition $0 \rightarrow 1$), since $\Delta \, m_r = + 1$ or $m_r' = m_r + 1$].

Relations (5), (6) and (7) giving the frequencies of the lines in the three branches are very similar to the empirical formula of Deslandres:

$$v = A + B\,m + C\,m^2.$$

If the moments of inertia in the initial and final states were equal (which would be the case if there was no electronic transition, hence a case of rotation-vibration)

$$I = I',\ B = B'\ \text{and}\ B - B' = c = 0.$$

Under these conditions, relation (6) for the Q branch loses its meaning, since $v_Q = a$, while (5) and (7) are reduced to

$$v_P = a - 2b\,m_r'$$
$$v_R = a + 2b\,(m_r' + 1)$$

which show that the spacing of the lines in the two branches P and R is uniform, as is to be expected.

Actually, however, the moments of inertia in the initial and final states differ markedly in the case of electronic transition, so that $I \neq I'$, $B \neq B'$ and $c \neq 0$. Consequently the terms in $m_r'^2$ do not vanish in relations (5), (6) and (7) and their effect becomes very pronounced for the higher values of m_r'.

Fortrat diagram

The complex rotational structure of the three branches can he best studied by means of a graphical representation, first used by Fortrat and hence known as the *Fortrat diagram.* This is obtained as follows:

Certain fixed values are given to a, b and c; then assigning different integral values 0, 1, 2, 3 etc. to m_r', the corresponding values for v_P, v_Q and v_R are calculated. The values of v thus obtained are plotted against the corresponding m_r' values, taking the former as abscissae and the latter as ordinates.

Taking first the simple case of $I = I'$ and $c = 0$, the relations to be used are
$$v_Q = a,$$
$$v_P = a - 2b\,m_r'$$
and
$$v_R = a + 2b\,(m_r' + 1)$$

The Fortrat diagram for this case is shown in Fig. 251. $v_Q = a$ for all values of m_r'; but v_P and v_R, *i.e.,* the P and R branches extend asymmetrically on either side of the Q line, intersecting it at $m_r' = 0$ and -1 respectively. The corresponding rotational lines of the two branches are obtained by projecting the points of intersection of the two branches with the horizontals at $m_r' = 1, 2, 3$, etc. on the v axis; these lines, indicated below the Fortrat diagram, are equi-spaced and would represent the rotational lines of a rotation-vibration band.

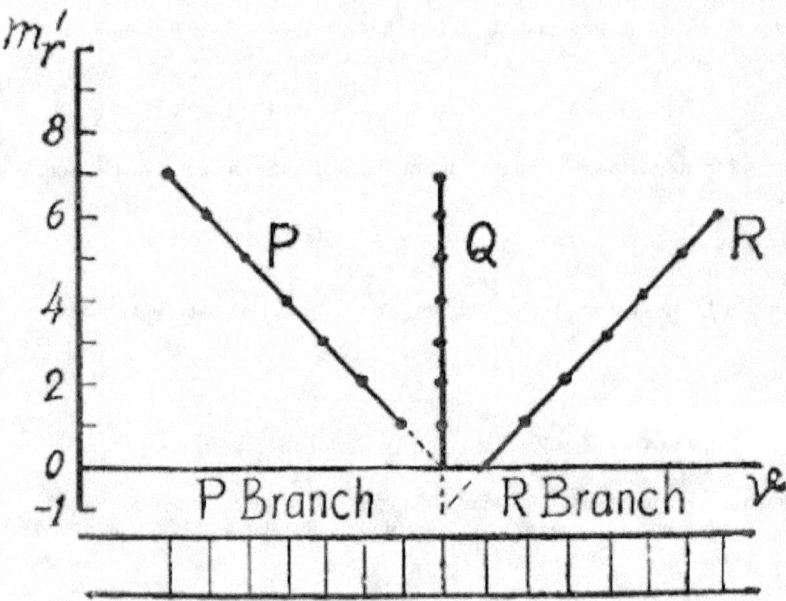

Fig. 251 Fortrat diagram (c=.0).

Considering the more complicated case, proper to electronic transition, i.e., where $c \neq 0$, two things might happen, *viz.* either $c < 0$, *i.e.,* negative or $c > 0$, *i.e.,* positive.

If $c < 0$, B - B' = c is negative, or B' > B or I > I', *i.e.;* the moment of inertia of the molecule in the initial state is greater than that in the final state. The relations to be used are:

$$v_P = a - 2b\, m_r' - c\, m_r'^2$$
$$v_Q = a - c\, m_r' - c\, m_r'^2$$
$$v_R = a + 2b\, (m_r' + 1) - c\, (m_r' + 1)^2$$

The Fortrat diagram corresponding to this case is shown in Fig. 252.

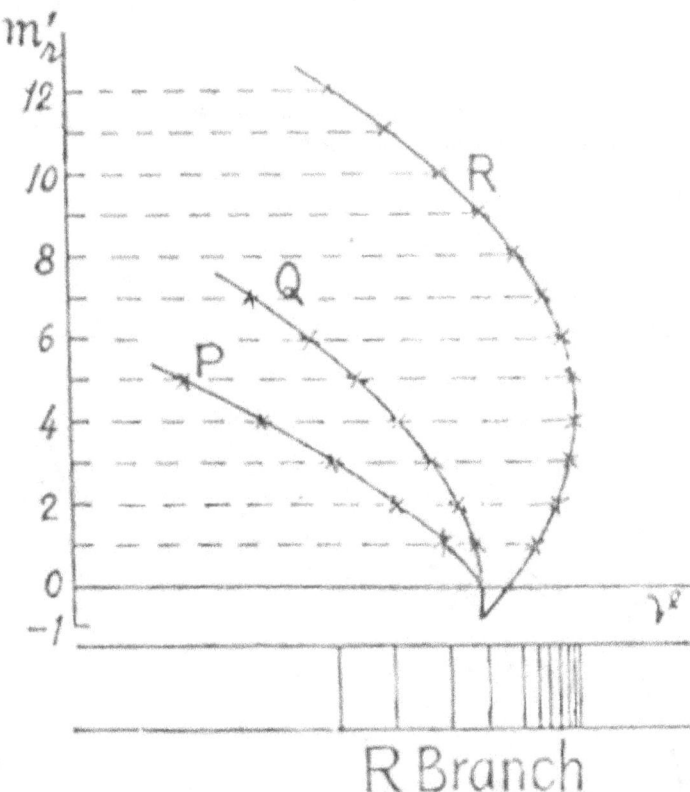

Fig. 252. Fortrat diagram *(0 < 0)*.

Three curves representing the three branches are parabolic in form, P and Q intersecting at $m_r' = 0$ and R and Q at $m_r' = -1$; the P and Q branches are always inclined in the lower frequency direction, while the R branch first rises in the higher frequency direction and then bends back towards the lower frequency at higher values of m_r'. This is to be expected from the nature of the relation used. Owing to the negative value of c and the preponderance of the quadratic terms for high values of m_r', the R branch is bound to bend back; for small values of $(m_r' + 1)$ the positive linear term $[+ 2b (m_r' + 1)]$ will predominate, but when m_r' reaches high values, the negative quadratic term $[- c(m_r'+1)'^2]$ will become more important, since $2b = B + B'$, while $c = B - B'$.

Projecting the points of intersection of the parabola of the R branch with the horizontals at $m_r' = 1, 2, 3$, etc. on the v-axis, we get the rotational lines of this branch, as shown below the Fortrat diagram. We see that there is a crowding of the lines at the right end, the so-called band-head. The band extends from this head in the direction of lower frequencies. The rotational lines of the P and Q branches, when drawn in a similar manner, will interpose themselves between the lines of the R branch. This type of band with the head in the R branch is said to be *shaded off towards the violet* (since, on account of the crowding of the lines at that side, the intensity will be great there, producing the impression of "shading") or *degraded towards the red* (on account of the diminution of intensity due to scarcity of lines on that side).

If, on the other hand, $c > 0$, *i.e.,* positive, $I < I'$ and the formulae show that the Q and R branches are always inclined in the higher frequency direction, while the P branch first rises in the lower frequency direction and then bends back towards the higher frequency at higher values of m_r', since the varying sign in the terms involving m_r' occurs in the relation

$$v_P = a - 2b\, m_r' + c\, m_r'^2$$

of the P branch and not in that of the R branch. The rotational lines of the branches can be drawn as in the previous case. Here the band-head which appears on the P branch is on the red side and the band extends towards the violet. Such a band is, therefore, *shaded off towards the red* or *degraded towards the violet*.

It is to be noted that the circumstance that the lines run partly towards the band-head and partly away from it apparently disturbs the regularity of the sequence of the lines. The band head, with its changing positions, is not of very fundamental significance.

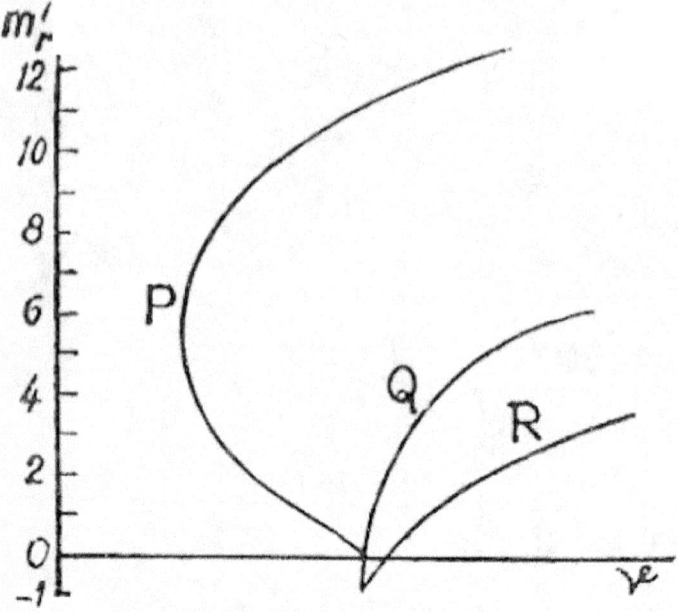

Fig. 253. Fortrat diagram {c > 0).

Distribution of intensity

In many electronic band spectra the maximum intensity lies at a m_r' value near to that at which the P or R branch turns back. Hence the intensity is at its greatest at the band-head. For this reason, measurements, such as the counting of the lines, were formerly referred to the band-head; but as we have remarked above, the position of the head has no great significance. Measurements are now made with reference to the zero *line*, i.e., the line for which $m_r' = 0$ in the Q branch and which is also called the *band origin*. As in the case of rotation-vibration spectra, here also Maxwellian distribution of intensity is to be expected and, in consequence, the minimum intensity should occur at the zero-point of the Q branch; on both sides of it, the intensity first increases and then decreases. If we count the lines from the band origin the intensities of the lines bearing the same number are equal in the R and P branches. But in the scale of frequency this symmetry becomes more or less unrecognizable. Since, in many cases, one of the two intensity maxima lies near the band-head, this produces the impression of strong channeling or "fluting", when the whole spectrum is viewed.

On account of the statistical nature of intensity distribution, the intensity of the rotational lines depends on the temperature. At low temperatures, only the levels of low quantum number are present in appreciable proportion; and at a given temperature, the preferred levels are lower, the greater the energy difference between them, i.e., lower for molecules with smaller moments of inertia. For this reason, in the H_2 molecule, even the first rotational level ($m_r' =1$) is relatively high, several hundred cm^{-1} as compared with about 20 cm^{-1} for HCl. Hence, at low temperatures only a very small proportion of H_2 molecules possesses even one rotational quantum. This means that rotation is practically absent and with it the corresponding degree of freedom. Thus, at moderately low temperatures, the H_2 molecule behaves, as regards specific heat, as a monatomic gas.

Testing the validity of the theory proposed

The cyanogen bands in the violet region provide a typical example for testing the theory of the rotational structure developed above. In these bands, the laws for the P and R branches are verified extremely well and the Q branch is not present, which simplifies the analysis very much.

Heurlinger has subjected the cyanogen bands to a very exact and close study. In several of the bands, approximately hundred lines in each can be counted. Taking the group of five bands at about $\lambda = 3800 A°$, the first band-head lies at 3884 A°, the second at 3871 A°, the third at 3862 A°, etc. The partial band belonging to each head resolves itself into a P and a R branch. Heurlinger has determined the zero lines belonging to the three band-heads mentioned above and has calculated the constants $2b$ and c; they are 3.843 cm⁻¹ and 0.0673 cm⁻¹ respectively. The value of c is very much less than that of b, as expected from the theory, since $b = (B + B')/2$, while $c = B — B'$. Further, the smallness of c indicates that B and B' are nearly equal, *i.e.,* the moments of inertia of the molecule in the initial and final states are nearly the same. Under these circumstances, it is sufficient to evaluate an average moment of inertia $I_m \approx I \approx I'$.

$B \approx B' \approx h /8 \pi^2 I_m c$, in wave number scale

Therefore,

$I_m \approx h /8 \pi^2 B c$
$= 6.55 \times 10^{-27} / [8 \pi^2 \times (1/2) \times 3.843 \times 3 \times 10^{10}]$
$= 1.44 \times 10^{-39}$ gm. cm²

Since the carrier of the cyanogen bands is the CN molecule, substituting the known masses of the C and N atoms in the relation
$I_m = [m_1 m_2 / (m_1 + m_2)] d^2$
we obtain $d = 1.16 \times 10^{-8}$ cm which is the distance between the atoms in the CN molecule. This is of the order of magnitude to be expected from other considerations. Thus the proposed theory is confirmed by experimental data.

(b) *Vibrational structure*
In order to explain the vibrational structure in a band system, we must consider the contribution due to changes in vibrational energy implied in the v_v component of relation (4). As already stated, under the influence of the alterations in the electronic configuration, the vibrations become markedly disturbed, so that the fundamental frequency v_o will be different in the initial and final states; likewise the coefficient x of anharmonicity will not be the same in the initial and final states.

Let m_o, v_o and x represent the vibrational quantum number, the fundamental frequency and the measure of anharmonicity respectively of the initial state;
m_o'. v_o' and x' are similar quantities referring to the final state.

The vibrational energy of the initial state is given by

$E_v = [(m_v + 1/2) — x(m_v + 1/2)^2] h v_o$

omitting higher terms of the series. Similarly, the vibrational energy of the final state is

$E_v' = [(m_v' + 1/2) — x' (m_v' + 1/2)^2] h v_o'$

Applying Bohr's frequency condition,

$$v_v = (E_v - E_v') / h$$
$$= v_0 \{ (m_v + 1/2) - x(m_v + 1/2)^2 \} - v_0' \{ (m_v' + 1/2)$$
$$- x' (m_v' + 1/2)^2 \} \qquad \ldots (8)$$

$$= v_0(m_v + 1/2) - v_0 x(m_v + 1/2)^2 - v_0'(m_v' + 1/2) + v_0' x' (m_v' + 1/2)^2$$

Adding and subtracting $v_0 \, m_0'$, we can arrange the different terms in the order of magnitude as

$$v_\nu = (m_v - m_v')v_0 + (m_v' + 1/2)(v_0 - v_0') - \{ (m_v + 1/2)^2 \, v_0 x$$
$$- (m_v' + 1/2)^2 \, v_0' x' \} \dots (9)$$

In this equation (9), *the first term* is the *principal* one which depends on the quantum transition of $(m_v - m_v')$ alone. It will have different values for the many possible values of m_v and m_v'.

If $(m_v - m_v') = \Delta m_v = 1$, we have the "fundamental" vibration;
If $\Delta m_v = 2$, the "first harmonic", and so forth. It is these different quantum transitions that give rise to the *different groups* of *bands* in a band system.

The *second term* is small compared with the first, since $(v_0 - v_0')$ is a small quantity; it depends upon the absolute value of the quantum number m_v' of the final state. By keeping the principal term fixed, *i.e.,* fur a given value of Δm_v and varying the value of m_v' we obtain a number of *partial bands in a particular group.*

The *third term is,* in general, still smaller, since the coefficients x and x' are individually small, being functions of the slight anharmonicitics in the linkage of the atoms in the molecule. It is this term which causes the *unequal spacing of the band groups, as well as of the band-hands in each group.*

Testing the validity of the theory proposed
In the violet system of the cyanogen bands, which has been subjected to close analysis, it has been possible to identify clearly *four groups,* the first three containing five band-heads each, while the fourth only three, as shown in Fig. 254.

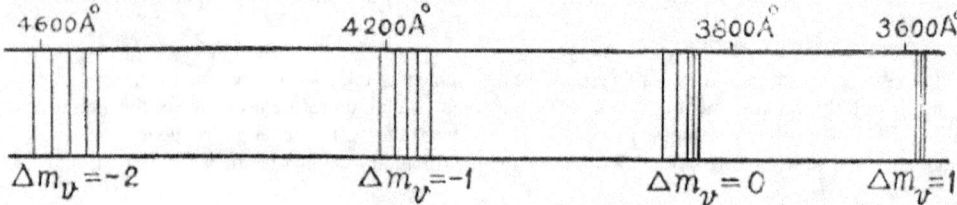

Fig. 254. Brand-heads of the cyanogen spectrum.

The four groups correspond to the quantum transitions $\Delta m_v = -2, -1, 0$ and $+1$ respectively, as predicted by the theory. The third group occurring at $\lambda = 3884$, which we have already considered in connection with the rotational structure and which corresponds to $\Delta m_v = 0$, contains five partial bands, for which, from left to right, the quantum numbers of the final state m_v' are 0, 1, 2, 3 and 4 respectively, again in agreement with the theory. Since $\Delta m_v = 0$ for this group, the five bands in it correspond to transitions $0 \to 0, 1 \to 1, 2 \to 2, 3 \to 3$ and $4 \to 4$. Considering the positions of the band-heads in it, viz., 3884, 3871, 3862, 3855 and 3850, we see that they are not equally spaced. The same is found for the band-heads in the other groups. Likewise the groups are not equally spaced either. The first group is at about $\lambda = 4550$ A°, the second at $\lambda = 4200$ A°, the third at $\lambda = 3880$ A° and the fourth at $\lambda = 3580$ A°. This unequal spacing of the groups as well as of the band-heads is due, according to the theory, to the anharmonicity of the linkage expressed by the third term of equation (9).

A more accurate and final test is obtained by considering the positions of the band origins or zero lines. For, in this case, the influence of anharmonicity does not come in and hence the regularity of the vibrational structure must appear. The position of the zero line in each of the partial bands can be accurately measured from the experimentally obtained band spectrum of cyanogen. On the other hand, the same may be calculated from the theory as follows:

Taking the Q branch, the frequency in it is given by the relation

$$v_Q = a + c \, m_r' + c \, m_r'^2$$

Now, since the zero line corresponds to $m_r' = 0$, the frequency of the zero line is given by

$$\nu_{zero\ line} = a = \nu_e + \nu_v$$

Substituting the value of v o from equation (S),

$$\nu_{zero\ line} = \nu_e + \nu_o\ [(m_v + 1/2) - x(m_v + 1/2)^2]\ -\nu_o{}'\ [(m_v{}' + 1/2) - x{}'\ (m_v{}' + 1/2)^2]$$

The constants ν_e, ν_o, $\nu_o{}'$, x and x' are chosen in such a way that there is perfect agreement with the most accurately measured zero line. Giving different values of m_v and $m_v{}'$ keeping one of them constant first and varying the other, then the other constant while the first is varied, it is possible to calculate the frequencies of all the zero lines of the different groups in the cyanogen band. Comparing these values with the experimentally measured ones, there is excellent agreement between the two. It is also found that the corresponding zero lines of the different groups are equally spaced, which brings out the regularity or equal spacing of the vibrational structure.

(c) *Electronic structure*
For a complete interpretation of the electronic band spectrum, we have still to consider the contribution to the frequency due to changes in the electronic energy. This can be done by the analysis of v e component in relation (4). This was achieved only long after the bands themselves and the band-heads had been investigated and it was Mullikan who was the first to start researches, both experimental and theoretical, on this point. Other workers, such as Mecke, Birge, Hund, Wood, and Heisenberg have collaborated in this work and have obtained very important and valuable results, We can only state them very briefly here:

(i) *Several band systems for one and the same molecule*
Since the electronic frequency of a molecule may assume several different values corresponding to different electronic transitions, the molecule can exhibit several band systems, which may even lie in quite separate spectral regions, like the violet and red cyanogen bands.

(ii) *Far-reaching analogy between molecular and atomic spectra*
This was first discovered, in 1925, by Mullikan. The analogy is found as regards:

(a) *electronic configuration, i.e.,*

the electrons of a molecule form one coherent system, as in the case of an atom, with closed shells containing a fixed number of electrons, as determined by Pauli's principle and the electrons outside the closed shells acting as the optical electrons of the molecule responsible for the observed band spectra.

But as already remarked the formation of a molecule from atoms with completed K shells leave these shells intact. Thus, for example, taking a diatomic molecule with 13 electrons, there are two closed K shells, one for each atom and a common L shell of eight electrons, so that the thirteenth electron would remain over as an optical electron. The spectra of such molecules must be analogous to the alkali spectra. That such is actually the case has been demonstrated by Mullikan, Birge and others with BeF, BO, CN, etc., molecules. Similarly, spectra of diatomic molecules with 21 electrons, *e g.,* MgF, AlO, SiN, etc., exhibit relationships with the alkali spectra. A direct experimental proof that the K shells of the atoms of these molecules are intact is furnished by their X-ray K spectra, which are practically identical with those of the free atoms.

(b) *multiplicity of spectral lines.* In many band spectra all the lines are found to be double; in some they are triple. This multiplicity is frequently an isotopic effect (as we shall see in the next section), but not always, since the multiplet structure is present even in cases where the isotopic effect is excluded. Hence the observed multiplicity must be due at least partly to the fine structure of the energy levels belonging to a given electronic configuration. Meeke was able to show that

molecular spectra also obey the alternation rule, so that with even number of electrons in the molecule the term multiplicity is odd and conversely.

The rules for a systematic treatment of molecular electron terms have been developed by Hund. In this connection is may be noted that the quantum numbers referring to the electronic structure of molecules are usually designated by Greek letters corresponding to those used for the atom (e.g., λ for l, σ for S term, π for P term) although the symbol s is retained for the electron spin.

(c) *fluorescence.* Like atoms, molecules also are found to be the agents in the fluorescence phenomena. This effect has been carefully investigated by Wood in the case of the iodine molecule.

PECULIARITIES OF CERTAIN BAND SPECTRA

(1) *The band spectra of elementary diatomic molecules*
A particular significance is attached to the band spectra produced by such molecules and among them that of hydrogen occupies a special position.

> The spectrum of the hydrogen molecule does not resemble the usual band spectrum, but appears as a *"many-line spectrum",* stretching from the infra-red to the ultra-violet, the band-heads being absent altogether.

The *Fulcher bands* have been known the longest of all, one in the red and one in the green, both having only a few lines: four bands discovered by Croze have about twelve lines each. But the sequence of the lines in the bands is so widely separated that the lines appear, at first sight, to have no relation among themselves. Merton, however, has tested these lines as regards their behavior with varying pressure, temperature, etc. and shown that they belong to the same bands.

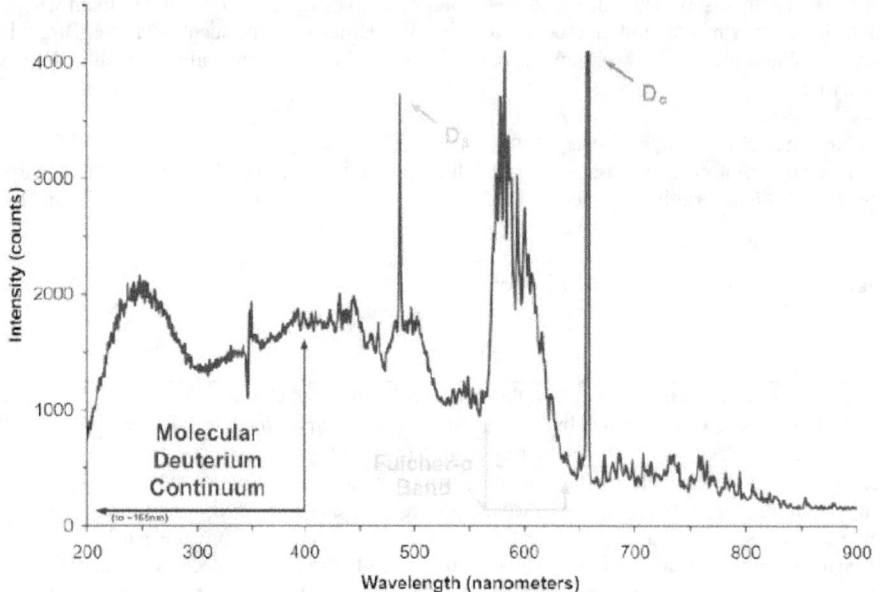

Spectrum of a deuterium lamp.
Balmer lines are marked with a "D" since this is deuterium not hydrogen (deuterium alpha on the right at 656 nm and the deuterium beta line on the left at 486 nm). The continuum emission on the far left from 400 nm to 200 nm does not actually decrease in intensity from 250 nm to 200 nm and this is an artefact of the decreased CCD photodetector efficiency at those wavelengths. The continuum actually increases in intensity all the way down to about 165 nm where the Lyman band then dominates emission at lower "vacuum UV" wavelengths. Note that the tallest peak of the Fulcher band and the D-alpha line are saturated to show more detail on less intense lines. This spectrum is not calibrated for intensity.

The reason for such an effect with the H_2 molecule is naturally due to the small moment of inertia which renders the intervals between adjacent lines ($d\nu = h/4\pi^2 I$) in a band proportionately so great that they present the appearance of not grouping together into a band. This serves as a limiting case in the general theory of band spectra. We have an instructive intermediate stage, between the many-line spectrum of hydrogen and the ordinary hand spectrum, in the spectrum of helium, discovered by Goldstein and Curtis and measured for the first time by Fowler. Whereas in the hydrogen spectrum the band character seems to have disappeared entirely, in the helium spectrum it can still be recognized, but by no means so strikingly as in the case of the cyanogen bands.

(2) *Alternating intensity in hand spectra*

Another peculiarity of the band spectra of diatomic molecules consisting of like atoms, *e.g.,* H^1, O^{16}, etc., is the alternating intensity of the rotational lines, *i.e.*, either every second line is missing or the lines are alternately strong or weak.

Heisenberg correctly ascribed this phenomenon to a *ware mechanical resonance effect* of the two identical atoms and showed that if the nuclear spin vanishes, alternate lines drop out, whereas if there is a resultant spin, the lines will alternate in intensity. We shall consider this effect, more in detail, when we deal with nuclear spin. The absence of every second line has been experimentally confirmed in the He^4, C^{12}, O^{16}, S^{32} molecules which, in consequence, have no nuclear spin. On the other hand;

molecules that consist of two isotopes of the same atom, such as $O^{16,18}$ or $Cl^{35,37}$ do not exhibit, either experimentally or theoretically, anomalies in intensity in their spectra.

Conclusion

Band spectra research has how developed into a science of considerable dimensions. For the interpretation of the extensive experimental data which have been collected, ingenious mathematical methods, such as group theory, consideration of symmetry, etc., have been called into service. The investigation and analysis of band spectra give valuable information regarding molecular structure, as we have already seen; they offer a very delicate and accurate means of studying the isotopic constitution of elements; they have formed also a convenient background for the study of the Raman effect. These we shall consider in the next two sections.

3. ISOTOPIC EFFECT IN MOLECULAR SPECTRA

Introduction

The isotopic effect in band spectra was first observed in 1919 by Imes who found that in the (2, 0) rotation-vibration band of HCl at 1.76μ the absorption lines were asymmetrical on the long wavelength side. This was interpreted later by Loomis and Kratzer as due to the presence of lines Of HCl^{37}. The theoretical foundations of the isotopic effect in band spectra were laid down in 1925 by Mullikan.

Since the three principal components of the total internal energy of the molecule, *vi:.*, the rotational and vibrational and electronic, depend upon the masses of the atoms that constitute the molecule, it follows that the isotopic nature of the atoms will affect the above-mentioned component energies, modify the frequencies of the spectral lines and thus make its appearance in the band spectra by supplementary lines and bands separated and deformed from the normal lines and bands. Thus, for instance, the doubling of each line in the rotation-vibration band spectrum of HCl can be shown to be due to the existence of two isotopes of Cl of masses 35 and 57.

The isotopic effect will make its appearance in all the three types of band spectra, but not to the same extent or degree of importance, since the dependence of the mechanism of production of each type on the masses of the atoms is not the same.

The isotopic nature of the constituent atoms will affect the pure rotation band through the intermediary of the moment of inertia, while in the case of rotation-vibration band, the chief agent transferring the isotopic nature to this type of band will be alteration in the vibrational frequency of the molecule. We shall first develop a simple theory for the isotopic effect in the three types of band spectra of diatomic molecules and then summarize the important results obtained in the study of the effect.

Theory of isotopic effect in band spectra

A. *The rotational effect*

In the case of pure rotational bands, we have seen that the frequency of the rotational line is inversely proportional to the moment of inertia of the rotating molecule. Now, since $I = \mu d^2$, where μ is the resultant mass and d the distance between the nuclei,

$$v_r \propto 1/ \mu d^2 \dots\dots\dots\dots\dots\dots\dots\dots\dots\dots(1)$$

The internuclear distance d will vary little on account of the isotopic nature of the atoms; so that d may be considered to be practically constant as far as the isotopic effect is concerned. Hence the change in frequency will be mainly due to the changes in μ, the resultant mass.

Taking the simplest case, where only one of the atoms of the diatomic molecule has two isotopes, if μ_1 and μ_2 be the resultant masses of the two isotopic molecules and v_{r1} and v_{r2} the corresponding frequencies of a rotational line in the two cases,

$$v_{r1} / v_{r2} = \mu_2 / \mu_1 \ldots\ldots\ldots\ldots\ldots\ldots\ldots \text{(2)}$$

Denoting the *relative frequency difference* between the rotational lines produced by the two isotopic molecules by dv / v,

$$dv / v = (v_{r1} - v_{r2}) / v_{r1} = 1 - v_{r2} / v_{r1} = 1 - \mu_1 / \mu_2 \ldots\ldots\ldots\ldots\ldots \text{(3)}$$

The isotopic displacement dv is therefore given by

$$dv = (1 - \mu_1 / \mu_2) / v \ldots\ldots\ldots\ldots\ldots \text{(4)}$$

This relation (4) shows that the isotopic effect (dv) depends upon two factors, viz., μ_1 / μ_2 and v. Now μ_1 / μ_2 differs little from unity, except in the case of very light molecules. The other factor v which expresses the change in rotational energy from one quantum state to another is a very small quantity, since the rotational energy E_r itself is a small quantity. Hence, the isotopic separation of pure rotational lines will be an extremely small quantity and will therefore escape direct observation, although it has been possible to recognize such separations in some cases.

The separation is usually of the order of magnitude of 0.1 cm^{-1}. Taking, for instance, the HCl molecule, both the constituent atoms have two isotopes, viz., H^1 and D^2, Cl^{35} and Cl^{37}. Hence, the ordinary HCl, used for the production of band spectrum, contains four different isotopic molecules, viz., H^1Cl^{35}, H^1Cl^{37}, D^2Cl^{35} and D^2Cl^{37}. Neglecting the isotopes of hydrogen for the present and considering only the isotopes of Cl, the resultant masses of the two isotopic molecules H^1Cl^{35} and H^1Cl^{37} can be determined as follows:

$$\frac{1}{\mu_{35}} = \frac{1}{m_{H^1}} + \frac{1}{m_{Cl^{35}}} = 1 + \frac{1}{35} = \frac{36}{35}$$

$$\frac{1}{\mu_{37}} = \frac{1}{m_{H^1}} + \frac{1}{m_{Cl^{37}}} = 1 + \frac{1}{37} = \frac{38}{37}$$

$$dv = \{ 1 - (\mu_1 / \mu_2) \} v = \left(1 - \frac{35}{36} \times \frac{38}{37} \right) v$$

$$= \left(1 - \frac{1330}{1332} \right) v = \frac{v}{666}$$

Expressed in wave number,

$$d\bar{v} = \frac{v}{666\, c} = \frac{\bar{v}}{666}$$

Since the pure rotational band lies in the region of 100 μ approximately,

$$\bar{v} = 1 / (100 \times 10^{-4})\ \text{cm}^{-1}$$

$$\therefore\quad d\bar{v} = \frac{10^4}{10^2 \times 666} = \frac{100}{666} = 0\cdot16\ \text{cm}^{-1}$$

The following points, are worthy of note:

(i) Since $d\nu / \nu$, equal to $(\nu_{35} - \nu_{37}) / \nu_{35}$ in the above consideration, is found to be a positive quantity, *i.e.*, $\nu_{35} > \nu_{37}$, the lines of H^1Cl^{35} will have shorter wavelengths, *i.e.*, the *lines of* H^1Cl^{37} *will be superposed on the longer wavelength side,* as is actually observed.

(ii) From the observed isotopic separation, one can work back to the *masses of isotopes* involved; thus here we have a means of measuring isotopic masses by a purely spectroscopic method; further, comparing the intensities of the isotopic lines, the relative abundance of the isotopes can also be estimated; finally, the presence of *rare isotopes* can be easily detected by this method.

In practice; however, the isotopic effect in pure rotation bands does not lend itself to these investigations by conventional methods, on account of the region in which this type of band occurs, viz., far infra-red.

(iii) *Microwave spectroscopy.* With the development of microwave spectrometers, in recent years, direct observation of the pure rotation absorption spectra of molecules has become possible. Further, the very high resolution and accuracy characteristic of the microwave technique have made measurement of the very small rotational isotopic effect practical, which means that masses of isotopes, even of heavy nuclei, can be determined with great precision. For example, in ICl, the shift in rotational frequency for the two isotopes Cl^{35} and Cl^{37} is only 5 per cent. With the limited resolution obtainable by infra-red techniques, the precision measurement of isotopic masses is very difficult, if not impossible. But with the microwave technique, absolute frequencies and frequency intervals in the region near 20,000 MHz (corresponding to the far infra-red) can be measured to an accuracy of 0.01 MHz. Hence the 5 per cent isotopic shift in IC1, which corresponds to about 1000 MHz, can be measured to an accuracy of 1 part in 10^5, which means that the mass of Cl^{37} can be determined to one twenty-thousandth of a mass unit relative to Cl^{35}.

It is interesting to note that the microwave spectrometer bids fair to become a keen competitor even to mass spectrographs of great precision, on account of its relative inexpensiveness, comparative insensitiveness to background and impurity effects, the small amount of the substance needed for analysis, etc. Already, valuable information has been obtained from microwave spectra for all the stable isotopes of Li, C, O, Si, S, Cl, K. Br, Ge, Se and Rb, as well as for a small but growing list of radioactive isotopes, such as H^3, S^{35}, Cl^{35}, Se^{75} and Sc^{79}. In certain cases, the accuracy of mass determination by microwave methods is as good as or even better than that obtained with the best mass spectrographs.

B. *The vibrational effect*
The frequency n of any oscillating body is given by $n = (1/2\pi)(f/m)^{1/2}$, where f is the restoring force per unit displacement and m the mass of the body. If f is kept constant,
$$n \propto 1/m^{1/2}$$

i.e., frequency varies inversely as the square root of the mass. Now, since isotopic molecules differ in mass, they will have different vibrational frequencies, the one of heavier mass having a lower frequency than that of the lighter.

Considering again the simple case of a diatomic molecule, in which one of the atoms has two isotopes, if μ_1 and μ_2 be the reduced masses and ν_{01} and ν_{02} the respective fundamental vibrational frequencies, then

$$\nu_{01} / \nu_{02} = (\mu_2 / \mu_1)^{1/2} \quad \ldots\ldots\ldots\ldots\ldots\ldots \quad (5)$$

provided the restoring force is constant, which is found to be the case, since the two atoms of a given diatomic molecule are bound with the same force, independent of their masses.

Since the vibrational frequency of the spectral line is proportional to the fundamental frequency, as seen from the relation

$$\nu_v = (m_{v1} - m_{v2}) \nu_o,$$

for a given quantum transition of vibrational energy,

$$\nu_{v1} / \nu_{v2} = \nu_{01} / \nu_{02} = (\mu_2 / \mu_1)^{1/2} = \rho \quad \ldots\ldots\ldots\ldots\ldots\ldots \quad (6)$$

where ν_{v1} and ν_{v2} are the frequencies of corresponding spectral lines;

$(\mu_2 / \mu_1)^{1/2}$
is usually replaced by ρ.

The relative frequency difference is given by

$$d\nu / \nu = (\nu_{\nu 1} - \nu_{\nu 2}) / \nu_{\nu 1} = 1 - \nu_{\nu 1} / \nu_{\nu 2} = 1 - (\mu_1 / \mu_2)^{1/2} \quad \dots\dots\dots\dots\dots\dots\dots (7)$$

This isotopic displacement is given by

$$d\nu = [1 - (\mu_1 / \mu_2)^{1/2}] \nu \quad \dots\dots\dots\dots\dots\dots\dots\dots\dots\dots\dots\dots\dots\dots(8)$$

This relation (8) shows that the vibrational isotopic effect will be much greater than the rotational effect, since E_v, on which ν depends, is very much greater than E_r ($E_v \gg E_r$). The value of $d\nu$ can be estimated from the reduced masses μ_1, and μ_2 and the position of the rotation-vibration band, *i.e.,* the near infra-red; it is of the order of 2 cm^{-1}, hence about twenty times greater than the rotational isotopic displacement.

Taking again the case of HCl and considering only the isotopes of Cl,

$$d\nu = [1 - (\mu_{35} / \mu_{37})^{1/2}] \nu$$
$$= [1 - (1330 / 1332)^{1/2}] \nu$$
$$= (1/1362) \nu$$

Since the rotation-vibration band of HCl is found in the region of 3.5 μ expressing the separation in wave numbers,

$$d\overline{\nu} = \overline{\nu} / 1362 = 10^4 / 3.5 \times 1362 \approx 2 \text{ cm}^{-1}$$

Thus, the theory predicts that in the case of rotation-vibration bands of HCl, on the low frequency side of each HCl^{35} line there should be a less intense satellite due to HCl^{37} about 2 cm^{-1} away; on the *low frequency side,* because HCl^{37} is heavier than HCl^{35} and *less intense,* because HCl^{37} molecules are less abundant than the HCl^{35} molecules, the proportions being approximately 1: 3, as known from other sources.

Experiment has completely confirmed these theoretical predictions. In 1919, Imes obtained the resolution of the HCl rotation-vibration bands (0 $\rightarrow$ 1) and (0 $\rightarrow$ 2), *i.e.,* $\Delta m_v = 1$ and 2, the second being the overtone of the first, occurring at $\lambda = 1.76$ μ, while the first is at $\lambda = 3.5$ μ. He found that the rotational lines broadened on the side of low frequencies; but the dispersion used by him being low, lines separated by 2 to 4 cm^{-1} or so could not be sharply resolved. The explanation of this broadening of each line was given independently by Loomis and Kratzer in 1920; they interpreted it as the vibrational isotopic effect. The approximate separation of the HCl^{35} and HCl^{37} components in the $\Delta m_v = 2$ band was construed by them as 4.5 cm^{-1}, in fair agreement with the theoretical. prediction. (In the theory given above, the band. corresponding to $\Delta m_v = 1$ was considered, which led to the value of 2 cm^{-1} for the isotopic separation; in the experimental data, the associated overtone band is considered, for which $d\nu$ will be twice as great.)

Since Imes's first observation, great improvements in the infrared technique have been made and it is now possible to resolve completely the isotopic components of the $\Delta m_v = 2$ band of HCl, as seen in Fig. 253, and measure their positions with great accuracy. The most precise measurements have been carried out in 1932, by Hardy and Sutherland, using Hardy's infra-red spectrometer.

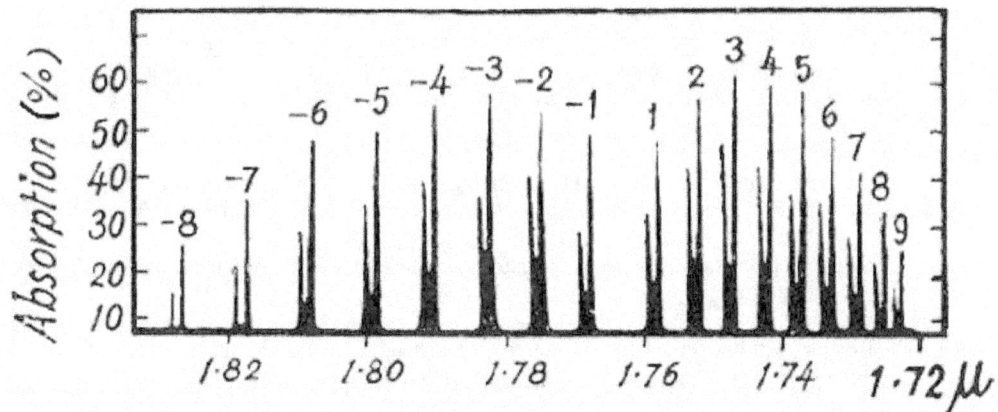

Fig. 255. Rotation-vibration band of HCl at 1.76μ showing the isotopic effect of HCl^{35} and HCl^{37}.

The isotopic differences for the lines of the R branch obtained are as follows:

m_r (number of line) :	2	3	4	5	6
Separation (cm.⁻¹) :	4·04	4·03	4·03	4·21	4·21

It is seen that the observed isotopic separation agrees very well with that predicted by the theory. Experiment confirms also the implication of the theory that the separation should be the same for all the lines belonging to the same vibrational transition, with a further indication that the vibrational effect is the chief source of displacement. But the accuracy of the experimental method is great enough to give significance to the gradual increase from .4.04 to 4.21 with increasing m_r values. This secondary effect is due to the associated rotational isotopic effect. For a single change in vibration there will be many values for the rotational frequency, increasing with the rotational quantum number m_r; accordingly the rotational displacement will get larger with higher values of m_r. When this effect is added to the constant vibrational displacement, the combined, displacement increases with the higher order of the lines, so beautifully illustrated by the results of Hardy and Sutherland.

C. *The electronic effect*
Our present knowledge of the isotopic effect would probably have not been so precise, if results from electronic bands have not been brought to bear upon the observations with the rotation-vibration bands.

In the case of the electronic band, just as there are three types of transitions which give the whole band structure, so also there will be three different manifestations of the isotopic effect:

(i) *The electronic isotopic effect*
The electronic transition will be slightly different for different isotopes of one or both the atoms of a diatomic molecule. The theory of this effect has not been fully worked even for the simplest of diatomic molecules. It is, however, obvious that the effect must be extremely small, as seen from the relation for electronic transition:

$$\bar{v} = (2\pi^2\, mE^2 e^2 \,/\, ch^3)\, [M\,/(M + m)]\, (1/n_1{}^2 - 1/n_2{}^2)$$

For the isotope of mass M_1, the relation will contain the fraction $M_1\,/(M_1 + m)$ and for that of mass M_2, $M_2\,/(M_2 + m)$, m being the mass of the electron. Both these fractions will differ little from unity, so that the difference in frequency due to the two isotopes will be extremely small. In practice, it has proved to be outside the range of experimental observations. Presumably, in the electronic band spectra of isotopic molecules, each line has a sort of hyperfine structure of separation less than 0.01 cm⁻¹.

(ii) *The vibrational isotopic effect*
Here, the theory is the same as that developed for rotation-vibration spectra.

The electronic effect being negligible, the total isotopic effect in electronic band spectra will be the algebraic sum of the vibration and rotation isotopic effects, exactly as in the rotation-vibration bands.

The vibrational effect, which is much larger, has been much more important in the history of band spectra. It is a constant for the lines of a given band, but increases linearly from band to band with the interval v_v.

(iii) *The rotational isotopic effect*

This is much smaller than the vibrational, hence it is usual to determine isotopic masses from a study of vibrational effect alone, *i.e.,* by measuring the displacement of lines of small m_r values, for which the rotational shift is negligible.

The accurate theory of the isotopic effect is much more complicated than the simple one, given here, for the following reasons:

(a) The subdivision of the isotopic shift into three parts is a convenient fiction; in practice, terms occur which are due to the interaction of one type of motion with another.

(b) The actual enharmonic feature of the vibrations has not been taken into consideration.

(c) The additional moment of inertia due to the peripheral electrons and the contraction of molecules in the case of heavy isotopes have not been taken into account.

Results obtained from the experimental study of the isotopic effect in band spectra

A good number of isotopic molecules have been studied either with rotation-vibration bands or with electronic bands and the results obtained both as regards masses and relative abundances of isotopes agree very well with those of the mass spectrum analysis.

The band spectra method is certainly a more sensitive method than the mass spectrograph method in the case of rare isotopes. The very important results obtained in this field of research may be summarized as follows:

(i) The *study of rotation-vibration bands* of HCl has established beyond doubt the two isotopes of Cl (35 and 37), while it has disproved the existence of Cl^{39}, which was at one time considered possible. We have already described the experiments of Hardy and Sutherland in connection with the isotopes of Hardy, Barker and Dennison. have been able to establish the existence of the two isotopes of hydrogen H^1 and D^2 by their researches on the rotation-vibration bands of HCl, in confirmatory evidence of the discovery of deuterium in the study of isotopic effect in atomic spectra by Urey, Brickwedde and Murphy. Hardy and his collaborators argued that if D^2 existed in the proportion suggested by the discoverers, there would be an appreciable amount of D^2Cl molecules in any volume of HCl and hence it should be possible to find the rotation- vibration bands of D^2Cl. The value of $\rho = (\mu_2 / \mu_1)^{1/2}$ for D^2Cl and H^1Cl being very large, the vibrational separation must be great. The position of the centre of the $(1 \rightarrow 0)$ band for H^1Cl being 3.46 μ, the centre of the corresponding band for D^2Cl is readily found to be 4.8 μ. Using a cell, seven metres long, Hardy and his co-workers found absorption in the calculated region. The structure of the band and the agreement with the theoretical prediction left no doubt that the band was actually the $(1 \rightarrow 0)$ band of D^2Cl^{35}. Most of the lines in the band were accompanied by weak satellites due to D^2Cl^{37}. Determining the value of ρ experimentally, the mass of D^2 could be calculated. The final value obtained was 2 01367 ± 0.0001, in good agreement with Bainbridge's mass spectrum value, 2.01353 ± 0.00006.

(ii) The *analysis of the electronic band spectrum* has led chiefly to the discovery of several new rare isotopes. Sometimes, of course, the discovery of an isotopic spectrum has been a matter of chance, *i.e.,* instead of going out to look in a calculated position for a spectrum of a given structure, workers found satellite bands which later were proved to be isotopic components. But the general method of procedure is to construct a theoretical expression from the principal band involving ρ, then to see whether the observed lines agree with theory or not, and finally to evaluate ρ, as described in the investigation of Hardy and others on deuterium. Some of the interesting results thus obtained are those on the isotopes of oxygen, nitrogen and carbon.

Oxygen

Close to the well-known red atmospheric bands of oxygen, Dieke and Babcock observed a weaker band of very much the same structure. Giauque and Johnston showed that these lines could be interpreted as due to $O^{18}O^{16}$, the calculated isotopic shifts agreeing extremely well with the observed separations. There were also twice as many rotational lines in the 0 180'6 band, as in the $O^{18}O^{16}$ band, as required by the theory of symmetrical and anti-symmetrical terms. Babcock proved also that another system of weak lines near to the main band was due to $O^{17}O^{18}$. The rareness of the atomic spectra of O^{17} and O^{18} is then naturally due to the weak intensity of the corresponding bands, their ratio as compared with O^{16} being

$$O^{16}: O^{17}: O^{18}: 1: 1/1250: 1/10,000$$

From the observed value of ρ, the masses of O^{17} and O^{18} have been calculated, assuming the mass of O^{16} as 16 and they are found to be 17.0029 and 18.0065. After these results upon the oxygen bands were reported, Naudé, studying the bands of NO, found structures corresponding to $N^{14}O^{17}$ and $N^{14}O^{18}$, which thus confirmed the existence of N^{17} and O^{18}.

Nitrogen
Along with the finding of the bands of the oxygen isotopes in NO, Naudé discovered also the bands of $N^{15}O^{16}$ in the same system. This was confirmed by Herzberg from a study of the bands of N_2.

Gerhard Herzberg (December 25, 1904 – March 3, 1999) Germany

Carbon
In the study of the so-called *swan* (C — C) *bands*, King and Birge discovered bands which they proved to be due to C^{13} C^{12}. The value of ρ gives for C^{13} a mass of 13 nearly, assuming that of C^{12} as 12. The isotope C^{13} was also proved to be present in the fourth. positive bands of CO and in the violet bands of CN.

The cases cited above are just the first fruits gathered from the interesting researches on the isotopic effect in band spectra; they do not, by any means, exhaust the list of cases studied. The study of the isotopic effect, both theoretical and experimental, has been extended to polyatomic molecules and interesting results have been obtained.

Note: Isotopic effect in atomic spectra
The isotopic effect is far less clearly marked as well as more complicated in principle in atomic spectra than in molecular spectra. There are two ways, one direct and the other indirect, in which the presence of isotopes may show itself in atomic spectra. The direct way is *the influence of the mass of the isotopic nuclei,* while the indirect one is *the effect of the spin of the nuclei,* on the electronic transition. We shall here consider only the first, reserving the second to a later chapter dealing with nuclear spin.

The *isotopic mass effect*

For the one-electron system, the principal spectral lines are given by Bohr's equation:

343

$$\tilde{\nu} = \frac{2\pi^2 \mathrm{E}^2 e^2}{ch^3} \cdot \frac{m\mathrm{M}}{\mathrm{M} + m} \left(\frac{1}{n_1{}^2} - \frac{1}{n_2{}^2} \right) \qquad \dots (1)$$

Now if the nucleus exists in two isotopic forms of masses M_1 and M_2, there will be satellite lines, which can be represented by

$$\tilde{\nu}_1 - \tilde{\nu}_2 = \frac{\dfrac{2\pi^2 \mathrm{E}^2 e^2}{ch^3} \left(\dfrac{\mathrm{M}_1 m}{\mathrm{M}_1 + m} - \dfrac{\mathrm{M}_2 m}{\mathrm{M}_2 + m} \right) \left(\dfrac{1}{n_1{}^2} - \dfrac{1}{n_2{}^2} \right)}{m\ (\mathrm{M}_1 - \mathrm{M}_2)\overline{\nu_2}/\mathrm{M}_1\mathrm{M}_2} \qquad \dots (2)$$

This relation (2) can be used with accuracy only in the case of hydrogen and hydrogen-like atoms, $e.g.,$ He^+, Li^{++}, etc. It ceases to apply to the many-electron systems. For two and three electron systems, the theory has been worked out by Hughes and Eckart and confirmed by experiments conducted on the two isotopes of Li of masses 6 and 7. The theory of the effect for atoms with more than three electrons has not yet been established and there are a few observations on the lines of neon and a few heavy atoms. The effect will be obviously very small and is likely to be confused with that due to nuclear spin.

The more important results obtained by the study of the isotopic effect in atomic spectra are:

(i) *Discovery of deuterium* by Urey and his co-workers, who were able to obtain the isotopic components of the Balmer lines in the hydrogen spectrum.

(ii) *Isotopic shift of spectral lines in more complicated atom.* It might be expected that with increasing atomic number the isotopic effect would become smaller, since the motion of the nucleus would become more and more unimportant. However, it has been found by Schüler and others that even for elements of rather high atomic number a noticeable isotopic effect is present, which is of the same order of magnitude as the nuclear spin effect—a fraction of a wave number. In general, it is not always easy to separate the two effects, viz., isotopy and nuclear spin. For this purpose, consideration of the intensity of the components is important, since obviously the various components of the isotopic structure will have intensities proportional to the relative abundances of the corresponding isotopes. An unambiguous decision is possible in the cases where separate isotopes are available, $e.g.,$ in lead, where the isotopes of 206 and 208 are obtained in an almost pure state from U and Th minerals. The existence of a hyperfine structure in the cases where the nuclear spin is zero is also a good proof for the reality of the isotopic effect.

Another means of differentiation is by the study of the Zeeman effect. For a pure isotopic effect, each of the individual components will show the Zeeman effect for the extra-nuclear electron quite independently of one another, whereas the hyperfine structure resulting from nuclear spin should show an essentially different Zeeman effect.

The fine structure of Li resonance line, which is not a simple doublet, was the first to be explained as due to the isotopic shift of Li^6 and Li^7 by Schüler. This interpretation has been verified by the intensity ratio of the corresponding lines in the hyperfine structure pattern, which agrees well with the abundance ratio of the two isotopes. Likewise, the lines of Ne, which consists of two components, were attributed to the two principal isotopes of Ne of masses 20 and 22, the intensity ratio agreeing with the abundance ratio (9:1). This interpretation was further confirmed by the fact, that the separated isotopes showed only one or the other component. After these initial attempts, it has been experimentally demonstrated that the phenomenon is of frequent occurrence with the heavier dements also. Schüler has found the effect in several elements, such as Zn, Hg, Sm, Th; etc. The mechanism of production of the observed shifts in these cases may be due to the different masses alone, as has been shown to be true at least in the case of light elements, such as H, Li. Ne. etc. But, in the case of heavier elements, the effect may have to be traced back to the change of nuclear radius with mass, as was shown by Pauling, Coudsmit and Bartlett.

4. RAMAN EFFECT

Sir Chandrasekhara Venkata Raman (7 November 1888 – 21 November 1970)

Discovery

In 1928, Sir C.V. Raman, while studying the scattering of light by liquids, with the intention of explaining and reproducing such natural phenomena as the blue of the sea and of the sky, found that

when a beam of monochromatic light was passed through organic liquids such as benzene, toluene, etc., the scattered light contained other frequencies in addition to that of the incident light.

Although such an effect had been predicted on theoretical grounds as early as 1923 by Smekal, Raman was the first to observe it experimentally.

The arrangement used by him was simple in design. A round-bottomed glass flask was filled, with dust-free toluene or benzene and the liquid strongly illumined by the 4358 line from a mercury arc, suitably filtered and concentrated by a lens. The scattered light was examined by means of a spectroscope placed transversely, *i.e.,* in a direction at right angles to that of the incident radiation. In the spectrum of the scattered light, a number of new lines was observed on both sides of the main line. Those on the low frequency side were more numerous and more intense than those on the high-frequency side. Most of these new lines were strongly polarized and their spacing was symmetrical about the main line. They are now generally referred to *as Raman lines,* or more specifically, those on the low-frequency side as *Stokes' lines* and those on the high as *anti-Stokes' lines.*

Raman, having thus observed a set of new discrete frequencies in the scattering process, contrary to the expectations of the classical theory, was able to establish that they constituted a new phenomenon, distinct from both the simple Rayleigh or coherent scattering and the more complex fluorescent scattering. He argued that

in Rayleigh's scattering no frequency change was caused, but only some of the already existing frequencies were selected, whereas in the phenomenon under study new lines of different frequencies appeared even when a single frequency was scattered.

Nor could the observed effect be assimilated to fluorescence, although there was a certain amount of superficial resemblance between the two, in so far as, with a given exciting line, new lines made their appearance in both.

For, in the *first* place, the frequencies in the fluorescent spectrum were always less than the incident frequency, while the Raman lines had frequencies both greater (anti-Stokes' lines) and less (Stokes' lines) than that of the incident line.

Secondly, the frequencies of the fluorescent lines were really independent of that of the exciting line, provided the latter was able to produce fluorescence; but in the case of the Raman fines, their frequencies were directly related to that of the incident light, since if the latter was varied the former changed at the same rate.

Thirdly, while the frequencies of the fluorescent lines were fixed by the nature of the scatterer, it was the frequency shifts of the Raman lines (known also as *Raman frequencies)* that were determined by the scatterer rather than the frequencies themselves.

Fourthly, the lines observed by Raman were strongly polarized unlike the lines in the fluorescent spectrum.

Finally, the most important fact that confirmed Raman's discovery as a new effect was that the Raman frequencies were either actual infra-red frequencies in the absorption spectrum of the scatterer or differences in such frequencies, which was not the case in fluorescent scattering. This fact proved further that the observed effect was a molecular phenomenon.

Hence, the modified frequencies observed in the scattering process were a new type of secondary radiation, to which the, name *Raman effect* was given and was initially considered as the *optical analogue of the Compton effect,* since both belong to the same category of *incoherent scattering* of radiation, which can be explained only on the basis of the quantum theory.

Almost simultaneously with Raman, Landsberg and Mandelstam in Russia, studying the light scattered by certain crystals, discovered the effect in the case of solids. It was soon found that many liquids, vapors, gases and transparent solids exhibited the effect which thus proved to be a universal phenomenon.

Experimental study
The Raman effect has been extensively studied by a great number of workers. The *general technique* used in these researches is to illumine the substance under investigation with an intense monochromatic source of light and photograph the scattered radiation by means of a spectrograph arranged in a transverse direction. But the technical details vary according to the nature of the substance under test, *i.e.,* liquid, solid or gas, the chief purpose being to obtain best and quick results.

Apparatus
The original simple arrangement of Raman was not quite efficient and required very long exposures of about hundred hours and more to obtain good records of the Raman spectrum. Hence, improvements were made as regards the container of the substance, the source of radiation, filter, spectrograph, etc.

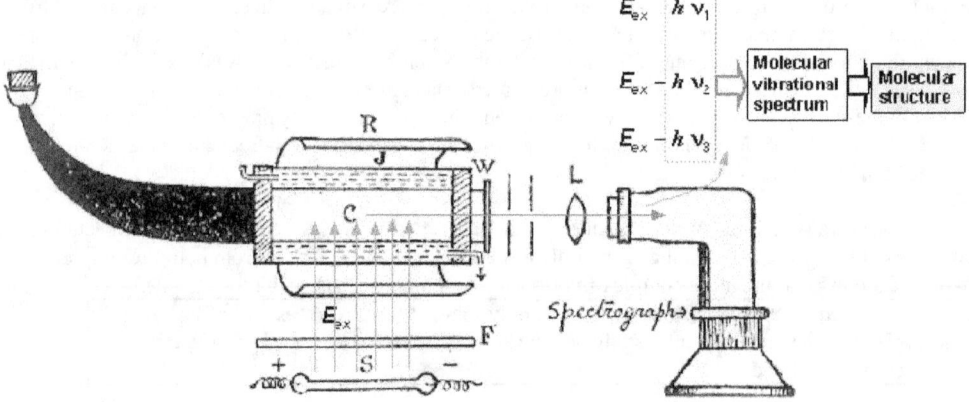

Fig. 256. Apparatus for the study of Raman effect.

The apparatus shown in Fig. 256 is the one first developed by Wood, and now ordinarily used in the study of the Raman effect in liquids.

The *container* C of the liquid to be investigated, called the *Raman tube,* consists of a glass tube of about 1 to 2 cms in diameter and 10 to 15 cms long, one end of which is drawn out into the shape of a horn and blackened outside to provide a suitable background, the other end being closed with an optically plane glass plate constituting the window W through which the scattered light emerges.

The container is surrounded by a water jacket J in which cold water is circulated to prevent overheating of the liquid due to the proximity of the exciting arc.

An ideal *source* S would be light from a helium discharge tube filtered by nickel oxide glass, giving a strictly monochromatic line of wavelength 3888 A°. But on account of the many technical difficulties involved in the construction and manipulation of this source, it is not widely used. The source ordinarily employed is the mercury arc, the next best available, from which it is possible to get single wavelengths by the use of suitable filters.

Thus, for instance, to obtain the 4358 line, slightly acidulated quinine sulphate solution contained in a novial glass vessel is used as filter, which cuts off all the other lines except 4358 A°. To get the 4046 line a solution of iodine in carbon tetrachloride contained in a novial glass cell is found to be a very satisfactory filter. The filter solution may be arranged either to surround the Raman tube or in front of the arc. The mercury arc is placed as close to the Raman tube as possible, which results in large intensity of the incident light.

A semi-cylindrical aluminum reflector R enhances the intensity of illumination still further.

The chief features of a *spectrograph,* suited for the study of the Raman spectra, are (1) large light-gathering power, (2) special prisms of high resolving power and (3) a short-focus camera. A lens L in front of the plane window W directs the scattered radiation noon the slit of the spectrograph which is carefully aligned along the axis of the Raman tube and screened from the direct rays of the arc. The intense Raman lines of a liquid such as CCl_4 can be photographed in about an hour with a small spectrograph, but the recording of the complete spectrum may require up to ten or fifteen hours, depending largely on the intensity of the incident light, the speed of the spectrograph and the intrinsic brilliance of the Raman lines. It may be noted that instruments of high resolving power such as gratings are not used with advantage, on account of the poor luminosity which necessitates long exposures.

Certain modifications in the experimental arrangements are necessary for the excitation of the Raman effect in solids and gases. In the case of substances which are available as *large and transparent solid blocks,* like gypsum, quartz, etc., a container is not required and the light from the exciting are can be directly focused on the material with a large condensing lens. With solids which are in the form of *loose crystals* or *powder,* the Raman effect can be obtained, as was first shown by Baer and Menzies, by reflecting light from crystal surfaces. But special precaution has to he taken to avoid the masking effect due to a large amount of direct light coming out of the container by repeated reflection at the crystal faces and entering the spectrograph. To achieve this, two techniques have been used, *one* the use of complimentary filters as developed by Ananthakrishnan at the Raman laboratory and *the other* a special type of spectrograph with two parts, each part having a prism and two lenses with a common slit in between the two, devised by Billroth, Kohlrausch and Reitz in Germany. Both give very good results with crystal powders and the latter even with very small quantities of the substance.

The intensity of the light scattered from *gases* is very weak, but this difficulty has been. overcome by intense illumination of the gas under high pressure and the use of spectrographs of great light gathering power. Wood employed a very long tube of HCl gas and obtained its Raman spectrum at atmospheric pressure using a specially made mercury arc, which was plated in contact with the gas tube, hollow, cylindrical reflectors enclosing both of them. The illumination produced in this way was very intense as the light from the arc was returned back and forth between the walls of the reflectors.

Rasetti was the first to develop the technique of exciting Raman effect in gases under high pressure, which shortened considerably the time of exposure. He used a thick walled quartz tube, 20 cms long and 2.2 cms internal diameter, which could withstand pressures of 10 to 15 atmospheres. With the 2537 line of the mercury arc as the exciting radiation, he was able to obtain the Raman spectra of several gases under pressure. Bhagavantam has constructed a Raman tube for gases which can stand pressures up to 50 atmospheres. It is made of transparent silica and enclosed in an outer steel tube for protection: He has been able to obtain with his apparatus good photographs of Raman spectra of gases in about 40 to 50 hours time of exposure using the 3650, 4046 and 4358 mercury lines.

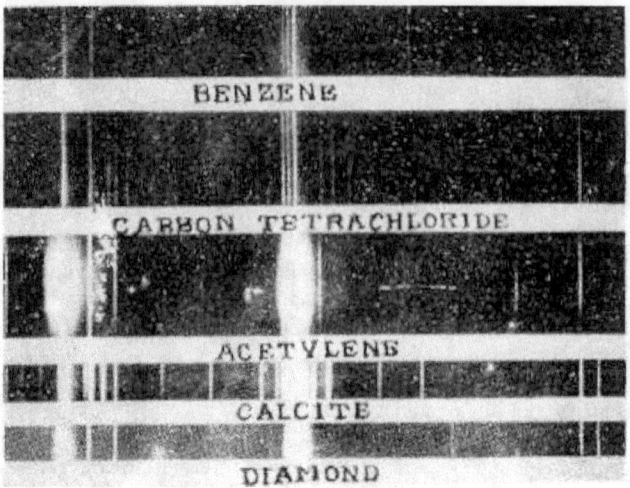

Raman spectra of different substances (Bhagavantam)

Suri Bhagavantam (October 14, 1909 - February 6, 1989)

Results

An idea of the type of Raman spectra produced by solids, liquids and gases may be formed by inspecting the photographs obtained with benzene, carbon tetrachloride (liquids), acetylene (gas), calcite and diamond (solids), given above. It is seen that a number of new lines and bands, exhibiting a variety of characters of intensity, width, polarization and fine structure are recorded on either side of the exciting radiation. There is also some unresolved continuous radiation, which generally appears as wings extending asymmetrically on either side of the parent line. This continuous spectrum shows great variations in intensity with different substances.

Each line in the incident spectrum, if of sufficient intensity, gives rise to its own set of lines or bands and associated continuous spectrum.

We shall now outline the main results obtained from researches made on the effect with such photographic records.

Raman effect in liquids

About a hundred liquids, so far examined, show the phenomenon in an unmistakable manner. The frequency shifts of the Raman lines produced with benzene correspond to an infra-red wavelength 3.27 μ, in which region benzene exhibits a strong band in its absorption spectrum. A close examination, however, of the infra-red absorption spectrum and the

Raman spectrum of benzene shows that none of the Raman lines are represented in infra-red absorption and *vice versa*. An interesting feature with CCl_4 is the triad of Stokes' and anti-Stokes' lines equally spaced on either side of the exciting 4358 line.

With very well purified water Dadieu and Kohlrausch found two broad bands instead of sharp lines at about $\lambda = 3\ \mu$.

Solutions of salts in water give the Raman spectra characteristic of the salts and the water. Gonesan and Veukateswaran found that the bands due to water in aqueous solutions of H_2SO_4, HCl and HNO_3 acids become sharper with increasing concentration. The similarities exhibited by solutions of carbonates of different metallic radicals and the similarity of the sulphates and nitrates appear to support the view that the characteristic frequencies are those of the ionized acid radical.

Raman effect in gases

The Raman spectrum of HCl gas obtained by Wood and those of CO and CO_2 by Rasetti were the first observations made with gases. Then many ether gases such as hydrogen, oxygen, nitrogen, ammonia, NO, N_2O, CS_2, etc., were studied. The frequency shift of the Raman line of HCl gas corresponds to $\lambda = 3.466\ \mu$, which is almost exactly the wavelength of the central missing line of the infra-red absorption band of HCl. Carbon monoxide gives a Raman line whose frequency shift is equal to the frequency of its infra-red band, while CO_2 gives a line whose shift is equal to the frequency difference of two of its infra-red hands.

Oxygen, hydrogen and nitrogen give a pattern of equally-spaced lines. The intensities of the individual lines alternate in the case of nitrogen and hydrogen, while with oxygen the alternate lines are absent, as seen in the adjacent figure, where the Raman spectra of oxygen and nitrogen, obtained by Rasetti, are reproduced. These peculiarities in intensity of alternate lines have led to very significant conclusions as regards molecular structure and nuclear spin. McLennan carried out a series of experiments on the Raman effect with liquid oxygen, nitrogen, hydrogen and nitrous oxide. His results with liquid oxygen suggested that the normal mode of vibration of the molecule is the one involved in the production of the four Raman lines observed, while those for liquid hydrogen supported the view that hydrogen at low temperatures must be regarded as a mixture of two distinct types of molecules known as the *para* and *ortho* hydrogen. The Raman spectrum of CO_2 which is similar to that of oxygen indicates a symmetric structure for its molecule, while that of N_2O where there is no alternation in intensity of the lines suggests an asymmetric molecular structure. Rasetti has observed a Raman line with NO whose frequency shift is 121 cm^{-1}, which has been classified as having an electronic origin.

Raman effect in solids

On account of the numerous experimental difficulties, comparatively a few solids only have been studied so far. Mandelstam, Landsberg, Baer and Menzies were the pioneer workers in this branch of Raman effect. Other investigators were Schaefer, Matossi and Aderhold, Miss Osborne, Cabannes and Canals, Nedungadi and Bhagavantam. The substances analyzed are gypsum, quartz, calcite, sodium nitrate, potassium nitrate, ammonium phosphate, ammonium chloride, diamond and a few others. The Raman lines obtained with crystals are sharp, becoming diffuse with rise of temperature.

In calcite ($CaCO_3$) two lines which are nearest to the parent line have been definitely identified with the oscillations of the crystal lattice; while the others are due to the vibrations of the CO_3 groups.

With gypsum ($CaSO_4.2H_2O$) which contains two molecules of water of crystallization, Krishnan found, in addition to the wavelengths which could be attributed to the SO_4 radical, three sharp lines at $\lambda = 2.8\mu$, 2.9μ and 3.0μ, which are evidently due to the water of crystallization and are practically in the same position as the components of the band observed with water.

Diamond exhibits a strong and sharp line of a comparatively large frequency shift, which has to be ascribed to a lattice oscillation.

In solid benzene, the intense continuous spectrum obtained with liquid benzene is replaced by bands.

Intensity of Raman lines

The experimental determination of the intensity of Raman line is beset with many difficulties on account of their extreme weakness. The intensity of a Raman line, when expressed as a fraction of the parent line, is usually a few hundredths in liquids and a few thousandths in gases. No accurate data are available as regards the absolute intensities of Raman lines in liquids or gases. More reliable results are obtained in the determination of relative intensities of a given set of Raman lines excited by a given parent line. To obtain good results, photographic plates which have great and uniform sensitivity in the region of investigation should be used. The densities obtained for the lines under comparison are measured with a microphotometer. On each plate, a set of calibration spectra are recorded, with the help of which density curves for each wavelength can be drawn, and the intensities corresponding to any density can be obtained from such curves. The measured intensities should be corrected for various causes of error, such as absorption in the body of the substance in the Raman tube, the oblique refraction at the prism surfaces in the spectrograph, etc.

In spite of the *numerous* experimental difficulties, results of fundamental importance have been obtained. Very faint Raman lines as well as very intense ones arc met with, *e.g.,* the band in NaCl recorded by Rasetti and the anti-Stokes' line' in diamond obtained by Bhagavantam are very faint, while the principal Raman line of diamond at 1332 and the line in benzene having a frequency shift of 992 are very intense. These variations in intensity as we pass from line to line and from substance to substance are of great significance in the study of molecular structure and chemical constitution.

Since the intensity of scattering increases with the fourth power of the frequency, it is advantageous to use, where the absorption of the substance under investigation permits, light of as short a wavelength as possible. Hence the ultra-violet mercury resonance line 2537 A° is often used. The Stokes' lines are always more intense than the corresponding anti-Stokes' lines. The anti-Stokes' lines grow more intense and, in addition, all the Raman lines move inward towards the parent line as the temperature is increased.

Polarization of Raman lines

Just as the Raman lines vary greatly in their intensities, so also their states of polarization. The fact that different lines are differently polarized is probably connected with their relative intensity.

The experimental arrangement used for determining the polarization of Raman lines is essentially the same as that described earlier but with the following modifications. The light from the source is concentrated by means of a condenser into the substance contained in the Raman tube. A suitably oriented double image prism whose function is to separate the vertical and horizontal components in the scattered light is placed in front of the slit of the spectrograph, so that two images, one above the other, are formed on the slit, which are simultaneously photographed.

The state of polarization of a Raman line is measured by a quantity known as the *depolarization factor which* is simply the ratio of the intensities of the horizontal and vertical components when the incident light is vertically polarized. This ratio is readily obtained from the traces photographed as described above, by one of the usual methods employed for comparing the intensities of two beams of the same wavelength. In order to get fairly accurate values of the depolarization factor, the following precautions should be taken:

(i) crystalline quartz should not be used for condenser, spectrographs or windows, since its optical activity complicates the phenomenon of polarization;

(ii) the window of the Raman tube through which the scattered light emerges should be strain-free and plane,

(iii) errors arising from oblique refraction at the prism surfaces, want of transversality in the incident beam and slit width should be eliminated.

Cabannes found that the Raman lines in crystals, such as quartz, are differently polarized, the intensity and depolarization of the lines depending upon the orientation of the crystal.

Menzies has investigated the polarization of the Raman lines in liquids, such as CCl_4, in directions perpendicular and obliquely forward to the incident beam and has shown that many of the observed facts could be accounted for by considering the initial and final directions of vibrations in the molecule involved to be parallel in the case of polarized lines, perpendicular in that of unpolarized lines and at an oblique angle for partially polarized lines. The following are some of the important results obtained:

(a) The depolarization factor varies from 0 to 0.86 for the vibrational Raman lines, while it has a constant value of 0.86 for the rotational lines.

(b) With circularly polarized incident light, part of the Raman line is circularly polarized in the reverse direction and part circularly polarized in the same direction as that of the incident light. Highly depolarized rotational lines exhibit reverse circular polarization.

(c) Sharp and strong lines are ordinarily characterized by low depolarization factors, while diffuse and weak ones by high depolarization factors.

(d) Corresponding lines in molecules having similar structures have nearly the same depolarization factors. This is to be expected as the polarization of a Raman line is mainly decided by the symmetry of the oscillation.

Nature of the Raman effect

From the many and varied experimental data, it is clear that the Raman effect is a molecular phenomenon. In the case of free molecules scattering light, three different kinds of Raman effect can therefore be distinguished, viz., a rotational effect, a vibrational effect and an electronic effect. A mixed "rotation-vibration" effect can also take place under certain circumstances.

Solids can exhibit yet another type of Raman effect in which the crystal lattice as a whole takes the place of the molecule.

Since the rotational energies involved are small relative to the vibrational energies, the pure rotational Raman lines lie correspondingly closer to the parent line and are often masked by the intense light of the parent line on the photographic plate. They are obtained separately only in the case of certain light gases, such as hydrogen, deuterium, oxygen, nitrogen, etc.

In heavier gases, the lines are much closer and instruments of greater resolving power are required to separate them. Most of the observed Raman lines and bands with moderate or large frequency shifts are due to the vibrational effect, corresponding to various normal modes of vibration of the molecule or the crystal lattice. The fainter lines may be due partly to overtones or combination tones. The lines arising from an oscillating crystal lattice are characteristic of only the solid state and are not present in liquids or gases.

Raman lines due to the electronic effect are rarely observed, as in the single case of NO obtained by Rasetti.

Analysis of the continuous spectrum under high dispersion reveals that it cannot be separated from the parent line and that its intensity is maximum near the parent line.. Although it is difficult to decide upon the exact nature and origin of the continuous spectrum in all cases, very probably it is due to an unresolved rotational effect. In the solid state, it is replaced by broad bands, while in the gaseous state by separate lines, which may be ascribed to the rotation of the molecule. In certain eases the broad bands which replace the continuous spectrum have been identified with the oscillations of the crystal lattice.

Relation between the Raman and infra-red absorption spectra

The rotational and vibrational Raman spectra closely resemble the far infra-red and near infra-red absorption spectra of molecules respectively. Further, in many cases the Raman frequencies are equal either to the actual infra-red frequencies of the absorption spectrum of the scatterer or to the differences of such frequencies. But there seems to be little or no correlation between the relative intensities of the Raman lines and the intensities of the corresponding infra-red bands. In addition, not all Raman lines have their corresponding infra-red bands; nor all infra-red bands their corresponding Raman lines. Thus the most intense Raman lines in benzene and toluene, whose frequency shifts are 10.3μ and 10.2μ respectively, have no corresponding intense infra-red bands. On the other hand, the intense absorption bands of benzene and toluene at wave-lengths 9.75μ and 6.86μ respectively have no corresponding Raman lines. Raman lines have also been observed corresponding to infra-red transitions which are forbidden by the selection rules, as in the case of HCl gas, already cited. In such cases, Raman effect provides a very valuable complement to the study of infrared absorption spectra in the detection of energy levels of molecules.

It is to be noted that the above consideration of Raman spectra in relation with band spectra leads to the following important conclusion:

Although, the ultimate result, as far as the molecule is concerned, is the same in both cases, yet the mechanism of production and the intervening laws are quite different in the two phenomena.

Theoretical explanation
The *classical electromagnetic wave theory* is able to give only a very elementary and imperfect explanation of the Raman effect as follows.

When a light wave represented by
$E_0 \cos 2\pi v_0 t$
falls on a molecule, it induces in the molecule an oscillating electric dipole moment,
$a \, E_0 \cos 2\pi v_0 t$
where a is called the deformability or polarizability of the molecule, since the induced dipole moment, in general, has a different direction from the exciting electric vector, chiefly in the case of anisotropic molecules.

The molecule then behaves like a Hertzian oscillator and radiates energy in the form of electromagnetic waves of frequency v_0. If we now suppose that some kind of mechanism, intrinsic in the molecule itself, such as its rotation or vibration, alters the amplitude of the emitted wave periodically with a frequency v, then the electric moment of the oscillator at any instant is given by
$a \, E_0 \cos 2\pi v_0 t \cos (2\pi v t + \delta)$,
which may be written as

$$(1/2) \, a \, E_0 \{ \cos [2\pi (v_0 + v) \, t + \delta] + \cos [2\pi (v_0 - v) \, t - \delta] \}$$

This expression shows that the emitted light will consist of two frequencies
$v_0 + v$ and $v_0 - v$.

If part of the induced moment is not affected by the internal state of the molecule, the incident frequency v_0 also will be present in the scattered light. Thus the appearance of lines with frequencies v_0 (parent line), $v_0 + v$ (anti-Stokes') and $v_0 - v$ (Stokes') in the Raman effect is accounted for.

Owing to the presence of the phase-factor δ, which varies arbitrarily from molecule to molecule, the three vibrations of the dipole moment and consequently the three frequencies in the scattered light are incoherent.

But this classical explanation is defective and incomplete in several respects.

First of all, it has been assumed that the vibration or rotation of the molecule is essential to the mechanism of Raman scattering. But it can be shown, using Boltzmann's law that molecules which are intrinsically in a state of vibration are very few at ordinary temperatures, their number decreasing rapidly at high frequencies. From this it would follow that the majority of the molecules in a scatterer does not contribute to the production of Raman lines. Such a conclusion is contradicted by the observed intense Raman lines of large frequency shifts.

Secondly, according to the classical theory, since both the Stokes' and anti-Stokes' lines arise apparently from the same molecule, they must have the same intensity, which is not true to facts, since the Stokes' lines are always much more intense than the corresponding anti-Stokes' lines.

Thirdly, no quantization of the rotational energy is contemplated in the proposed theory, which means that there is no possibility of calculating separately the intensities of the individual rotational Raman lines. Thus the special features met with in the intensities of the Raman lines find no adequate solution in the classical theory.

A *simple but more satisfactory explanation based* on *the quantum theory* of radiation was put forward by Prof. Smekal in 1923, which solved the problem of intensities as well.

According to the quantum theory a beam of monochromatic light of frequency v_o has its energy distributed in small bundles or quanta, each of energy hv_o. When such a light quantum or photon hits a molecule of the scatterer, three things might happen. The molecule might merely deviate the photon without absorbing its energy which would result in the appearance of the *unmodified line* in the scattered beam. The molecule might, on the other hand, absorb part of the energy of the incident photon, giving rise to the *modified Stokes' line,* whose frequency would evidently be less than that of the incident radiation. It may also happen that the molecule, itself being in an excited state, imparts some of its intrinsic energy to the incident photon and this would produce the *anti-Stokes' line,* of frequency greater than that of the incident radiation.

The mechanism of Raman scattering can therefore be analytically expressed as follows:

Treating the phenomenon as a collision between the photon and the molecule and applying the principle of conservation of energy, we may write

$$E_p + (\tfrac{1}{2}) mv^2 + hv_o = E_q + (\tfrac{1}{2}) mv'^2 + hv' \quad ....(1)$$

where E_p and E_q are the intrinsic energies of the molecule before and after collision respectively, m the mass of the molecule, v and v' its velocities before and after impact, v_o and v' the frequencies of the incident and scattered photon.

As the collision does not cause any appreciable change of temperature, we may assume that the kinetic energy of the molecule remains practically unaltered in the process and hence simplify the above relation as

$$E_p + hv_o = E_q + hv'$$

Therefore
$$h (v' + v_o) = E_p - E_q$$
or
$$v' = v_o + (E_p - E_q)/ h \quad .... (2)$$

In this expression,
if $E_p = E_q$, $v' = v_o$, which represents the unmodified line;
if $E_p < E_q$, $v' < v_o$, referring to Stokes' line and finally,
if $E_p > E_q$, $v' > v_o$, which corresponds to the anti-Stokes' line. Thus the fundamental feature of the Raman effect is accounted for.

Considering the molecule involved in the process, since the same property of possessing energy in quanta appears in molecules and atoms, as we have already seen, we may apply quantum principles to the change in the intrinsic energy of the molecule and write

$$E_p - E_q = n h v_m \quad .... (3)$$

where $n = 1, 2, 3$, etc., and v_m the characteristic frequency of the molecule.

Hence equation (2) reduces, in the simplest case of $n = 1$, to:

$$v' = v_o \pm v_m \quad .... (4)$$

This expression shows that the difference in frequency $(v_o - v')$ between the incident and scattered lines in the Raman effect corresponds to the frequency v_m characteristic of the molecule which thus verifies the experimentally observed fact, viz., that the Raman lines are symmetrically situated on either side of the parent line at intervals corresponding to the characteristic frequency of the molecule and hence to the infra-red absorption lines of the scatterer.

Turning now to the question of intensities of the lines, we know that the molecules of a material medium are distributed among a series of quantum states of energies E_1, E_2, E_3 ... Assuming that the statistical distribution of the molecules in these different quantum states is governed by Boltzmann's law, the number of molecules N_p in the particular states E_p is given by

$$N_p = C \, N \, g \, \exp(-E_p/kT) \, \ldots\ldots\ldots\ldots\ldots\ldots\ldots (5)$$

where C is a constant, N the total number of molecules, g_p the statistical weight of the state and k the Boltzmann's constant. One can draw from this expression the following conclusions as regards the relative intensities of the Raman lines:

(a) The Stokes' lines should be more intense than the corresponding anti-Stokes' lines. For, the former are caused by molecules of low energy value and hence are more numerous than those of higher energy. In fact, the smaller the value of E_p the larger the value of: $\exp(-E_p/kT)$ and in consequence the greater the number N_p. Thus, statistical distribution results in the Stokes' transitions being more frequent than the anti-Stokes', which means that the Stokes' lines will be more intense than the anti-Stokes' lines as experimentally observed.

(b) The intensity of the anti-Stokes' lines should increase as the temperature is raised. For they are produced by molecules which are of high energy value and will be comparatively few at ordinary temperatures. But, with increase of temperature the kinetic energy of molecules increases and more molecules will be raised to higher energy states by inelastic collisions. Referring N_p in equation (5) to the molecules in higher energy states, we see that their relative number will increase with rise of temperature, since as T increases the value of : $\exp(-E_p/kT)$ becomes greater. With the increase in the number of high energy molecules, that of low energy molecules will correspondingly diminish.

Thus the quantum picture of the Raman scattering explains adequately most of the experimentally observed characteristics, including the intensity, of the Raman lines.

It has a further advantage of fitting the Raman effect into the same picture as several other phenomena connected with the interaction between radiation and matter. In fact, from this point of view

the Raman effect can be treated as the optical analogue of the Compton effect.

The two effects resemble one another in the fact that there is partial absorption of the photon in both cases of scattering. They differ, however, in the result produced:

In the Compton effect the atom is ionized so that after the process not only a photon of altered frequency but also an electron leaves the atom, while in the Raman effect the molecule is merely excited, due evidently to the lower energy content of the incident photon and a radiation of altered frequency alone emerges from the process without any electron leaving the molecule.

The quantum theory interpretation of the Raman effect, though sound and satisfactory as far as it goes, yet is found to be too simple, as it does not cover other complicated features of the actually observed effect, such as the discrepancy between the Raman and infra-red absorption spectra, i.e., want of correlation between the two, the different conditions that determine the appearance of the rotational Raman lines alone, or the vibrational Raman lines alone, or both together, etc.

A *more complete theory has been proposed by Kramers and Heisenberg* in their treatment of the phenomenon of dispersion, worked out first (1925) from the point of view of the correspondence principle and later (1927) perfected by the use of quantum mechanics. As it is beyond the scope of this book to deal with this rather complicated theory, involving as it does, the calculus of matrices, we shall rest content with some of the important; conclusions arrived at, in relation to the Raman effect.

Hendrik Anthony Kramers (17 Dec 1894 - 24 Apr 1952)

Considering a molecule in the quantum state k, the perturbation due to a light wave of frequency v gives rise to harmonic components in the electric moment of the system having frequencies expressed by

$$v^*_{lk} = v — v_{lk}$$

where l represents another state of the system and v_{lk} the frequency corresponding to a transition of the scattering molecule from state k to state l.

When the electric moment associated with the transition involved has the same direction as the electric vector of the light wave, the amplitude of the harmonic component v^*_{lk} in the electric moment and hence the intensity I of the scattered radiation is proportional to:

$$\Sigma_r \, a_{lr} a_{kr} \left[1/ (v_{rl} + v) + 1/ (v_{rk} - v) \right]$$

where $a_{lr} a_{kr}$ are numbers which determine the transition probabilities between the states r and l and r and k. If a_{lr} or a_{kr} = 0, the corresponding transition is forbidden. The special feature of the above expression is that it does not contain the probability of the transition ($k \rightarrow l$).

From this it follows that the selection rules that govern the Raman and the infra-red absorption spectra are quite different.

The condition for observing a transition of the molecule from a state k to another state l as a Raman line is not that this transition ($k \rightarrow l$) should occur in direct absorption or emission (which means $a_{kl} \neq 0$) but that both the k and l states should combine with a third intermediate state r (or with more than one), such that the transitions ($k \rightarrow r$) and ($r \rightarrow l$) can take place in ordinary absorption or emission. In other words, if the selection rule that holds for infra-red absorption spectra, i.e., for transitions ($k \rightarrow r$) and ($r \rightarrow l$) is given by $\Delta J = \pm 1$, then the selection rule for Raman spectra, i.e., for the transition ($k \rightarrow l$) is given by $\Delta J = 0$ or ± 2. Hence it is that, although every Raman line corresponds to a definite line in the spectrum of the molecule and the frequency of a Raman line can be considered as the difference between the frequencies of two lines, one of which must be an absorption line, yet the intensities in the two cases may be quite different.

The reason for this difference in the two phenomena is to be traced, ultimately, to the different conditions that are required in the two cases. In the direct absorption effect, the presence of an electric moment or an alteration in that moment is essential, while in the Raman effect other factors like optical anisotropy, alteration of the mean polarizability or mean refractivity of the molecule, prevail.

Thus a *rotational Raman line* occurs only when the molecule involved is optically anisotropic, its intensity being the greater, the greater the anisotropy. Hence it is that electrically non-polar but optically anisotropic molecules like H_2, O_2, N_2, etc., give rise to pure rotational Raman lines, while they have no corresponding far infra-red absorption lines.

A *vibrational Raman line* occurs when the mode of oscillation involved causes a change in the mean polarizability of the molecule, its intensity being greater, the greater the above mentioned change. This is why molecules like CO_2, CS_2, NO_3, etc., in which symmetric oscillations occur, cause no change in electric moment, but on account of the large variation in the mean polarizability involved, give rise to intense vibrational Raman lines which do not correspond to any infra-red absorption lines. Rasetti, studying the Raman effect in nitrogen and oxygen has succeeded in verifying the selection rule $\Delta J = 0, \pm 2$. Similarly the analysis of the Raman spectra of HCl gas *by* Wood and Dieke has led to the same conclusion as regards the selection rules involved in the origin of Raman lines.

It may be noted that the theoretical values obtained for the relative intensities of the individual Raman lines as well as for their depolarization factors have also been proved to be in good agreement with experimental results.

Importance of the Raman effect

Over and above the great theoretical interest attached to it as a further confirmation of the quantum theory of radiation, the Raman effect is of immense practical importance on account of its many useful applications in Physics and Chemistry:
1. The universal nature of the phenomenon;
2. The relative simplicity of the experimental technique as compared with that of infra-red measurements;
3. The ease with which the effect can be controlled, making it appear as a part of the visible spectrum, which can be chosen at will (its position depending only on the choice of the incident frequency) and
4. The complementary character of the spectra obtained with reference to those of infra-red absorption (the former containing lines which are forbidden in the latter)

All those practical factors make the Raman effect a very convenient and powerful tool of research in problems concerning the intimate structure of matter, chiefly as regards the constitution of molecules, their number, arrangement and motion, in gaseous, liquid and solid states.

We shall here summaries some of the important applications which have led to sure and significant results.

APPLICATIONS OF THE RAMAN EFFECT IN PHYSICS

(i) Molecular structure

The Raman effect has been put to very great use in the study of the structure of molecules. Raman spectra are, in general, determined by those factors which affect most the nature of vibration, viz., the number of atoms in the molecule, the masses of the atoms and the strength of the chemical bonds between the atoms.

Taking a *diatomic-molecule*, its natural frequency of vibration is given by

$$\nu_o = (1/2\pi) (F/\mu)^{1/2},$$

where F is the restoring force per unit displacement and μ the resultant mass. This means that
(a) there is only one vibration frequency in diatomic molecules,
(b) a molecule containing light atoms should have a higher frequency than one containing heavier atoms and
(c) a molecule in which the force binding the atoms is great should have higher characteristic frequency than one in which the force is weak. This force depends on the nature and strength of the inter-atomic bonds. Thus a diatomic molecule with a double bond should have a higher frequency than one containing only a single bond. The Raman lines are also expected to appear with great intensity when the bond is of the covalent type (homopolar or electrically non polar molecules) while the reverse is the case when the bond is of the electrovalent type (polar or heteropolar molecules). The reason for this is to be found in the fact that

the Raman lines essentially depend upon the symmetry of the molecules and the extent to which the polarizability is affected by the oscillations.

In covalent molecules the binding electrons remain common to the nuclei, so that the polarizability of the molecule is considerably modified by nuclear oscillations and this variation in polarizability gives rise to Raman lines.

In electrovalent molecules the binding electrons definitely change over from one nucleus to the other in the formation of the molecule so that the polarizability of the molecule is little affected by nuclear oscillations, which means no Raman lines will occur.

Considering *polyatomic molecules* which are constituted by more than two atoms, the Raman spectra will naturally be much more complex.

First of all, more than one characteristic frequency is to be expected: *e.g.,* triatomic molecules will, in general, have three such frequencies.

Secondly, the arrangement of atoms, such as symmetrical or unsymmetrical, linear or non-linear symmetry, etc., is an additional factor by which the intensity of the Raman lines is essentially determined. Hence, from the number and intensity of the observed lines in the Raman effect, in conjunction with infra-red data, it is possible to draw important conclusions about molecular structure.

Theory leads to the following important rule, known as *the rule of mutual exclusion.*

For molecules with a centre of symmetry, transitions that are allowed in the infra-red are forbidden in the Raman spectrum and *vice versa.*

This rule implies that for molecules without a centre of symmetry there are transitions that can occur both in the infra-red and the Raman effect. It does not however, imply that all transitions that are forbidden in one must occur in the other. For, some transitions may be forbidden in both.

The following examples may serve to illustrate these remarks.

Diatomic molecules

Some of the diatomic molecules studied are H_2, O_2, D_2, N_2, HCl, HBr and HI. The first four are homonuclear molecules, *i.e.,* constituted with identical atoms, while the last three heteronuclear molecules, *i.e.,* made up of non-identical atoms. In all these cases there is only one vibration frequency and its value obtained from Raman spectra is given below. The restoring force per unit displacement F in each case, calculated with the help of the relation $v_0 = (1/2\pi)(F/\mu)^{1/2}$, is also given.

	H_2	D_2	N_2	O_2	HCl	HBr	HI
v_0 (cm.$^{-1}$)	4156	2993	2331	1556	2880	2558	2233
$F \times 10^{-5}$	5·1	5·3	22·4	11·4	4·8	3·8	2·9

It is readily seen from such a tabulation that (1) the heavier the molecule the lower is the vibration frequency and (2) the values of F, which is a measure of the binding strength, may be roughly classified in the ratio of 3: 2: 1, thereby indicating the existence of three different types of bonds, triple, double and single, between the atoms in the molecule.

Triatomic molecules

Among the several triatomic molecules analyzed, we shall consider here only three, viz., CO_2, N_2O and H_2O as typical cases illustrating some special points of interest in molecular structure.

Carbon dioxide (**CO₂**) is one of the most frequently and thoroughly studied molecules. It has two very strong bands in its infra-red absorption spectrum at 668 and 2349 cm^{-1}, while only one strong band in its Raman spectrum at 1389 cm^{-1}. As none of these occurs both in the Raman and infra-red spectrum, it follows from the rule of mutual exclusion that the molecule of CO_2 must have a centre of symmetry. For triatomic molecules this implies that the molecule is linear and symmetric. Hence the molecular structure of CO_2 is O—C—O. This symmetric structure is confirmed by the rotational Raman spectrum of CO_2, which consists only of alternate (odd) levels, like that of O_2. The three bands stated above, two in the infra-red spectrum and one in the Raman spectrum, represent the three fundamental frequencies of CO_2. Carbon disulphide (CS_2) also belongs to this class.

Nitrous oxide (**N₂O**). This molecule has the same number of electrons as CO_2 and one might therefore expect it to have a linear symmetric structure. But analysis of the infra-red and Raman spectra of N_2O shows clearly that this molecule, although linear, is not symmetrical. The three fundamental frequencies of N_2O are 2224, 1285 and 589 cm^{-1}. All the three appear in the infra-red absorption and two of them, viz., 2224 and 1285, appear in the Raman spectrum. 589 has, however, not been recorded, due to weak intensity. At any rate, the two Raman lines appear also in the infra-red, which argues to the absence of the centre of symmetry in the molecule. For, if there was a centre of symmetry, only one fundamental should appear in the Raman spectrum and this fundamental should not appear in the infra-red, according to the rule of mutual exclusion. Hence the molecule has the asymmetrical structure N — N —O. This conclusion is confirmed by the rotational Raman spectrum of N_2O, which consists of both (odd and even) sets of lines without any alternation in intensity. Other molecules of this type are HCN, ClCN, BrCN, ICN, etc.

Water (**H₂O**). The observation of a strong pure rotational spectrum and its structure, as well as that of the rotation-vibration spectrum lead unambiguously to the conclusion that H_2O, though symmetrical, is not linear but bent (Fig. 257).

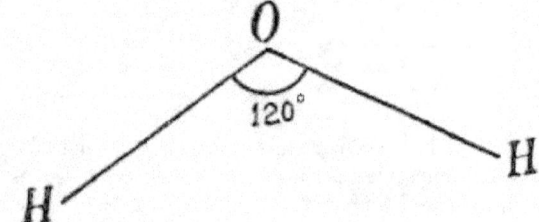

Fig. 257

According to theory, all triatomic molecules of bent symmetrical structure should give rise to three Raman lines all of which are represented in the infra-red spectrum. There are two very strong bands at 1595 and 3756 cm^{-1} in the infra-red absorption spectrum, which are very probably two of the three fundamental frequencies. There is also evidence to show that the bend in the system is 120°. In the Raman spectrum, two frequencies have been recorded at 1665 and 3605 cm^{-1}, which correspond roughly with the infra-red frequencies. There is some difficulty in recording the Raman lines of H_2O, which are very weak, due evidently to the fact that the moments of the light hydrogen atoms do not cause appreciable variation in the polarizability tensor. Water gives some other extra bands which have to be ascribed to polymerized molecules—$(H_2O)_2$ and $(H_2O)_3$. D_2O, S_2O and H_2S have similar bent symmetrical structures.

Thus the study of Raman spectra of different substances enables one to classify them according to their molecular structure.

(ii) Nature of the liquid state

Considering the gaseous state, it can be readily shown that the translatory motions of the molecules are such that the Rayleigh scattering will be spread out in accordance with Maxwell's law. On the other hand, for the solid state, Brillouin, disregarding molecular structure and replacing the solid by a homogeneous elastic continuum, so that the translatory motions may be resolved into a system of elastic waves traversing the medium in all directions, arrived at an expression showing that the scattered light should merely consist of two displaced components depending upon the velocity of the waves and the angle of scattering. Now, as regards the intermediate liquid state, Raman and Ragavendra Rao, in their study of the Raman spectra of a number of liquids, found that while Brillouin's theory was verified, a central line brighter than the displaced components was seen. With increase of temperature the central line grew in intensity at the expense of the neighboring Brillouin components which themselves broadened out and tended to merge into the central line. The two Brillouin components were found to be completely polarized in transverse scattering, while the central line was also nearly fully polarized in non-polar liquids. In strongly associated liquids, like formic acid, the central line showed appreciable depolarization, while in viscous liquids the Brillouin components were faint.

The experimental finding of the Brillouin components is characteristic of the elastic continuous medium. The presence of the central line, however, which is not to be expected in Brillouin's theory, can be accounted for solely on the basis of random type of movements, as in a gas. The central component is always brighter and this readily follows from the fact that gaseous Rayleigh scattering is always much more intense than the corresponding scattering from liquids or solids. The relative intensities of the central and Brillouin components vary from liquid to liquid and this may be taken to show how the thermal energy is shared between the two types in each liquid.

The observations of Raman spectra have thus led to an important conclusion regarding the liquid state, viz., that it is characterized by both types of thermal energy, one in which it is organized in the manner of elastic waves and the other in which it is akin to random types. In other words, the thermal motion of liquid molecules, which goes to make up its heat content, is neither wholly disorganized as in the case of the molecules of a gas, nor wholly organized in the form of elastic waves as in a solid.

(iii) Crystal physics

The analysis of crystal structure by X-ray diffraction methods concerns itself with the determination of the position and arrangement of the atoms or molecules which scatter the X-rays. It does not tell us anything about the forces binding the atoms or molecules together in the lattice, which determine the physical properties of the crystal, and hence of the solid state, since most solids are crystalline in structure.

The Raman effect offers us a new method of studying crystals, which, in a sense, is complementary to X-ray diffraction. For, it furnishes us with just the kind of information that X-rays do not give concerning the strength and nature of the forces which hold the crystal together. Furthermore, this information is of a precise character, as the Raman lines are very sharp and can be measured very accurately. The frequencies deduced from the Raman spectra and the known positions and masses of the atoms in crystals enable us to determine the binding forces in the crystals.

Although the study of Raman effect in crystals has been initiated only recently and some important results have already been obtained. The case of diamond, studied by Ramasamy and Bhagavantam, shows clearly that it is not necessary for a crystal to contain molecules in order that it should exhibit Raman effect, for diamond is a non-polar substance in which it is impossible to identify any particular group of atoms as forming an ion or molecule.

The Raman spectra of a series of crystalline nitrates analyzed by Krishnamoorthi and Gerlach established that the frequency characteristics of the NO_3 ions are notably influenced by the presence of the cations. This suggests that the assumption of complete ionization in the solid crystal may be invalid.

The Raman effect in crystals has been studied also with reference to two other fundamental problems in crystal physics, viz., the existence of a *"mosaic" structure* and *thermal agitation* in crystals. From the Raman spectra of crystals, the existence of discrete monochromatic infra-red frequencies has been clearly demonstrated. In a quantitative study of the phenomenon, Raman has shown that

a crystal has, in general, $(24p - 3)$ fundamental modes of atomic vibration with monochromatic frequencies, where p is the number of interpenetrating atomic lattices of the crystal.

Of this number $(3p - 3)$ modes may be ascribed as oscillations of the p interpenetrating lattices, while the remaining $21p$ modes are oscillations with respect to each other of various important planes of atoms in the crystal. The number of monochromatic frequencies is shown to be considerably reduced when the crystal has a high degree of symmetry. But the number of frequencies and the geometry of the modes have been fully worked out from considerations of symmetry.

For instance, diamond has been shown to have eight fundamental frequencies, rocksalt nine, and fluorspar fourteen. The fine structure of the infrared spectrum of crystals actually observed is readily and adequately accounted for on the basis of the above-mentioned. analysis.

(iv) Nuclear physics

The Raman effect has been applied also in the study of certain aspects of nuclear physics, such as the *spin, statistics* as well as the *isotopic* constitution of the nucleus. The rotational Raman lines of homonuclear diatomic molecules like H_2, O_2, D_2, N_2, etc., characterized by their alternating intensities, have been most fruitful in giving precise results.

Thus, for instance, in the case of *hydrogen*, the Raman lines arising from odd rotational levels are three times more intense than those coming from transitions between even levels, winch can be shown to be due to the fact that
the hydrogen nucleus has a spin of half a unit, obeying Fermi-Dirac statistics and that the statistical weight of the odd rotational levels is three times that of the even rotational levels.

With *deuterium,* the stronger of the alternating Raman lines correspond to transitions between even levels and their intensities are twice that of the weaker ones arising from odd levels. These observed data lead to the results that
the deuterium nucleus has a spin of one unit, conforming to Bose-Einstein statistics, and that the statistical weight of the even levels is twice that of the odd levels.

In the case of *oxygen* the entire rotational spectrum consists of lines arising from transitions between odd levels only, while those from even levels are completely missing, which means that
the nuclear spin of oxygen is zero.

With *nitrogen* the intensities of the Raman lines alternate, the more intense lines corresponding to transitions between even levels, unlike hydrogen but similar to deuterium. This leads to the conclusion that
the nitrogen nucleus has a spin of one unit and obeys Bose-Einstein statistics.

These results are in complete agreement with those derived by other methods. The substitution of an isotopic atom for the original one in the molecule is expected to alter the symmetry of the molecule and the effective mass. Many isotopic molecules have been studied from this standpoint and it has been found that the phenomenon of alternating intensities in the rotational Raman lines, well exhibited by the spectra of H_2 and D_2, is absent in the spectrum of HD showing that the substitution of D for H has destroyed the symmetry.

APPLICATIONS OF THE RAMAN EFFECT IN CHEMISTRY

The applications of Raman effect in chemistry, in all its three different branches, inorganic, organic and physical, are so vast and ever-increasing, that it is impossible to go into a detailed study here. We can just mention a few of the important results obtained.

(i) Inorganic chemistry. Chemical constitution and valency bonds
Numerous facts point to the relation existing between Raman effect and the chemical valency bonds. The study of the Raman effect in a large number of the halides by Krislinamoorthi has established beyond doubt that
Raman lines appear with great intensity in covalent compounds while the reverse is the case in electrovalent compounds.
Thus, for instance, NaCl, KCl, etc., which are electrovalent compounds, exhibit no Raman lines corresponding to their infra-red frequencies, whereas HCl, the chlorides of non-metals, etc., which are covalent compounds, give vibrational Raman lines.

It is of interest to note that the vibrational Raman line is observed in.HCl, gas and liquid, but not in aqueous solution of HCl, which consists of hydrogen and chlorine ions. In aqueous solutions of salts, where ionization is complete, one finds no Raman lines characteristic of the salt as a whole but only those corresponding to the molecules of the separated ions and water. These facts confirm what we have stated above about the dependence of the intensity of the Raman lines on the chemical nature of the linkage, *viz.,*
Raman effect appears most strongly when the atomic structure under consideration is bound by covalent bonds and tends to weaken when the bonds change their character and pass into electrovalent linkages.

The fact that most of the organic compounds, characterized by covalent linkages, give rise to strong Raman spectra may be considered as further evidence in support of the above conclusion.

The strength of chemical bonds is also duly reflected in the Raman effect so that, calculating the force constants with the help of the Raman frequencies, the chemical bonds can be classified into triple, double and single Complex compounds, mixed molecules and water of crystallization are some of the other important problems which have been tackled with the help of the Raman effect.

(ii) Organic chemistry
The Raman effect has been found of great value in organic chemistry also. It has been used in determining the presence or absence of specific linkages in a molecule, identifying impurities of certain types, estimating quantitatively the relative proportions in which the constituents of a mixture are present, fixing up the structure of important molecules and studying the different types of isomerism. It is interesting to note that the chemist's classification of organic substances into *aliphatic* or open chain compounds and *aromatic* or closed chain compounds is clearly reflected in the character of the respective Raman spectra.

(iii) Physical chemistry
Very many interesting problems appertaining to this branch of chemistry, such as the transition from crystalline to amorphous state, electrolytic dissociation, hydrolysis, etc., have been investigated with the Raman effect. Some of the more important results obtained are:

(a) Although there is a general correspondence between the Raman spectra obtained in the crystalline and amorphous states of a substance, the latter state gives rise to broad and diffuse bands in the place of sharp lines produced by the former.

(b) Raman effect affords an ideal method for investigating the phenomenon of *electrolytic dissociation*, as spectra can be obtained both for the pure substance and for the aqueous solution of it at different concentrations. The observed positions and the variation of intensity of the lines enable one to determine directly the nature and number of ions produced and thus decide whether the dissociation is complete or only partial.

(c) The problem of *hydrolysis* is similar to that of electrolytic dissociation. If a salt, possessing a Raman spectrum different from that due to the base or the acid, is hydrolyzed by water, the degree of hydrolysis can be readily estimated by measuring the relative intensities of a set of lines characteristic of either the base, the acid or the salt.

Thus the Raman effect has come to stay as a powerful weapon in the hands of scientists for the analysis of those intricate phenomena which reveal the intimate structure of matter.

5. SOLID STATE SPECTROSCOPY

Introduction
The investigation of the solid state of matter may be said to occupy an important place in Modern Physics, both experimental and theoretical. From the study of bulk properties in the macroscopic state of solids, it has been shown that solids, in general, are characterized by their elasticity of volume and shape. As regards their internal structure, it has been established that they are essentially *crystalline, i.e..* they are aggregates of minute crystals packed together in a more or less random fashion. Many metals and alloys in the solid state are thus *microcrystalline* or *polycrystalline.* It is usual to distinguish the polycrystalline bodies from *single crystals* which may ho defined as solid objects of uniform chemical composition, which as they occur in nature or are formed in the laboratory, have a very regular structure, *i.e.,* bounded by plane faces, the interrelation of which exhibits a typical symmetry: *e.g.,* quartz, rock-salt etc. It is evident that the study of crystals is bound to throw light on the intrinsic nature of the solid state and hence the introduction of a now branch in Physics, known as *Crystal Physics,* which comprises both theoretical crystallography and analysis of crystal structure by X-rays, Raman effect and neutrons. We have already seen how Raman effect is utilized to analyze crystal structure. The use of neutrons for the same purpose will be mentioned later in the section dealing with neutrons. Here we shall limit ourselves to X-rays.

The basic structure of the solid state has been recently analyzed in the following three different ways:

(i) fine structure- of the X-ray absorption edges,
(ii) X-ray emission spectra of solid targets,
(iii) modified X-ray reflections from solid crystals.

It is to be noted that although the spectroscopic theory developed for free gaseous atoms. could be utilized in dealing with X-ray spectra of solids, yet, in cases, where the outer shells of the atom are involved, modifications will have to be made since these shells will be greatly disturbed by the interactions of the atoms. A brief description of the above-stated three lines of investigation will now be given.

A. Fine structure of the X-ray absorption edges

We have already seen that, in the case of free atoms, due to the intervention of the outer part of the atom, i.e., when the electron from an inner shell, instead of being ejected completely from the atom, is stopped in some outer vacant shelf, a series of resonance absorption lines can arise with the absorption edge as their series limit. The arrangement of the electrons in the outer part of an atom, however, will be considerably altered when the atom is forced into close proximity with many other atoms, as in a solid. Consequently, we may expect the observed fine structure of absorption edges with solid absorbers to be materially different from that found for a gaseous absorber, but nevertheless, some type of resonance absorption is to be expected. Just as an interpretation of the fine structure of the absorption edges in a gaseous absorber may give us information about the internal structure of a free atom, so may a similar phenomenon with a solid absorber reveal details about the basic structure of a solid.

The fine structures of the K edges of several solids have been obtained by Beeman and Friedman (1939), Trischka (1945) and others. A typical absorption curve showing the K-edge of Cu is reproduced from the work of Beeman and Friedman (Fig. 258).

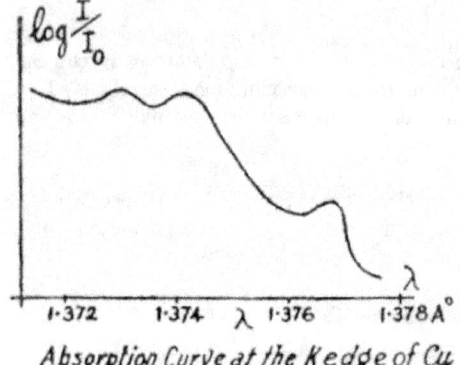

Absorption Curve at the K edge of Cu

The curve shows obvious indications of resonance absorption at the **X** limit, but no distinct lines stand out. Theoretical considerations lead to the conclusion that *an electron removed from the interior of an atom must be transferred to an electronic shell that belongs to the entire solid rather than to any individual atom* and the energies associated with such transitions are distributed over broad bands instead of being confined to a discrete set of values. The theory, however, has not yet been developed to the extent of giving a clear interpretation of the details of the observed structure.

B. X-ray emission spectra of solid targets

Skinner (1940) and Cady and Tomboulian (1941) have studied the emission spectra from solid targets, in the long wavelength region, from 5 to 180 A°. Mica crystal spectrograph was used up to about 20 A° and ruled grating spectrograph for higher wavelengths. Attention was chiefly directed on the lines that involve transitions of the valence electrons which belong to the solid as a whole. The effect of the interaction between neighboring atoms in a solid is equivalent to replacing the sharp levels of the free atom by bands. Levels associated with vacancies deep in the interior of the atom, such as the K levels in heavy atoms, are only very slightly broadened. But levels corresponding to changes in the outermost atomic electrons will broaden out appreciably. Hence the higher members in the series of X-ray lines, as observed when they are emitted by free atoms, become, in the case of solid emitters, broad bands which may overlap. The shape of such bands, produced when the valence electrons associated with the solid as a whole drop into vacancies in the inner shells of atoms, will give information about the structure of the solid. Thus Cady and Tomboulian analyzing the shape of the X-ray bands emitted in the "valence → L" transitions for aluminum, have been able to show that *the energy distribution of the valence or conduction electrons in a metal is of the Fermi-Dirac type.*

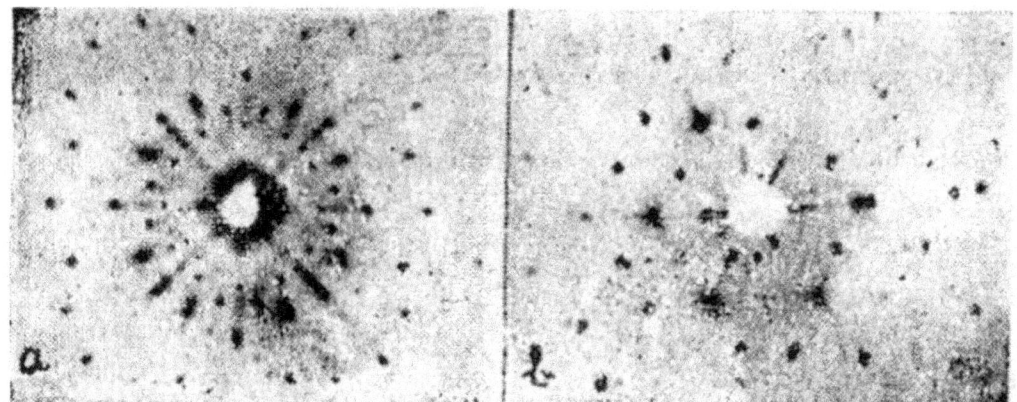

Modified X-ray reflections from crystals. (*a*) Sylvine, (*b*) Benzil (Lonsdale and Smith)

C. Modified X-ray reflection from solid crystals

Well-exposed photographs taken of X-rays after they have traversed stationary single crystals, as in the Laue method, frequently show, in addition to the usual clearly defined Laue spots, a fainter background pattern of diffuse streaks, spots and halos, which do not apparently correspond in position to the Bragg reflection from any planes of the stationary crystals, as seen in the adjacent photo.

Although such effects were first observed by Friedrich as early as 1913, it is only in the past few years that any serious attempt has been made to study them, on account of their probable connection with other phenomena met with in the study of crystal structure, such as scattering of light by crystals with altered frequency, luminescence and absorption spectra at low temperatures, the anomalous variation of specific heat of solids at low temperatures, etc., and of the consequent importance of these background secondary reflections of X-rays from crystals in the building up of a cogent theory for the solid state of matter.

Wadlund (1938), Preston (1939), Laval and Magnin in France (1939), Raman and his collaborators in India (1940) and Lonsdale, Knaggs and Smith in England (1940) have made extensive researches on the different aspects of the phenomenon, such as the azimuth effect, temperature influence, etc. They employed X-rays from Cu and Mo targets, unfiltered or monochromatized, and different crystals, both inorganic and organic, as well as metallic, such as diamond, quartz, KCl, NaCl, $NaNO_3$, $CaCO_3$, naphthaline, benzil, sorlic acid, Al, powdered Ag, etc. The results obtained may be summarized as follows:

(i) The diffuse pattern does not seem to be affected by the nature of the surface of the crystal or by any crystal imperfection. It resembles in many respects the Laue pattern of an oscillating or rotating crystal. This is a general fact applying to inorganic and organic crystals of any system and to all wavelengths of radiation used. It is not only seen in Laue photographs, where it is, in general, well separated from the regular Laue pattern, but also in powder photographs, where it forms a background to the selective Bragg reflection.

(ii) The diffuse halo surrounding the central spot does not appear on monochromatic photographs; the radial streaks on ordinary Laue photographs are almost due to diffuse reflection of unfiltered 'white' radiation; the diffuse spots are always obtained in monochromatic Laue photographs, thereby indicating that they are due to diffraction of characteristic X-rays.

(iii) The shape and size of the diffuse spots vary considerably, being very frequently not circular or even elliptical. They are more closely related to the shape of the corresponding Bragg spots than to those of the Laue spots. The positions of the diffuse spots are very nearly, sometimes even exactly, those of Bragg reflections from crystal planes, which are not necessarily, however, quite at their proper reflecting angle. The diffuse spots seem thus to be related to the structure and nature of the crystal planes to which they correspond and hence may give valuable help in structure determination.

(iv) The intensity of the diffuse spots depends on their positions with respect to the Laue spots, increasing as the distance diminishes. A very important factor affecting the intensity is the *crystal temperature. At high temperatures* (up to 500°C) some diffuse spots increase enormously both in intensity and size and non-radial streaks appear, which were not previously visible. Other diffuse spots are not so strongly affected. The Laue spots usually become fainter with rise of temperature. At *low temperatures* (liquid-air) the diffuse spot pattern nearly disappears in many crystals. The Laue spots become more profuse and the general background clearer. But the case of diamond seems to depart from this general rule under certain conditions. The Indian scientists, working with (1, 1, 1) crystal planes of diamond, have shown that the intensity of the diffuse spots is independent of temperature.

Although many of these background effects may be accounted for as due to imperfections in the specimen, strains, etc., yet there still remain streaks and spots, much too regularly and systematically organized and depending on too many factors such as the nature of the incident beam, orientation, symmetry and temperature of the crystal, which therefore demand a more serious explanation than mere bulk or macroscopic defects stated above. It is admitted by all that while the primary reflections of X-rays from crystals, which give rise to the ordinary Laue spots, are due to the *static, space-periodic* arrangement of the atoms, the secondary reflections which produce the background diffuse pattern are to be ascribed to the *dynamic, time-periodic* nature of crystal lattices that vibrate about equilibrium positions within certain limits. But as regards the origin and nature of these vibrations, a lively controversy has arisen between eminent scientists, such as Max Born, Lonsdale and others on the one hand and Raman and his co-workers on the other.

The English scientists attribute the observed effect to purely thermal agitation of the atoms in the crystal and try to explain the different characteristics of the background pattern on the semi-classical basis of elastic vibrations in the solid constituting a continuous spectrum with a limiting maximum frequency, first suggested by Debye in the interpretation of the specific heats of solids.

Max Born, with his so-called "cyclic lattice" theory has given a rather complicated mathematical treatment of the problem and obtained a scattering law which represents the positions of the diffuse maxima by a formula similar to that of Bragg, giving at the same time the dependence on temperature.

But the basic idea underlying his treatment seems essentially to be the same as that of Debye, viz., that all the possible vibrations of a crystal lattice are analogous to elastic vibrations, having the same distribution of phase waves in respect to the wavelengths and relation to the size and orientation in space of the crystal. Herein probably lies the chief drawback of the otherwise erudite theory of Born. For, the reason why the Debye formula for specific heat of solids fails to represent the experimental data in several cases is precisely due to the fundamental assumption that the atomic vibrations constitute a continuous spectrum of all possible elastic wavelengths with an arbitrarily chosen limiting wavelength, depending on the external dimensions of the crystal.

A crystal, in the first place, cannot be treated as a continuum having a uniform density and elasticity, since it has a discrete atomic structure, uniform neither in electron density nor in mass distribution.

Secondly, the atomic vibrations of the crystal lattice which depend on the individual masses of the atoms and on their arrangement in the fine structure of the crystal cannot be assimilated to vibrations of the elastic type which are governed by the macroscopic properties and external dimensions of the solid.

Thirdly, the contribution of the low frequency elastic type vibrations to the thermal energy will be, under normal conditions, relatively of minor importance.

Fourthly, the Raman effect in crystals leads to the existence of discrete and monochromatic vibrations of the crystal lattice, which fall in the infra-red region, hence relatively of much higher frequencies than the low elastic vibrations of Debye's thermal spectrum.

Finally, the effect of all vibrations of the elastic type of low frequency can be only a diffuse scattering without any geometrical arrangement of spots, the intensity being much smaller and varying much less rapidly with temperature than what is actually observed.

The Indian scientists, under the leadership of Raman, interpret the observed phenomenon as primarily a quantum effect of the interaction of the incident X-ray photon with the atoms of the crystal vibrating with discrete monochromatic frequencies, in a manner analogous to the Raman effect in the optical region. It is, therefore, not essentially of thermal origin, though for low-lattice frequencies or at high temperatures it may be thermally influenced. Hence they have named the background pattern "quantum or modified X-ray reflections", not satisfied with the nomenclature of "diffuse X-ray reflections" used by the English scientists. According to them, the incident X-ray quantum exciting the infra-red monochromatic frequencies of the crystal lattice produces dynamic stratifications in the crystal, which scatter the modified incident quantum, thereby giving rise to dynamic reflections with altered frequency. The discrete frequencies of the vibrations of the crystal lattice as well as their modes, though induced by the incident quantum will depend on the internal structure of the crystal, i.e., the geometric grouping of the atoms and the forces which they exert on each other, hence on the symmetry characters also. If the number of the lattice cells be sufficiently large, as is usually the case even for a sub-microscopic crystal, these frequencies and modes will be independent of the external size and shape of the crystal, unlike the elastic vibrations of a continuous nature. The vibrations must, moreover, be pictured as having an identical frequency, amplitude and phase in all the cells of the lattice, as only then the motion will be truly monochromatic.

Basing himself on these postulates, Raman has been able to account for the simultaneous appearance of the background pattern (as modified geometric reflections) and the Laue pattern (as unmodified geometric reflections) by the same lattice planes of the crystal. He has also derived a quantitative expression for the geometric direction of the dynamic X-ray reflection, known as *Raman and Nath formula*:

$$2d \sin [(\varphi + \theta)/2] \sin [\eta + (\varphi - \theta)/2] = n \lambda \sin \eta$$

where
d is the distance between the static lattice planes,
φ the glancing angle of the dynamic reflection,
θ the glancing angle of the incident beam with respect to the static planes and
η the angle which the phase waves make with the same static planes. This relation has enabled him to put to experimental test the theory proposed.

As regards the intensity of the dynamic reflections, it is proved that under the most favorable conditions, the modified reflections have intensities which are of the same order of magnitude as the unmodified reflections, though as a rule definitely weaker. The dynamic reflections should appear not only more intense but also more sharply defined as the crystal setting approaches the position in which the static and dynamic reflections coincide. Considering the influence of temperature on the phenomena, an expression for the effect of thermal agitation that would increase the intensity of the pattern, known as the *temperature factor,* is derived, from which it is shown that the intensity can be affected by temperature only when the frequency of vibration of the lattice is very low or the temperature is very high. Under all other circumstances, the intensity is solely determined by the discrete energy of vibration of the crystal lattice.

This quantum theory explanation seems to be in accordance with the modern general treatment of the problem of interaction between radiation and matter in the microscopic atomic state. The different conclusions drawn from it as regards the position and intensity of the background spots as well as the temperature effect seem to answer fairly well to the experimentally observed facts. But its chief merit lies in the fact that it gives a more satisfactory explanation of the problem of specific heats of solids than the older Debye theory. Supposing that according to the ideas of this new theory, the part of the thermal energy associated with the elastic vibrations is practically negligible and that the bulk of the thermal energy is due to monochromatic infra-red vibrations, the specific heat of a crystal can be expressed as the sum of a number of Einstein terms, each one of which corresponds to a discrete frequency of the crystal lattice. These frequencies are taken from the available spectroscopic data, with the appropriate weightage. The specific heats thus calculated are found to be in good agreement with the observed values for a number of solids over a wide range of temperatures.

CHAPTER 13

MODERN PROBLEMS OF RADIOACTIVITY

The Nucleus of the Atom

In the study of the atom, after having analyzed the peripheral electronic structure and properties, our attention must naturally be directed on the very kernel of the atom, the nucleus, in which the essential individuality of the atom resides.

Within recent years, nuclear physics has made rapid advance, thanks to the intense and ingenious researches made by many eminent physicists both on the experimental and theoretical sides. The first real foundation of the modern conception of the nucleus was laid by Rutherford who discovered, in his experiments on the large angle scattering of alpha particles by matter, that all the positive charges of the atom and practically the whole mass of the atom were concentrated in a very small region of dimensions of about 10^{-12} cm. and less, situated at the centre of the atom. A more comprehensive idea of the structure and properties of the nucleus was then gained steadily from many independent lines of investigation, such as a more detailed study of the phenomena of natural radioactivity, artificial disintegration, cosmic rays, etc. Theoretical interpretations of these experimental data have also been made with such success that a very satisfactory understanding of the nucleus is now well under way of realization.

An exhaustive account of this wide subject cannot be given within the compass of this book. We shall, therefore, content ourselves with a brief survey of the main experimental methods and results, supplemented by relevant theoretical conclusions. The three following chapters will be devoted to the study of:

Some important modern problems of natural radioactivity,
Artificial transmutations of elements and
Cosmic rays

All of those have furnished valuable information concerning the nucleus. Then, in the last chapter, a short account of

The *structure, and properties of the nucleus,*

viz., its constitution, size, mass, charge, quantum states and statistics, spin and magnetic moment, stability and nature of the forces acting in it, will be given.

Introduction to radioactivity

It was early realized that *radioactivity is essentially a nuclear phenomenon.* For, over and above the general reason which we have already stated, viz., the independence of radioactivity from any external influence, physical or chemical, the following particular and additional considerations indicate that the rays emitted by radioactive substances have their origin in the nucleus:

(i) Certain radioactive elements emit α-particles which are identified with helium nuclei. Now, since outside the nucleus nothing except electrons is provided in the accepted atom model, the α-particle must be supposed to proceed out of the nucleus itself.

(ii) Considering the radioactive elements which emit β-particles identified with electrons, although the above line of argument cannot be directly used, as there are many difficulties in accepting the actual existence of electrons in the nucleus, still there are other reasons to believe that the real β-rays or *disintegration electrons* as they are called, do not come from the extranuclear family of electrons but from some other source. For, it is possible to extract electrons out of the various peripheral shells of the atom; but when this is done, the resulting "ionized" atom promptly takes in other electrons to fill up the gap and reverts to its original state. This is not the case, however, with the radioactive elements which emit β-particles. Neither can one extract by external force these electrons, nor does the atom that has lost a β-particle revert to its original state. The change is *irreversible* and the atom is altered for good, giving rise to a new different atom. Hence, it follows that the β-particles which cause the disappearance of their very *parent* atoms cannot originate from the extranuclear group of electrons.

(iii) As regards the nuclear origin of the γ-rays identified with electromagnetic radiation, we have the following evidence. The γ-rays emitted by some radioactive substances are of extremely short wavelength. Now, the emission of radiation from the extranuclear structure, *i.e.,* caused by the return of the deeply ionized atoms to their normal states, is ordinarily classified as X-rays.

The shortest X-ray which the extranuclear structure can radiate is that emitted by uranium.

Any radiation of shorter wavelength, as least some of the γ-rays are, must therefore come from the nucleus.

After this first surmise about radioactivity as a nuclear property, researches were made to determine more precisely the nature aid circumstances of the origin of the three radiations of radioactivity. They form one of the most interesting problems of the radioactive phenomenon. A second and even more important investigation, intimately related to the first, is concerned with the spectra of these rays. The *origin* and *spectra* of the radiations from radioactive substances will therefore be considered in this chapter.

RADIATIONS FROM NATURAL RADIOELEMENTS

Alpha disintegration

According to current ideas of nuclear constitution with protons and neutrons, the α-particles made up of two protons and two neutrons, *probably pre-exist in some form or other, though not as individual units in the nucleus before their emission,*

since α-particles alone are ejected, not protons or neutrons, in natural radioactivity, while in artificial disintegrations they are emitted as readily as protons and neutrons.

Further, the α-particles inside the nucleus are not lying in any confused and idle manner, but are *arranged in distinct energy levels in active motion* as is shown by the experimentally observed long range α-particles and γ-ray spectra of which we shall speak presently.

But a serious difficulty arises against the emission of α-particles by radioactive nuclei due to the existence of an electrostatic *potential barrier* surrounding the nucleus. For, the energy of any α-particle inside the nucleus cannot be greater than the height of the potential barrier, or it would not stay there at all, which means that a activity could have never been detected, contrary to observation. On the other hand, if the energy of the α-particles less than the height of the potential barrier, the α-particle should never be able to get out.

Classical mechanics, which considers material particles as solely corpuscular, is quite incapable of explaining this difficulty. But in wave mechanics, which attributes to every material particle a wave aspect also, the difficulty disappears, since an electrostatic potential barrier, though very high, cannot completely hinder the passage of a wave through it. For, there is a certain probability of penetrating through the barrier, however small that probability may be, for all positive energies of the material particle.

More precisely, the Schrödinger wave function ψ, whose square measures the density of probability of the particle in a certain region, does not vanish in the regions where the potential energy V is higher than the total energy E and where the particle in the classical model would have a negative kinetic energy; instead, although decreasing exponentially with the distance the wave function ψ maintains a finite value provided E is greater than zero. This statement can be established by the following considerations:

The apparently paradoxical behavior of the particle in this case has a complete analogy in certain phenomena connected with the reflection of light. If a beam of light falls on the boundary between two media at an angle of incidence greater than the critical angle, then according to geometrical optics, total reflection will occur, *i.e.,* all the light will be reflected at the surface of separation and no disturbance will enter the second medium.

According to the wave theory, however, the process of total reflection is much more complicated. On this theory, the disturbance in the second medium is not every where zero, but within the space of few wavelengths decreases exponentially to become entirely negligible at greater distances. This 'forbidden' penetration of the disturbance cannot be described in terms of ray of light at all, the lines representing the directions of energy-flow being curved and returning to the surface again. If now, the second medium be confined to a thin sheet, of thickness less than the range of penetration of the disturbance, and if it is immediately followed by the first medium, a small fraction of the disturbance which has penetrated the sheet will emerge from it and enter the first medium. This transmission of energy is obviously in contradiction to the prediction of geometrical optics; it is, however, established by experiment.

The change from classical mechanics to wave mechanics introduces an exactly similar possibility for the transmission of a particle through a potential barrier which would otherwise be insurmountable.

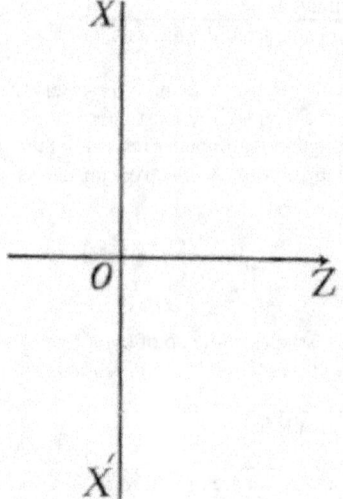

Fig. 259

Let us now consider analytically this type of *reflection of De Broglie* waves associated with material particles. Let a beam of material particles, say electrons, moving along the Z-axis, enter a uniformly opposing field of force (to the right of XOX') from free space (to the left of it), XOX' representing the boundary and O the origin (Fig. 259). For the sake of simplicity, we assume the beam to be infinite along the X-Y plane and of uniform density in that plane. The wave function associated with a material particle can be represented by

$$\psi = \psi_o \ \exp(2\pi i v t) \exp(-2\pi i \ \overline{v} \ z) \qquad \ldots\ldots\ldots\ldots \ \ldots(1)$$

On account of the assumptions made above, ψ will be a function of z, t only. Omitting the time factor, the incident wave is represented by

$$\psi = \exp(-2\pi i \ \overline{v} \ z) \qquad \ldots\ldots\ldots\ldots \ \ldots(2)$$

Considering the wave equation, appropriate to the case, viz.,

$$\partial^2 \psi / \partial z^2 + (8\pi^2 m / h^2)(E - V) \ \psi = 0 \ \ldots\ldots\ldots\ldots\ldots(3)$$

the initial conditions of the problem give:

For $z < 0$, $V = 0$, while for $z > 0$, V has a finite value, proportional to the distance from the origin 0, which may be represented by $V = az$, where a is a constant.

The wave number $\overline{v} = 1/\lambda = mv/h = [2m (E - V)]^{1/2} /h$.

If $z < 0$, $V = 0$ and $\overline{v} = (2m E)^{1/2} /h.$.

If $z > 0$, V steadily increases; hence $\bar{v}$ decreases and becomes zero when $V - E = az$; this means that at a distance $z = E/a$, $\bar{v} = 0$. At this point, the particle is said to be turned back, *i.e.*, reflected, according to classical mechanics.

If $V > E$ or $z > E/a$, $\bar{v} = i\,[2m\,(V - E)]^{1/2}\,/h$, $\bar{v}$ is imaginary and is interpreted, according to classical ideas, as indicating the absence of the particle.

Wave mechanics, however, leads to a different conclusion. The wave equation (3), being a linear equation of the second order, has two independent solutions, *viz.*,

$$\bar{v} = \pm i\,[2m\,(V - E)]^{1/2}\,/h,$$

for points beyond $z > E/a$. Applying these solutions in equation (2), it is readily seen that for

$\bar{v} = + i\,[2m\,(V - E)]^{1/2}\,/h$, ψ goes on increasing exponentially for $z > E/a$. This means that the probability of finding the particle beyond the point of reflection rapidly increases, which is evidently impossible; hence this solution does not correspond to reality. On the other hand, for

$\bar{v} = - i\,[2m\,(V - E)]^{1/2}\,/h$, ψ decreases continuously, also in an exponential manner, which means that the probability of finding the particle beyond the point of reflection rapidly diminishes; this solution alone can represent the actual state of affairs. In such a case, even when v is imaginary, ψ is real, though very small; the point of reflection ($z = E/a$) marks the beginning of an exponential decrease of the wave function ψ. Thus, according to wave mechanics, at the point $z = E/a$, although the particles are reflected as in classical mechanics, yet unlike in classical mechanics a few of them must get beyond, since ψ does not quite vanish, however far we go along the Z-axis.

This additional refinement of wave mechanics may now be used to explain α-disintegration, *i.e.*, how α-particles *leak through* the potential barrier, though they may not have the energy to clear the whole height of the barrier.

In solving the problem of α-disintegration, we must first of all fix up the nature of the wave that will represent the facts of such a disintegration.

Let ψ be the wave function that describes the position and velocity of an α-particle. Initially, the probability that the α-particle is in the nucleus is very much greater than that it will be in an equal volume outside; hence the amplitude of the wave function in the nucleus must be much greater than that of the wave function outside. After disintegration, the α-particle is outside the nucleus, and it may move in any direction away from the nucleus, but with a definite velocity v. These facts are described by a spherical wave diverging from the nucleus, of wavelength h/mv.

The energy of the α-particle, both before and after disintegration, will be equal to its kinetic energy in free space; hence the frequency of the waves is equal to $(1/2)\,mv^2/h$. Further, the probability of the α-particle remaining in the nucleus becomes smaller and smaller as the time increases, *i.e.*, the amplitude of the wave function in the nucleus becomes steadily less and less with the advance of time.

Next, we must enquire into the nature of the field of force in the neighborhood of the nucleus in which the α-particle moves. At large distances, we know that there is a repulsive electrostatic force (Coulomb's) due to the charge on the nucleus. The potential of the α-particle in this field is $(Ze \times 2e)\,/r$, where Ze is the nuclear charge, $2e$ the charge on the α-particle and r the distance of the α-particle from the nucleus.

At small distances from the nucleus, however, the field must change into an attractive field of some kind which enables the α-particle to remain in the nucleus before disintegration. Let us assume, therefore, that the potential energy of the α-particle plotted against the distance r is somewhat of the form shown in Fig. 260; it is known as the *potential barrier curve*.

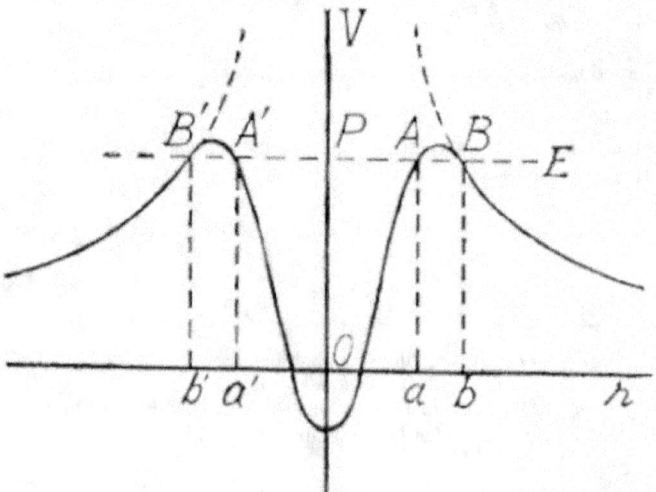

For distances greater than r_0, corresponding to the maximum of the curve, the field is repulsive; in this region, as the distance decreases the potential energy increases and reaches a maximum at r_0.

For distances smaller than r_0; the field becomes attractive; the potential energy decreases and at the centre of the nucleus is negative but finite; (we shall deal with this potential barrier in more detail latter).

We know from experiments on the scattering of α-particles that r_0 is certainly less than 3×10^{-12} cm; the *height of the potential barrier* is therefore greater than $4e^2/r_0$, (the factor 4 arising from the double charge of the α-particle) and this is equal to 14×10^{-6} erg. But the energy of the α-particle is found to be only 6.6×10^{-6} erg, *i.e.*, less than half the height of the barrier, If, therefore, the α-particle is originally inside the nucleus, according to the laws of classical mechanics, it can never get out.

For, let E be the total energy of the α-particle, represented by the line PAB (Fig. 260). Then, according to the classical theory, at points A and B (or at A' and B') the α-particle would have zero kinetic energy since they are points of reflection. If placed at B (or B'), it would fly away from the nucleus due to the repulsive force acting there; if placed at A (or A') it would oscillate between the points A and A'. It is not *possible for the α-particle to be between the lines Aa and Bb (or A'a' and B'b')*, where the kinetic energy is negative, since at points A and B (or A' and B'), all the kinetic energy has already been converted into potential energy. Hence if the α-particle is initially inside the nucleus it cannot get out, *i.e.*, α-disintegration is impossible, if the energy E of the α-particle is less than the height of the potential barrier; likewise if the α-particle is initially outside the nucleus, it cannot get into it, i.e., artificial disintegration by α-particle is impossible, unless the energy of the α-particle is greater than the top of the potential barrier.

But in wave mechanics such a difficulty disappears. For, the behavior of the particle is determined by the behavior of the wave function of wavelength $h/[2m(V-E)]^{1/2}$ at each point. Between the lines Aa and Bb (or Aa' and Bb') the wavelength becomes imaginary; in other words, the wave suffers total reflection at the walls of the nucleus. It can vibrate, therefore, only in one of the series of normal mode with definite frequencies, Thus, the α-particle within the nucleus must have one of a series of discrete energies, just like the electron in the atom.

In the region AB (or A'B'), although the wavelength becomes imaginary, the wave function is real and the amplitude of the wave does not vanish at A or A' but begins to fall off exponentially, as we have proved above. At B the wave amplitude will have become very small, but not zero. Now the wavelength becomes real again and so a wave of small amplitude leaks out of the nucleus.

Right outside the field of the nucleus, there will be a spherical wave of very small amplitude and of wavelength $h/(2m E)^{1/2}$. A wave leaking out of the nucleus means that the amplitude of the wave inside the nucleus must slowly die away, since, by analogy with other waves, where the amplitude determines the energy, if energy is flowing away, the amplitude must die down. The amplitude of the wave inside the nucleus dying away indicates, as we have seen, the probability that the α-particle has escaped from the nucleus.

Since the square of the amplitude $|\psi|^2$ gives the probability of occurrence of the particles, it follows that the particles penetrate through te region of the barrier, AB or A'B', with a frequency which is given by $|\psi|^2$. This will diminish extremely rapidly with increasing thickness of the barrier at the height which corresponds to the energy of the α-particle. Hence, *swift α-particles have a much greater probability of getting out of the nucleus than allow ones.* The higher the top of the barrier and the wider the barrier, the slower is the rate at which the wave packet streams outward and the smaller is the chance per second that the α-particle gets out of the nucleus.

The decay constant λ

The probability of escape is a measure of the decay constant since the more probable the escape, the shorter is the, mean life of the α-particle inside the nucleus. The decay constant is defined as the probability of disintegration, (*i.e.,* escape from the nucleus) in unit time. At time t = 0, when the α-particle is well within the nucleus, the wave function vanishes outside the nucleus, whereas inside it represents a vibration of frequency E/h and unit amplitude, given by

$$\int\int\int |\psi|^2 \, dx.dy.dz = 1$$

But, as pointed out above, a small wave will instantly leak through the potential barrier. Now, if a/r be the amplitude of the wave that flows out of the nucleus, then the probability that an α-particle crosses an element of area ds at a distance r from the nucleus in time dt is given by

$$(a/r)^2 \; ds \, . \, dt \, . \, v$$

where v is the velocity of the α-particle. The probability that it passes a sphere of radius r surrounding the nucleus is

$$(a/r)^2 \; 4\pi^2 \, r^2 \, . \, dt \, . \, v = 4\pi^2 \, a^2 \, . \, dt \, . \, v$$

This is the probability that a disintegration takes place in time dt. Hence the decay constant is given by

$$\lambda = 4\pi^2 \, a^2 \, . \, dt \, . \, v$$

This relation states a fundamental law of α-disintegration, viz., *the mean life τ of α- radioactive atom between two disintegrations is inversely proportional to the energy E of the emitted α-particle,* since $\lambda \propto 1/\tau$ and $E \propto v$; it also provides a theoretical justification of the empirical Geiger-Nuttall relation

$$\log \lambda = A + B \log R$$

since the range R of an α-particle is directly proportional to its energy E.

But a more precise quantitative treatment, given independently by Gamow and by Gurney and Condon, in 1928, leads to the more exact but somewhat complicated formula:

$$\log \lambda = \log (h/4mr_0^2) - 8\pi^2 \, e^2 \, (Z-2) / h\tau + 16\pi e \, [(Z-2) \, r_0]^{1/2} / h$$

This theoretical formula differs from the empirical law of Geiger- Nuttall in the fact that it is not linear in v but in 1/v; it also depends on the atomic number Z and the radius of the nucleus. It can be shown that the empirical Geiger-Nuttall law is only a first approximation, its comparative accuracy being due to the fact that Z, v and r_0 are nearly constant for all the radioactive elements. Further, on account of the large factor in the second term, in which v appears, the range of values of the decay constant is extremely wide.

It may be noted that the theory developed is not perfect in all respects. For one thing, to represent the interaction of a particle with the rest of the nucleus by means of a potential field is a very rough approximation and for another, it is probably incorrect to consider the α-particle existing as such in the nucleus.

Beta disintegration

371

In the original conception of nuclear constitution with protons and electrons, the emission of β-particles by radioactive nuclei was easily understood since those unstable nuclei could be legitimately assumed to part with their electrons while trying to reach a more stable state,

But in the actually accepted opinion of the composition of the nucleus with protons and neutrons, the *β-particles do not pre-exist in the nucleus but are created just at the moment* of their emission.

It is further believed that the β-*radioactivity is due to the transformation of neutrons into protons or vice versa inside the nucleus.* When a neutron transforms itself into a proton an electron is emitted:
$n \rightarrow p^{+} + e^{-}$.
In the converse process of a proton transforming itself into a neutron, a positive electron or positron, the counterpart of the electron, is emitted:
$p^{+} \rightarrow n + e^{+}$.

Perhaps it is more correct to say that the proton and neutron represent two quantum states of one and the same fundamental heavy particle which is subject to transformations from one state to the other, with the emission or absorption of energy and the consequent liberation of negative charge (electron) or positive charge (positron).

This simple but exquisite explanation of the origin of β-rays was evolved only after laborious and ingenious researches which have not yet reached completion. There were *two* main *difficulties* to be solved.

The *first* one arose from the very nature of the β-particles which are nothing but electrons, pure and simple, like those surrounding the nucleus. Now, since the radioactive atoms, like other atoms, can emit electrons from the peripheral electronic structure under the influence of radiation by a process, such as the photoelectric effect, there is a possibility of confusing the two types of electrons, the one arising from the nuclear change and the other from the extranuclear family as a photoelectric emission, chiefly given the fact that electrons, whatever be their genesis, are the same in mass and charge.

Moreover, in the case under study, the admixture of the two kinds of electrons is real, as the radioactive elements which exhibit β-ray disintegration usually emit also γ-rays. A portion of this high frequency radiation gets absorbed in the outer electronic structure of the disintegrating atom itself and as a result of this, electrons from the different shells, K, L, M, etc., are ejected. This process, known as the **internal conversion of γ-rays,** has been subjected to direct test and proved to be true. Further if the *γ-rays* are of different frequencies, since each of these frequencies can give rise to as many different speed electrons as there are shells and sub-shells in the electronic structure of the atom, complex groups of electrons of discrete and characteristic energies will be ejected from the radioactive element along with the real disintegration electrons, and it is not easy to distinguish between the two. As a matter of fact, *internal conversion electrons or secondary β-rays,* as they are now called, *were initially mistaken for the primary* β-rays, until carefully devised researches clearly differentiated the two phenomena.

Hahn and Meitner using the magnetic spectrograph method were able to show that

the *internal conversion electrons always consist of groups, whereas the disintegration electrons have a continuous energy spectrum.*

The photographic plate which records the magnetic spectrum of the β-rays emitted by a radioactive substance clearly demonstrates a continuous background within certain limits, produced by the disintegration electrons, on which is superposed the line spectrum formed by the groups of secondary electrons.

Chadwick using a Geiger counter instead of the photographic plate in the magnetic spectrograph (*cf.* Fig. 267) and counting the number of electrons of different energies produced by a radioactive substance was able to prove that the *'groups' formed only a small portion of the total emission.* The experimental data gave curves of the type shown in Fig. 261.

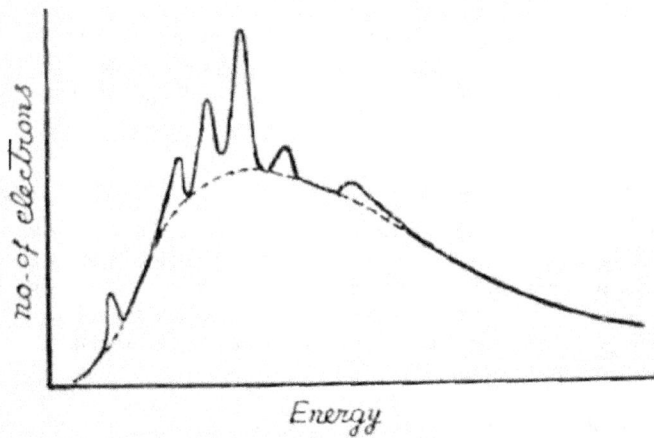

Fig. 261.

It is seen that the primary β-ray emission constitutes an extensive continuum indicated by the dotted line upon which a few sharp lines are superposed. The integrated intensity of the continuous part is much greater than the sum total of the intensities of the individual peaks. In the case of RaB, for instance, the total intensity of the individual lines is found to be only about 30% of the total emission.

Further researches were made to confirm this interpretation which distinguished the two kinds of electrons met with in β-ray emission. Thus, for instance, Rutherford, Robinson and Rawlinson replaced the fine wire source of radon of the magnetic spectrograph method by a small glass tube containing radon, around which could be wrapped thin metal foils. The γ-rays emitted by the source liberated secondary electrons from the metal sheath. The foils were of sufficient thickness to stop most of the primary electrons and to straggle widely even the most penetrating ones among them so that no traces due to the primary β-rays could appear on the plate. In spite of this, a definite line spectrum was obtained. This clearly labeled the electrons producing the observed line spectrum as purely secondary, since there was no chance for any primary β-rays to interfere.

Gurney, replacing the Geiger counter by a Faraday cylinder in the magnetic spectrograph, was able to measure the number of β-particles given off per second by different radioactive substances and established that *in a β-disintegration one disintegration. electron is ejected per atom in consonance with theory, while the number of the accompanying internal conversion electrons is much smaller.* Thus, it is found that

RaB, where γ-rays also are emitted, gives off 1.25 electrons per atom approximately, while RaE which has no γ-emission, ejects only one electron per atom,

The excess (0.25) in the former case is evidently due to the secondary electrons emitted by the γ-rays.

A *second* and more serious difficulty came from the *continuous nature of the primary β-ray spectra* which we shall consider below when dealing with β-ray spectra. For the present, it may be noted that this difficulty, if unsolved, is powerful enough to veto the interpretation of the origin of β-rays of radioactive nuclei given above, since a transition between two well-defined neutron and proton quantum states in the nucleus can give rise to the liberation of only a discrete and not to **a** continuously varying amount of energy.

Gamma-ray emission

In general, a body excited and rendered unstable by some means, either external or internal, emits radiations. For instance, a body heated above the temperature of its surroundings emits heat radiation. An atom when excited emits also radiations comprised between infra-red and X-rays, the various phenomena occurring according to the degree and nature of the excitation. Similarly, a nucleus excited, say during a disintegration, can emit radiations. In every one of these cases, when the excited body returns to its normal state, the excess of energy is given out in the form of electromagnetic radiations. All these radiations are propagated in space with the same velocity, viz., that of light. According to the classical electromagnetic theory these radiations take place in a continuous fashion, which is what is ordinarily observed in the macroscopic bulk state. But the quantum theory has formulated a discrete discontinuity in the way in which radiations are emitted, which is actually observed in the microscopic atomic state.

According to this general principle and given the fact that the γ-rays are electromagnetic in nature like other radiations, one might say that these rays emitted by radioactive nuclei are merely the excess of energy radiated away in the form of quanta by the excited nucleus returning to a more stable state. That this is actually the case is established from the following experimental facts:

(i) *The γ-ray emission usually accompanies the α- or β-disintegration of a radioactive nucleus.*

Owing to the departure of an α-particle or a β-particle from the nucleus, the latter becomes excited and hence transition can take place from this excited state to another of lower energy with the consequent emission of the excess of energy as a γ-ray. If there are several states lower than the excited one, transitions may take place to these various states, giving rise to γ-rays of different energies, as is proved to be the case by the experimentally observed γ-ray spectra. It is to be noted that γ-rays are not emitted in every α- or γ-disintegration; certain atoms are known to disintegrate without emitting γ-rays: *e.g.,* RaE emits β-particles without γ-ray emission.

(ii) *The frequent emission of γ-rays in artificially provoked nuclear reactions.*

It is the rule rather than the exception that artificial nuclear disintegrations lead to excited states of the resultant nucleus. The change brought about by a nuclear reaction manifests itself not only by the departure of a particle from the nucleus but also by the rearrangement of the remaining particles in the nucleus with the emission of γ-rays. The time taken for this emission is normally very small, of the order of 10^{-12} sec., unless some other circumstances introduce a delay.

The origin of the *γ-rays* having thus been fixed, investigations were made to determine the exact moment of their emission, *i.e.,* whether before or after the disintegration particle has left the nucleus.

In the case γ-rays which accompany β-disintegration, Rutherford, Ellis and Wooster,. experimenting upon the secondary β-rays and X-rays produced by the γ-rays have shown that the γ-rays are emitted *only after* the parent nucleus has changed, *i.e.,* after the disintegration electron has been ejected. This means that the resulting product nucleus is left in an excited state, which therefore returns to the normal by the emission of γ-rays.

As regards the γ-rays associated with α-disintegration, there are reasons to believe that the γ-rays may be emitted *either before or after* the departure of the α-particle, although the latter is the more common occurrence. We shall deal with this point more in detail when we speak of α-ray spectra, to which it is closely related.

It may be noted that as the emission of γ-rays involves no change in nuclear charge or mass, being analogous in this respect to the emission of radiation by atoms, there is a tendency nowadays to restrict the term "radioactivity" to the emission of α- or β-particles alone. But, it might be argued against this opinion that although light quanta are never considered as a constituent part of an atom, their mass in the form of the energy of the electromagnetic field makes an appreciable contribution to the total mass of the system.

SPECTRA OF THE RADIATIONS FROM NATURAL RADIOELEMENTS

A close analysis of the three radiations emitted by radioactive substances shows that all of them have a spectral distribution as regards their energy, but each one a special type of its own.

Thus the α-rays have a fine *structure,* the β-rays a *continuous spectrum* and the γ-rays highly *monochromatic lines.* The study of the α- and γ-ray spectra is, in a certain sense. more important than that of the β-ray spectra, since the former two furnish valuable data about the internal structure of the nucleus, viz., the existence of *nuclear energy levels,* while the latter, on account of its continuous nature, gives no such direct information of what is happening inside the nucleus.

But the analysis of β-ray spectra appears to be the more important from another point of view, since, in spite of the great difficulties encountered in the theoretical interpretation, it has been able to make the enriching contribution of a new and novel entity called the *neutrino* to nuclear physics.

The γ-ray spectra are the most easily studied and interpreted of the three, while a proper understanding of the α-ray spectra involves a knowledge of the γ-ray spectra. Hence, in giving an account of the spectra of the radiations from natural radioelements it is advantageous to deal with first the γ-rays, their the α-rays and finally the β-rays.

GAMMA RAY SPECTRA

The γ-ray spectra, in a few cases, have been studied *directly* by means of crystal gratings (Rutherford and Andrade, Thibaud, Frilley). Although this direct method has the advantage of leaving no doubts about the interpretation of the observed γ-ray wavelengths, it is hardly suited for γ-rays of very short wavelengths, the crystal being too coarse to produce diffraction effect with them. Consequently *indirect methods* based on measurements of the secondary electrons produced by the γ-rays have been developed. There exist three such methods; *viz.,*

(i) the *natural secondary β-ray or the conversion electron method,*
(ii) the *excited secondary electron or the photoelectron method* and
(iii) the *Compton electron method.*

The first two employ a magnetic spectrograph, while the third a Wilson's cloud chamber for the experimental technique. As the first method is the most important of the three and has provided almost all the available information about γ-ray spectra it will be described somewhat in detail, while the other two will be more briefly stated.

Lise Meitner (7 or 17 November 1878 – 27 October 1968)

The natural secondary β-ray method

This method, first worked out independently by Ellis and Meitner, in 1921, is based on the phenomenon of the internal conversion γ-rays, which, as we have already noted, can be regarded as a special case of the photoelectric effect, in so far as the γ-ray quantum issuing from the nucleus is absorbed in the extranuclear electronic structure of the disintegrating atom itself and thereby liberates the electrons found there. But according to the wave-mechanical theory, this interpretation is shown to be imperfect, as there is also another possibility of a direct mechanical transmission of energy from the excited nucleus to one of the peripheral electrons of the same atom. Whatever be the theoretical explanation of the process, the effect produced is the same as in the external photoelectric effect, but with two additional advantages, viz.,

(i) the probability of the yield of electrons being greater on account. of the *internal* nature of the phenomenon and

(ii) the spectral lines produced by the internal conversion electrons being better defined and sharper than those obtained in the external photoelectric effect, on account of the better elimination of absorption effects.

Principle of the method

With every γ-ray emitted by the nucleus, provided it is intense enough to act upon the various electronic shells of the disintegrating atom, a number of secondary β-rays will be emitted having energies

$h\nu - W_K$,
$h\nu - W_{LI}$,
$h\nu - W_{LII}$, etc.
where
ν is the frequency of the γ-ray and
W_K, W_{LI}, W_{LII} are the binding energies of the different shells involved.

These β-rays are called natural to distinguish them from those produced in the external photoelectric effect and *secondary* to differentiate them from the primary disintegration electrons.

Measuring the energies of these secondary electron groups with a magnetic spectrograph and using the above relations the energies and hence the frequencies of the primary γ-rays, can be readily deduced, remembering that the binding energies W_K, W_{LI} , etc., refer to the product atom and not to the parent atom, since the γ-rays, as has already been shown, come from the excited product nucleus and not from the original nucleus that disintegrates.

Apparatus

The experimental arrangement, first designed by Ellis and Meitner and modified by subsequent investigators, is similar to the magnetic spectrograph used by Maurice de Broglie and Robinson for X-rays. In Ellis's experiments the radioactive substance which was the source of γ-rays, say RaB. was placed on a fine wire at S near a thick lead block B (Fig. 262). A photographic plate PP was suitably arranged on the lead block which protected it from any direct radiations from the source. The whole apparatus, enclosed in a highly .evacuated box A, was placed in a uniform magnetic field acting perpendicular to the plane of the figure. The conversion electrons, emitted at S and deflected through circular paths under the influence of the magnetic field, were passed through a fairly wide slit W and made to strike the photographic plate. All the electrons having the same energy would describe circular paths of the same radius and would therefore be focused approximately at the same point on the plate. When the plate was developed a number of fairly sharp line traces were found at different points L_1, L_2, etc. The adjacent photo, where only a part of the spectrum of RaB is shown, was obtained by Ellis. It is seen that the lines produced by the secondary conversion electrons are superposed on the continuous spectrum due to the primary β-rays from RaB.

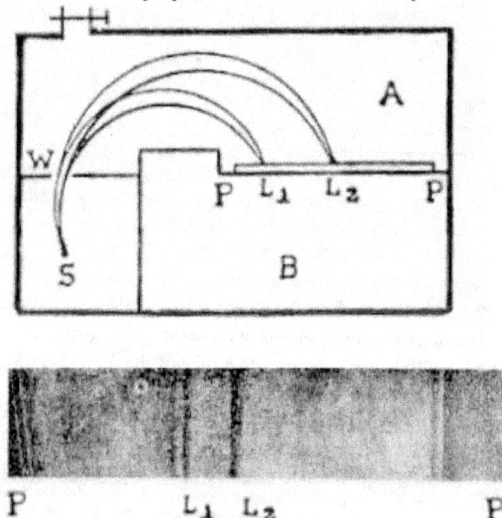

Fig. 262. Magnetic spectrograph of Ellis for the study of γ-ray spectra.
Film developed: Part of the secondary β-ray spectrum of RaB (Ellis)

Determination of the frequencies of the γ-rays from the photographic traces

The electron energies corresponding to the different traces were first determined by knowing the strength of the magnetic field and the diameters of the corresponding circular paths, given by SL_1, SL_2, etc., the relativistic formula being used on account of the high velocities of the electrons. Next, paying attention to the fact that the γ-rays come from the product nucleus, RaC in the present case, the electrons ejected by these γ-rays must come from the electronic shells of this type of atom whose atomic number is 83: Hence the binding energies of the shells involved could be obtained from tables prepared with experimental data on the X-ray spectra of bismuth (Z = 83) which is isotopic with RaC. It was then possible to identify the electronic shells to which the electrons forming the traces originally belonged.

The sum of the kinetic energy of any group of electrons which had produced a given trace and the binding energy of the corresponding shell was equal to the energy by of the exciting γ-ray, which when divided *by h* gave the frequency ν. When several β-ray lines arising from a single γ-ray were observed, there could be no doubt about the interpretation. When a number of γ-rays of different energies were, however, simultaneously involved, the analysis became somewhat more difficult, but could be achieved with a little and care, as was actually done by Ellis and Meitner.

As an illustration of the method of procedure, the data of one of the experiments of Ellis on the γ-rays emitted by RaB is given in the table below.

Determination of γ-ray energy from the lines of the conversion β-ray spectrum of RaB→RaC.

Line No.	β-ray energy in eV $\times 10^5$	Conversion shell	Binding energy of the shell in eV $\times 10^5$	γ-ray energy in eV $\times 10^5$
1	0·3674	L_1	0·1634	0·5308
2	0·3737	L_{11}	0·1567	0·5304
3	0·3963	L_{111}	0·1338	0·5301
4	0·4885	M_1	0·0399	0·5284
5	0·4910	M_{11}	0·0368	0·5278
6	0·4966	M_{111}	0·0317	0·5283
7	0·5190	N_1	0·0093	0·5283
8	0·5264	O	0·0020	0·5284
			Mean γ-ray energy $= 0·5291 \times 10^5$ eV	

It is seen from this tabulated result that a γ-ray of energy 0.5291×10^5 eV from RaB gives rise to a group of eight secondary β-ray lines. As RaB emits several γ-rays of different energies, the actual interpretation of experimental data is evidently more complicated.

It. will be noticed that in the present case conversion in the K shell does not take place as the binding energy of this shell (0.9×10^5 eV) is much greater than the energy of the γ-quantum involved (0.53×10^5 eV). The other γ-rays of RaB, however, eject electrons from the K shells on account of their high energy values.

Wherever possible, the frequencies of the γ-rays determined by this method have been checked by independent measurements with the crystal grating method and good agreement has been established.

The excited secondary electron method

This method based on the external photoelectric effect consists in a magnetic analysis of the photoelectrons produced when the γ-rays are made to fall upon a sheet of metal. The experimental technique is essentially the same as in the previous method except for the special disposition of the source at S (Fig. 262). The fine wire which carries the radioactive substance is replaced by a very small glass tube, the inside walls of which are coated with the radioactive substance, the source of γ-rays. The glass tube is wrapped by a metal foil of tungsten, platinum, lead, etc., which serve as the source of the secondary photoelectrons. It is these which are used in the magnetic spectrograph and their traces on the photographic plate are obtained in the usual way.

Ellis was the first to employ this method, but others also have worked with it. The adjacent photo obtained by Thibaud with photoelectrons excited in a lead foil by the γ-rays of RaB demonstrates the magnetic spectrum of the secondary electrons recorded in this method.

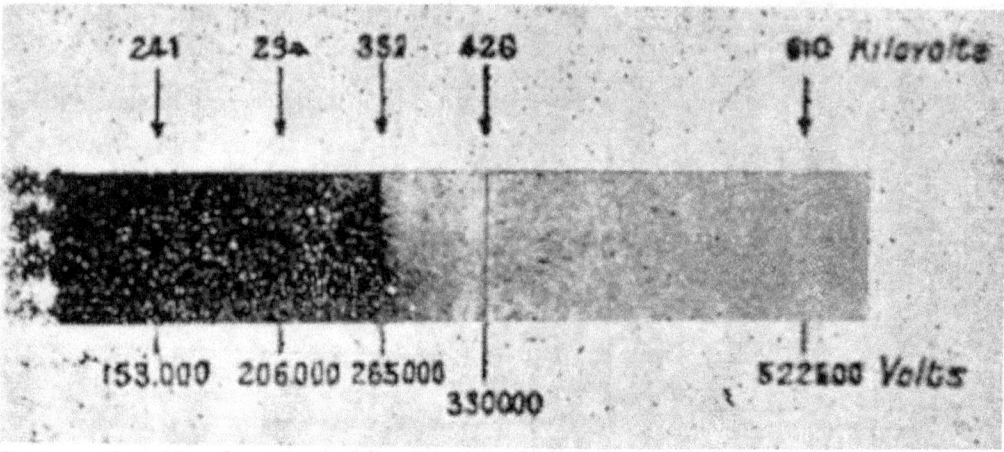

Spectrum of secondary β-rays produced by the γ-rays of RaB in a lead foil (Thibaud)

A number of traces is seen of which three are more intense. The continuous background due to primary β-rays will be practically eliminated in this method, as they are absorbed in the glass walls and the metal foil at the source, but there is still a background due to Compton electrons resulting from scattering of γ-rays. The traces are more diffuse than those obtained in the other method on account of the slight slowing down of the photoelectrons which arise at a certain depth below the surface of the metal foil.

But this method has a decided advantage over the previous one, in that it lends itself to an easy determination of the frequencies of several γ-rays at once, wherever possible. Let us illustrate this point taking the case of RaB cited above. The three intense lines in the photographic plate correspond to energies of 1.53×10^5, 2.06×10^5 and 2.65×10^5 electron volts. The same three lines are recorded with different metal foils but are shifted in position in such a direction as to indicate that the corresponding electrons have less energy the greater the atomic weight of the metal used. The change in the energy of the electrons from one metal to another corresponding to a given line agrees closely with the difference of the binding energies of the K shells of the two atoms. Hence it follows that the electrons causing the three lines are expelled from the K shells of the respective atoms by three different γ-ray frequencies.

Applying Einstein's photoelectric equation $h\nu = E_K + W_K$, where E_K is the measured kinetic energy of the photoelectron and W_K the energy required to remove the electrons from the K shell of the atom, the energy by of the incident γ-rays is obtained, assuming the value of W_K known from the X-ray spectrum of the metal used. Thus for lead, W_K being equal to 0.88×10^5 eV the energies of the three lines are 2.41×10^5, 2.94×10^5 and 3.52×10^5 eV. Data with different metals lead to the same values of the γ-ray energies, although W_K is different for each case. *Thus there are three lines in the gamma ray spectrum of RaB having the above given energies.* The other lines recorded in the photo are due to photoelectrons ejected from the other shells L, M, etc. This does not mean that there are no other lines in the γ-ray spectrum of RaB.

The Compton electron method

This method, first employed by Skobelzyn, depends upon the observation of the Compton electrons produced in the gas of a Wilson's cloud chamber and the determination of their energies by means of an applied magnetic field. The source of γ-rays is placed at some distance from the chamber and a canalized beam of γ-rays is sent into the chamber through a screened window. While traversing the chamber the γ-rays are scattered by the air in it and Compton electrons are ejected.

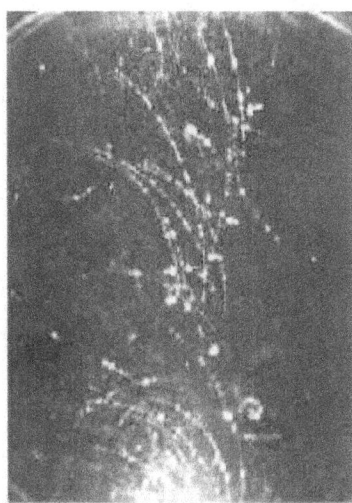

Compton electrons produced by the γ-rays from Th C") (Source: Skobelzyn, Leningrad)

If a gamma ray collides with an electron, it may give up some energy to the electron. This photograph shows the tracks of electrons scattered in this way. Arthur Compton studied this effect in 1923. It was one of the first pieces of evidence that gamma rays could behave like particles.

The expansion chamber being placed in a suitable magnetic field, these electrons describe circular, sometimes even spiral paths, which can be photographed with a fairly strong illumination of the chamber at the moment of expansion. One such photo of the tracks of the Compton electrons produced by the γ-rays of Th (C + C") taken by Skobelzyn, is reproduced here. From a photographic record of this type, the energy W corresponding to each electronic track can be determined by knowing the strength of the magnetic field and the radius of curvature of the circular path; the angle φ which the initial direction of the electron track makes with that of the primary beam can also be determined. Then using Compton's relation,

$$W = 2\alpha \, h\nu \, / \, [\, 1 + 2\alpha + (1 + \alpha)^2 \tan^2 \varphi]$$

where $a = h\nu/m_0c^2$, the energy $h\nu$ of the exciting γ-ray can be deduced.

This method has confirmed the existence of sharp line spectra in the γ-rays emitted by radioactive substances. It permits also more reliable estimates of the intensities of the γ-rays than the other two methods. Counting the Compton electrons produced by each γ-ray group of a given frequency ν, the number of primary quanta for the different values of the frequency can be deduced. This gives the distribution of intensity in the γ-ray spectra. The number of γ-ray quanta whose energies are contained in a small constant interval can be represented by a graph of the type shown in Fig. 263, which refers to the γ-rays from ThC and ThC". It is seen that among the γ-rays of ThC + ThC" there exists a group of very high energy, of 2.65 MeV, which is in complete agreement with the results obtained by the internal conversion method.

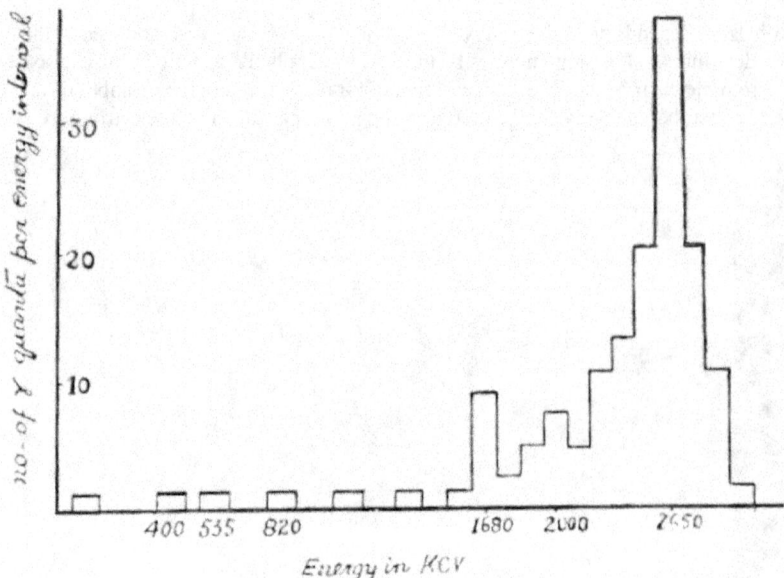

Fig. 263. γ-ray spectra of ThC + ThC" obtained from the study of Compton electrons.

The chief drawbacks of this method are:

(i) energies are not measured with the same high degree of accuracy as in the other methods, and

(ii) only the stronger lines in the spectrum are easily detected.

It may be noted that all the three methods require fairly strong sources for their success; hence they are applicable chiefly to the γ-rays emitted by natural radioelements.

Results

Meitner and Hahn located forty-nine lines in the β-ray spectrum of radioactinium. Of these, thirty-seven could be attributed to the action of one or the other of twelve γ-ray frequencies. They also observed twenty lines in the spectrum of actinium X and explained fourteen of them by postulating seven frequencies. The spectrum of RaD consists of only five lines whose measurements show that they may be supposed to consist of electrons ejected from L_I, L_{II}, L_{III}, M_I, and N_I, shells of the product atom RaE by a single γ-ray of energy 0.0472 MeV. Likewise the three lines of the spectrum of Ra may be ascribed to a single γ-ray of energy 0 . 189 MeV expelling electrons from the K, L and M I , shells of the product nucleus Rn. The number of γ-rays estimated for RaB → RaC is five, for RaC → RaC' nine and so on. The γ-ray spectra thus far mapped out consists from one to fourteen frequencies. The highest frequency γ-ray having an energy of 2.65 MeV occurs in the spectrum of ThC", while the next in merit with an energy of 2.2 MeV is found in the γ-ray spectrum of RaC.

The substances which emit γ-rays emit also characteristic X-rays since the internal conversion of the γ-rays leaves the atoms in excited states, which in returning to the normal state should emit characteristic X-rays. These X-rays have been observed and measured in several cases and afford additional checks on the assignment of the origin of the ejected electrons. Often, however, instead of the emission of an X-ray photon, the emission of another electron by a secondary photoelectric effect known as auto-ionization or *Auger effect* is observed. The presence of these tertiary electrons is bound to increase the difficulty of interpretation of the natural secondary β-ray spectra.

Nuclear energy levels

The study of γ-ray spectra has led to a very important result concerning the nucleus, viz. the existence of discrete energy levels in the nucleus. For, as was first suggested by Ellis, the different monochromatic frequencies actually observed in

380

the γ-rays emitted by a radioactive substance can be adequately accounted for only on the supposition of a number of excited quantum energy levels in the nucleus. Many attempts have been made to construct systems of nuclear energy levels based on the simple additive relationships found in the measured frequencies of the γ-rays, but with only a partial success.

In the case of the β-emitters the suggested schemes are highly dubious, since the primary β-spectrum, being continuous, offers no indication of the quantum levels, so that these have to be built solely upon data of the γ-ray spectrum using the above-stated relationships. In the case of the α-emitters, one feels on surer ground, on account of the inter-relations existing between α-ray and γ-ray spectra, which could be established beyond doubt in several cases, as we shall see presently in the study of α-ray spectra.

ALPHA RAY SPECTRA

As a result of the measurement of velocities of the α-particles in connection with the experimental determination of their E/M value, it was concluded that all the α-particles emitted by a given radioactive substance came forth from their parent atoms with a single speed and hence with the same kinetic energy, except for slight variations due to the *straggling effect* caused by statistical fluctuations of the collisions in the absorbing medium. This belief continued for quite a long time and the Geiger and Nuttall empirical law was based upon it. The existence of an extremely small number of particles having much higher ranges than that of the main group and hence called *long-range,* α-particles, discovered by Rutherford in one or two radioactive substances such as ThC' and RaC', even in the early stages of the study of α-rays, was treated as an anomaly.

But with the discovery of a real *fine structure* in the α-rays, belonging to the main group itself, by S. Rosenblum, in 1930, the situation changed entirely and the long cherished idea of single speed of α-particles from a given α-emitter had to be given up. Fresh and more delicate researches were undertaken, which resulted in the definite establishment and satisfactory interpretation of the long range and the fine structure characteristics of α-rays. This revision of idea was extremely fruitful, as it led to important discoveries concerning the internal structure and activity of the nucleus, especially to the existence of quantum energy levels in it. We shall give here a brief summary of the experimental study made on α-ray spectra and of the theoretical explanation of the results thus obtained, leading up to the construction of nuclear energy levels.

The **long range α-particles** were *first discovered by the scintillation method.* Rutherford and Wood, in 1916, measuring the ranges of the α-particles from ThC' by the use of absorbing screens and a spinthariscope found a very few particles with a high range of 11.6 cms, among the main group of range 8.6 cms. A few years later (1919) similar long range particles were observed by Rutherford among the α-rays emitted by RaC' using the same arrangement. This method, in spite of its singular delicacy and value, is wearisome and taxing, since all the observations are ocular and there is no permanent record left behind except in the observer's memory or notes.

In more recent years, *three other more efficient methods have* been used in the study of these long range part ides.

The first of these is the **Wilson's cloud chamber** by means of which the tracks of the α-particles are photographed. Study of such photos gives information regarding the number of particles involved and their ranges. Now and then, on the same photograph, one or more long range particles are recorded among an enormous number of normal particles, as seen in the one reproduced here. But it is a long, tedious and costly work to obtain a sufficient number of such photographs required for plotting really good distribution in range curves.

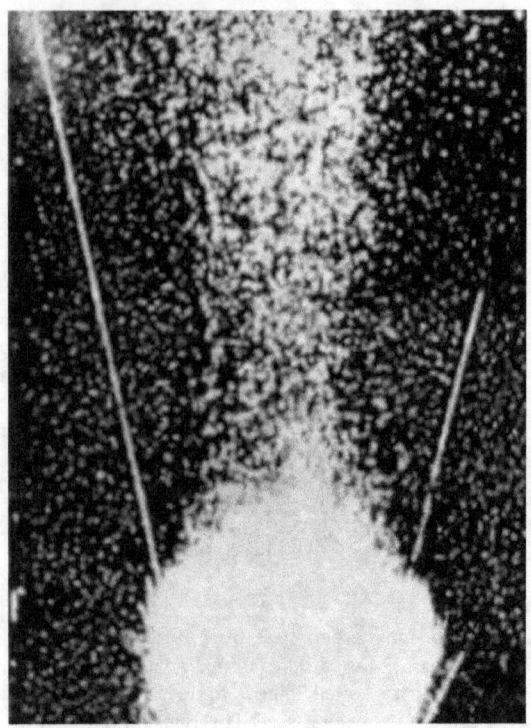

Long range a-particles from RaC'.

The **second,** known as the **differential ionization chamber method,** was devised by Rutherford, Ward, Lewis and Wynn Williams in 1931. The differential ionization chamber is a very effective and sensitive device for detecting single charged particles and hence well suited to the study of the extremely small number of the long range α-particles, almost lost in the overwhelming majority of the normal group. It consists of a pair of very shallow ionization chambers separated by a metal foil, of negligible thickness, which acts at the same time as the negative electrode of one of the chambers and the positive electrode of the other. The charge collected by this special electrode is then the difference between the ionizations in the two chambers which can be measured by connecting the electrode to an electrometer through an amplifier.

If the chambers are traversed by a particle which is yet far from the end of its range, the difference in ionization will be small and even imperceptible; if by a particle which is approaching its maximum ionizing power (the peak of Bragg's curve), the difference will be appreciable and of one sign; if by a particle which is coining to the end of its range, the difference will again be appreciable but of the opposite sign. Such an arrangement is therefore sensitive above all to particles which are nearing the ends of their ranges. With this delicate detector of single particles almost at the ends of their ranges very accurate data for the plotting of the distribution-in-range curve can be easily obtained by recording the readings of the electrometer corresponding to different thicknesses of the absorbing screen interposed between the source of the α-particles and the differential chamber.

Rutherford and his co-workers analyzed, in this range α-particles of RaC' and obtained a curve of the *Fig.* 264.

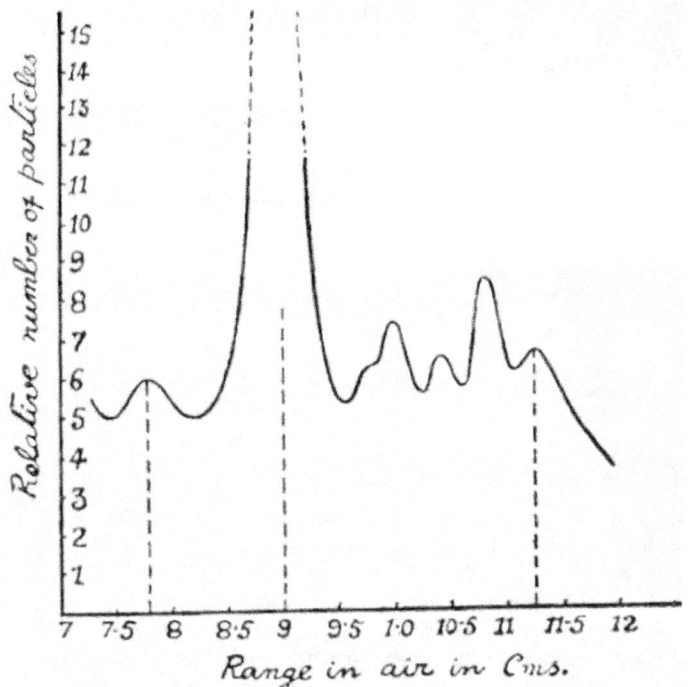

Fig. 264. Long range α-particles from RaC differential ionization chamber method

The range, which is given by the thickness of the absorber, is expressed in air equivalent; the relative number of particles referred to unit time is obtained from the electrometer readings. The curve shows a high peak (not completed in the figure) at a range of 9 cms.; the left of it, there is a much lower peak at a range of 7.8 cms; to the right a wavy curve with four distinct maxima which the experimenters interpreted as a superposition of. seven peaks. Even the tallest of the peaks just mentioned (9 cms.) is nothing compared with the one that would correspond to the principle group of RaC' at 6.9 cms, range (not shown in the figure). The estimated abundances of the 7.8 cms., 9 cms., and 6.9 cms, groups are in the ratio of 1: 44: 2 x 10^6. Yet the differential chamber is able to detect them; herein lies, its great merit.

A **third** method of investigating the long range particles is the **magnetic spectrograph with annular magnet,** devised also by Rutherford and his colleagues. As this apparatus is very effective in the analysis of the fine structure of the normal particles as well, we shall speak of it presently.

The **fine structure of β-rays has been studied by the magnetic spectrograph semi-circular focusing method.** But the method has to be adopted to obtain a perceptible resolution of the fine structure whose order of magnitude is not much greater than the variations due to the straggling effect. This is achieved by the following two devices:
(i) Use of a very strong and extensive magnetic field; with the fields actually employed it is possible to separate particles whose velocities differ by only 0.02 per cent.

(ii) High vacuum along the path of the α-rays to minimize the straggling effect. Rosenblum was the first to obtain these conditions and thereby succeeded to place in clear evidence the fine structure of α-particles. He used the huge magnet of the Academy of Sciences at Bellevue, Paris, with solid circular pole-pieces 75 cms in diameter and 6 cms apart, capable of producing an essentially uniform field of 25,000 gauss over a region of some 35 cms in diameter. After passing through collimating slits, the α-particles were deflected through a semi-circle by the magnetic field and made to fall on a photographic plate. High vacuum was maintained in the semi-circular box which contained the source at one end of its diameter and the plate at the other.

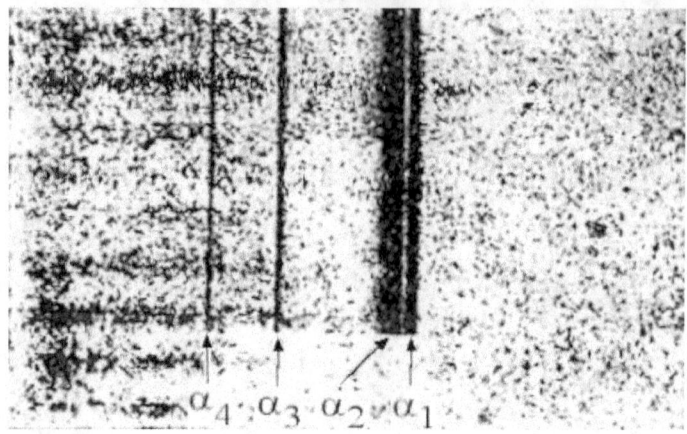

Fine structure of α-particles of ThC (Rosenblum)

In this arrangement, all the groups of particles of different speeds, deviated simultaneously in circular paths, each of its own particular radius, fell upon the plate producing what looked like an optical line spectrum.

The photographic record thus obtained by Rosenblum with the α-particles of ThC, the first to be analyzed, is reproduced here. Four lines are seen in it, but some plates after long exposures have shown as many as five. The lines marked α_1 and α_2 are very intense as they correspond to ranges very near the mean range and hence the number of particles having those ranges is very great. The two other feeble lines marked α_1 and α_2 are due to particles of lower range and their number is necessarily small; α_4 is not recorded at all on account of its still smaller number. To have a concrete idea, the relative numbers of particles that go to produce α_1 , α_2 , α_3 , α_4 and α_5 are in the ratio 27.2: 69.8: 1.8: 0.16: 1.1. The respective velocities of these groups as calculated in the usual way with the magnetic spectrograph data are 1.7108, 1.7053, 1.6651, 1.6445 and 1.6417 x 10^9 cms/sec and the corresponding energies, including the recoil energy of the residual nucleus, are 6.201, 6.161, 5.873, 5.728, 5.709 MeV. *These sharp traces clearly indicate that the normal α-particles from ThC are emitted with several discrete velocities,* though closely spaced (hence the name 'fine structure'), and not with velocities varying continuously over a narrow range, as would be expected from the statistical straggling effect.

Rutherford and his associates in the Cavendish laboratory modified in 1933 the technique of the magnetic spectrograph used in the study of α-ray spectra in the following ways:

(a) Instead of solid disc-shaped pole-pieces, narrow circular rings were used, which reduced the amount of magnetizable metal and, in consequence, the weight of the apparatus to a great extent.

(b) The source of α-particles was placed in the narrow annular space between the faces of the rings at one side, while on the diametrically opposite side, the photographic plate was replaced by a simple ionization chamber connected through an amplifier to a Wynn-William thyratron scaling system, by means of which individual particles arriving at quick succession could be counted accurately.

(c) Arrangement was made to vary the magnetization of the rings over sufficiently wide limits, so that the field strength could be adjusted step by step in order to bring group after group of the α-particles to the ionization chamber. Although this entailed the accurate determination of the field strength at the different stages, which was no easy work, yet the very possibility of choosing any desired field strength combined with the very sensitive electronic detector increased the efficiency of the apparatus in the investigation of not only extremely weak groups but also the rare long range particles.

(d) The annular space and everything within it was evacuated, being walled in by means of a ring band.

The apparatus constructed with these modifications, viz., a fixed radius of curvature, a variable magnetic field and an electronic detector, is shown in the adjacent photo. The rings were 40 cms in radius, 5 cms broad and 1 cm, apart. The field strength was adjustable up to 12,000 gauss, which was sufficient for α-particles of range up to and even beyond 11.5 cms. It was claimed that the arrangement could measure velocities accurate to 1 in 10,000.

Annular magnetic spectrograph used by Rutherford for the analysis of α-ray spectra.

Rutherford, Lewis and Bowden with this annular magnetic spectrograph were able to analyze the *fine structure;* of the α-particles from several radioactive substances with great accuracy. Plotting the different values of the field strength against the corresponding readings of the detector, they obtained curves which showed a peak for every group. From the value of the field strength corresponding to a peak and the radius of the ring pole-pieces, the energy of the group responsible for the peak could be readily computed. With ThC the results of Rosenblum were confirmed and thus the fine structure of the α-ray spectrum was definitely established.

As regards the *long range* particles, working with RaC', all the peaks indicated by the differential ionization chamber were clearly separated, a hump which had suggested two groups was resolved into three maxima, and an extra group was discovered. Thus it was shown that there were in all 12 groups of long range particles from RaC'. Only two groups of long range particles were found with ThC', one at 9.7 cms, and the other at 11.54 cms. which agreed well with the results obtained by the measurement of more than 500 cloud chamber tracks of long range particles of ThC'. This fact indicates also the advantage of the magnetic spectrograph method over the cloud chamber method from the point of view of economy of cost and labor.

More recently, in 1940, Roy Ringo, in America, has constructed a magnetic spectrograph for α-ray analysis, which claims to have overcome certain inherent defects of the methods of Rosenblum and Rutherford, such as the immobility of the Bellevue magnet, which prevents its use in conjunction with reactions produced by artificially accelerated particles, the background effect in the detector of the annular magnet method caused by contamination and cosmic rays, which necessarily limits the accuracy of results, the impossibility of repetition of experiment in quick succession, arising from the fact that the source and detector are placed in a highly evacuated region inside the magnetic field in both the methods, etc.

In Roy Ringo's apparatus, diagrammatically shown in Fig. 265, the particles are deflected through an angle of 60° by a curved magnet. The source of α-particles S as well as the detector, a photographic plate P, are placed outside the magnetic field at a distance of about 40 cms The distance between the pole-pieces is $1\frac{3}{26}$ " and there is an adjustable slit A at the entrance to the field.

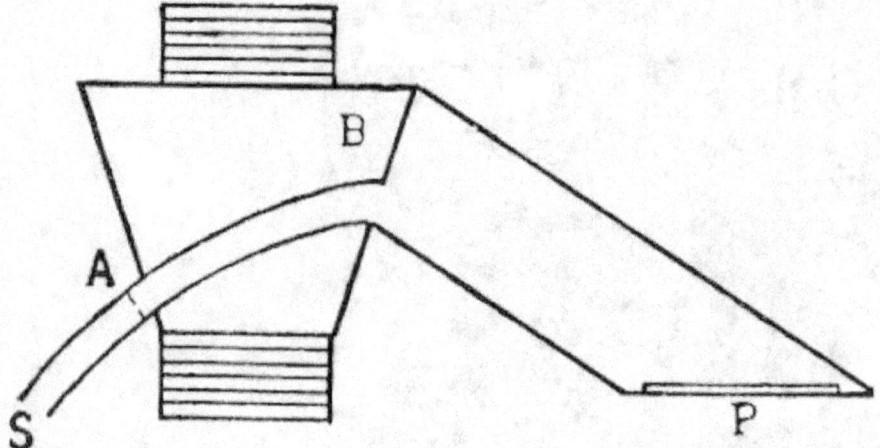

Fig. 265. Roy Ringo's magnetic spectrograph for the study of α-ray spectra. .

The whole apparatus was enclosed in a brass vacuum case B inside which the pressure is reduced to 10^{-3} mm. The arrangement weighing 1200 lbs is mounted on a movable truck.

The salient features of this spectrograph, according to its author, are a great resolving power as high as 1300 and the unerring detection of even single α-particles from their traces by a microscope examination of the plate. It is also claimed that in 1 in every 70,000 of the emitted particles could be got inside the apparatus.

Results
A great majority of the twenty-five known natural a-emitters has already been analyzed. The α-ray spectra of some of them obtained by Rosenblum are reproduced here. The results of these experimental researches load to the conclusion that α-ray spectra may be divided into three categories:

(i) Those that consist of a *single line* as in the case of Rn, RaA, Po, etc., which in consequence appear to have no *fine structure*.
(ii) Those that have two or more closely packed but distinct components of more or less the, same intensity and hence show a *well defined fine structure:* e.g., ThC, An, AcC, AcX, RdAc, the last one having as many as eleven components.

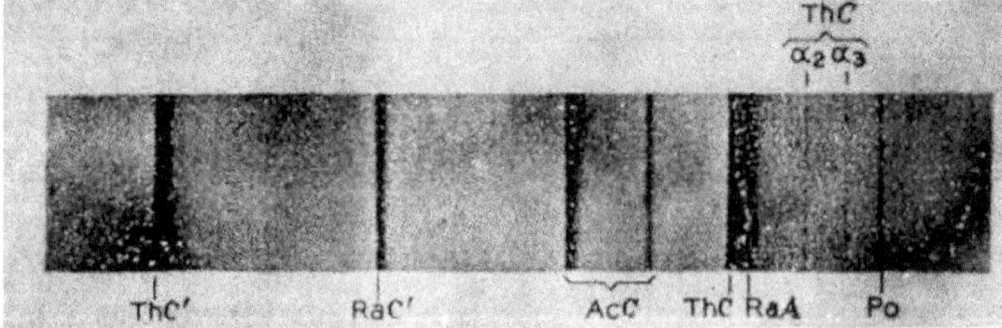

α- ray spectra of several elements (Rosenblum)

The difference in range of adjacent groups is often only a few tenths of a. millimetre and never more than a few millimetres. All of them occurring within the region of straggling, it is not surprising that they had escaped observation for a long while.

(iii) Those having not only a main group but also others of much higher energy, the latter containing, however, only an extremely small fraction of the total number of particles. These exhibit the *long range* phenomenon. So far, only two sure cases belonging to this class have been found, *viz., the extremely short-lived* Rae' and ThC'. This means that the ejection of a long range particle is a rare occurrence. It is interesting to note that the longest range particles from both RaC' and The' have the same range 11.5 cms about, corresponding to an energy of 10.5 MeV. These represent the fastest particles ejected by natural radioelements.

Interpretation of results

Gamow was the first to give a simple and logical explanation of the "fine structure" and "long range" phenomena met with in α-ray spectra and the experimentally observed inter-relations between the α-ray and γ-ray spectra on the basis of **nuclear energy levels.**

The problem to be solved is this: Since all α-particles must evidently come from the nucleus, why should they be ejected from a given radioactive substance, which presumably is made up of the same kind of identical nuclei, with a variety of discrete energies !

As the idea of energy levels has been applied with great success in the explanation of the structure and properties of the extra-nuclear part of the atom, it is but natural to attempt an interpretation of these phenomena in terms of discrete energy levels within the nucleus. Let us suppose that there do exist in the nucleus a number of discrete energy levels.

The fine *structure can then be explained by postulating that the nucleus resulting from the α-disintegration is left excited in different quantum states.* If the energy released in the disintegration is in excess of an excitation energy of the resulting nucleus there is always the possibility that it will be found in an excited state, the energy of the emitted α-particle being smaller just by this amount. In these cases, the spectra of the α-rays will show several discrete groups corresponding to the different quantum levels of the product nucleus. This excited nucleus will return to its normal state almost immediately by releasing the excess energy in the form of γ-rays which must evidently belong to the product nucleus and not to the original nucleus. If all these suppositions are correct, the scheme of the energies of the various groups of α-particles turned upside down will give the scheme of levels of the product nucleus, and the γ-rays emitted by that nucleus will fit in that level scheme. These consequences of the proposed theory have been tested in several cases and found true.

As a specific example, let us consider the fine structure of the α-rays from The. We know that a ThC atom emits an α-particle to form ThC" which has a γ-activity. Experiments show that the α-particles ejected by ThC exhibit a fine structure of five components α_1, α_1, α_2, α_3, α_4, α_5 in the decreasing order of energy. When the most energetic α_1 group is emitted, the resulting nucleus ThC" would be formed in its lowest energy state. In the case of the emission of the other groups, ThC" would be left in correspondingly higher excited states. The energy of these excited states are readily obtained by the differences between the energies of the group α_1 and of the others. From the numerical values of the energies of the five groups, already given, the required energy differences are 0, 0.04, 0.328, 0.473, 0.492 MeV. Turning these values, upside down we get the energy level scheme of ThC", as shown in Fig. 266.

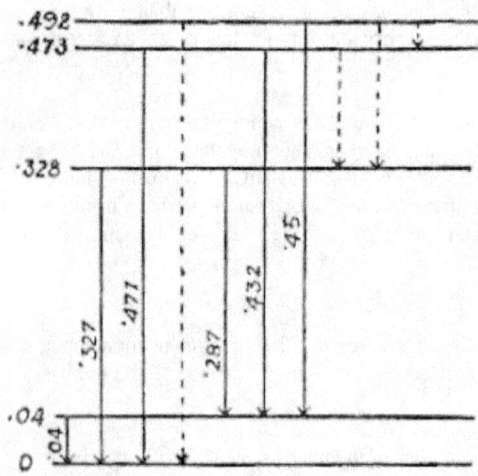

Fig. 266. Energy level scheme of ThC'.

Out of the 10 possible combinations between the five levels, six transitions, represented, by continuous lines, have been observed as γ-rays within the limits of experimental accuracy. The other four theoretically possible γ-rays, represented by dotted lines, have 'not been observed, either due to their feeble intensity or some nuclear selection rule which forbids certain transitions. It may be noted, however, that there are other γ-lines actually observed with ThC", which are not indicated by this method of analysis of observed fine structure; but they also can be accounted for by the introduction of an additional level at 0.629 MeV (not shown in figure).

The long range particles emitted by RaC' and ThC' can also be interpreted in terms of nuclear energy levels, provided it is assumed that the nucleus before the α-disintegration can exist in different quantum states. It is possible that the nucleus which is going to emit an α-particle, is already excited as a result of its own formation from a preceding disintegration. In such a case two alternative processes can take place:

(i) a γ-ray may first be emitted, bringing the nucleus to its normal state, and this may be followed by normal α-disintegration, or

(ii) the α-particle may be emitted taking with it the excess energy of the excited nucleus..

The second process will evidently give rise to an α-particle whose energy is much higher than the normal particle, *i.e.,* to a long range particle. The probabilities of the excited nucleus to follow one way or the other are inversely proportional to the respective mean lives. Since the mean life of an excited nucleus with respect to γ-ray emission is extremely short, of the order of 10^{-12} sec., hence much shorter than the usual mean lives even for high energy α-particle emission, the first process must be the more common occurrence. In spite of this fact, the alternative process of α-emission actually takes place in a few cases such as RaC' and ThC'. This must then be attributed to the extremely short mean lives of these nuclei with respect to α-emission, which are actually of the order of magnitude of 10^{-5} and 10^{-9} sec., respectively, in fact, the shortest lived among the radioactive atoms. Further, such a process can take place only with anomalously fast or long range α-particles on account of the law of inverse proportionality that exists between mean life and range in the case of α-particles. The extremely small number of long range particles observed is therefore due to the *rare* occurrence of the process involved; it may also be partly due to the fact that not all nuclei are necessarily left in excited states.

If this line of argument is correct, the extra energy of the long range α-particle over the normal in one process and the energy of the γ-ray emitted in the alternative process must be equal. Further, if there are several groups of long range particles, to each of them must correspond a different excitation energy of the nucleus and hence each one o f them must indicate a discrete energy level. With the data derived from the groups of long range particles, the nuclear energy levels can therefore be drawn. These levels, conveniently combined, must account for the actually observed γ-ray spectra, belonging evidently to the initial nucleus and not to the product nucleus as in the previous case (fine structure).

When these tests are applied to the two known cases of long range α-particle emitters, ThC' and RaC', a satisfactory confirmation of the theoretical predictions is obtained. Thus, for instance, taking the case of ThC', it is formed by β-disintegration of ThC, which might well leave it in an excited state. ThC' transforms itself into ThD by the emission of α-particles. Some of the exceedingly short lived ThC' atoms may disintegrate, ejecting an α-particle and becoming ThD *before* the excited ThC' nucleus has lost its excess energy by γ-radiation and thus give rise to long range particles. Experiment shows that two such long range groups are emitted by ThC'.

This means that the ThC' nucleus can be initially left in at least two excited states. Three transitions between the two excited states and the normal state are possible and hence three lines would be expected in the γ-ray spectrum of ThC'. Of these, two have been actually observed, whose energy values agree closely with those calculated from the data obtained for the two long range groups. In a similar manner RaC' can be studied, although the analysis becomes more complicated on account of the many long range groups and γ-ray spectrum associated with them.

Gamow's theory, therefore, not only gives an essentially correct explanation of the origin of the fine structure and long range α-particles as well as of the γ-rays of α-emitters, but also a convincing proof for the actual existence of discrete, quantized energy levels within the nucleus.

BETA RAY SPECTRA

The main interest in the study of β-rays proper lies in the fact that these rays have *a continuous energy spectrum* which has raised a very difficult problem in nuclear physics.

Before we consider the ingenious attempts made to solve this knotty point, we shall describe briefly the experimental methods in use for obtaining the velocity distribution of the primary β-rays.

There exist *two chief methods* for determining the energies of the β-particles emitted by radioactive substances, *viz.*,
(i) *the magnetic spectrograph* and
(ii) the *Wilson's cloud chamber.*

The magnetic spectrograph method

The basic principle of this method has already been explained in several places, in connection with the photoelectric effect of X-rays, γ-ray spectra and α-ray spectra. The experimental technique in the case of β-rays, however, assumes the following three different forms:

(a) *The photographic plate*

In this technique the β-rays given off by the radioactive substance under study are bent by a *constant magnetic field* in such a way that they traverse semi-circular paths and then fall on a photographic plate. β-particles of different velocities are thereby focused at different places on the plate.

From the variation in the intensity of the continuous background along the plate the distribution of velocities can be evaluated approximately. Hence this method gives only a rough estimate of the primary β-ray spectrum. A photo of the magnetic spectrum of β-rays of RaE obtained by J. d'Espine using this method is reproduced here.

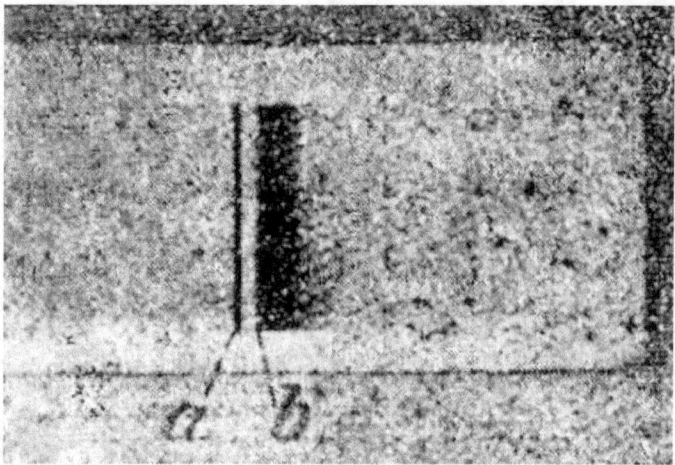

Magnetic spectrum of β-rays from RaE *(a)* central ray; *(b)* the beginning of a continuous band. Emax =1.3 MeV (J. d' Espine)

(b) *The Geiger tube counter*

In this method, first used by Chadwick, the photographic plate is replaced by a Geiger electronic tube counter, by means of which β-particles entering it can be counted accurately. The experimental arrangement is shown in Fig. 267.

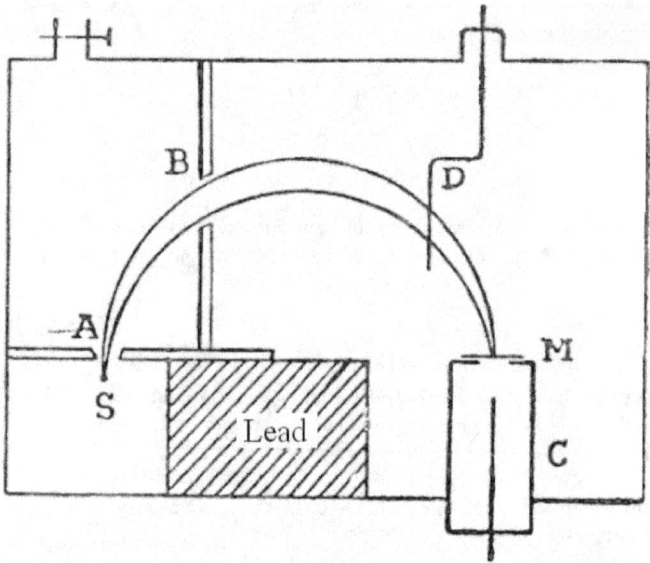

Fig. 267. Magnetic spectrograph with Geiger counter for detecting β-particles.

The source S is the radioactive substance coated on a fine wire.
A and B are diaphragms used for limiting the beam of β-particles.
C is a Geiger counter covered with a thin sheet of mica M, which is arranged in the plane which the plate formerly occupied.
D is a movable screen by means of which the β-rays can be intercepted and prevented from reaching the counter.
The lead block protects the counter from parasitical radiations.

In performing the experiment the strength of the magnetic field is varied step by step and for each value of the field strength the number of β-particles of a definite velocity traversing the circular path of constant diameter SM and entering the counter is determined. In this manner the velocity distribution of the β-rays can be obtained.

(c) *The Faraday cylinder*

Gurney, in order to minimize the errors due to secondary electrons, replaced the counter by a Faraday cylinder and, altering the magnetic field as described above, was able to explore the whole spectrum in successive steps and obtain very reliable data for the energy distribution of the β-particles. With this arrangement he was also able to establish that the number of nuclear primary electrons was one per disintegration, as already noted.

These last two techniques, where β-rays of definite energies could be closely analyzed in a direct and efficient manner, have been used by many other workers, such as Schmidt, Madgwick, Sargent, Scott, Lymann and O'Conor. The chief experimental difficulties of the method are:
- need of good vacuum in the apparatus and of high resolving power,
- maintenance of uniform magnetic field in the effective part,
- the very troublesome scattering effects from the walls and slits
and
- want of proper focusing,

all of which have their due share in vitiating the results.

The cloud chamber method
The paths of individual β-particles from a radioactive substance, curved under the influence of a constant magnetic field, are photographed with the help of a cloud chamber. With a record of the tracks thus obtained, as illustrated in the adjacent photo, the energies of the β-particles can be determined from a knowledge of the field strength used and a measure of the radii of curvature of the tracks.

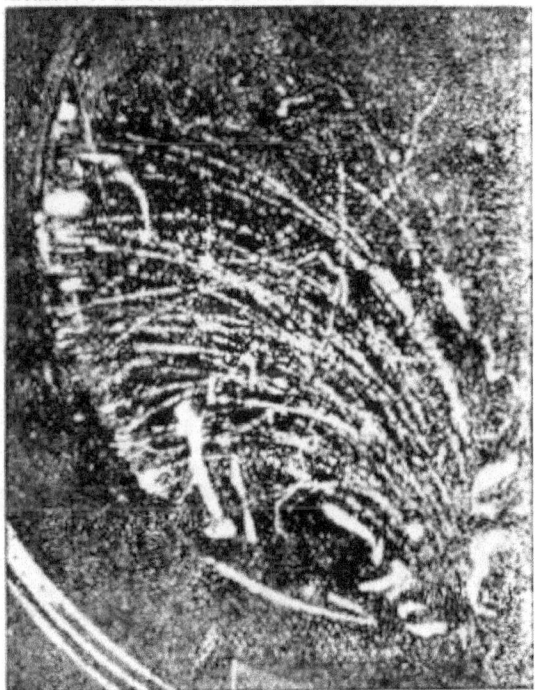

Cloud chamber photo of β-rays from RaE deflected and dispersed by a magnetic field (M. Lecoin)

Counting also the number of particles which have the same radii of curvature and hence same energy, the distribution-in-energy curve can be drawn provided a sufficient number of tracks are available.

This method, besides being direct, is also spectacular, because the particles studied leave a permanent trace of their behavior under well-defined conditions so that they could be analyzed at leisure. But from the experimental side, the major difficulties encountered are:

- the distortion of curvature due to disturbing convection currents inside the cloud chamber,
- the maintenance of uniform temperature during operation,
- the fluctuations of the magnetic field,
- proper illumination of the chamber required for taking good photos of the tracks
and
- correct measurement of the curvature of the tracks.

As regards the results obtained, it suffers from the common difficulty of all statistical estimations in that a large number of tracks must be photographed to arrive at a fairly accurate conclusion.
Many contradictory results may be obtained when different thicknesses of the radioactive substance are used.

Further, if the specimen studied emits γ-rays also, as is very often the case, secondary electrons of those γ-rays will be mixed up with the primary disintegration electrons, and will, if the γ-ray energy is greater than the β-ray upper limit, add a tail to the distribution curve which may lead to a false estimate of the maximum energy of the β-particles.

In spite of the many difficulties, this method has been employed with great success in the study of the β-ray spectra of natural and artificial radioelements by several workers, such as Champion, Kurie, Lecoin, Richardson and Paxton, Turin and Crane, etc.

Results
The distribution-in-energy curves have been drawn fairly accurately for several radioactive substances. Some of these obtained by Gurney with the magnetic. spectrograph are shown in Fig. 268.

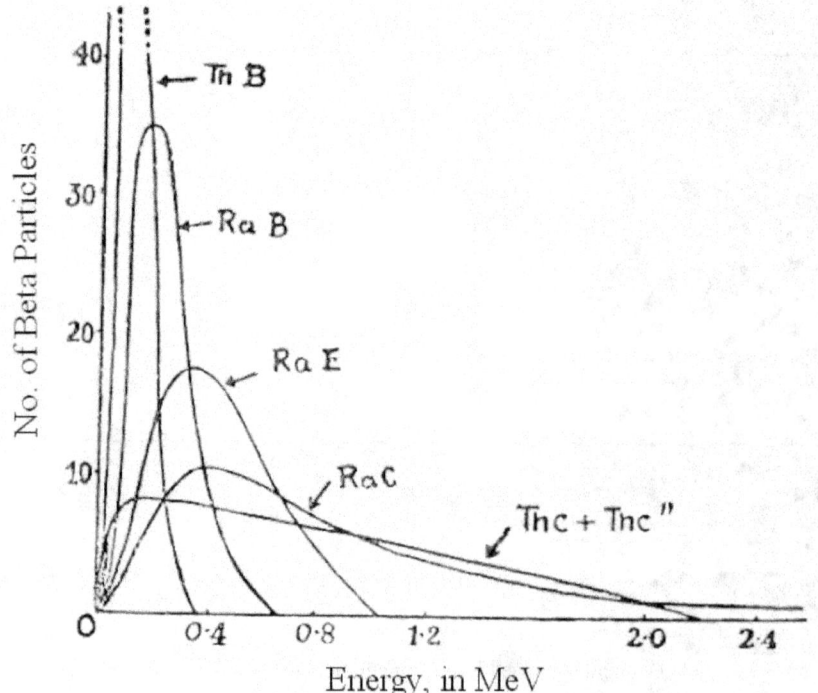

Fig. 268. Primary β-ray spectra.

It is seen that all, the curves are similar, being something like the familiar Maxwellian
velocity distribution curves, first rising to a maximum and then decreasing to zero at a well defined energy. This upper limit of the energy, which varies with the different β-emitters, can now be fixed with certainty, thanks chiefly to the researches carried out by Sargent. It is, when expressed in MeV, 0.36 for ThB, 0.65 for RaB, 1.05 for RaE, 1.80 for ThC'', 2.20 for ThC, 2.25 for MsTh$_{II}$, 2.32 for UX$_2$ and 3.15 for RaC. This definite maximum energy of the β-particles from a radioactive substance is probably the most important result concerning the β-ray spectrum, since according to present theories, it gives the total disintegration energy. It may be noted that although there is evidence that some materials may emit a few β-rays with energies as great as 8 or 11 MeV, still the great majority of the β-rays have energies limited to within about 3 MeV. Hence, in general, the energies of β-particles are less than those of the normal z-particles, which range between 4 and 9 MeV about.

Unlike the well-defined upper limit, the exact shape of the curve in the very low energy region is not at all certain. Results obtained are inconsistent, which is evidently due to considerable experimental difficulties. To investigate the low speed β-particles, one would have to use the source alone in order to prevent the scattering effect due to the very material that supports it. Consequently, it has not been possible to ascertain whether there is a definite inferior limit like the upper one, or whether the distribution extends to zero energy, which is, however, generally presumed, as indicated by the curves passing through the origin in the above figure.

It is certain that all the curves have a definite maximum, the height and position of which depend upon the substance emitting the β-rays. It is, in general, situated at about one-third of the upper limit.

The essential feature of these experimental curves is evidently their *continuous character* which constitutes one of the most difficult problems of nuclear physics. We shall now consider the attempts made to solve it.

Theoretical interpretation of the β-ray continuum

The real difficulty of interpreting the observed β-ray continuum arises only when it is admitted that the continuous characteristic of the experimental curves means that *the β-particles are actually ejected from the nucleus with a continuous distribution of energies.* But this implication appears certain for the following reasons:

(a) Every radioactive β-emitter gives rise to a continuous spectrum, whether it emits or not a line spectrum. This is true both for natural and artificial radioelements, as experimentally established.

(b) In all cases, where reliable measurements have been made, the number of continuous β-particles ejected per unit time is equal to the number of atoms disintegrating during the same interval. This shows that the β-particles responsible for the continuous spectrum arise from the nucleus, since according to theory there can be only one electron per disintegration.

(c) Sargent has shown from carefully gathered experimental data that there exists an approximate linear relation between the limiting maximum energy in the continuous β-spectrum and the decay constant of the source, analogous to Geiger-Nuttall's law in the case of α-rays; the greater the decay constant the greater also is the maximum energy. Now, since the decay constant refers exclusively to the disintegration of the nucleus, it follows that at least the upper limit of the continuous curve must be attributed to the nuclear electrons.

All these experimental findings, considered together, prove definitely that the β-particles constituting the continuous spectrum are really of nuclear origin.

Having thus established that the β-particles are emitted by the nucleus itself with continuously variable energy, an extremely embarrassing situation is created, which may be expressed as follows:

> While all the nuclear phenomena, such as the α- and the γ-ray spectra, show the existence of nuclei in perfectly definite quantum energy states, how is it that the β-ray spectra alone, manifesting a continuous distribution of energies, appear to deny the existence of such discrete energy states in the nucleus ?

The difficulty may be stated in yet another way.

As regards the origin of the β-rays, it is now generally admitted that the disintegration electron arises as a result of the transition of the nucleus from the neutron quantum state to the proton quantum state. This implies that the energy of β-disintegration must have a definite and constant value equal to the difference of the energies of the initial neutron and the final proton quantum states involved. On the other hand, in the β-ray spectrum we actually observe a continuous variation from the total energy of disintegration as maximum to zero as minimum, which means that the energy of the initial state minus the energy of the final state is not equal to the release of energy during β-disintegration. *But according to the principle of conservation of energy it is to be expected that the β-rays should be expelled either all with identical speed or at least a discontinuous group,* as in the case of the fine structure of α-rays.

How, then, to reconcile the fact of a continuous β-spectrum with the existence of discrete quantum states in nuclei without violation of the law of conservation of energy?

This *difficulty* concerning energy conservation *is supplemented by another equally serious one arising from the point of view of conservation of momentum.*

Experimental researches show that every nucleus has an angular momentum (nuclear spin), which is a half integral value of $h/2\pi$ for nuclei of odd mass number and a whole integral value of $h/2\pi$ for nuclei of even mass number.

This is just what would be expected if all nuclei were constituted with neutrons and protons, each with spin ½.

The β-particles, like all electrons, has also a spin ½; but when it leaves a nucleus of odd mass number, the remaining nucleus still retains a half-integral spin, since its mass number remains odd, the β-disintegration effecting no change in the mass number.

Evidently this goes against the law of conservation of momentum, since a half cannot be removed from a half integral quantity, the balance still remaining half integral. The same is true if the parent nucleus has an integral spin.

Several solutions have been suggested, but only one of them is now generally accepted as the most satisfactory. Some of the attempted but improbable solutions, which can be easily dismissed are:

(i) *The pre- and post-disintegration energies may not be the same but different for different nuclei in a given element,* so that the energy available for disintegration need not show a rigorous constancy but may vary continuously from zero to an arbitrary maximum. But this would mean, side by side, that the attendant nuclear process, viz., α-ray, γ-ray emission, etc., must be different for different nuclei, which, however, is contrary to experimental facts.

(ii) *Granting that during disintegration, the total energy (equal to the observed maximum energy) is transferred to the primary or disintegration electron, part of it may be subsequently lost through collisions with the outer electrons of the atom in a statistical fashion giving rise to the continuous Maxwellian-like velocity spectrum.*

But experiments rule out such an assumption. For, over and above the indirect experimental evidences in favor of the nuclear origin of the very continuous nature of β-ray spectrum, already stated, experiments performed by Ellis and Wooster and by Meitner and Orthmann directly disprove this hypothesis. These authors determined the average energy of disintegration per nucleus in the case of RaE, which is almost a pure β-emitter, by measuring the total heat energy produced by a known amount of the element with a microcalorimeter whose walls were thick enough to absorb all the emitted β-particles.

If the continuous β-spectrum were due to secondary scattering caused later by the external electrons of the atom, then the total energy should remain inside the calorimeter and one should expect the energy per nucleus measured by the heating effect to be equal to the observed maximum given by the upper limit of the continuous β-spectrum curve, which is 1.05 MeV for RaE. The actually measured energy was, however, only 0.34 MeV, which closely agreed with the average value deduced from the distribution curve.

Thus it appears certain that the original energy of the β-particles is, on the average, the same as that which one finds these particles possess when they get out of the atom. They have not, therefore, all been ejected with an energy corresponding to the upper limit and then lost varying amounts in collisions in the manner supposed. In other words, the continuity of the β-spectrum is a property of the disintegrating nucleus

(iii) *The discrepancy between the constant amount of energy released and the continuously varying energy observed might be explained by assuming that two electrons which are simultaneously emitted share the total constant disintegration energy in the observed continuous fashion.*

But Gurney's experiment, clearly demonstrating that the number of electrons in the spectrum of a β-emitter is exactly one per disintegration, excludes such a possibility.

With the elimination of these easily conceived but hardly proved hypotheses we are left with the choice between two final and drastic explanations, the one proposed by Bohr and the other by Pauli.

Bohr suggested that since the differences in energy and momentum between the initial nucleus and the product nucleus are not equal to the energy and momentum involved in β-decay, *the laws of conservation of energy and of momentum do not hold for this nuclear process.* He tried to justify his opinion by saying that in so far as the behavior of *light* particles in the nucleus could not be described by means of modern wave mechanics, but required a fundamentally new theory, there was no *a priori* reason for clinging to the laws of conservation of energy and of momentum.

But very serious difficulties have been raised against this hypothesis both from the experimental and theoretical points of view. For instance, Sargent's determination of well-defined upper limits for β-rays throws its weight on the side of the conservation laws. All other experimentally observed nuclear phenomena, without exception, even the γ-rays, which may be considered *lighter* than the electrons, abide by the conservation laws, and are, therefore, in favor of the general validity of these laws. From a purely theoretical point of view, the non-conservation hypothesis not only upsets the basis of the present nuclear theories, but also leads, as Landau has shown, to a contradiction of the general laws of gravitation worked out on relativity principles. There are, therefore, very few adherents of this hypothesis.

Pauli, on the other hand, proposed *a solution* of the difficulty, which, with the aid of a new-particle, called the **neutrino**, *safeguards the universal validity of the conservation laws, even in the case of β-. disintegration.* He assumed that the emission of a β-particle by a nucleus is occasioned by the transition of the nuclear heavy particle from the neutron to the proton state with the *simultaneous creation of an electron-neutrino pair.* These latter then escape with a constant discrete total energy equal to the difference between the energies of the original and final nuclei, thus conforming to the principle of conservation of energy. They share also the total momentum involved in such a way that it is conserved.

The observed continuous energy distribution of β-rays merely expresses the manner in which the total energy is shared between the electron and the neutrino. The upper limit of the continuous β-spectrum thus corresponds to the case where the neutrino is emitted with zero energy, while the whole energy is carried away by the electron; at the maximum point of the curve the two share more or less equal energies; in the low energy portion of the curve the neutrino gets a greater share of the energy and the electron correspondingly less.

Pauli's hypothesis is thus able to explain satisfactorily the continuous nature of the β-ray curves without, at the same time, violating the laws of conservation of energy and of momentum. Hence, it is now generally regarded as the best among the different solutions so far suggested.

Fermi, developing the idea of Pauli, has worked out a quantitative theory of β-emission on the basis of the simplest hypothesis concerning the interaction energy between heavy and light particles (neutrons and protons, electrons and neutrinos) and of the general principles of quantum mechanics. The calculated distribution in momentum of the electrons, according to Fermi's theory, is given by

$$N (\eta_e) \, d \eta_e = C \, \eta_e^2 \, \eta_v^2 \, d\eta_e$$

Where

N is the number of electrons of momentum η_e expressed in units of mc^2,
η_v the momentum of the neutrinos in the same units and
C a factor approximately independent of the momentum for low atomic numbers.

The theoretical distribution curve obtained from the above expression agrees fairly well with the experimental β-ray curves, with this difference, however, that the former exhibits a symmetry about the half-energy value, while the latter shows a pronounced shift of the maximum distribution to about one-third of the limiting energy.

To obtain a better agreement with experimental data, Konopinski and Uhlenbeck introduced a modification in the original Fermi theory by proposing a different type of interaction between the electron-neutrino field and the heavy particle. Of course, this achieved the end in view and gave theoretical curves that agreed closely with the experimental ones, but authors have discussed which of the two, either Fermi's or K — U theory, represents actual fact, At any rate, we have here a *theoretical justification* of Pauli's hypothesis of the neutrino.

The *crux of Pauli's explanation is evidently centred* on *the actual existence of the neutrino.* According to theory, this particle can have *no charge,* as all the charge is taken by the electron, its *mass* must be *extremely small,* since there is no change in mass number in β-disintegration; theoretical estimate based on the position of the maximum in the β-ray continuum as well as on the upper limit of the curve strongly suggests even a *zero rest-mass;* it must have a *spin* 1/2 , like its companion the electron, because only then the angular momentum will be conserved in the β-process; but it has been shown that it can have only a *very small magnetic moment,* much smaller than that of the electron. The name *neutrino* (a miniature neutron) represented by the symbol v, has been adopted, because the new particle is without charge like the neutron, while perhaps without mass unlike the neutron. It is, however, more akin to a photon or light quantum than to a material particle, although it has no associated electromagnetic field and hence cannot interact with matter. Some authors have therefore preferred to call it **"ergon"** or mere carrier of energy and momentum.

These peculiar properties of the neutrino make direct detection of the particle almost impossible. Possessing no charge, it cannot produce sufficient ionizations to be detected, say in a cloud chamber; similarly, having no mass, the possibility of observing the nuclei set in motion by it, as in the case of the neutrons, is also excluded. But, in recent years, indirect experimental evidence of its existence is becoming more and more convincing. The basic principle used in the search for the neutrino is to study the *recoil of disintegrating atom ejecting β-rays.* In natural radioactive elements whose atoms are all very heavy, the track of the recoiling residual atom is too short to be measured even in a cloud chamber operated at the lowest possible pressure. But with the modern technique of artificial transmutation, light β-emitters can be produced and with these the recoil would certainly be observable.

If then, in such cases, the direction of the tracks of the 1β-particle and of the residual atom are not exactly opposite, the assumption of a third invisible particle would be preferable to giving up the law of conservation of momentum. Hence, when experiments conducted to measure the energy and momentum of the recoiling nucleus and the emitted β-particle lead to the conclusion that neither momentum nor energy is conserved in the individual disintegrations, if only these two particles are involved. the existence of a third particle, the neutrino, is definitely indicated. The experiments of Leipunski (1936), Crane and Halpern (1939), and Allen (1942) performed in this direction seem to be in favor of the existence of the neutrino.

Thus, when lithium is bombarded with deuterons, a radioactive isotope of beryllium is produced according to the reaction

$$_3Li^6 + {}_1D^2 \rightarrow {}^*_4Be^7 + {}_0n^1$$

$_4Be^7$ is less stable than $_3Li^7$ as it contains a proton more; but the difference is too small for expelling a positron (β^+ decay). But $_4Be^7$ captures one of the innermost extra-nuclear electrons (K shell) and is transformed into $_3Li^7$ with a decay constant of about 43 days. In this case, the recoil energy of the resulting atom depends solely on the emitted neutrino. From the mass difference of $_4Be^7$ and $_3Li^7$ (corresponding to 370 KeV) the recoil energy to be expected is about 58 eV; the experimental results obtained by Allen are in fair agreement with this value.

A careful study of the beta spectrum of H^3 (triton) showed that the mass of the neutrino is considerably less than that of an electron. It has not been possible to make an exact determination of the mass of the neutrino, but it may be stated definitely that it does not exceed 1/1000 of that of an electron.

The recent experiments of Reines and Cowen (1956) give a more direct proof of the existence of the neutrino. Near an atomic reactor where is an intense flux of neutrinos and the inverse β-process can be used for their detection. In this process a neutrino is captured by a proton which then decays into a neutron and a positron. Simultaneous detection of the neutron and positron has given convincing proof that the neutrino exists as an elementary particle.

Pauli's theory, which has the great merit of solving one of the greatest riddles of nuclear phenomena, in a very simple and satisfactory manner, can, therefore, be considered as having received the necessary experimental confirmation.

BIBLIOGRAPHY

General
Max Planck, *Survey of Physics*, 1925.
F. H. Newman. *Recent Advances in Physics*, 1934.
E. N. da C. - Andrade, The *Structure of the Atom*, 1934.
A. E. Ruark and H. C. Urey, *Atoms, Molecules and Quanta*, 1930.
G. Castelfranehi, *Recent Advances in Atomic Physics*, (2 Vole.), 1932.
A. Hass, *Theoretical Physics*, (2 Vols.), 1929.
Harnwell and Livingood, *Experimental Atomic Physics, 1933.*
F. K. Richtmyer, *Introduction to Modern Physics*, 1934.
H. A. Wilson, *Modern Physics*, 1944.
Max Born, *Atomic Physics*, 1954.
G. F. M. Jauneey, *Modern Physics*, 1946.
G. Joos, *Theoretical Physics*, 1950.
J. D. Stranathan, *The Particles of Modern Physics*, 1942.
R. A. Milliken, *Electron (+ and -), Protons, Photons, Neutrons, Mesotrons and Cosmic Rays*, 1946.
E. Grimsehl, *Physics of the Atom*, 1949.
H. Semat, *Introduction to Atomic Physics*, 1947.
C. T. Chase, *The Evolution of Modern Physics*, 1947.
S. Tolansky, *introduction. to Atomic Physics*, 1948.
F. K. Richtmyer and E. H. Kennard, *Introduction to Modern Physics*, 1946.
F. K. Richtmyer, E. H. Kennard and T. Lauritsen, *Introduction to Modern Physics*, 1956.
Oldenberg, *Introduction to Atomic Physics*, 1949.
W. Finklenburg, *Atomic Physics*, 1950.
S. Dushman, *Fundamentals of Atomic Physics*, 1951.
F. W. Van Name, *Modern Physics*, 1952.
C. Kittol, *Introduction to Solid State Physics*, 1954.
R. S. Shankland, *Atomic and Nuclear Physics*, 1955.

Chapter 1 -Discharge of Electricity through Gases
J. S. Townsend, *Electricity do Gases*, 1914.
J. J. Thomson and G. P. Thomson, *Conduction of Electricity through Oases*,
Vol. I (1928). Vol, II (1932)..
K. G. Emedeus, *Conduction of Electricity through Oases*, 1928.
K. K. Darrow, *Electrical Phenomena in Oases*, 1932.
A. AL Tyndall, Mobility *of Positive Ions in Gases*, 1938.

Chapter 2 and 3-The Electron
R. A. Millikan, *The Electron*, 1926.
0. W. Richardson, *The Electron Theory of Matter*, 1916,
D. Grimes, *Meet the Electron*, 1944.
E. C. Stoner, *Magnetism*, 1930.
J. H. Van Vleck, *The Theory of Electric and Magnetic Susceptibilities*, 1932.
F. Bitter, *Introduction, to Ferromagnetism*, 1937.
O. W. Richardson, *The Emission of Electricity from Hot Bodies*, 1921.
A. L. Riemann, *Thromionic Emission*, 1934.
H. S. Allen, *Photoelectricity*, 1925.
A. L. Hughes and L. A. du Bridge, *Photoelectric Phenomena*, 1932.
K. H. Spring, *Photons and Electrons*, 1950.
J. Stokley, *Electrons in Action*, 1946,
L. R. Koller, *Physics of Electron Tubes*, 1937.
F. E. Terman, *Fundamentals of Radio*, 1938.
W. L. Everitt, *Fundamentals of Radio and Electronics*, 1958
A. L. Albert, *Fundamental Electronics and Vacuum Tubes*, 19-18.
L. B. Arguimban, *Vacuum tube circuits and transistors*, 1956

Admiralty, *Handbook. of Wireless Telegraphy,* (2 Vols.) 1939.

A. A. Ghirardi, *Radio Physics* Course, 1933.

J. H. Rayner, *Modern Radio Communication.* 1947

W. F. Lovering, *Radio Communication,* 1958

J. L. Hornung, *Radar Primer,* 1948.

Radar School, *Principle of Radar,* 1946.

E. C. Pollard and J. M. Sturtevant, *Microwaves and Radar Electronics,* 1918.

W. Gordy, W. V. Smith & R. F. Trambaru o, *Microwave Spectroscopy,* 1953.

AL W. P. Strand berg. *Microwave Spectroscopy,* 1954.

H. Motz, *Electromagnetic problems of Microwave theory,* 1951.

Campbell and Ritchie, *Photocells,* 1934.

Zworykin and Wilson, *Photocells,* 1934.

-0. A. Briggs, *Sound Reproduction,* 1950.

G. F. Jones, *Sound Film Reproduction,* 1936.

V. K. Zworkyin and E. G. Ramberg, *Photoelectricity and its applications,* 1949.

H. K. Henisch, *Metal Rectifiers,* 1949.

J. T. Mac Gregor Morris and J. A. Henly, *Cathode Ray Oscillography,* 1936.

J. H. Rayner, *Cathode Ray Oscillography,* 1945.

W. C. Eddy, *Television The Eyes of Tomorrow),* 1945.

J. H Boyner, *Television,* 1934.

M. G. Scroggie, *Television,* 1935.

A. Dinsdale, *First Principles of Television,* 1932.

G. V. Dowding, *Practical Television,* 1935.

E. F. Burton and W. H. Kohl, *The Electron Microscope,* 1946.

Chapter 4 - Positive-Rays

F. W. Aston, *Mass Spectra and Isotopes,* 1933,

J. J. Thomson, *Rays of Positive Electricity,* 1923.

Reviews of Modern Physics, *Nuclear Physics* C., July 1937.

Chapter 5-X-Rays

A. H. Compton and S. K. Allison, *X-Rays in Theory and Experiment.* 1935.

Maurice de Broglie, *X-Rays,* 1925.

0. W. G. Kaye, *X-Rays,* 1926.

W. H. Bragg and W. L. Bragg, *X-Rays* and *Crystal Structure.* 1918.

M. J. Buerger, *X-Ray Crystallography,* 1949.

B. L. Worsnop, *X-Rays,* 1930.

W. H. Zachariassen, *Theory of X-Ray Diffraction in Crystals,* 1915.

R. W. James, *X-Ray Crystallography,* 1930.

K. Lonsdale, *Crystals* and *X-Rays,* 1948.

M. Siegbahn, *Spectroscopy of X-Rays,* 1925.

A. J. C. Wilson, *X-Ray Optics,* 1949.

G. L. Clark, *Applied X-Rays,* 1940.

Chapter 6 -Radioactivity

E. Rutherford, *Radioactivity,* 1905,

W. H. Bragg, *Studies in Radioactivity,* 1912.

E. Rutherford, J. Chadwick and C. D. Ellis, *Radiations from Radioactive Substance,* 1930.

G, Newsy and F. A. Paneth, *A Manual of Radioactivity, 1938.*

K. K. Darrow, *Bell Telephone, Some Contemporary Advances in Physics-XII Radioactivity,* 1927.

Madame Pierre Curie, *Radioactivity,* 1935.

Chapter 7-Relativity

A. Einstein, *Relativity : Special and General Theory,* 1922.

L, Silberstein, *The Theory of Relativity, 1914.*
H. Schmidt, *Relativity and the Universe, 1921.*
A. Hass, *Introduction to Theoretical Physics.* (Vol. II), 1929.
Sir James Jean?, *Through Space and Time, 1934.*
W. H. M. C. Crea, *Relativity Physics, 1935.*
A. S. Eddington, *Space, Time and Gravitation, 1921,*
E. Cunningham, *The Principle of Relativit , 1921.*
Max Born, *Einstein's Theory of Relativity, 1922.*
H. Dingle, *The Special Theory of Relativity, 1940.*
G. J. *Whitrow, The Structure of the Universe, 1949.*
G. Y. Rainich, *Mathematics of Relativity, 1950.*
A. Einstein, *The Meaning of Relativity, 1950.*
L Barnett, *The Universe and Dr. Einstein, 1952.*

Chapter 8-Quantum Theory of Radiation
F. Reiche, *The Quantum Theory, 1930.*
W. Heisenberg, *The Physical Principles of the* Quantum *Theory, 1930.*
L. Infeld, *The World in Modern Science, Matter and Quanta, 1934.*
W. Heitler, *Quantum Theory of Radiation, 1936.*
G. Temple, *An Introduction to Quantum Theory, 1931.*
Louis de Broglie, *Matter and Light, 1939.*
J. Jeans, *The New Background of Science, 1947,*

Chapter 9-Wave Nature of Matter
G. Birtwistle, *New Quantum Mechanics, 1928.*
D. Bohm, *Quantum Theory, 1952.*
J. Frenkel, W*ave Mechanics, 1934.*
E. Schrödinger, *Four Lectures on Nave Mechanics, 1929.*
N. F. Mott, *Wave Mechanics, 1930.*
A. Sommerfeld, *Wove Mechanics, 1930.*
Louis de Broglie, An *Introduction to the Study of Wave Mechanics, 1930.*
H T. Flint. *W woe Mechanics, 1931.*
F. G. Kemble, *The Fundamental Principles or Quantum, Mechanics, 1937.*
P. A. M. Dirac, *The principles of Quantum Mechanics, 1947.*
W. Wilson, *Relativity and Quantum Dynamics, 1040.*
V. Rojansky, *Introduction to Quantum Mechanics, 1950.*
K. R. Dixit, *The Elements of Wave Mechanics and Quantum Mechanics, 1953.*
G. K. T. Conn, *The Wave Nature of the Electron, 1944.*
R. Beeching, *Electron Diffraction, 1946.*
G. T. Thomson and W. Cochrane, *Theory and Practice of Electron Diffraction,1939.*

Chapter 10-Quantum Statistics
J. Rice, *introduction to Statistical Mechanics J or Students of Physics and Physical Chemistry, 1930.*
E. H. Kennard, *Kinetic Theory of Gases, 1938.*
M. N. Saha and B. *N.* Srivastava, *A Treatise on Heat, 1953.*

Chapter 11-Peripheral Electronic Structure of the Atom
N. Bohr, *The Theory or Spectra and Atomic Constitution, '1924.*
E. C. Stoner, *Magnetism and Atomic Structure,• 1926.*
A. Haas, *Atomic. Theory, 1927.*
A. Sommerfeld, *Atomic Structure and Spectral Lines, 1934.*
G. Herzberg, *Atomic Spectra and Atomic Structure, 1937.*
H. E. White, *introduction to Atomic Spectra, 1934.*
F. K. Richtmyer, *Introduction to Modern Physics, 1934.*
0. L. Padding and S. Goudsmit, *The Structure of Line Spectra, 1930.*
G. K. T. Conn, *The Nature of the Atom, 1944.*

F. 0. Rice and E. Teller, *The Structure of Matter*, 1949.

S. Tolansky, *High Resolution Spectroscopy*, 1947.

Chapter 12-Molecular Spectra

L. Kroenig, *Bond Spectra and Molecular Spectra*, 1930.

R. C. Johnson, *Introduction to Molecular Spectra*, 1949.

P. Debye, *The Structure of Molecules*, 1932.

K. B. Ramanathan, *Molecular Scattering of Light*, 1923.

8. Bhagavantham, *Scattering of Light and Raman Effect*, 1940.

G. Herzberg, *Infra-red and Raman Spectra of Polyatomic Molecules*, 1945.

G. B. B. Sutherland, *Infra-red* and *Raman Spectra*, 1935.

Chapter 13-Modern Problems of Radioactivity

K. K. Darrow, *Bell Telephone Series XII, Radioactivity*, 1927.

F. Rasetti, *Elements of Nuclear Physics*, 19 7.

G. Gamow, *Atomic Nuclei and Nuclear Transformations*, 1937.

J. M. Cork, *Radioactivity and Nuclear Physics, 1947.*

Chapter 14 -Artificial Transmutation

K. K. Darrow, *Bell Telephone* Series *XXII, Transmutation, 1931.*

N. Feather, *An Introduction to Nuclear Physics*, 1936.

Reviews of Modern Physics, *Nuclear Physics, C, July, 1937,*

W. B. Mann, *The Cyclotron*, 1940.

D. H. Wilkinson. *Ionizations Chambers and Counters*, 1950.

J, B. Birks, *Scintillation Counters, 1. 954.*

S. C. Curran, Luminescence *and the Scintillation Counter*, 1953.

Curran and Craggs, *Counting Tubes : Theory and Applications*, 1949.

J. 0. Wilson, *The Principles of Cloud Chamber Technique, 1951.*

C. F. Powell and G. P. S. Occhialini, *Nuclear Physics in Photographs*, 1947.

H. Yagoda, *Radioactive Measurements with Nuclear Emulsion*, 1949.

J. B. Hoag and S. A. Kor, The *Electron* and *Nuclear Physics, 1949-*

P. B. Moon, *Artificial Radioactivity*, 1949.

E. Pollard and W. L. Davidson, *Applied Nuclear Physics, 1946.*

J. B. Rajam, *Nuclear Isomerism* (Thesis), 1939.

K. Mendelssohn, *What is Atomic Energy ?* 1946.

J. De Ment and H. C. Dake, *Uranium* and *Atomic Power*, 1945.

J. J. O'Neill, *The Almighty' Atom,* 1945.

G. Gamow, *Atomic Energy*, 1947.

W. E. Stephens, *Nuclear Fission and Atomic* Energy, 1948.

J. Cockcroft, *Development and Future of Atomic Energy*, 1950.

W. L. Lawrence, *The Hell Bomb*, 1951.

Chapter 15 Cosmic Rays

R. A. Millkan, *Cosmic Rays,*

L..Janossy, *Cosmic Rays,* 1948.

J. G. Wilson, *About Cosmic Rays,* 1948.

D. J. X. Montgomery, *Cosmic Rays Physics*, 1949.

L. Leprince Ringuet, *Cosmic Rays,* 1950.

B. Rossi, *High-Energy Particles*, 1952.

F. C. Frank and D. R. Rexworthy, *Cosmic Radiation*, 194.9.

A. Dauvillier, *Les Rayons Cosmiques,* 1954.

R. Marshals, *Meson Physics,* 1952.

H. A. Bathe and F. De Hoffman, *Mesons* and *Fields*, 1955.

J. G. Wilson, *Progress* in *Cosmic Ray Physics,* Vols. I (1952), II (1954) and III (1956).

Chapter 16-Structure and Properties of the Nucleus

Reviews of Modern Physics. *Nuclear Physics, A and B,* April, 1936 and April, 1931.
K. K. Darrow, *Bell Telephone Series, The Nucleus-I,* II, III, IV, 1933-35.
D. Halliday, *Introductory Nuclear Physics,* 1950.
F. Bitter, *Nuclear Physics,* 1950.
S. Tolansky, Hyperfine Structure in Line *Spectra and Nuclear Spin,* 1948.
N. F. Ramsey, *Nuclear Moments,* 1953.
Reviews of Modern Physics, Jul?, 1946.
S. Devons, *Excited States of Nuclei,* 1949.
0. R. Frisch, *Progress in Nuclear Physics,* 1950.
F. Fermi, *Nuclear Physics,* 1951,
W. Heisenberg, *Nuclear Physics,* 1952.
J. M. Blatt and V. F. Weisskopi, *Theoretical Nuclear Physics,* 1952.
H. I. Bathe, _Elementary Nuclear Theory,* 19 7.
C. Wentzel, *Quantum Theory of Fields,* 1949.
W. Pauli, *Meson Theory of Nuclear Forces,* 1948.
R. E. Marshak, *Meson Physics,* 1952.
R. D. Evans, *The Atomic Nucleus,* 1955.

www.ingramcontent.com/pod-product-compliance
Lightning Source LLC
Chambersburg PA
CBHW081255170526
45165CB00011B/3310